DATE DUE

Immunoglobulin Genes
Second edition

Dedicated to the memory of Georges Kohler

Immunoglobulin Genes
Second edition

Edited by

T. Honjo
Kyoto University, Japan

F. W. Alt
*Howard Hughes Medical Institute Research Laboratories,
The Children's Hospital, Boston*

ACADEMIC PRESS

Harcourt Brace & Company, Publishers
London, San Diego, New York, Boston, Sydney, Tokyo, Toronto

ACADEMIC PRESS LIMITED
24–28 Oval Road
LONDON NW1 7DX, UK

U. S. Edition Published by
ACADEMIC PRESS INC.
San Diego, CA 92101, USA

This book is printed on acid-free paper

Copyright © 1995 ACADEMIC PRESS LIMITED

All rights reserved

B lympohocyte tolerance in the mouse, pages 365–378, and Immunoglobulin variable region gene segments in human autoantibodies, pages 379–396, are US government works in the public domain and not subject to copyright.

No part of this book may be reproduced or transmitted in any form or by any means, electronic or mechanical including photocopying, recording, or any information storage and retrieval system without permission in writing from the publisher

A catalogue record for this book is available from the British Library

ISBN 0 12 053640 4

Typeset by Phoenix Photosetting, Chatham
Printed and bound by Hartnolls Ltd, Bodmin, Cornwall

Contents

Contributors	vii
Foreword	xi
Preface to first edition	xv
Preface to second edition	xvii

PART I: B CELLS

1 **B-cell differentiation in humans** — 3
P D Burrows, H W Schroeder Jr and M D Cooper

2 **The role of B cell and pre-B-cell receptors in development and growth control of the B-lymphocyte cell lineage** — 33
F Melchers

3 **The maturation of the antibody response** — 57
C Milstein and C Rada

4 **Development of B-cell subsets** — 83
S M Wells, A M Stall, A B Kantor and L A Herzenberg

5 **Regulation of B-cell differentiation by stromal cells** — 103
E A Faust, D C Saffran and O N Witte

6 **Antigen receptors on B lymphocytes** — 129
M Reth

PART II: ORGANIZATION AND REARRANGEMENT OF IMMUNOGLOBULIN GENES

7 **Immunoglobulin heavy chain loci of mouse and human** — 145
T Honjo and F Matsuda

8 **The human immunoglobulin κ genes** — 173
H G Zachau

9 **Immunoglobulin λ genes** — 193
E Selsing and L E Daitch

10	**The variable region gene assembly mechanism** A Okada and F W Alt	205
11	**Regulation of class switch recombination of the immunoglobulin heavy chain genes** J Zhang, F Alt and T Honjo	235
12	**Generation of diversity by post-rearrangement diversification mechanisms: the chicken and the sheep antibody repertoires** J-C Weill and C-A. Reynaud	267
13	**Immunoglobulin heavy chain genes of rabbit** K L Knight and C Tunyaplin	289
14	**The structure and organization of immunoglobulin genes in lower vertebrates** J P Rast, M K Anderson and G W Litman	315

PART III: IMMUNOGLOBULIN GENE EXPRESSION

15	**Ig gene expression and regulation in Ig transgenic mice** U Storb	345
16	**B-lymphocyte tolerance in the mouse** D Nemazee	365
17	**Immunoglobulin variable region gene segments in human autoantibodies** K N Potter and J D Capra	379
18	**Regulation of immunoglobulin gene transcription** K Calame and S Ghosh	397
Index		423

Contributors

FW Alt
Howard Hughes Medical Institute Research Laboratories, The Children's Hospital, 300 Longwood Avenue, Boston, MA 02115, USA

MK Anderson
Department of Molecular Genetics, All Children's Hospital, 801 Sixth Street South, St Petersburg, FL 33701-4899, USA

PD Burrows
Division of Developmental and Clinical Immunology, Departments of Microbiology, Medicine and Pediatrics, University of Alabama at Birmingham, 263 Wallace Tumor Institute, UAB Station, Birmingham, AL, USA

K Calame
Columbia University College of Physicians and Surgeons, Department of Microbiology, 701 West 168th Street, New York, NY 10032-2704, USA

JD Capra
Southwestern Medical School, Department of Microbiology, 5323 Harry Hines Boulevard, Dallas, TX 75235, USA

MD Cooper
Howard Hughes Medical Institute, Department of Pediatrics and Microbiology, University of Alabama at Birmingham, 263 Wallace Tumor Institute, UAB Station, Birmingham, AL 35294, USA

LE Daitch
Tufts University School of Medicine, Department of Pathology, 136 Harrison Avenue, Boston, MA 02111, USA

EA Faust
Howard Hughes Medical Institute Research Laboratories, University of California, Los Angeles, 5-748 MacDonald Building, 10833 Le Conte Avenue, Los Angeles, CA 90024-1662, USA

S Ghosh
Howard Hughes Medical Institute, Yale University, 295 Congress Avenue, 154 BCMM, New Haven, CT 06510, USA

LA Herzenberg
Stanford University School of Medicine, Department of Genetics, Immunogenetics and Cell Sorting Laboratory, Stanford, CA 94305-1662, USA

T Honjo
Faculty of Medicine, Department of Medical Chemistry, Kyoto University, Yoshida, Sakyo-ku, Kyoto 606, JAPAN

AB Kantor
Department of Microbiology, College of Physicians & Surgeons, Columbia University, New York, NY, USA

KL Knight
Loyola University Chicago, Stritch School of Medicine, Department of Microbiology and Immunology, 2160 South First Avenue, Maywood, IL 60153, USA

GW Litman
Department of Molecular Genetics, All Children's Hospital, 801 Sixth Street South, St Petersburg, FL 33701-4899, USA

F Matsuda
Kyoto University, Center for Molecular Biology and Genetics, Shogo-in Kawaharacho 53, Sakyo-ku, Kyoto 606, JAPAN

F Melchers
Basel Institute for Immunology, Grenzacherstrasse 487, Postfach, CH-4005 Basel, SWITZERLAND

C Milstein
Laboratory of Molecular Biology, Medical Research Council, Hills Road, Cambridge CB2 2QH, UK

D Nemazee
National Jewish Center, Department of Pediatrics, 1400 Jackson Street, Denver, CO 80206, USA

A Okada
Howard Hughes Medical Institute Research Laboratories, The Children's Hospital, 300 Longwood Avenue, Boston, MA 02115, USA

KN Potter
Southwestern Medical School, Department of Microbiology, 5323 Harry Hines Boulevard, Dallas, TX 75235, USA

C Rada
Laboratory of Molecular Biology, Medical Research Council, Hills Road, Cambridge CB2 2QH, UK

JP Rast
Department of Molecular Genetics, All Children's Hospital, 801 Sixth Street South, St Petersburg, FL 33701-4899, USA

M Reth
MPI fur Immunbiologie, Postfach 79011, D-7800 Freiburg, GERMANY

C-A Reynaud
Institut Necker, Faculté de Medicine, Necker-Enfants Malades, 156 rue de Vaugirard, 75730 Paris cedex 15, FRANCE

DC Saffran
Howard Hughes Medical Institute Research Laboratories, University of California, Los Angeles, 5-748 MacDonald Building, 10833 Le Conte Avenue, Los Angeles, CA 90024-1662, USA

HW Schroeder, Jr
Division of Developmental and Clinical Immunology, Departments of Microbiology, Medicine and Pediatrics, University of Alabama at Birmingham, 263 Wallace Tumor Institute, UAB Station, Birmingham, AL, USA

E Selsing
Tufts University School of Medicine, Department of Pathology, 136 Harrison Avenue, Boston, MA 02111, USA

AM Stall
Department of Microbiology, College of Physicians & Surgeons, Columbia University, New York, NY, USA

U Storb
University of Chicago, Department of Molecular Genetics and Cell Biology, 920 East 58th Street, Chicago, IL 60637, USA

C Tunyaplin
Loyola University Chicago, Stritch School of Medicine, Department of Microbiology and Immunology, 2160 South First Avenue, Maywood, IL 60153, USA

J-C Weill
Institut Necker, Faculté de Medicine, Necker-Enfants Malades, 156 rue de Vaugirard, 75730 Paris cedex 15, FRANCE

SM Wells
Department of Microbiology, College of Physicians & Surgeons, Columbia University, New York, NY, USA

ON Witte
Howard Hughes Medical Institute Research Laboratories, University of California, Los Angeles, 5-748 MacDonald Building, 10833 Le Conte Avenue, Los Angeles, CA 90024-1662, USA

HG Zachau
Institüt für Physiologische Chemie der Universität München, München, Germany.

J Zhang
The Rockefeller University, 1230 York Avenue, Box 167, New York, NY 10021, USA.

Foreword

When Behring and Kitasato discovered antibodies back in 1890, they could hardly have imagined that these factors were prime examples of both the chemical and biological bases of molecular recognition. The specificity of antibodies was soon recognized as a biological puzzle by Ehrlich, who in 1905 proposed the first (selective) theory of antibody specificity. Even so, Ehrlich considered it inconceivable that there could be specific substances ready to recognize and neutralize toxins that the animal species had never encountered before. So he produced the most ingenious idea that the toxins were the ones which by pure coincidence were capable of recognizing 'side chains', located on the surface of cells, and required for the utilization 'of foodstuffs'. Ehrlich therefore made the remarkable prediction of the existence of receptors on the cell surface, but could not conceive the capacity of the organisms to recognize the unknown and to learn to improve such recognition. While he laid the foundations of what was going to be a major preoccupation of immunologists for a long time, namely quantitative immunochemistry, it was the role of Landsteiner to demonstrate that antibodies were indeed capable of recognizing substances, naturally occurring or otherwise, which the animal had never seen before, through his classical studies with haptens. This remarkable property of the immune system became a dominating intellectual challenge to basic immunologists. The additional conviction that such specific recognition was capable of further improvement through a process of maturation of the immune response, demonstrated by Heidelberger and Kendall in the late 1930s, was an added complication for which no rational explanation could be proposed at the time.

At first it was proposed that antigens act as a form of template, around which the antibody is synthesized or folded. Prominent proponents of such an 'instructive' hypothesis were Haurowitz and Pauling. But this soon ran into difficulties, for a number of reasons. A major one was the progress in the understanding of the molecular basis of protein structure and folding and, more generally speaking, the molecular basis of biological specificity.

The new ideas and the developments of new methods which characterize the birth of molecular biology, had a direct impact on immunology. The emerging techniques of protein chemistry were immediately applied to the studies of antibodies by Rodney Porter, soon after his PhD supervisor, Sanger, developed the methods which culminated with the demonstration that proteins had defined amino acid sequences. This early inroad into the protein chemistry and early amino acid sequences of the N-terminus of antibodies, in the early 1950s, gave support to the instruction theories, in that no heterogeneity could be discovered. On the contrary, a single N-terminal sequence could be discerned in rabbit immunoglobulins, leading to calculations into the highly improbable possibility of different molecules having identical N-terminal sequences. This may sound strange today, when we have become so accustomed to the idea of protein families, and closely related tandem arrays of genes. However, in those days, the idea that genes could arise by gene duplication only gained acceptance in the 1960s, following Braunitzer's comparison of the α and β chains of haemoglobin. That antibodies were indeed heterogeneous was very difficult to demonstrate, and was the result of a variety of studies coming from different directions that slowly built up into an inescapable conclusion at a later date. Earliest among them was the discovery of idiotypes made by Oudin in the early 1960s. The clonal selection theory proposed by Burnett and inspired by an alternative selective theory made by Niels Jerne, provided a very sound theoretical basis to the generation of specificity through protein microheterogeneity.

The common structural architecture of antibody molecules made up of two heavy and two light chains could be established in the early 1960s by Edelman and Porter, with the heterogeneous population of antibodies, because of the very fact that they represented an invariant character of all molecules. The critical element which revealed the essential character of the antibody diversity did not come from studies of antibodies themselves, but from myeloma proteins. Myeloma proteins have been known for a very long time, so much so that what turned out to be the light chains of myeloma proteins were discovered by Bence-Jones in 1847. The relationship between human myeloma proteins and antibodies arose largely from the careful antigenic analysis performed by Henry Kunkel, and extended to the mouse counterparts from the mouse plasmacytomas discovered by Michael Potter. It was the structural analysis of such molecules which brought about a further understanding of the underlying diversity within the frame of a general common architecture. The early peptide maps of human and mouse Bence-Jones proteins performed, respectively, by Putnam and colleagues and by Dreyer and his colleagues, were quickly superseded by the demonstration in 1965 by Hilschmann and Craig that such light chains consisted of a common segment and a variable segment. The existence of allotypic markers which appeared to be localized in the V region of rabbit heavy chains, and shared by IgG, IgM and IgA (the Todd phenomenon), and my own demonstration that variable segments of human kappa light chains consisted of at least three non-allelic sets of V_κ regions in association with a single C_κ region, provided experimental evidence that the variable and constant domains must be encoded by separate genes, as proposed by Dreyer and Bennett.

The comparison between the rapidly expanding sequences of V segments of myeloma proteins disclosed the existence of the hypervariable regions, which were to be called complementarity determining regions (CDRs) by Kabat, to imply that those were the residues directly involved in the antigenic recognition. The generalized common architecture, predicted by our analysis of disulphide bonds, and conceptualized by the domain structure proposal of Edelman, with the hypervariable segments predictably located at the tips of the Y-shaped molecule (as seen in electron micrographs) received confirmation from the structural studies which followed the crystallization of Fab fragments of myeloma proteins by Poljak and Nisonoff. This exciting period characterized by studies of myeloma proteins using protein chemistry techniques, was to be enriched by the first glimpse of the somatic hypermutation, which resulted from the comparison of different lambda chains made by Cohn, Weigert and co-workers.

All this was soon to be superseded by the application of the DNA recombinant technology, in myelomas and in the newly derived hybridomas. The spectacular confirmation of the two genes/one polypeptide made at the DNA level by Tonegawa in 1976 led, within a period of less than 10 years, to our present understanding of the genetic arrangement and rearrangement of the antibody genes. Further success was provided by the attack on the problem of the T-cell recognition system. The long-drawn out controversy concerning the T-cell receptor was finally solved, to close the chapter of basic understanding of the genetic nature of the origin of diversity and of the structures involved in antigen recognition. The connection between the major histocompatibility complex and immunology became established and the observations of Zinkernagel and Doherty (which later culminated with the crystallographic explanation of the phenomenon) opened a new area with the first glimpse of the molecular bases of T–B cell collaboration.

The complexity of cell–cell and receptor–ligand interactions were soon to become a major focus of research. The International Workshops of Human Leucocyte Differentiation Antigens (now organizing its sixth conference) gave impetus and introduced order in efforts towards the dissection and characterization of the relevant cell surface molecules and receptors. Other techniques of molecular and cell biology, like transgenic and knockout mice at the one extreme, and gene amplification and fast DNA sequencing techniques at the other, provided new approaches to reassess old problems and to uncover and investigate previous intractable questions. For example, the bases of antigen recognition and of the maturation of the immune response have been generally clarified to uncover unique molecular mechanisms and interactions which underpin the processes.

As in years gone by, the impact of technological advances is not only revolutionizing but also reshaping the way we think about immunology. The pace of progress and change is faster and faster all the time. I wonder how many young colleagues realize that there are still a good number of active scientists who in their own youth, were asking the question: does a single amino acid sequence code for the world of antibodies? and were yet to learn that lymphocytes were not only large or small but also T or B.

C. Milstein

Preface to first edition

Since Kitasato and Behring discovered antibodies in animal serum in the late 19th century, the structure, function and expression of antibodies or immunoglobulins have posed exciting and important questions in immunology. There is no doubt that immunoglobulins are essential molecules in the immune system since most infectious diseases can be prevented or cured by appropriate specific antibodies.

Protein chemical studies on the immunoglobulin structure showed, firstly, that the light chain of an immunoglobulin molecule is composed of variable and constant regions. This discovery, however, served only as the vanguard to further questions: How can single polypeptides with variable and constant regions be synthesized? How are the immunoglobulin genes organized? How can so many variable regions be produced by a limited number of genes?

We had to wait until the next major technical development, namely DNA cloning, to elucidate the basic framework of dynamic rearrangement of the immunoglobulin genes. During the period up to this discovery, a variety of models were proposed, most of which did, however, turn out to be partially correct. A new technology had to become available to solve questions which emerged from immunology, but are fundamental to molecular genetics in the eukaryote. The development of the recombinant DNA techniques allowed us to explore the above questions in a straightforward manner.

During the past decade, we have accumulated an enormous amount of information on the organization, structure, rearrangement and expression of immunoglobulin genes in a variety of organisms. Studies on immunoglobulin genes have had a great impact not only on immunology but also on molecular biology in general. Such studies have provided many precedents for new concepts in eukaryotic molecular biology: exon–intron organization, differential splicing, site-specific as well as region-specific recombination, gene deletion and somatic mutation are examples.

This book provides up-to-date overviews of various aspects of immunoglobulin genes by authors who have actively participated in the accumulation of our knowledge on this subject. The editors hope that this book will serve as a prelude to further advancement and look forward to new developments in the field.

Tasuku Honjo
Frederick W. Alt
Terry H. Rabbitts

Preface to second edition

Since the first edition of *Immunoglobulin Genes* appeared five years ago, we have accumulated a large amount of knowledge about the structure and function of immunoglobulin (Ig) genes and the mechanisms that regulate their assembly and expression. We now have substantial information concerning the organization and structure of the human Ig heavy chain and κ light chain loci. In particular, the complete human Ig heavy-chain variable region (V_H) locus has been isolated in the form of linked cosmids and yeast artificial chromosomes; all of the V_H gene segments have been precisely arrayed within the locus, the 5′ end of which is linked to the telomeric sequence of human chromosome 14. A fundamental question in immunology has been the nature of the enzymatic machinery involved in the assembly of Ig variable region genes, and the related TCR variable region genes, from their germ-line V_H, D and J_H segments and the mechanisms by which this process (V(D)J recombination) is regulated. This area has witnessed remarkable progress following the seminal discovery of the Recombination Activating Genes (RAG) 1 and 2, the simultaneous expression of which generates V(D)J recombination activity in any mammalian cell type. The RAG gene discovery has facilitated numerous ongoing studies which have identified novel genes which encode products involved in the V(D)J recombination and DNA double strand break repair processes, at least one of which has been implicated as the target of the Murine Scid mutation. An additional major advance has been the rapid development of the gene targeted mutation technology which has been particularly useful to address questions concerning the regulation of Ig gene assembly and expression. 'Knock-out' experiments which have focused on various sequences within the Ig gene loci have elucidated both expected and unexpected functions of various exons and sequences.

In spite of the rapid progress in the aforementioned areas, there still remain many fundamental areas of investigation which are wide open for future in-depth studies. The molecular and cellular mechanisms that regulate assembly of Ig heavy and light chain genes remain a subject of intense investigation. In particular, the long debated phenomena of allelic and isotype exclusion and the mechanisms that regulate this process are still not well understood. Despite considerable progress in the V(D)J recombination field, a similar level of mechanistic understanding of the class switch recombination and somatic hypermutation processes awaits the further development of model systems and the identification of involved molecules. The function of Ig as a surface receptor and mechanisms and processes involved in its signaling has been another rapidly emerging area of new discoveries and progress, but much remains to be done. Finally, we expect that the complete nucleotide sequences of the Ig heavy and light chain loci will be determined by the time we complete the next edition of this book. Such information, coupled with the increasing ability to perform targeted genetic manipulations in animals, should allow very rapid advances with respect to elucidation of mechanisms that regulate Ig gene assembly and expression.

The second edition of *Immunoglobulin Genes* provides updated reviews of various aspects of the structure, function, and regulation of Ig Genes, and the related aspects of B-lymphocytes development and function, by authors who have actively participated in the accumulation of this knowledge. We hope that this book will continue to serve as a reference to past progress and as a stimulus for future studies. We look forward to many exciting developments in Ig gene research during the next five years.

Tasuku Honjo
Fred Alt
June, 1995

PART I: B CELLS

1

B-cell differentiation in humans

Peter D. Burrows*, Harry W. Schroeder Jr* and Max D. Cooper*†

Division of Developmental and Clinical Immunology, Departments of Microbiology, Medicine and Pediatrics, University of Alabama at Birmingham; and †Howard Hughes Medical Institute, Birmingham, Alabama, USA

Introduction	3
B-cell differentiation antigens	8
B-cell ontogeny	17
Development of the antibody repertoire	17
Immunodeficiency diseases	22
Conclusions	25

INTRODUCTION

Our current understanding of B cell development is derived from comparative studies in humans and non-primates, including rodents and non-mammalian species such as birds. This chapter provides an overview of the developmental highlights of human B cell differentiation that emphasizes features whose conservation in multiple species underscores their importance, while at the same time illustrating aspects of the process that are unique in humans.

B-CELL DIFFERENTIATION ANTIGENS

The discovery of the functionally distinct T and B lineages of lymphocytes that are responsible for cell-mediated and humoral immunity was originally made in chickens, and it was immediately apparent that the pathogenesis of primary immuno- deficiencies and lymphoid malignancies in mammals could be most easily explained by defects in either T or B cell development or both. Early attempts to develop serologic reagents to distinguish B and T lymphocytes relied on cross-species immunization with cells or cell-derived proteins, followed by absorption of the resulting heteroantisera to the point of acceptable specificity. The success of this strategy was variable, with the notable exception of antisera to immunoglobulins, which soon became the most reliable and readily detectable marker for B cells and their various differentiation stages.

Immunoglobulins and immunoglobulin genes

Immunoglobulins are composed of two identical heavy and two identical light chains held together by disulfide bonds as well as by extremely stable non-covalent interactions. The amino-terminal portion (100–110 amino acids) of both heavy and light chains varies among different immunoglobulins and forms the antigen-binding portion of the molecule. Each variable domain is the product of a complex series of gene rearrangement events. Starting with a human germline repertoire that includes fewer than 300 elements, combinatorial joining of V_H, D_H and J_H gene segments (for heavy chains) and of V_L and J_L gene segments (for light chains) coupled with combinatorial association of the resulting heavy and light chains can yield more than 10^6 different antibody molecules. Non-germline encoded nucleotides (N regions) can be added at the site of gene segment joining, a process catalyzed by the enzyme terminal deoxynucleotidyl transferase (TdT) (Desiderio et al., 1984; Gilfillan et al., 1993; Komori et al., 1993). TdT is detected in B lineage cells prior to production of μ heavy chains, and the lack of N region addition in murine V_L–J_L junctions has been attributed to the absence of TdT protein in cells at the time of light chain gene rearrangement. Recent data have demonstrated N region addition in a significant subpopulation of human κ chains (Lee et al., 1992; Victor and Capra, 1994), suggesting either that TdT expression is not always extinguished in pre-B cells prior to light chain gene rearrangement, or that light chain gene rearrangement may precede heavy chain gene rearrangement in some B cell precursors (Kubagawa et al., 1989). N regions can be inserted between V and D, between Ds in D–D fusions, between D and J, and between V and J in light chains (Sanz, 1991; Yamada et al., 1991; Lee et al., 1992; Victor and Capra, 1994), and every codon added as a result of N region addition increases the potential diversity of the repertoire 20-fold. Thus, more than 10^{10} different heavy and light chain V(D)J junctions can be generated at the time of gene segment rearrangement, yielding a potential preimmune repertoire in excess of 10^{16} different antibodies.

Outside of the variable regions the remainder of the heavy and light chains are constant in amino acid sequence except for occasional polymorphic residues (allotypes) that vary in the population and have no known biologic consequences. The different antibody classes, IgM, IgD, IgG_{1-4}, IgE and $IgA_{1\ and\ 2}$, have different immunologic effector functions that are determined by the constant region of the heavy chain. Each constant region is encoded by separate gene segments in the immunoglobulin H-chain locus on chromosome 14q32.33, which also contains two non-functional pseudo constant region genes arranged as follows: $C_\mu, C_\delta, C_\gamma 3, C_\gamma 1, \psi\epsilon, C_\alpha 1, \psi\gamma, C_\gamma 2, C_\gamma 4, C_\epsilon$ and $C_\alpha 2$ (Honjo et al., 1989). All immunoglobulin classes can exist as either transmembrane or soluble proteins, a choice regulated at the level of RNA cleavage/polyadenylation (Mather et al., 1984; Guise et al., 1989; Staudt and Lenardo, 1991). Light chains, which have no transmembrane segment, are of two types, κ and λ, these are encoded in unlinked loci located respectively on chromosomes 2p12 and 22q11.1–q11.2. A notable difference among species is that approximately 60% of human serum immunoglobulin contains κ light chains, while in mice and horses this figure is 95% and 5%, respectively. The relative serum levels correlate well with the frequency of κ- and λ-bearing B cells. The paucity of λ-containing immunoglobulin in mice is influenced by two factors: the simplicity of the locus and the fact that recombination signal sequences (RSS) required for $V_\lambda \rightarrow J_\lambda$ gene segment joining are less efficient rearrangement substrates than the κ RSS (Ramsden and Wu, 1991). Horses have a relatively complex κ light chain gene locus (Ford et al., 1994), so that the preponderance of λ-bearing immunoglobulins in this species must have another explanation.

Surface immunoglobulin (sIg)-associated proteins

Only three amino acids of the carboxyl terminus of μ and δ heavy chains extend beyond the transmembrane domain and into the cytoplasm of B lineage cells. None of these amino acids is a serine or tyrosine, potential substrates of protein kinases involved in signaling through the B cell receptor (BCR). A similar situation exists for the T cell receptor (TCR) αβ and γδ heterodimers, where it has been shown that a complex of proteins (CD3) associates with the TCR during its assembly, chaperones the receptor to the cell surface, and there transduces a signal following antigen receptor aggregation (Jorgensen et al., 1992; Malissen and Schmitt-Verhulst, 1993). Two similar disulfide-linked proteins have been found to be non-covalently associated with sIg in mice, humans and other vertebrate species (Reth et al., 1991; Cambier and Campbell, 1992; Reth, 1992). These molecules were discovered independently as proteins and cDNAs and thus have several names, Igα/mb-1 and Igβ/B29, and have also recently been assigned CD designations, CD79a and CD79b. Since Igα and Igβ are the most well-recognized names, these are used in this chapter. The BCR is thus a multimeric protein complex, with sIg providing antigen specificity and the Igα/β heterodimer conferring signal transduction capability on the receptor (DeFranco, 1993; Nakamura et al., 1993; Desiderio, 1994; Peaker, 1994). As for the TCR, the entire BCR complex must be assembled for it to be expressed on the cell surface.

The cytoplasmic tails of Igα/β contain the antigen receptor homology motif (Reth, 1992), which consists of paired tyrosines and leucines (D/EXXYXXL(X)$_{6-8}$YXXL) and which is also present in several other signal transducing molecules in the immune system. These include the ζ chain, which is a component of both the TCR and the Fcγ receptor III; the CD3 γδε and η chains; and the γ and β chains of the high-affinity Fc receptor for IgE on mast cells, FcεRI (Weiss and Littman, 1994). Phosphorylation of tyrosines in this motif allows SH2 domain-mediated association of cytoplasmic protein tyrosine kinases, initiating a cascade of events that culminate in modulation of gene expression at the transcriptional level. (A human disease caused by a defective cytoplasmic protein kinase is described in the section on immunodeficiency.) The eventual outcome of antigen recognition by the BCR depends on the stage of differentiation of the B cell, the nature of the antigenic stimulus, and the availability of T cell help. B cells may thus be activated to proliferate and differentiate, rendered unresponsive or anergic, or induced to die by apoptosis.

Stages of B-cell development

Based on immunoglobulin gene rearrangement status and expression, B lineage cells can be divided into four general stages of differentiation: progenitor (pro-B) cells, precursor (pre-B) cells, B cells and plasma cells.

Pro-B cells do not express immunoglobulin proteins but may actively transcribe their immunoglobulin heavy chain gene segments before their rearrangement. These sterile transcripts may contain isolated V_H gene segments or C_μ domains. Other sterile transcripts initiate upstream of the J_H proximal $D_H Q52$ gene segment, transcribe through $J_H 1$, and splice appropriately to C_μ. Germline transcription is thought to reflect the accessibility of an immunoglobulin locus to the recombination enzymes involved in V(D)J joining (Yancopolous and Alt, 1986). Recombination in the heavy chain locus typically begins with $D_H \rightarrow J_H$ joining in pro-B cells. Promoter sequences are found upstream of most D_H gene segments, and DJC_μ transcripts are easily detected (Schroeder and Wang, 1990; F. Mortari and H.W. Schroeder, unpublished observations). In the mouse, a conserved ATG translation start site exists upstream of the second of three potential D_H reading frames (rf2) (Gu *et al.*, 1991), and a truncated D_μ protein has been described in virally transformed pre-B cell lines (Yancopoulus and Alt, 1986). However, open reading frames are uncommon upstream of human DJC_μ transcripts, and D_μ proteins have not been described in humans.

Pre-B cells are the earliest cells of B lineage to produce conventional immunoglobulin proteins. They were originally defined as being devoid of sIg while containing small amounts of μ heavy chains in the cytoplasm (reviewed by Burrows and Cooper, 1990). Light chains are not synthesized by these cells because the immunoglobulin gene rearrangement processes in the κ or λ loci usually begin later (Yancopolous and Alt, 1986). The newly synthesized μ chains associate with BiP/GRP78 (Ig *b*inding *p*rotein/*g*lucose *r*egulated *p*rotein of molecular mass 78 kDa) (Haas and Wabl, 1983; Munro and Pelham, 1986) and other resident endoplasmic reticulum (ER) proteins that serve as quality control monitors for heterodimeric proteins including immunoglobulin and the major histocompatibility complex (MHC) class II α/β chains (Kelley and Georgopoulos, 1992; Doms *et al.*, 1993; Gaut and

Hendershot, 1993; Bergeron et al., 1994; Cresswell, 1994). These incompletely assembled immunoglobulin molecules are retained with BiP in the ER until the noncovalent BiP–μ heavy chain interaction is dissociated by light chains. In light of a stringent mechanism that prevents free μ chains from being expressed on the cell surface, it was therefore surprising when pre-B cell lines derived from patients with acute lymphocytic leukemia were found to be dimly positive for immunofluorescence staining with anti-μ antibodies (Findley et al., 1982). Studies in mice and humans later revealed that these were not free μ chains but instead were associated with a complex of two or three proteins with molecular masses ranging from 16 to 22 kDa (Pillai and Baltimore, 1988; Bauer et al., 1988; Kerr et al., 1989). These proteins behave as surrogate or pseudo-light chains (ψLC).

The genes encoding the ψLC proteins, λ5 and V_{preB} were cloned in mice by Sakaguchi, Kudo, and Melchers (Sakaguchi and Melchers, 1986; Kudo and Melchers, 1987), who were performing a systematic screen of a subtracted cDNA library in search of genes whose expression was restricted to pre-B cells. The human counterpart genes, 14.1 and V_{preB} (Chang et al., 1986; Bauer et al., 1988; Hollis et al., 1989; Bossy et al., 1991; Schiff et al., 1991), are located within the λ light chain locus on chromosome 22. The organization of the λ5 and V_{pre-B} genes and their predicted protein sequences show considerable homology to conventional λ light chains, the λ5 gene containing J_λ- and C_λ-like sequences and the V_{pre-B} genes including a V_λ-like sequence. A critical difference between conventional and ψLC genes is that λ5 and V_{preB} gene rearrangement is not required for their expression, and mRNA encoding the ψLC can be detected in the earliest B cell precursors prior to the onset of immunoglobulin gene rearrangement.

The ψLC proteins may disrupt the BiP–μ heavy chain interaction in the ER of pre-B cells, allowing the μ chains to form a complex with the ψLC which can then leave the ER, trafficking first to the Golgi where the μ chains are glycosylated, and eventually to the cell surface. This earliest form of cell surface immunoglobulin is similar to conventional sIg in that it is non-covalently associated with the two other proteins, Igα and Igβ, that are required for transport of the μ–ψLC complex to the cell surface. However, surface μ–ψLC differs from conventional sIg in that it is expressed at much lower levels, and inter-clonal variability is contributed solely by the heavy chain, λ5 and V_{pre-B} being monomorphic proteins. An important question concerns when in development the ψLC receptors are expressed on the cell surface. Analysis of λ5 and V_{pre-B} mRNA in B lineage cell lines has indicated that the genes are expressed throughout the pro-B and pre-B cell stages (Sakaguchi and Melchers, 1986; Kudo and Melchers, 1987; Bauer et al., 1991; Schiff et al., 1991), but the issue of cell surface protein expression required the availability of monoclonal antibodies that recognize ψLC. Even with these reagents the analysis has been complicated by the very low levels of ψLC that are expressed on the cell surface. Our studies with human pre-B cells indicate that although cytoplasmic expression of ψLC occurs throughout pre-B cell differentiation, cell surface expression is confined to late stage pre-B cells (Nishimoto et al., 1991; Lassoued et al., 1993). Cross-linkage of the μ–ψLC complex on pre-B cell lines modulates these receptors and leads to signal transduction as measured by Ca^{2+} flux, but has no obvious effect on cell growth or differentiation (Takemori et al., 1990; Misener et al., 1991; Bossy et al., 1993; Lassoued et al., 1993; Nakamura et al., 1993). The striking reduction in pre-B and B cell production by λ5-deficient mice (Kitamura et al., 1992) suggests, however, that the μ–ψLC receptors transduce a signal that allows pre-B cell survival and B cell conversion.

Following a successful V→$J_κ$ or V→$J_λ$ gene rearrangement, IgM monomers are expressed on the surface of the *immature B cell*, non-covalently associated with Igα and Igβ. This multimeric complex serves as the BCR for antigen, with specificity provided by the *immunoglobulin* and signal transduction capabilities provided by Igα/β. Cells at this early stage in B cell development are exquisitely sensitive to clonal deletion if their receptor recognizes multimeric self-antigens with sufficient avidity (Cooper et al., 1980; Hasbold and Klaus, 1990; Goodnow, 1992; Nemazee, 1993).

The next modification of immunoglobulin expression by B lineage cells is the production of cell surface IgD (sIgD), endowing the *mature B cell* with two classes of antigen receptors, both with the same specificity since the variable regions of the μ and δ heavy chains are identical. This is accomplished by alternate splicing and cleavage/polyadenylation of a primary RNA transcript derived from the complex transcriptional unit: 5′–Leader–VDJ–$C_μ$–$C_δ$–3′ (Blattner and Tucker, 1984). Messenger RNA encoding four different types of heavy chain, $μ_m$, $μ_s$, $δ_m$ and $δ_s$, the membrane and secreted forms of μ and δ heavy chains respectively, can be derived from this locus. The simultaneous expression of IgM and IgD is conserved in rodents and primates, but the advantage of having two classes of sIg, each with identical specificity, has been a matter of speculation for many years. Homologous recombination in embryonic stem cells has been used to create mutant 'knock-out' mice unable to make IgD (Roes and Rajewsky, 1993). No dramatic effects on either B cell development or immune responses occurred, but subtle defects in affinity maturation of antibodies were observed during a secondary immune response. Why IgD should be important in this process is unclear.

The mature B cell can be triggered to proliferate and differentiate in either a T cell-independent fashion, by antigens such as carbohydrates that contain multiple repetitive antigenic determinants, or with the help of T cells and their cytokine products (Paul and Seder, 1994). The latter response is typical of those to protein antigens, and differs from a T-independent response in several important ways including the switching of immunoglobulin isotypes, production of high-affinity antibodies by somatic mutation, and generation of a long-lived pool of *memory B cells* (Tsiagbe et al., 1992; Berek, 1993; Gray, 1993; Kelsoe, 1994; Sprent, 1994). The proliferating B cells are relatively large lymphoblasts. The expression of IgD begins to decline at this stage, although terminal differentiation of IgD-switched cells occurs in humans with the production of measurable amounts of serum IgD (~30 µg ml^{-1}) (Kuziel et al., 1989; Kerr et al., 1991).

Immunologic memory is functionally defined for antibody responses as an increase in both the quantity and quality of antibodies produced following secondary challenge with a T cell-dependent antigen. The quantitative increases are also characterized by a switch to production of non-IgM class antibodies, and substantial increases in antibody affinity are the basis for the qualitative changes. This affinity maturation, originally measured by serologic assays, is the result of somatic hypermutation of rearranged variable region gene segments in responding B cells (Tonegawa, 1983). The cellular basis of this response has been elucidated through a combination of classical immunohistochemical and contemporary molecular techniques, although the mechanism for the selective hypermutation of antigen receptor genes in B cells [and perhaps in antigen-responsive T cells (Zheng et al., 1994)] remains an intriguing enigma. Both somatic mutation and isotype switching occur in a T cell-dependent fashion in germinal centers of secondary lymphoid organs (Berek et al., 1991; Jacob et al., 1991; Leanderson et al., 1992; Liu et al., 1992; Kuppers et al., 1993). These are sites of intense proliferation, during which essentially random mutations are introduced into the immunoglobulin variable regions. This is followed by a quiescent stage during which cells whose receptors bind antigen with sufficient affinity are selected for survival, and those which no longer bind antigen, or do so at too low an affinity, undergo apoptotic cell death. These two stages of B cell development occur in morphologically distinct compartments of the germinal center (Liu et al., 1992; Nieuwenhuis et al., 1992; MacLennan, 1994). Proliferation occurs in the 'dark zone', while antigen selection occurs in the 'light zone', a region rich in follicular dendritic cells, which expose the intact antigen on their surface where it mediates affinity-based selection of B cells, and in tingible body macrophages, which engulf apoptotic lymphocytes. Cells at various stages along this complex developmental pathway, which begins with a virgin B cell and ends with death, terminal differentiation, or long-term survival, can now be isolated based on cell density and the differential expression of a relatively limited panel of cell surface markers. Pascual et al. (1994) have followed the progression of cells through this pathway in human tonsils and isolated five fractions, Bm1–5, that are sequentially related. The immature high-density follicular mantle cells (Bm1 and Bm2), which surround the germinal centers, express IgD, while low-density germinal center cells (Bm3 and Bm4) are IgD negative and CD38 positive. The Bm1 and Bm2 population is separable based on activation status: Bm1 is $CD23^-$ and Bm2 is $CD23^+$. Bm3 corresponds to the dark zone centroblasts and expresses the glycolipid (globotriaosylceramide) CD77 antigen, while Bm4 contains the light zone centrocytes and is CD77 negative. Memory B cells in the tonsil, many of which are IgG positive, are negative for both CD38 and IgD.

Surviving B cells exit the germinal center and either differentiate into plasma cells, most of which are found in the bone marrow or lamina propria of mucosal tissues, or enter the recirculating memory cell pool. The phenotypic characteristics of these circulating memory cells are controversial, as are the related issues of whether persistence of antigen is required for chronic restimulation and maintenance of memory, and whether memory resides in the switched population of B cells or in an IgM/±IgD population also derived from the germinal center. One defining feature of memory B cells generated during a T cell-dependent immune response is the presence of somatic mutations (Jacob et al., 1991; Berek and Ziegner, 1993). Analysis of blood, tonsil and lymph node B lineage cells has shown that the variable regions of IgM⁺/IgD⁺ mature 'virgin' B cells have few if any somatic mutations, a situation that persists throughout life (Klein et al., 1993, 1994). IgG⁺ B cells from germinal center and blood have undergone significant mutational changes in both V_H and V_L genes (Andris et al., 1992; Ikematsu et al., 1993; Klein et al., 1993; Kuppers et al., 1993; Varade and Insel, 1993; Pascual et al., 1994), and the B cells are thought to be recirculating memory B cells derived from the germinal center. Mutations are also observed in IgM⁺/IgD⁻ cells (Klein et al., 1994) and the percentage of mutations is equivalent in this subpopulation irrespective of the cell source. However, the mutation frequency is lower than that seen in isotype-switched B cells, implying that these cells resided for a limited period of time in the specialized environment of the germinal center, joining the recirculating pool before isotype switching and accrual of high levels of somatic mutation. If one accepts the notion that isotype-switched B cells in the circulation are memory cells, data obtained from IgA⁺ cells (Irsch et al., 1994) indicates that they are resting lymphocytes [negative for CD23, IL-2Rα (CD25), and transferrin receptor (CD71)] of relatively uniform phenotype. The cells express the B lineage antigens CD19, CD20, CD21, and CD22, but not CD10, CD38, or CD77. This phenotype appears identical to that of memory cells isolated from germinal centers. These IgA⁺ cells have undergone switch rearrangement on both heavy chain alleles so that reversion to an IgM⁺ B cell is no longer a possibility.

The effector cells of the B lineage, terminally differentiated *plasma cells*, have an abundant cytoplasm and a highly developed rough ER and Golgi apparatus, features typical of a secretory cell. The most mature plasma cells lack sIgM since the reaction to generate the membrane or secretory forms of μ chain has been shifted to favor μ_s (9 : 1 $\mu_s : \mu_m$). This is in contrast to resting B lymphocytes, which have a scant cytoplasm that is nearly devoid of membranous organelles and a $\mu_s : \mu_m$ ratio of ~1 : 9 (Mather et al., 1984; Guise et al., 1989; Staudt and Lenardo, 1991). Plasma cell lines are unable to express sIgM for another reason: they no longer synthesize the Igα component of the BCR and thus any μ_m that is produced cannot be transported to the cell surface as sIgM. However, relatively immature plasma cells may continue to express Igα (Mason et al., 1991), and still express sIg. Although we have been describing development of IgM-secreting plasma cells, similar changes in morphology and mRNA content occur in plasma cells secreting other immunoglobulin isotypes.

Cluster of differentiation antigens and B cell development

The advent of monoclonal antibody (MAb) technology coupled with improved multicolor flow cytometry has revolutionized the serologic identification of lymphocyte subpopulations. Many of these antibodies have been made against human cell surface proteins that are highly immunogenic in mice, the major species for MAb production. In fact it seemed that a surfeit of antibodies was being produced, all having different names and different apparent specificities depending on the laboratory of origin and the assay system. This problem was solved by the creation of international workshops in which MAb were collected and tested for their pattern of reactivity with human leukocytes. Antibodies showing similar reactivity were placed into groups and given a cluster of differentiation (CD) number (Barclay et al., 1993). Genes encoding many of the CD antigens have now been cloned, often using the MAb that defined them. Still more CD antigens have been described, with provisional CD designations up to CD130 being assigned at the most recent workshop in 1993 (Schlossman et al., 1994). Many of these CD antigens are preferentially expressed on B lineage cells and have been shown to be important in their differentiation. A few proteins of clear relevance to B cell differentiation are described below.

CD34

CD34 is a highly glycosylated 110–120 kDa protein that is expressed by immature hemopoietic cells of all lineages and on the pluripotent hemopoietic stem cell (Greaves et al., 1992), so named for its ability to generate all hemopoietic lineages (Uchida et al., 1993). This latter feature of CD34 has great potential clinical significance since it provides a marker for isolating stem cells that can be used in bone marrow transplantation and as targets for gene therapy. The earliest recognizable cells committed to the B lineage can be identified by joint expression of CD34 and the pan B cell marker CD19.

CD40

CD40 is expressed throughout B cell development except in plasma cells, but is also expressed on other cell types involved in immune system development and function (Fuleihan et al., 1993a; Armitage, 1994; Banchereau et al., 1994). These include follicular dendritic cells, basophils, interdigitating cells in the T cell-rich areas of secondary lymphoid organs, and certain thymic epithelial cells. CD40 is a 45–50 kDa glycoprotein member of the nerve growth factor receptor superfamily, which includes the tumor necrosis factor α receptor. CD40 is involved in the non-MHC-restricted cognate interactions between B and T cells, especially those interactions in germinal centers leading to secondary immune responses featuring isotype switching and somatic mutation (Callard et al., 1993; Fuleihan et al., 1993a; Kehry and Hodgkin, 1993; Lederman et al., 1993; Gray et al., 1994). Its role in this process is crucial, as will be discussed in the section on immunodeficiency.

CD45

CD45, originally termed leukocyte common antigen, is a highly glycosylated protein of 180–240 kDa that is abundantly expressed on the surface of all hemopoietic cells except erythrocytes (Koretzky, 1993; Law and Clark, 1994;

Trowbridge and Thomas, 1994). Different isoforms of CD45 can be generated by alternate splicing to include or exclude various combinations of exons 4, 5, 6, and 7, and these isoforms may be differentially expressed by T and B lineage cells. The mRNA encoding the B cell isoform, CD45R (B220 in mice), contains all four of these exons. An important clue to CD45 function was suggested when the cDNA was cloned and found to encode a protein tyrosine phosphatase. The phosphatase domain is located in the cytoplasmic portion of the protein where it was predicted to counter the effects of protein tyrosine kinases, many of which are involved in proliferative responses. Most functional studies of CD45 have focused on T cells and have recently been reviewed (Penninger et al., 1993; Chan et al., 1994; Thomas, 1994). A surprising finding has been that CD45 does not generate an antiproliferative signal, but rather is crucial for activation of mature T cells. An important role for CD45 in T cell development was also revealed with gene knockout mice in which there are normal numbers of double positive ($CD4^+/CD8^+$) thymocytes, but few single positive thymocytes or peripheral T cells (Kishihara et al., 1993). Even though CD45 is normally expressed at all stages of the B cell lineage, the frequency of B cells was unaffected in the CD45-deficient mice. The B cells in these mice were not entirely normal, since they could be induced to proliferate with lipopolysaccharide (LPS) but not with anti-μ. LPS stimulation occurs independently of the BCR, while anti-μ is thought to mirror the normal physiologic triggering that occurs via BCR aggregation by antigen. The signal transduction defect in B cells from CD45-deficient mice has not yet been identified, and a counterpart human deficiency has not been described.

CD10

CD10 was called CALLA (common acute lymphoblastic leukemia antigen) because of its original identification on this precursor B cell malignancy. The 100 kDa protein, encoded by a large 80 kb gene on chromosome 3 in humans, is a cell surface enzyme, neutral endopeptidase, which cleaves small peptides at hydrophobic residues (LeBien and McCormack, 1989). It belongs to a family of metallopeptidases that are widely distributed in many tissues, but which display lineage and differentiation stage-specific expression among hemopoietic cells (Shipp and Look, 1993). Other examples include CD13, aminopeptidase N, which is expressed on myeloid cells; CD26, dipeptidyl peptidase IV, which is expressed primarily on T cells; and BP-1, aminopeptidase A, which is found on pro-B and pre-B cells and newly formed B lymphocytes (Wu et al., 1990). These ectoenzymes may activate or inactivate biologically active peptides that can modulate hemopoietic cell development or function. Among the known substrates of CD10 are the enkephalins, bradykinin, oxytocin, substance P, and fMet-Leu-Phe, but whether these are biologically relevant to B cell development is unknown. Salles et al. (1992, 1993) have demonstrated that inhibition of CD10 enzymatic activity enhances murine B cell development in vitro and in vivo, suggesting that CD10 may inactivate a peptide growth factor. CD10 is expressed on early precursors of both B and T cells and some bone marrow stromal cells, as well as the brush border epithelium of kidney and intestine. CD10 is not expressed on peripheral B cells, except for germinal center centroblasts and centrocytes. This is an interesting exception, since it appears that lymphoid cells that express CD10 are engaged in modification of their antigen receptor genes; thus CD10+ pre-B and pre-T cells are undergoing V(D)J rearrangement and somatic mutation and isotype switching are occurring in germinal center B cells.

CD19

CD19 is a 95 kDa glycoprotein member of the immunoglobulin gene superfamily with a large (~240) amino acid cytoplasmic domain (Tedder *et al.*, 1994). It is a B lineage marker whose expression begins prior to CD10 in lineage-committed precursors and is extinguished at the terminally differentiated plasma cell stage. Recent studies have identified CD19 as a key signaling molecule that may function at several B cell differentiation stages (Uckun *et al.*, 1993; Law and Clark, 1994; Peaker, 1994; Tedder *et al.*, 1994). Antibody crosslinking of CD19 induces a cascade of signaling events that include calcium mobilization and activation of phospholipase C and tyrosine and serine protein kinases, association with phosphoinositide (PI)3-kinase, and increased DNA binding activity of nuclear factor-κB NF-κB. In mature B cells, antibody crosslinking of CD19 inhibits subsequent activation via the BCR. On the other hand, when CD19 and the BCR are coligated, the cell becomes hypersensitive and able to proliferate in response to much lower concentrations of anti-μ than are required for optimal stimulation by BCR crosslinking alone. The analogous *in vivo* situation may arise from the ability of CD19 to associate with several other transmembrane glycoproteins on the B cell surface including Leu13, CD81, and the complement receptor 2 (CD21) to form a complex capable of mediating signal transduction. The biologically relevant crosslinking of CD19 and sIg is thought to be mediated by antigen that has fixed complement and thus can bind to both CD21 and antigen-specific sIg. This activation pathway is not essential but may be especially important when concentrations of antigen are limiting. A central role for CD19 in pre-B and pro-B cell signaling is also suggested by the finding that CD19 is physically and functionally associated with *src* family kinases in these cells. CD19 is phosphorylated following anti-CD19 treatment, and also following crosslinking of other cell surface receptors including sIgM, CD40, and CD72, suggesting that the CD19 pathway may intersect with other signaling pathways in B lineage cells.

CD19 also functions in homotypic adhesion, a reaction that can be dramatically demonstrated by incubating CD19-expressing cells with certain anti-CD19 MAb (Wolf *et al.*, 1993). The significance of homotypic adhesion *in vivo* is unclear, but heterotypic adhesion could also occur since follicular dendritic cells also express CD19 and may provide differentiation signals. There has also been speculation that an unknown counter-receptor for CD19 may exist on bone marrow stromal cells, and recent evidence suggests that CD77, a glycolipid antigen expressed on germinal center B cells, may bind CD19 (Maloney and Lingwood, 1994).

A potentially important role for CD19 in very early development of B lineage cells has also been recognized. For example, the stromal cell-derived cytokine interleukin (IL)-7 upregulates expression of CD19 on progenitor B lineage cells (Wolf et al., 1993). Recently we observed that IL-7 downregulates mRNA for the recombinase activating genes, *RAG*-1 and *RAG*-2, and for TdT, the nuclear enzyme responsible for inserting N sequences into V(D)J joins. CD19 ligation completely negates the IL-7-mediated downregulation of *RAG* gene expression without affecting the TdT response (Billips et al., 1995a,b). These findings suggest that immunoglobulin gene rearrangement may be modulated by antagonistic signals generated through CD19 and the IL-7 receptor.

CD20

CD20 is another pan B cell marker with an expression pattern similar to CD19. CD20, a molecule of 33–37 kDa, appears slightly later in B cell development and is expressed only on B lineage cells (Clark and Lane, 1991). The structure of CD20 is interesting in that it spans the membrane four times, and both the N and C termini of the protein appear to reside in the cytoplasm. It is a non-glycosylated protein but can undergo phosphorylation at both serine and threonine residues, thus accounting for its size heterogeneity. There are some indications that CD20 may be a calcium channel.

CD21

CD21 is a 145 kDa complement receptor (CR2) for the complement cleavage fragments C3d and C3dg (Fingeroth et al., 1984). During complement activation, C3b becomes covalently attached to antigen–antibody complexes via the classical pathway, or to the surface of certain pathogens by the alternative pathway. Bound C3b (~170 kDa) undergoes a sequential series of regulated proteolytic cleavages by complement family proteases to generate C3dg, a 40 kDa fragment that remains covalently attached to the activator surface. C3dg is eventually cleaved by serum proteases to yield soluble C3g and bound C3d (30 kDa).

CD21 also serves as a receptor for Epstein–Barr virus (EBV), a member of the herpesvirus family that can transform human B cells *in vitro* and is associated with malignant B cell lymphomas and smooth muscle cell tumors in immunosuppressed individuals. EBV has also been strongly implicated in the pathogenesis of Burkitt's lymphoma and nasopharyngeal carcinoma (Klein, 1994; Lee et al., 1995; McClain et al., 1995). CD21 is expressed on B cells, follicular dendritic cells, and a subset of thymocytes and epithelial cells (Fearon and Ahearn, 1990). Structurally, the extracellular portion of CD21 consists of 15 or 16 copies of a short consensus repeat (SCR), a motif that is also found in other complement and complement regulatory proteins including the C3b receptor (CR1, CD35), C4-binding protein, decay accelerating factor (DAF), and membrane cofactor protein (MCP). The binding sites for C3d and EBV are located in the two SCRs nearest the N terminus. CD21 can be involved in B cell activation via its association with CD19 (Fearon, 1993), and may also be important in binding of immune complexes to follicular dendritic cells.

CD21 was originally thought to be a marker only of mature B cells since its expression was undetectable by immunofluorescence on pre-B cells. The discovery that EBV could transform pre-B and even pro-B cells that had no immunoglobulin gene rearrangements (Katamine et al., 1984; Gregory et al., 1987; Kubagawa et al., 1988) led us to re-examine this issue. The OKB7 monoclonal antibody inhibits binding of both EBV and C3d to CD21 since the binding sites are overlapping. We found that this MAb also inhibited EBV transformation of pro-B and pre-B cells, indicating that these cells express enough of the receptor to allow viral entry (only a few molecules per cell may be sufficient) but that the levels of CD21 are too low to be detected by conventional methods (H. Kubagawa and P.D. Burrows, unpublished observations).

CD22

CD22 is a heterodimer of 130 and 140 kDa chains that has an interesting expression pattern during B cell development. It is found in the cytoplasm of pro-B, pre-B and B cells, and only appears on the cell surface in a subset of B cells (Clark and Lane, 1991; Clark, 1993). Perhaps there is a CD22-associated protein that is only synthesized in these B cells, analogous to the situation with Igα/β, which remain in the cytoplasm of pre-B cells until immunoglobulins are produced. The CD22⁺ B cell subset undergoes a Ca^{2+} flux and proliferates in response to anti-µ, and the expression of CD22 then ceases following activation; however, CD22 is again expressed on circulating IgA⁺ memory B cells (Irsch et al., 1994).

CD38

CD38 is a type II transmembrane glycoprotein whose distribution is dependent on cell lineage and differentiation and activation stages (Malavasi et al., 1994). In addition, there appear to be significant species differences in the expression of human and murine CD38 (Lund et al., 1995). In humans, expression tends to be highest among precursor cells: medullary thymocytes express CD38, but circulating T cells do not unless activated. Most natural killer (NK) cells constitutively express CD38 as do monocytes, although at lower levels than myeloid precursors in bone marrow. An immunoreactive CD38 protein is also found on red blood cells. Among B lineage cells, CD38 is found on early precursors but is not expressed by circulating B lymphocytes. CD38 is re-expressed *in vivo* in germinal centers on proliferating centroblasts and on centrocytes. Both normal and malignant plasma cells are CD38 positive, but memory B cells, which are also germinal-center-derived, no longer express this antigen. Antibody crosslinking of CD38 may lead to signal transduction, as evidenced by proliferation of T lineage and NK cells, but requires the presence of accessory cells and IL-2. The function of CD38 is unclear, although it may promote adhesion to endothelial cells and, most strikingly, CD38 shows predicted structural homology to cyclic ADP-ribosyl cyclase. Recombinant murine CD38 possesses this enzymatic activity (Howard et al., 1993), which converts nicotinamide adenine dinucleotide to cyclic ADP-ribose, an intracellular second messenger involved in calcium mobilization. The catalytic domain of CD38, however, is normally extracellular, and it is not yet clear whether substrate is available at that site, whether any cyclic ADP-ribose generated there exerts a physiologic effect, or if other mechanisms such as internalization of CD38 are required for its biologic activity. From a practical point of view, CD38 expression has been useful in dissecting the subpopulations of B cells responding to antigen exposure in germinal centers.

Overview of B-cell differentiation antigens

Figure 1 summarizes the expression pattern of some of the CD antigens discussed in this section, and illustrates the value of these markers for dissecting human B cell development. Multicolor flow cytometry easily allows separation of lymphoid cells

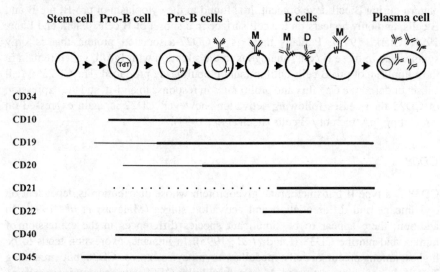

Fig. 1. Development of human B lineage cells. TdT is found in the nucleus of pro-B cells, and the expression of cytoplasmic μ heavy chains marks the onset of the pre-B cell stage. The surface receptor illustrated on the late stage pre-B cell is composed of μ heavy chains plus the ψLC complex and is non-covalently associated with the Igα/β heterodimer. The receptors on B cells are composed of either μ (M) or δ (D) heavy chains disulfide bonded to either κ or λ light chains; both sIgM and sIgD also associate with Igα/β. See text for description of immunoglobulin isotype switching and memory cell formation. Expression of differentiation antigens is illustrated by horizontal lines, the thickness of which is proportional to expression levels. See text for details of CD antigen expression.

on the basis of up to three independent markers of differentiation and their relative expression levels. The immunoglobulin genotype and phenotype and developmental potential of the cells as they progress along this differentiation pathway can then be elucidated, allowing abnormalities in this process to be readily recognized.

B-CELL ONTOGENY

Pre-B cells can be found in the fetal liver during week 8 of gestation, immature sIgM$^+$ B lymphocytes are detected by week 9 of gestation, whereas mature sIgM$^+$/sIgD$^+$ B cells appear in week 12 (Cooper, 1987). Isotype-switched B cells soon follow, and plasma cells secreting IgG or IgA antibodies may be detected by 20 and 30 weeks of gestation, respectively (Gathings et al., 1981). During the second trimester, production of B lineage cells shifts to the bone marrow. At birth, IgM$^+$/IgD$^+$ cells comprise approximately 20% of the circulating mononuclear cells, whereas IgG$^+$ and IgA$^+$ B cells are present in lower frequencies. Newborns can respond to prenatal infection with the production of IgM, IgG, and IgA antibodies (Stiehm et al., 1966).

Most of the serum immunoglobulin in the fetus comes from the mother. Active transport of maternal IgG across the human placenta begins around 20–21 weeks of gestation (Kohler and Farr, 1966), and the fetus acquires levels higher than that of the mother by birth. This maternally derived IgG is catabolized with a half-life of approximately 1 month so that the serum IgG concentration declines after birth. It reaches trough levels at 3–6 months, by which time the infant is beginning to make sufficient amounts of its own IgG, the levels of which slowly build up to adult values by about 6 years of age. Adult levels of serum IgM are typically reached around 1 year of age, whereas IgA does not reach adult values in serum until around puberty. Secretory IgA in mucosal secretions develops much more rapidly, attaining adult levels by 1 or 2 years (Husband and Gleeson, 1990; Smith and Taubman, 1992).

DEVELOPMENT OF THE ANTIBODY REPERTOIRE

The ability to respond to antigen develops in a programmed fashion

Antibodies can be generated with specificity against a limited range of antigens at an early stage in embryonic development (Silverstein, 1977). Premature infants weighing as little as 1500 g (26–28 weeks' gestation) can respond to vaccination with bacteriophage $\phi\chi 174$ (Uhr et al., 1962). Fetuses can respond to syphilis and toxoplasmosis by week 29 of gestation (Silverstein, 1977), and up to 0.6% of the cord blood IgG can derive from the infant (Maretensson and Fudenberg, 1965). The human neonate is unable to mount satisfactory responses to a number of microorganisms and vaccines (Stein, 1962). Almost all neonates are able to respond to tetanus toxoid before the first 6 months of life and to diphtheria toxin around 6 months of age (Cooke et al., 1948). However, children under the age of 2 years are often unable to respond to bacterial polysaccharides (Stein, 1962; Paton et al., 1986).

Development of the V_H antibody repertoire

The developmental delay in achieving a humoral response to certain antigens is surprising, given the stochastic mechanisms that underlie generation of the antibody repertoire. However, the fetus may not generate the full range of antibody diversity present in the adult. The restriction may correlate with preferential use of certain V_H gene segments and limited TdT expression, which thus limits the addition of N nucleotides in V(D)J junctions of fetal antibodies.

The consequences of limiting V utilization or reducing N region addition affect the three-dimensional structure of the antibody and its antigen-binding site (Kabat et al., 1991; Padlan, 1994). Each immunoglobulin variable domain contains three intervals of sequence hypervariability (termed complementarity determining regions, CDRs) separated from each other by four intervals of relatively constant sequence (termed frameworks, FRs). FRs 1, 2, and 3 and CDRs 1 and 2 are encoded by their respective V gene segments, and FR4 is contributed by the carboxy terminus of the J. Unlike these germline-encoded intervals, CDR3 is somatically generated by V(D)J joining and N region addition. The six juxtaposed CDR intervals (CDRs 1, 2, and 3 of the heavy and light chain, respectively) create an antigen-binding surface of approximately 25–30 nm^2 (Padlan, 1994). The heavy and light chain CDRs 1 and 2 form the outside borders of this antigen-binding surface and surround the paired heavy and light CDR3 intervals which form the center of the antigen-binding site (Kirkham and Schroeder, 1994). An average protein antigen may interact with about one-third of this surface (Padlan, 1994).

Located at the tip of the long arm of chromosome 14 are approximately 90 V_H gene segments, only half of which are potentially functional. Based on the extent of nucleotide similarity, these gene segments can be grouped into seven distinct families (Kirkham and Schroeder, 1994). These gene segments are identified by family and by position relative to the D_H locus (Matsuda et al., 1993). For example, V6-1 is the only member of the V_H6 family and is located immediately proximal to D_H, whereas V1-69 belongs to the multisequence V_H1 family and is 69 gene segments upstream from D_H. Members of the seven families are interspersed throughout the V_H locus (Schroeder et al., 1988; Matsuda et al., 1993). Gene segments belonging to the same family share similar FR structures and primarily diverge in the sequence and structure of their CDRs (Kirkham and Schroeder, 1994). Limitations in V_H family utilization lead to constraints in the underlying structure of the variable domains and affect antigen binding indirectly, whereas limitations in V_H gene segment utilization lead to direct constraints in the composition of the antigen-binding site itself.

All of the V_H families are used from week 8 of gestation through adult life. Within a family, however, individual gene segments are over-represented in fetal repertoires. For example, the V_H3 family contains more than 20 functional members, yet two gene segments, V3-23 and V3-30.3, contribute to more than 25% of fetal μ transcripts (Schroeder *et al.*, 1987, 1994; Schroeder and Wang, 1990; Raaphorst *et al.*, 1992; Cuisinier *et al.*, 1993). Studies in the mouse (Yancopolous *et al.*, 1984; Perlmutter *et al.*, 1985) and rabbit (Knight and Becker, 1990) have suggested that physical proximity to the heavy chain constant genes and their associated enhancer regions is correlated with increased use of V_H gene segments during fetal life. In one study that examined two unrestricted cDNA libraries generated from human 104- and 130-day gestation fetal liver mononuclear cells, gene segments from throughout the locus, ranging from V6-1 to V1-69, were detected (Schroeder *et al.*, 1987, 1994; Schroeder and Wang, 1990). However, although the sample size was small, 40% of the V_H gene segments expressed in 130-day fetal liver were upstream of V5-51, the most telomeric V_H gene segment detected in C_μ^+ transcripts from the 104-day fetal liver ($P < 0.02$, chi-squared). Most notably V1-69 (51p1) was not detected in the 104-day sample. Further support for the influence of physical proximity derives from analysis of the single member V_H6 family. Transcripts containing V6-1 were detected in fetal liver of 7 weeks' gestation (Cuisinier *et al.*, 1989), up to 10% of transcripts from second trimester fetal liver contained this gene segment (Schroeder *et al.*, 1994), whereas in adult life V_H6 contributed to less than 2% of the repertoire (Berman *et al.*, 1991; Huang and Stollar, 1993; Van Es *et al.*, 1993).

The proximity hypothesis has been further tested through analysis of the human V_H5 and $V_H1/7$ families (Cuisinier *et al.*, 1989; Pascual *et al.*, 1993). In some haploid genomes, the V_H5 family contains only two functional gene segments (V_H32 and V5-51), providentially located approximately one-fourth and two-thirds of the distance to the most distal V_H gene segment, V7-81. At 10 and 11 weeks' gestation, both V_H5 gene segments were used at equal frequency. However, in week 8–10 fetal liver samples, 9 of 12 $V_H1/7$ μ transcripts (75%) used V1-2, the second most DJC-proximal V_H gene segment, whereas none used V1-69 (Schroeder *et al.*, 1994). Thus, in human as in mouse and rabbit, chromosomal position may play a role in preferential V utilization early in fetal life.

In spite of these relative differences, it should be emphasized that from the earliest stages of gestation examined, the repertoire is enriched for a small subset of V_H3 gene segments, including V3-30.3 and V3-23 (Schroeder *et al.*, 1987, 1994; Schroeder and Wang, 1990; Raaphorst *et al.*, 1992; Cuisinier *et al.*, 1993). Indeed, in individuals who lack V3-30.3, V3-23 is the most prominent component of the repertoire (Hillson *et al.*, 1992). V3-23 is of particular interest because it shares the greatest similarity in sequence and structure with gene segments preferentially used in murine fetal life (Kirkham and Schroeder, 1994). Moreover, V3-23 is highly utilized in cord blood as well as in the peripheral blood of adults (Mortari *et al.*, 1993; Stewart *et al.*, 1993). Antibodies incorporating V3-23 can encode pathogenic autoantibodies directed against DNA as well as antibodies protective against

Haemophilus influenzae, a specificity that arises after the age of 2 years in humans (Dersimonian *et al.*, 1990; Adderson *et al.*, 1991). Thus, although the relative frequency of use of some V_H gene segments varies with gestational age, a significant component of the repertoire is used from 8 weeks of gestation through adult life. If differences in the antibody repertoire contribute to the programmed acquisition of antigen specificities during ontogeny, then other components of the antigen-binding site must play a significant if not a major role.

In addition to V_H, the antigen-binding site is the product of the light chain repertoire, the varied combinations that can be made between heavy and light chains, and the composition of the heavy chain CDR3. Little is known of the development of the light chain repertoire, but considerable evidence suggests that CDR3 plays a major role in controlling the diversity of the antibody repertoire during ontogeny.

Controlled diversification of heavy chain CDR3

The heavy chain CDR3 (HCDR3) is generated by VDJ joining and N nucleotide addition. HCDR3 is found in the center of the antigen-binding site, and virtually all antigen–antibody interactions studied at the structural level reveal that HCDR3 is directly involved in binding to the antigen. The composition and structure of HCDR3 thus play a critical role in defining the antigen specificity of the antibody (Kirkham *et al.*, 1992; Padlan, 1994).

In the mouse, although utilization of individual D_H and J_H gene segments appears similar between fetal and adult life (Gu *et al.*, 1990; Feeney, 1992), the potential diversity of fetal HCDR3 is severely restricted due to limited N region addition. In humans, the diversity of the fetal CDR3 repertoire is also limited, but some of the mechanisms leading to these restrictions differ from those seen in mouse.

D_H and J_H utilization in humans differs as a function of developmental age. D_H utilization, J_H utilization and the distribution of HCDR3 lengths in a representative sample of first and second trimester fetal liver μ transcripts are compared with those observed in cord blood in Fig. 2. The J_H-proximal D_HQ52 gene segment contributes to a plurality of fetal transcripts, but is rare in the cord blood and in adult peripheral blood. In contrast, D_HDXP and DLR containing transcripts are frequent in cord blood and in peripheral blood, but rare in fetal tissues. Similarly, the fetal repertoire is enriched for use of J_H3 and 4, whereas cord blood and adult peripheral blood C_μ^+ transcripts are enriched for use of J_H 4, 5 and 6. Finally, D–D fusion is rare in fetal tissues, but common in cord blood and adult peripheral blood (Schroeder *et al.*, 1987, 1994; Nickerson *et al.*, 1989; Schroeder and Wang, 1990; Sanz, 1991; Yamada *et al.*, 1991; Mortari *et al.*, 1992; Raaphorst *et al.*, 1992; H.W. Schroeder, unpublished observations).

In addition to differences in D_H and J_H utilization, the extent of N region addition also increases with the age of the human fetus (Schroeder *et al.*, 1987; Sanz, 1991; Yamada *et al.*, 1991; Mortari *et al.*, 1992; Cusinier *et al.*, 1993). Unlike mouse,

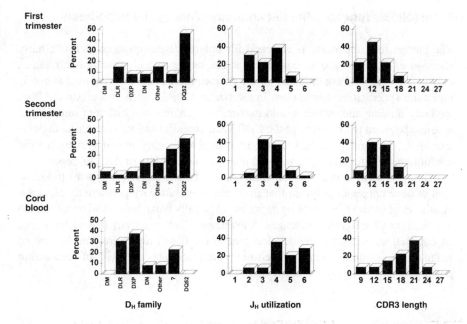

Fig. 2. D_H and J_H utilization and CDR3 length distributions during human ontogeny. D_H utilization, J_H utilization and the distribution of HCDR3 lengths in a representative sample of first and second trimester fetal liver μ transcripts are compared with those observed in cord blood.

which lacks TdT during fetal life, TdT can be detected in the human fetal liver as early as 8 weeks' gestation (J. Kearney, personal communication). However, while N regions are present at the V→D junction in the majority of first and second trimester C_μ^+ transcripts, N region addition at the D→J junction varies dramatically depending on the identity of the D_H gene segment. Nine of eleven $D_HQ52 \rightarrow J_H$ rearrangements had N additions with an average of six nucleotides per join (i.e. two codons), whereas only four of nine non-$D_HQ52 \rightarrow J_H$ rearrangements had N additions, averaging only one nucleotide per join. In contrast, N region addition in non-$D_HQ52 \rightarrow J_H$ cord blood transcripts averaged seven nucleotides per join (Mortari et al., 1992). D_HQ52 is the shortest human D_H specifying at most four codons, whereas D_H DXP and DLR gene segments can encode up to 10 codons. J_H 3 and 4 are the shortest J_H gene segments, J_H5 and 6 the longest. The lengths of H chain CDR3s in fetal transcripts averaged 11–12 codons and were all less than 18 codons long. Due to increased use of the longer D_H and J_H gene segments, the increased number of D–D fusions, and the increased extent of N region addition, cord blood transcripts contained an average of 17 codons, ranging between 6 and 24 codons in length. In humans, as in mice, the fetal repertoire thus lacks long CDR3 domains, although the human CDR3 intervals are less likely to be germline in sequence and different mechanisms are used to achieve the limitations in length and structure.

Form follows function: the flat antibody-binding site hypothesis

The human fetal repertoire is enriched for polyreactive sequences of low affinity (Guigou et al., 1991). It is possible that the sequence and structural limitations exhibited by HCDR3 are the basis for the characteristic features of the fetal antibody repertoire specificities. Limitations in the size of CDR3 results in a relatively 'flat' antibody-binding site, which would maximize the number of different interactions possible between the residues of the CDR3 and potential antigens, resulting in polyspecificity. However, the lack of topographical variability would prevent a tight conformational fit, correlating with the observed low affinity of the fetal repertoire.

In humans, as in mouse, diversification of the H chain repertoire appears to follow a strict developmental program that may reflect evolutionary pressure to express a plastic set of antibodies, enabling an immunologically naive individual to respond to a wide range of antigenic challenges. A low level of self-reactivity may also serve as an essential stimulant for the development of the B cell repertoire, but the cost of maintaining a multi-reactive repertoire could include the risk of generating potentially deleterious self-reactive antibodies.

IMMUNODEFICIENCY DISEASES

Impressive progress has been made recently in identifying the gene defects responsible for several inherited human immunodeficiencies, raising hope that this information can be translated into effective therapy or cure for these diseases in the foreseeable future.

X-linked agammaglobulinemia (XLA)

XLA is due to an intrinsic defect in the differentiation of B lineage cells (Rosen et al., 1995). Affected boys have near-normal numbers of pro-B and pre-B cells, but very few B lymphocytes and their plasma cell progeny. Because of their inability to produce antibodies, these patients have repeated infections, primarily with bacterial pathogens. They are protected from infections by monthly intravenous injections of pooled immunoglobulin from normal donors. Deficiency in a cytoplasmic protein tyrosine kinase has recently been found to be the cause of XLA (Tsukada et al., 1993; Vetrie et al., 1993). This src family kinase is called Bruton's tyrosine kinase (BTK) to honor Ogden Bruton, the physician who first diagnosed a boy with recurrent bacterial infections as lacking serum immunoglobulin and then successfully treated him with injections of pooled human immunoglobulin. Since BTK is expressed in all cells of B lineage and in myeloid cells as well, it is remarkable that defective BTK causes a bottleneck only in the differentiation of early B lineage cells. Recent data suggest defective reception of signals that are transduced by IL-7 and by CD19 ligation in pro-B cells from XLA patients (Billips et al., 1995a,b). IL-7 normally downregulates *RAG* and TdT gene expression in pro-B cells, an effect that is abolished for *RAG* but not TdT by CD19 ligation. However, TdT expression in pro-B cells from XLA patients is insensitive to IL-7 down-modulation. *RAG* gene expression remains sensitive to IL-7 but cannot be rescued by CD19 ligation. Thus essential signals transduced by the IL-7 receptor and CD19 may be aborted by the BTK defect in XLA progenitor B cells.

Hyper-IgM immunodeficiency (HIM)

HIM patients have high levels of IgM and sometimes IgD, but markedly reduced levels of the other immunoglobulin isotypes (Rosen et al., 1995). It has recently been discovered that this immunodeficiency is caused by a defect in a gene encoding the ligand for CD40 (Allen et al., 1993; Aruffo et al., 1993; DiSanto et al., 1993; Fuleihan et al., 1993b; Korthauer et al., 1993). The CD40L, a 39 kDa glycoprotein that is transiently expressed on activated T cells (primarily CD4$^+$ T cells), is involved in non-MHC restricted interaction of T cells with B cells, all of which express CD40 (Armitage, 1994; Banchereau et al., 1994). This interaction between antigen-specific helper T cells and B cells in germinal centers can rescue the B cell from apoptotic death, and induces proliferation, isotype switching, somatic hypermutation of V region genes, and plasma cell differentiation. Activated T cells in HIM patients either fail to express CD40L or express a mutated version that cannot bind to CD40 on B cells, thus accounting for the defect in isotype switching and affinity maturation. Given that CD40 is expressed at all stages of B lineage differentiation, it is interesting that a defective CD40L causes no abnormalities during earlier stages of development. This suggests that redundant pathways exist in pro-B and pre-B cells, or that another ligand for CD40 is important in early development. CD40-deficient mice also have normal numbers of B cells, indicating that CD40 is dispensable for early development of murine B lineage cells, but these mice have immunodeficiency patterns similar to those of the patients with CD40L gene defects (Castigli et al., 1994; Kawabe et al., 1994).

A minor group of HIM patients have normal CD40L expression, but their B cells display defective CD40 signal transduction (Conley et al., 1994). HIM may thus result either from genetic defects in CD40, CD40L components, or their signal transduction pathways.

X-linked severe combined immunodeficiency (XSCID)

XSCID features the absence of T cells and deficient antibody production despite the presence of B cells in normal numbers (Rosen et al., 1995). The faulty humoral immunity in these patients is due both to the lack of T cells and functional abnormality of their B cells. The gene for the common γ (γ_c) chain of the IL-2 receptor (IL-2R) has been mapped to the XSCID locus (Xq13), and mutations of the γ_c chain gene identified as the cause of XSCID (Noguchi et al., 1993a). The high-affinity IL-2R, which is important in T and B cell proliferation, is composed of IL-2R α, β chains and the γ_c chain. An intermediate-affinity IL-2R containing the IL-2Rβ and γ_c chains is constitutively expressed by resting T cells, neutrophils, and NK cells (Taniguchi and Minami, 1993; Liu et al., 1994). Although IL-2 is clearly an important cytokine for T cells (Paul and Seder, 1994), IL-2 knockout mice have apparently normal T cell development and function (Schorle et al., 1991). The severity of the XSCID phenotype is explained by the fact that the IL-2R γ_c chain is also a functional component of the receptors for IL-4, IL-7, IL-9, and IL-15 (Kondo et al., 1993, 1994; Noguchi et al., 1993b; Russell et al., 1993; Zurawski et al., 1993; Giri et al., 1994). In mice, IL-7 is important in T cell development and is crucial for B cell generation; thus, deletion of the γ_c chain gene aborts development of both T and B cells in this species (T. DiSantos and K. Rajewsky, personal communication). In contrast, the human XSCID γ_c mutations have no demonstrable effect on the development of B cells, thus implying the existence of a γ_c replacement chain in the human receptor for IL-7 or another growth factor for human B cell progenitors. The latter possibility is favored by the finding that IL-7 has minimal growth promoting potential for human pro-B or pre-B cells (Saeland et al., 1991; Billips et al., 1995a; Dittel and LeBien, 1995). These differences between mice and humans emphasize the importance of species differences in cellular requirements for individual cytokines and the potential for redundant cytokine pathways.

Wiskott–Aldrich syndrome

Wiskott–Aldrich syndrome (WAS) is an X-linked disorder characterized by thrombocytopenia, eczema, bloody diarrhea and both B and T cell dysfunction (Rosen et al., 1995). The inability of WAS patients to respond to polysaccharide antigens is an early hallmark of the immunodeficiency, and the progressive loss of T cells occurs later. An increased risk of developing lymphoreticular tumors and leukemias may be secondary to the immunodeficiency. CD43, also known as leukosialin and sialophorin, was an early candidate for the defective gene in WAS (Remold-O'Donnell et al., 1987). This 95–135 kDa mucin-like molecule is found on many hemopoietic cells including thymocytes, T cells, activated B cells, plasma cells, monocytes, macrophages, granulocytes, and platelets, and on hemopoietic stem cells (Barclay et al., 1993). Although expression of this heavily O-glycosylated molecule is clearly abnormal in WAS, chromosomal mapping localized the CD43 gene to 16p11.2, a location inconsistent with the X-linkage of the disease. A positional cloning strategy has recently been used to isolate the gene, *WASP* (WAS protein), that is mutated in WAS patients (Derry et al., 1994). *WASP* is expressed in lymphocyte and megakaryocyte cell lines in keeping with the thrombocytopenia and immunodeficiency seen in WAS. The function of the protein encoded by *WASP* is presently unknown, but sequence homologies suggest that it may be a transcription factor or may interact with SH3 domains of intracellular signal transduction proteins.

IgA deficiency and common variable immunodeficiency

IgA deficiency (IgA-D) is the most common primary immunodeficiency, occurring with a frequency of approximately 1 in 600 individuals of European ancestry (Schaffer et al., 1991; Strober and Sneller, 1991). The clinical consequences of IgA-D are highly variable. Many IgA-D individuals are healthy, while others have recurrent infections, gastrointestinal disorders, autoimmune disease, and malignancies. The ultimate health or illness of the patient is dependent on a number of poorly defined variables in this heterogeneous syndrome. For example, while most IgA-D individuals have normal levels of total serum IgG, many are deficient in IgG_2 and IgG_4 subclass antibodies. IgA-D and another heterogeneous immunodeficiency syndrome called common variable immunodeficiency (CVID) often occur in members of the same family and may represent polar ends of the same disease spectrum. Progression of IgA-D to CVID is sometimes seen in members of these families. CVID features a deficiency in all immunoglobulin classes as a consequence of defective terminal differentiation of B cells, which these patients produce in normal numbers (Rosen et al., 1995).

IgA deficiency is relatively infrequent in individuals of Asian ancestry, e.g. in Japan it occurs in approximately 1 in 18 500 (Schaffer et al., 1991; Volanakis et al., 1992). Recent studies have mapped IgA-D/CVID disease susceptibility to MHC haplotypes that are more common to European populations. Our studies predict linkage to a class III gene, possibly the C4A complement gene (Volanakis et al., 1992) while others have proposed linkage with an MHC class II gene, perhaps the HLA-DQβ chain gene (French and Dawkins, 1990). Individuals with a 'Sardinian haplotype' are informative in this regard in that they have undergone a crossover that splits the class II and class III regions in the disease-associated MHC haplotype. Analysis of a group of individuals homozygous for the Sardinian haplotype suggests that even though this haplotype contains the class II-linked susceptibility gene for insulin-dependent diabetes mellitus and celiac diseases, IgA-D is not seen in individuals who are homozygous for this allele (S. Powis, F. Cucca and J. Volanakis, unpublished observations). This study favors the idea of a non-MHC class II susceptibility gene for IgA-D/CVID, but the challenge remains to pinpoint the precise genetic basis and precipitating environmental factors for this immunodeficiency spectrum.

CONCLUSIONS

The human immunoglobulin heavy and light chain gene loci have been cloned in their entirety, and the details of the B cell developmental pathways in which these genes are rearranged and expressed are rapidly coming into sharper focus. This continuing explosion of information has been generated through complementary studies in humans and mice, principal among the many informative animal models.

Immunodeficient patients, whose defects have contributed much to our present understanding of the structural and functional aspects of the immune system, are likely to begin receiving their just rewards in the form of reparative gene therapy in the foreseeable future.

REFERENCES

Adderson, E.E., Shackelford, P.G., Quinn, A. and Carroll, W.L. (1991). *J. Immunol.* **147**, 1667.
Allen, R.C., Armitage, R.J., Conley, M.E., Rosenblatt, H., Jenkins, N.A., Copeland, N.G., Bedell, M.A., Edelhoff, S., Disteche, C.M., Simoneaux, D.K., Fanslow, W.C., Belmont, J and Spriggs, M.K. (1993). *Science* **259**, 990.
Andris, J.S., Ehrlich, P.H., Osterberg, L. and Capra, J.D. (1992). *J. Immunol.* **149**, 4053.
Armitage, R.J. (1994). *Curr. Opin. Immunol.* **6**, 407.
Aruffo, A., Farrington, M., Hollenbaugh, D., Li, X., Milatovich, A., Nonoyama, S., Bajorath, J., Grosmaire, L.S., Stenkamp, R., Neubauer, M., Roberts, R.L., Noelle, R.J., Ledbetter, J.A., Francke, U., and Ochs, H.D. (1993). *Cell* **72**, 291.
Banchereau, J., Bazan, F., Blanchard, D., Briere, F., Galizzi, J.P., van Kooten, C., Liu, Y.J., Rousset, F. and Saeland, S. (1994). *Annu. Rev. Immunol.* **12**, 881.
Barclay, A.N., Birkeland, M.L., Brown, M.H., Beyers, A.D., Davis, S J., Somoza, C. and Williams, A.F. (1993). 'The Leukocyte Antigen Facts Book'. Academic Press, London.
Bauer, S.R., Kudo, A. and Melchers, F. (1988). *EMBO J.* **7**, 111.
Bauer, S.R., Kubagawa, H., MacLennan, I. and Melchers, F. (1991). *J. Cell. Biol.* **102**, 1558.
Berek, C. (1993). *Curr. Opin. Immunol.* **5**, 218.
Berek, C. and Ziegner, M. (1993). *Immunol. Today* **14**, 400.
Berek, C., Berger, A. and Apel, M. (1991). *Cell* **67**, 1121.
Bergeron, J.J., Brenner, M.B., Thomas, D.V. and Williams, D.B. (1994). *Trends Biochem. Sci.* **19**, 124.
Berman, J.E., Nickerson, K.G., Pollock, R.R. Barth, J.E., Schuurman, R.K.B., Knowles, D.M., Chess, L. and Alt, F.W. (1991). *Eur. J. Immunol.* **21**, 1311.
Billips, L.G., Lassoued, K., Nunez, C., Wang, J., Kubagawa, H., Gartland, G.L., Burrows, P.D. and Cooper, M.D. (1995a). *Ann. N.Y. Acad. Sci.*, in press.
Billips, L.G., Nunez, C.A., Bertrand, F.E., Stankovic, A.K., Gartland, G.L., Burrows, P.D. and Cooper, M.D. (1995b). *J. Exp. Med.*, in press.
Blattner, F.R. and Tucker, P.W. (1984). *Nature* **307**, 417.
Bossy, D., Milili, M., Zucman, J., Thomas, G., Fougereau, M. and Schiff, C. (1991). *Int. Immunol.* **3**, 1081.
Bossy, D., Salamero, J., Olive, D., Fougereau, M. and Schiff, C. (1993). *Int. Immunol.* **5**, 467.
Burrows, P.D. and Cooper, M.D. (1990). *Semin. Immunol.* **1**, 189.
Callard, R.E., Armitage, R.J., Fanslow, W.C. and Spriggs, M.K. (1993). *Immunol. Today* **14**, 559.
Cambier, J.S. and Campbell, K.S. (1992). *FASEB Journal.* **6**, 3207.
Castigli, E., Alt, F.W., Davidson, L., Bottaro, A., Mizoguchi, E., Bhan, A.K. and Geha, R. S. (1994). *Proc. Natl Acad. Sci. USA* **91**, 12135.
Chan, A.C., Desai, D.M. and Weiss, A. (1994). *Annu. Rev. Immunol.* **12**, 555.
Chang, H., Dmitrovsky, E., Hieter, P.A., Mitchell, K., Leder, P., Turoczi, L., Kirsch, I.R. and Hollis, G.F. (1986). *J. Exp. Med.* **163**, 425.
Clark, E.A. (1993). *J. Immunol.* **150**, 4715.
Clark, E.A. and Lane, P.J. (1991). *Annu. Rev. Immunol.* **9**, 97.
Conley, M.E., Larche, M., Bonagura, V.R., Lawton, A.R., III, Buckley, R.H., Fu, S.M., Coustan-Smith, E., Herrod, H.G. and Campana, D. (1994). *J. Clin. Invest.* **94**, 1404.
Cooke, J.V., Holowach, J., Atkins, J.E. and Powers, J.R. (1948). *J. Pediatr.* **33**, 141.

Cooper, M.D. (1987). *N. Engl. J. Med.* **137**, 1452.
Cooper, M.D., Kearney, J. F., Gathings, W. E. and Lawton, A. R. (1980). *Immunol. Rev.* **52**, 29.
Cresswell, P. (1994). *Annu. Rev. Immunol.* **12**, 259.
Cuisinier, A.M., Guigou, V., Boubli, L., Fougereau, M. and Tonnelle, C. (1989). *Scand. J. Immunol.* **30**, 493.
Cuisinier, A.M., Gauthier, L., Boubli, L., Fougereau, M. and Tonnelle, C. (1993). *Eur. J. Immunol.* **23**, 110.
DeFranco, A.L. (1993). *Curr. Opin. Immunol.* **6**, 364.
Derry, J.M.J., Ochs, H.D. and Francke, U. (1994). *Cell* **78**, 635.
Dersimonian, H., Long, A., Rubinstein, D., Stollar, B.D. and Schwartz, R.S. (1990). *Int. Rev. Immunol.* **5**, 253.
Desiderio, S. (1994). *Curr. Opin. Immunol.* **6**, 248.
Desiderio, S.V., Yancopoulos, G.D., Paskind, M., Thomas, E., Boss, M.A., Landau, N., Alt, F.W. and Baltimore, D. (1984). *Nature* **311**, 752.
DiSanto, J.P., Bonnefoy, J.Y., Gauchat, J.F., Fischer, A.D. and Basile, G. (1993). *Nature.* **361**, 541.
Dittel, B.N. and LeBien, T.W. (1995) *J. Immunol.* **154**, 58.
Doms, R.W., Lamb, R.A., Rose, J.K. and Helenius, A. (1993). *Virology* **193**, 545.
Fearon, D.T. (1993). *Curr. Opin. Immunol.* **5**, 341.
Fearon, D.T. and Ahearn, JM. (1990). *Curr. Top. Microbiol. Immunol.* **153**, 83.
Feeney, A.J. (1992). *J. Immunol.* **149**, 222.
Findley, H.W., Cooper, M.D., Kim. J.H., Alvarado, C. and Ragab, A.H. (1982). *Blood* **60**, 1305.
Fingeroth, J.D., Weiss, J.J., Tedder, T.F., Strominger, J.L., Biro, P.A. and Fearon, D.T. (1984). *Proc. Natl Acad. Sci. USA* **81**, 45.
Ford, J.E., Home, W.A. and Gibson, D.M. (1994). *J. Immunol.* **153**, 1099.
French, M.A. and Dawkins, R. L. (1990). *Immunol. Today* **11**, 271.
Fuleihan, R., Ramesh, N. and Geha, R.S. (1993a). *Curr. Opin. Immunol.* **5**, 963.
Fuleihan, R., Ramesh, N., Loh, R., Jabara, H., Rosen, R.S., Chatila, T., Fu, S.M., Stamenkovic, I. and Geha, R.S. (1993b). *Proc. Natl Acad. Sci. USA* **90**, 2170.
Gathings, W.E., Kubagawa, H. and Cooper, M.D. (1981). *Immunol. Rev.* **57**, 107.
Gaut, J.R. and Hendershot, L.M. (1993). *Curr. Opin. Cell Biol.* **5**, 589.
Gilfillan, S., Dierich, A., Lemeur, M., Benoist, C. and Mathis, D. (1993). *Science* **261**, 1178.
Giri, J.G., Ahdieh, M., Eisenman, J., Shanebeck, K., Grabstein, K., Kumaki, S., Namen, A., Park, L.S., Cosman, D. and Anderson, D. (1994). *EMBO J.* **13**, 2822.
Goodnow, C.C. (1992). *Annu. Rev. Immunol.* **10**, 489.
Gray, D. (1993). *Annu. Rev. Immunol.* **11**, 49.
Gray, D., Dullforce, P. and Jainandunsing, S. (1994). *J. Exp. Med.* **180**, 141.
Greaves, M.F., Brown, J., Molgaard, H.V., Spurr, N.K., Robertson, D., Delia, D. and Sutherland, D.R. (1992). *Leukemia* **1**, (suppl. 6), 31.
Gregory, C.D., Kirchgens, C., Edwards, C.F., Young, L.S., Rowe, M., Rabbits, T.H. and Rickinson, A.B. (1987). *Eur. J. Immunol.* **17**, 1199.
Gu, H., Forster, I. and Rajewsky, K. (1990). *EMBO J.* **9**, 2133.
Gu, H., Kitamura, D. and Rajewsky, K. (1991). *Cell* **65**, 47.
Guigou, V., Guilbert, B., Moinier, D., Tonnelle, C., Boubli, L., Avrameas, S., Fougereau, M., and Fumoux, F. (1991). *J. Immunol.* **146**, 1368.
Guise, J.W., Galli, G., Nevins, J.R. and Tucker, P.W. (1989). In 'Immunoglobulin Genes' Academic Press, London. (T. Honjo, F.W. Alt and T.H. Rabbits, eds), p. 275.
Haas, I.G. and Wabl, M.R. (1983). *Nature* **306**, 387.
Hasbold, J. and Klaus, G.G.B. (1990). *Eur. J. Immunol.* **20**, 1685.
Hillson, J.L., Oppliger, I.R., Sasso, E.H., Milner, E.C. and Wener, M.H. (1992). *J. Immunol.* **149**, 3741.
Hollis, G.F., Evans, R.J., Stafford-Hollis, J.M., Korsmeyer, S.J. and McKearny, J.P. (1989). *Proc. Natl. Acad. Sci. USA* **86**, 5552.
Honjo, T. Shimizu, A. and Yaoita, Y. (1989). In 'Immunoglobulin Genes' Academic Press, London (T. Honjo, F.W. Alt and T.H. Rabbitts, eds), p. 123.

Howard, M., Grimaldi, J.C., Bazan, J.F., Lund, F.E., Santos-Argumedo, L., Parkhouse, R.M.E., Walseth, T.F. and Lee, H.C. (1993). *Science* **262**, 1056.
Huang, C. and Stollar, B.D. (1993). *J. Immunol.* **151**, 5290.
Husband, A.J. and Gleeson M. (1990). In 'Ontogeny of the Immune System of the Gut' (T.T. MacDonald, ed.) CRC Press, Boca Raton, Fl.
Ikematsu, H., Harindranath, N., Ueki, Y., Notkins, AL. and Casali, P. (1993). *J. Immunol.* **150**, 1325.
Irsch, J., Sigrid, I., Radl, J. Burrows, P.D., Cooper, M.D. and Radbruch, A.H. (1994). *Proc. Natl Acad. Sci. USA* **91**, 1323.
Jacob, J., Kelsoe, G., Rajewsky, K. and Weiss, U. (1991). *Nature* **354**, 389.
Jorgensen, J.L., Reay, P.A., Ehrich, E.W. and Davis, M.M. (1992). *Annu. Rev. Immunol.* **10**, 835.
Kabat, E.A., Wu, T.T., Perry, H.M., Gottesman, K.S. and Foeller, C. (1991). Sequences of Proteins of Immunological Interest, 5th edn, p. xiii–xcvi. USPHS, Bethesda, MD. **57**, 1.
Katamine, S., Otsu, M., Tada, K., Tsuchiya, S., Sato, T., Ishida, N., Honjo, T and Ono, Y. (1984). *Nature* **309**, 369.
Kawabe, T., Naka, T., Yoshida, K., Tanaka, T., Fuijwara, H., Suematsu, S., Yoshida, N., Kishimoto, T. and Kitutani, H. (1994). *Immunity* **1**, 167.
Kehry, M.R. and Hodgkin, P.D. (1993). *Semin. Immunol.* **5**, 393.
Kelley, W.L. and Georgopoulos, C. (1992). *Curr. Opin. Cell Biol.* **4**, 984.
Kelsoe, G. (1994). *J. Exp. Med.* **180**, 5.
Kerr, W.G., Cooper, M.D., Feng, L., Burrows, P.D. and Hendershot, L.M. (1989). *Int. Immunol.* **1**, 355.
Kerr, W.G., Hendershot, L.M. and Burrows, P.D. (1991). *J. Immunol.* **146**, 3314.
Kirkham, P.M. and Schroeder, H.W., Jr (1994). *Semin. Immunol.* **6**, 347.
Kirkham, P.M., Mortari, F., Newton, J.A. and Schroeder, H.W. (1992). *EMBO J.* **11**, 603.
Kishihara, K., Penninger, J., Wallace, V. A., Kündig, T.M., Kawai, K., Wakeham, A., Timms, E., Pfeffer, K., Ohashi, P.S., Thomas, ML., Furlonger, C., Paige, C.J., and Mak, T.W. (1993). *Cell* **74**, 143.
Kitamura, D., Kudo, A., Schaal, S., Müller, W., Melchers, F. and Rajewsky, K. (1992). *Cell* **69**, 823.
Klein, G. (1994). *Cell* **77**, 791.
Klein, U., Küppers, R. and Rajewsky, K. (1993). *Eur. J. Immunol.* **23**, 3272.
Klein, U., Küppers, R. and Rajewsky, K. (1993). *J. Exp. Med.* **180**, 1983.
Knight, K.L. and Becker, R.S. (1990). *Cell* **60**, 963.
Kohler, P.F. and Farr, R.S. (1966). *Nature* **210**, 1070.
Komori, T., Okada, A., Stewart, V. and Alt, F.W. (1993). *Science* **261**, 1175.
Kondo, M., Takeshita, T., Ishi, N., Nakamura, M., Watanabe, S. Arai, K.I. and Sugamura, K. (1993). *Science* **262**, 1874.
Kondo, M., Takeshita, T., Higuchi, M., Nakamura, M., Sudo, T., Nishikawa, S.-I. and Sugamura, K. (1994). *Science* **263**, 1453.
Koretzky, G.A. (1993). *FASEB J.* **7**, 420.
Korthauer, U., Graf, D., Mages, H.W., Briere, F., Padayachee, M., Malcolm, S., Ugazio, A.G., Notarangelo, L.D., Levinsky, R.J. and Kroczek, R.A. (1993). *Nature* **361**, 539.
Kubagawa, H., Burrows, P.D., Grossi, C.E., Mestecky, J. and Cooper, M.D. (1988). *Proc. Natl Acad. Sci. USA* **85**, 875.
Kubagawa, H., Cooper, M.D., Carrol, A.J. and Burrows, P.D. (1989). *Proc. Natl Acad. Sci. USA* **86**, 2356.
Kudo, A. and Melchers, F. (1987). *EMBO J.* **6**, 2267.
Kuppers, R., Zhao, M., Hansmann, M.L. and Rajewsky, K. (1993). *EMBO J.* **13**, 4967.
Kuziel, W.A., Word, C.J., Yuan, D., White, M.B., Mushinski, J.F., Blattner, F.R. and Tucker, P.W. (1989). *Int. Immunol.* **1**, 310.
Lassoued, K., Nunez, C.A., Billips, L.G., Kubagawa, H., Monteiro, R.C., LeBien, T.W. and Cooper, M.D. (1993). *Cell* **73**, 73.
Law, C.-L. and Clark, E.A. (1994). *Curr. Opin Immunol.* **6**, 238.
Leanderson, T., Kallberg, E. and Gray, D. (1992). *Immunol. Rev.* **126**, 47.
LeBien, T.W. and McCormack, R.T. (1989). *Blood* **73**, 625.
Lederman, S., Yellin, M.J., Covey, L.R. and Clearny, A.M. (1993). *Curr. Opin. Immunol.* **5**, 439.

Lee, E.S., Locker, J., Nalesnik, M., Reyes, J., Jaffe, R., Alashari, M., Nour, B., Tzakis, A. and Dickman, P.S. (1995). *N. Engl. J. Med.* **332**, 19.
Lee, S.K., Bridges, S.L., Jr, Koopman, W.J. and Schroeder, H.W., Jr (1992). *Arthritis Rheum.* **35**, 905.
Liu, J.H., Wei, S., Ussery, D., Epling-Burnette, E., Leonard, W.J. and Djeu, J.Y. (1994). *Blood* **84**, 3870.
Liu, Y.J., Johnson, G.D., Gordon, J. and MacLennan, I.C.M. (1992). *Immunol. Today* **13**, 17.
Lund, F., Solvason, N., Grimaldi, C. Parkhouse, R.M.E. and Howard, M. (1995). *Immunol. Today*, in press.
McClain, K.L., Leach, C.T., Jenson, H.B., Joshi, V.V., Pollock, B.H., Parmley, R.T., DiCarolo, F.J., Chadwick, E.G. and Murphy, S.B. (1995). *N. Engl. J. Med.* **332**, 12.
MacLennan, I.C. (1994). *Annu. Rev. Immunol.* **12**, 117.
Malavasi, F., Funaro, A., Roggero, S., Horenstein, A., Calosso, L. and Mehta, K. (1994). *Immunol. Today* **15**, 95.
Malissen, B. and Schmitt-Verhulst, A.M. (1993). *Curr. Opin. Immunol.* **5**, 324.
Maloney, M.D. and Lingwood, C.A. (1994). *J. Exp. Med.* **180**, 191.
Maretensson, L. and Fudenberg, H.H. (1965). *J. Immunol.* **94**, 514.
Mason, D.Y., Cordell, J.L., Tse, A.G.D., Van Dongen, J.J.M., Van Noesel, C.J.M., Micklem, K., Pulford, K.A.F., Valenski, F., Comans-Bitter, W.M., Borst, J., and Gatter, K.C. (1991). *J. Immunol.* **147**, 2474.
Mather, E.L., Nelson, K.J., Haimovich, J. and Perry, R.P. (1984). *Cell* **36**, 329.
Matsuda, F., Shin, E.K., Nagaoka, H., Matsumura, R., Haino, M., Fukita, Y., Taka-ishi, S., Imai, T., Riley, J.H., Anand, R., Soeda, E. and Honjo, T. (1993). *Nature Genet.* **3**, 88.
Misener, V., Downey, G.P. and Jongstra, J. (1991). *Int. Immunol.* **3**, 1129.
Mortari, F., Newton, J.A., Wang, J.Y. and Schroeder, H.W., Jr (1992). *Eur. J. Immunol.* **22**, 241.
Mortari, F., Wang, J.Y. and Schroeder, H.W., Jr (1993). *J. Immunol.* **150**, 1348.
Munro, S. and Pelham, H.R.B. (1986). *Cell* **46**, 291.
Nakamura, T., Sekar, M.C., Kubagawa, H. and Cooper, M.D. (1993). *Int. Immunol.* **5**, 1309.
Nemazee, D. (1993). *Curr. Opin. Immunol.* **5**, 866.
Nickerson, K.G., Berman, J.E., Glickman, E., Chess, L. and Alt, F.W. (1989). *J. Exp. Med.* **169**, 1391.
Nieuwenhuis, P., Kroese, F.G.M., Opstelten, D. and Seijen, H.G. (1992). *Immunol. Rev.* **126**, 77.
Nishimoto, N., Kubagawa, H., Ohno, T., Gartland, G.L., Stankovic, A.K. and Cooper, M.D. (1991). *Proc. Natl Acad. Sci. USA* **88**, 6284.
Noguchi, M., Masayuki, Y.H., Rosenblatt, H.M., Fillipovich, A.H., Adelstein, S., Modi, W.S., McBride, O.W. and Leonard, W.J. (1993a). *Cell* **73**, 147.
Noguchi, M., Nakamura, Y., Russell, S.W., Ziegler, S.F., Tsang, M., Cao, X. and Leonard, W. J. (1993b). *Science* **262**, 1877.
Padlan, E.A. (1994). *Mol. Immunol.* **31**, 169.
Pascual, V., Verkruyse, L., Casey, M. L. and Capra, J.D. (1993). *J. Immunol.* **151**, 4164.
Pascual, V., Liu, Y.-J., Magalski, A., de Bouteiller, O., Banchereau, J. and Capra, J.D. (1994). *J. Exp. Med.* **180**, 329.
Paton, J.C., Toogood, I.R., Cockinton, R.A. and Hansman, D. (1986). *Am. J. Dis. Child.* **140**, 138.
Paul, W.E. and Seder, R.A. (1994). *Cell* **76**, 241.
Peaker, C.J.G. (1994). *Curr. Opin Immunol.* **6**, 359.
Penninger, J.M., Wallace, V.A., Kishihara, K. and Mak, T.W. (1993). *Immunol. Rev.* **135**, 183.
Perlmutter, R.M., Kearney, J.F., Chang, S.P. and Hood, L.E. (1985). *Science* **227**, 1597.
Pillai, S. and Baltimore, D. (1988). *Curr. Top. Microbiol. Immunol.* **137**, 136.
Raaphorst, F.M., Timmers, E., Kenter, M., Van To, M., Vossen, J.M. and Schuurman, R.K. (1992). *Eur. J. Immunol.* **22**, 247.
Ramsden, D.A. and Wu, G.E. (1991). *Proc. Natl Acad. Sci. USA* **88**, 10721.
Remold-O'Donnell, E., Zimmerman E., Kenney, C. and Rosen, F.S. (1987). *Blood* **70**, 104.
Reth, M. (1992). *Annu. Rev. Immunol.* **10**, 97.
Reth, M., Hombach, J., Wienands, J., Campbell, K.S., Chien, N., Justement, L.B. and Cambier, J.C. (1991). *Immunol. Today* **12**, 196.
Roes, J. and Rajewsky, K. (1993). *J. Exp. Med.* **177**, 45.
Rosen, F.S., Wedgewood, R.J. and Cooper, M.D. (1995). *N. Engl. J. Med.*, in press.

Russell, S.M., Keegan, A.D., Harada, N., Nakamura, Y., Noguchi, M., Leland, P., Friedmann, M.C., Miyajima, A., Puri, R.K., Paul, W.E. and Leonard, W.J. (1993). *Science* **262**, 1880.
Saeland, S., Duvert, V., Pandrau, D., Caux, C., Durand, I., Wrighton, N., Widerman, J., Lee, F. and Banchereau. (1991). *Blood* **78**, 2229.
Sakaguchi, N. and Melchers, F. (1986). *Nature* **329**, 579.
Salles, G., Chen, C.Y., Reinherz, E.L. and Shipp, M.A. (1992). *Blood* **80**, 2021.
Salles, G., Rodewals, H.R., Chin, B.S. and Reinherz, E.L. (1993). *Proc. Natl Acad. Sci. USA* **90**, 7618.
Sanz, I. (1991). *J. Immunol.* **147**, 1720.
Schaffer, F.M., Monteiro, R.C., Volanakis, J.E. and Cooper, M.D. (1991). *Immunodeficiency Rev.* **3**, 15.
Schiff, C., Nilili, M., Bossy, D., Tabilio, A., Falzetti, F., Gabert, J., Mannoni, P. and Fougereau, M. (1991). *Blood* **78**, 1516.
Schlossman, S.F., Boumsell, L., Gilks, W., Harlan, J.M., Kishimoto, T., Morimoto, C., Ritz, J., Shaw, S., Silverstein, R.L., Springer, T.A., Tedder, T.F. and Todd, R.F. (1994). *Immunol. Today* **15**, 98.
Schorle, H., Holtschke, T., Hunig, T., Schimpl, A. and Horak, I. (1991). *Nature* **352**, 621.
Schroeder, H.W., Jr and Wang, J.Y. (1990). *Proc. Natl Acad. Sci. USA* **87**, 6146.
Schroeder, H.W., Jr, Hillson, J.L. and Perlmutter, R.M. (1987). *Science* **238**, 791.
Schroeder, H.W., Jr, Walter, M.A., Hosker, M.H., Ebens, A., Willems VanDijk, K., Lau, L., Cox, D.W., Milner, E.C.B. and Perlmutter, R.M. (1988). *Proc. Natl Acad. Sci. USA* **85**, 8196.
Schroeder, H.W., Jr, Mortari, F., Shiokawa, S., Kirkham, P.M., Elgavish, R.A. and Bertrand, F.E., III (1995). *Ann. N.Y. Acad. Sci.*, in press.
Shipp, M.A. and Look, A.T. (1993). *Blood* **82**, 1052.
Silverstein, A.M. (1977). In 'Development of Host Defense' (M.D. Cooper, ed.), pp. 1–10. Raven Press, New York.
Smith, P.J. and Taubman, M.A. (1992). *Crit. Rev. Oral Biol. Med.* **3**, 109.
Sprent, J. (1994). *Cell* **76**, 315.
Staudt, L.M. and Lenardo, M. (1991). *Ann. Rev. Immunol.* **9**, 373.
Stein, K.E. (1962). *J. Infect. Dis.* 165, S49.
Stewart, A.K., Huang, C., Stollar, B.D. and Schwartz, R.S. (1993). *J. Exp. Med.* **177**, 409.
Stiehm, E.R., Ammann, A.J. and Cherry, J.D. (1966). *N. Engl. J. Med.* **275**, 971.
Strober, W. and Sneller, M.C. (1991). *Ann. Allergy* **66**, 363.
Takemori, T., Mizugushi, J., Miyazoe, I., Nakanishi, M., Shigemoto, K., Kimoto, H., Shirasawa, Maruyama, T. and Taniguchi, M. (1990). *EMBO J.* **9**, 2493.
Tanaguchi, T. and Minami, Y. (1993). *Cell* **73**, 5.
Tedder, T.F., Shou, L.-J. and Engel, P. (1994). *Immunol. Today* **15**, 437.
Thomas, M.L. (1994). *Curr. Opin. Immunol.* **6**, 247.
Tonegawa, S. (1983). *Nature* **302**, 575.
Trowbridge, I.S. and Thomas, M.L. (1994). *Annu. Rev. Immunol.* **12**, 85.
Tsiagbe, V.K., Linton, P.J. and Thorbecke, G.J. (1992). *Immunol. Rev.* **126**, 113.
Tsukada, S., Saffran, D.C., Rawlings, D.J., Parolini, O., Allen, R.C., Klisak, I., Sparkes, R.S., Kubagawa, H., Mohandas, T., Quan, S., Belmont, J.W., Cooper, M.D., Conley, M.E. and Witte, W.N. (1993). *Cell* **72**, 279.
Uchida, N., Fleming, W.H., Alpern, E.J. and Weismann, I.L. (1993). *Curr. Opin. Immunol.* **5**, 177.
Uckun, F.M., Burkhardt, A.L., Jarvis, L., Jun, X., Stealey, B., Dibirdik, I., Myers, D.E., Tuel-Ahlgren, L. and Bolen, J.B. (1993). *J. Biol. Chem.* **268**, 21172.
Uhr, J.W., Dancis, J., Franklin, E.C., Finkelstein, M.S. and Lewis, E.W. (1962). *J. Clin. Invest.* **41**, 1509.
Van Es, J.H., Raaphorst, F.M., van Tol, M.J.D., Gmelig-Meyling, F.H.J. and Logtenberg, T. (1993). *J. Immunol.* **150**, 161.
Varade, W.S. and Insel, R.A. (1993). *J. Clin. Invest.* **91**, 1838.
Vetrie, D., Vorechovsky, I., Sideras, P., Holland, J., Davies, A., Flinter, F., Hammarstrom, L., Kinnon, C., Levinsky, R., Bobrow, M., Smith, C.I.E. and Bentley, D.R. (1993). *Nature* **361**, 226.
Victor, K.D. and Capra, J.D. (1994). *Mol. Immunol.* **31**, 39.
Volanakis, J.E., Zhu, Z.-B., Schaffer, F.M., Macon, K.J., Palermo, J., Barger, B.O., Go, R., Campbell, R.D., Schroeder, H.W., Jr and Cooper, M.D. (1992). *J. Clin. Invest.* **89**, 1914.

Weiss, A. and Littman, D.R. (1994). *Cell* **76**, 263.
Wolf, M.L., Weng, W.-K., Steiglbauer, K.T., Shah, N. and LeBien, T.W. (1993). *J. Immunol.* **151**, 138.
Wu, Q., Lahti, J.M., Air, G.M., Burrows, P.D. and Cooper, M.D. (1990). *Proc. Natl Acad. Sci. USA* **87**, 993.
Yamada, M., Wasserman, R., Reichard, B.A., Shane, S., Caton, A.H. and Rovera, G. (1991). *J. Exp. Med.* **173**, 395.
Yancopoulos, G.D. and Alt, F.W. (1986). *Annu. Rev. Immunol.* **4**, 330.
Yancopoulos, G.D., Desiderio, S.V., Paskind, M., Kearney, J.F., Baltimore, D. and Alt, F.W. (1984). *Nature* **311**, 727.
Zheng, B., Xue, W. and Kelsoe, G. (1994). *Nature* **372**, 556.
Zurawski, S.M., Vega, F., Huyghe, B. and Zurawski, G. (1993). *EMBO J.* **12**, 2663.

2

The role of B cell and pre-B-cell receptors in development and growth control of the B-lymphocyte cell lineage

Fritz Melchers

Basel Institute for Immunology, Basel, Switzerland

Introduction	33
Helper T-cell-dependent and -independent responses of mature, resting, surface Ig-positive B cells	35
Continuation and completion of the cell cycle of a proliferating B cell	39
Memory B-cell development and responses	42
Self-reactive B-cell mutants	43
Suppressive mechanisms involving the Ig molecule	44
Roles of Ig and Ig-like molecules in early phases of B-cell development	45
Functions of the μH chain–surrogate L chain pre-B receptor	48
The roles of Ig on immature B cells	50
Deletion versus anergy of B cells	51
Conclusions	53

INTRODUCTION

Antigen-specific receptors (B-cell receptors (BCR) or immunoglobulins (Ig) on B lymphocytes, and T-cell receptors (TCR) on T lymphocytes) control the specificity and clonal selection of immune responses of lymphocytes. This chapter considers the roles of Ig and Ig-like molecules on B-lineage committed lymphocytes. It is likely that TCR molecules on T-lineage committed lymphocytes follow much the same principles in their control of the development, growth and maturation of T lymphocytes.

Binding of antigen to pre-existing Ig receptors on lymphocytes determines the specificity of an immune response. More than 97% of all mature B lymphocytes produce only one of the many Ig receptors of the immune system and deposit it on the surface for recognition of antigen. Antigen binds by chemical complementarity with variable (V) regions of Ig molecules and thereby selects its specific lymphocytes. In a 'positive' immune response, binding of an antigen leads to proliferation of clones of antigen-specific lymphocytes and to maturation of effector cells within these clones (Ehrlich, 1904; Jerne, 1955; Kabat, 1956; Humphrey and Porter, 1957; Burnet, 1959; Sell and Gell, 1965).

Lymphocytes of the various lineages are generated from stem cells, first during embryonic development and later continuously throughout adult life. The development of B lymphocytes with their diverse repertoires of antigen-recognizing Ig receptors is ordered by a process of sequential rearrangements of gene segments encoding the antigen-binding V regions of Ig molecules (Tonegawa, 1983). Soon after the dividing and differentiating precursor (pre-) B lymphocytes have completed their gene rearrangements and have expressed Ig on their surface, they fall into a resting state, ready to respond to antigen or to die within a few days. Daily almost 1% of the total cellular pool of the immune system is newly generated; consequently, an equal number of lymphocytes die. The daily generation of cells is maintained by the proliferation of precursor states of lymphocytes, which is internally driven in the absence of foreign antigens (Gowans, 1957; Simonsen, 1957; Sprent, 1973; Osmond, 1986).

It was originally expected that the binding of antigen to Ig molecules on the cell surface of a mature B lymphocyte would, by itself, suffice to signal a lymphocyte to enter and complete the cell cycle, i.e. to proceed through G_1, S and G_2 phases and through mitosis. This view had to be altered when it was discovered that two other cell types—helper T cells and accessory (A) cells (macrophages)—have to cooperate with B cells in most of their responses and, correspondingly, that two determinants (the hapten and the carrier) were needed for an antigen to be immunogenic (Claman et al., 1966; Mosier, 1967; Rajewsky et al., 1969; Mitchison, 1971). Furthermore, it became evident that B cells could be activated polyclonally, under the apparent circumvention of binding to surface Ig, by mitogens such as lipopolysaccharides (LPS) irrespective of the antigen-binding specificity or the occupancy of the Ig (Andersson et al., 1972). Mitogens also activated B cells without T-cell help, inducing them to successive cell cycles and to maturation. It became clear later that they were still dependent in their responses on A cells (Chused et al., 1976; Corbel and Melchers, 1984). The action of cooperating cells and of mitogens in B-cell responses make it plausible that at least a part of the total proliferative and maturational response could be effected by antigen-non-specific, polyclonally stimulating factors (Andersson et al., 1972; Schimpl and Wecker, 1972; Waldmann and Munro, 1973).

The problem of B-cell growth control was further complicated by the finding that antigen-specific, T- and A-cell-dependent B-cell cooperation was restricted by the allelic forms of the major histocompatibility complex (MHC) genes, in particular of class II MHC genes, which are expressed as membrane glycoproteins on the surface of A and B cells (Katz et al., 1973; Rosenthal and Shevach, 1973). It became evident that control of growth and maturation of B cells must be effected by a series of signals given by different ligands from the two cooperating cells via different surface receptors on B cells (Bretscher and Cohn, 1970; Coutinho and Möller, 1975). Today it is clear that such MHC restriction of B-cell responses operates at two levels (Sprent, 1978; Andersson et al., 1980). At the first level, helper T lymphocytes recognize foreign antigen on A cells together with MHC class II molecules (see Figs 1 and 4). At the second level the same helper T lymphocytes recognize the same foreign antigen on B cells in the context of MHC class II molecules. For this to happen, antigen has to be taken up and degraded by A cells, and the breakdown products are presented on the surface in association with MHC proteins (Unanue and Cerottini, 1970; Chesnut and Grey, 1981, 1986; Lanzavecchia, 1985; for additional literature, see Allen, 1987). This recognition of processed antigen on A cells by helper T cells acts as a signalling receptor for T cells, while MHC plus processed antigen acts as a signalling receptor for A cells. As a consequence of this signalling, both cells produce antigen-non-specific growth and maturation factors, called lymphokines (for literature, see Howard and Paul, 1983; Kishimoto, 1985; Melchers and Andersson, 1984).

Virgin, mature, resting G_0-phase B cells, however, are refractory to the proliferation-propagating activity of lymphokines (Andersson et al., 1980; Melchers et al., 1980; Schreier et al., 1980). They only mature, without proliferation, to Ig-secreting cells when exposed to cytokines. Two interleukins, IL-2 and IL-5, were found to induce resting B cells to mature without proliferation (Lernhardt et al., 1987). In order to become susceptible to the proliferation–propagating activity of lymphokines, resting B cells have to exit from the G_0 phase or, as we have termed it, have to become 'excited'. It is at this point that Ig molecules control responses of lymphocytes.

In recent years it has become evident that stimulation of T and B lymphocyte proliferation and the maturation of clones of effector cells requires not only the binding of antigen to TCR and Ig, and the production and recognition of secreted soluble cytokines, but also is crucially dependent on cell to cell contacts and on signals mediated by these contacts. The interactions between B7–2 molecules on B cells and CD28 molecules on T cells, and those between CD40 ligand molecules on T cells and CD40 molecules on B cells, appear to be important, essential contacts and signalling devices in the collaboration between T and B cells (Noelle et al., 1992; Armitage, 1994; June et al., 1994).

HELPER T-CELL-DEPENDENT AND -INDEPENDENT RESPONSES OF MATURE, RESTING SURFACE IG-POSITIVE B CELLS

We distinguish two major ways by which a virgin, resting B cell can be positively activated to proliferate and mature into Ig-secreting cells (Figs 1 and 2). One needs the help of T lymphocytes; the other does not. Consequently, the responses as well as the antigen eliciting them are called T-dependent and T-independent. To a large degree, the same populations of B cells in the immune system are involved in these two types of responses (Andersson et al., 1979; Schreier et al., 1980).

Phase 1

Resting G_0 cell, refractory to cell-to-cell contact signals and lymphokine action

↓

Cross-linking of Ig (Others?)

↓

Phase 2

Excited B cell, susceptible to cell contact signals and lymphokine action

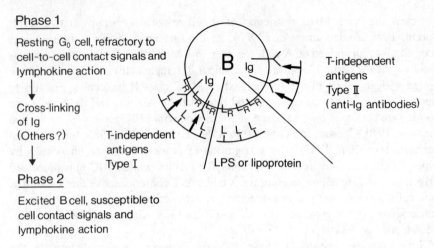

Fig. 1. Three modes of T-independent B-cell excitation from the G_0 resting state. The first is the cross-linking of surface Ig (—<) by repetitive determinants of T-independent antigens (↑, right); the second is the cross-linking of putative LPS receptors (L_R) by lipopolysaccharide (L) in micellar form with repetitive ligands (middle); the third is the cross-linking of surface Ig and of LPS receptors by repetitive haptenic determinants on LPS as a carrier (left). The first mode is called activation by T-independent antigen of type II, the last is activation by T-independent antigens of type I.

Surface Ig molecules play quite distinct roles in the control of T-independent B-cell responses (Sell and Gell, 1965; Kishimoto and Ishizaka, 1971; Kishimoto *et al.*, 1975; see also Möller, 1980), and it might be the chemical composition of the antigens that plays the distinguishing role. T-dependent antigens are usually polymeric, with repetitive determinants on one carrier. Hence, they have the capacity to cross-link several Ig molecules in the surface membrane of a B cell with specificity for the repetitive haptenic group (Fig. 1). Model polyclonal activators of this type are Ig-specific antibodies that are covalently coupled to polymeric carriers. It is at least questionable whether these repetitively presented antigens, immobilized to Sepharose beads as they often are, are even taken up by the B cells. Naturally occurring T-independent antigens are often not proteins but polysaccharides. Even if they are taken up by the B cells after multiple binding to surface Ig, it remains unclear what happens to them inside the cell. Processing, and reappearance of processed antigen, as appears to occur with T-dependent antigens (see below), is not very likely. Hence, cross-linking of surface Ig molecules by suitably spaced repetitive determinants is the most likely way to initiate the T-independent signalling to B cells that we have termed excitation (Fig. 1). IgD and IgM can both act as receptors for T-independent signalling.

T-dependent antigens are most often monomeric proteins, with only one or very few given determinants per molecule. This so-called haptenic group of the T-dependent antigen binds to surface Ig, is taken up, perhaps as an antigen–Ig–complement complex, and is processed by proteolytic cleavage to peptides (Fig. 2). The

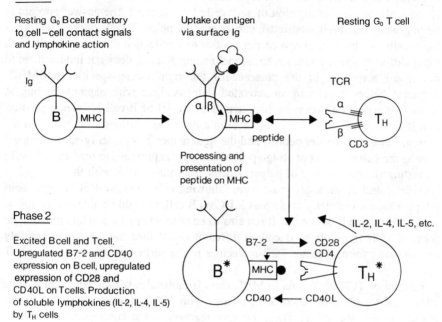

Fig. 2. The cellular cooperation of helper T cells (T_H) with resting, G_0 B cells in responses to T-dependent antigens. The antigen is taken up by binding to surface Ig, is processed and then presented as peptides (●) on MHC class II molecules. Presentation of MHC plus peptide on B cells to T-cell receptors (TCRα/β) on T helper cells establishes contact between the two cells, which is aided by CD4 making contact with MHC class II molecules. This recognition upregulates the expression of B7-2 and CD40 on B cells, and of CD28 and CD40 ligand on T cells. Lymphokine production and secretion is induced in T cells, and these lymphokines are bound to lymphokine receptors on B cells.

peptides then reappear on the surface of the hapten-specific B cell in association with MHC class II molecules. Peptides of T-dependent antigens are the so-called carrier determinants.

It is important to realize that hapten and carrier of an antigen need not be and, in fact, most often are not the same chemical entities. It follows that a series of haptens on one antigen can bind to a series of B cells with different surface Ig. However, processing by the different B cells may produce the same carrier determinant peptides that associate with the same self-MHC class II molecule, all recognized by one helper T cell with one carrier-specific MHC-restricted TCR. The peptide–MHC class II complex then binds to TCR molecules on the appropriate antigen (peptide-)specific, self-MHC class II restricted helper T cell. Hence, the excitation of a resting B cell in helper T-dependent activation can be dissected in two steps, first signalling via Ig, and then signalling via MHC plus peptide.

Can B-cell activation occur via signalling of the second receptor, the MHC class II molecule, without occupation of surface Ig by a hapten? The answer appears to be 'no' when self-MHC-restricted, carrier-specific helper T cells interact with resting B cells. Addition of excess carrier peptide to a collection of resting B cells with many different specificities, i.e. to normal resting B cells, does not induce them to polyclonal activation in the presence of the right carrier-specific, self-MHC-restricted helper T cells in an activated state. Antigen with haptens binding to surface Ig appears a necessary first step. This model of B-cell activation predicts that B cells should exit the resting state in a polyclonal fashion only when, in the presence of excess carrier peptide and the right helper T cells, an Ig-specific ligand such as the Fab fragment of an Ig-specific antibody is provided to bind to surface Ig on resting B cells. Instead of helper T cells, the soluble TCR with the appropriate specificity and in the appropriate form (allowing either monovalent or polyvalent interactions with carrier peptide plus MHC on B cells) should be able to function as a signalling ligand for B cells. It remains to be seen whether a soluble, monomeric form of TCR transmits to B cells a different signal than the repetitive, possibly cross-linking form of many TCR molecules in the surface membrane of helper T cells.

Binding of TCR on T cells to MHC class II molecules with peptides on B cells allows CD4 molecules to come in contact with MHC class II molecules to cooperate in the signalling. These binding reactions have consequences for both partners in this cellular cooperation (Fig. 2). Helper T cells upregulate the expression of CD28 and CD40 ligand, and begin to synthesize and secrete soluble lymphokines such as IL-2, IL-4 and IL-5, amongst others. B cells are now excited to upregulate the expression of B7–2 and CD40 on their surface (Armitage, 1994; June et al., 1994), as well as of receptors for the cytokines (Kishimoto et al., 1994; Paul and Seder, 1994).

It should be emphasized at this point that we know very little of the mechanisms that control proliferation and maturation of memory B cells (Cammisuli and Henry, 1978; Cammisuli et al., 1978), since they constitute a minor population within the total pool of B cells. This is also true for other B-cell subpopulations like the CD5-positive B cells (Okumura et al., 1982; Hayakawa et al., 1986). The schemes of B-cell controlling mechanisms developed in this chapter are applicable to the pool of short-lived, primary, virgin, surface IgM and IgD expressing, antigen-sensitive cells. However, it is reasonable to expect that the role of surface Ig of other classes and on other B-cell populations will be comparable to that of IgM of a primary B cell, i.e. to control the resting state of the cell.

In summary, two distinct roles of surface Ig are apparent in the signalling by T-dependent and T-independent antigens. Surface IgD and IgM function in both responses as hapten-specific receptors. T-dependent activation uses uptake of the Ig receptor inside the cell and needs additional signalling via surface MHC class II molecules. T-independent activation occurs by cross-linking of Ig in the surface membrane and is independent of the need for signalling via MHC class II, via B7–2 and via CD40.

CONTINUATION AND COMPLETION OF THE CELL CYCLE OF A PROLIFERATING B CELL

Once Ig has been cross-linked by T-independent antigens, or once Ig has been taken up, processed, presented on MHC class II molecules and recognized by helper T cells, cell–cell contacts and soluble lymphokines will then stimulate the excited B cell to proceed through the cell cycle and complete it with mitosis. Two control points in the cell cycle, called restriction points, where lymphokines act have been identified (Melchers and Andersson, 1984). A-cell-derived α factors, recently identified as an alternative transcription/translation product of the complement C3 gene (Cahen-Kramer et al., 1994), control the entry into S phase. Since the promotor which controls alternative C3 expression contains an NF-IL6-like element, it may well be that the alternative C3 is produced as a response to an acute phase reaction. It is likely that the complement receptor CR2 (CD21) acts as the signal-transducing receptor (Fearon, 1993). CR2 acts in concert with CD19, TAPA-1 and CD45 to modulate signals at this point in B-cell progression through the cell cycle (Fig. 3).

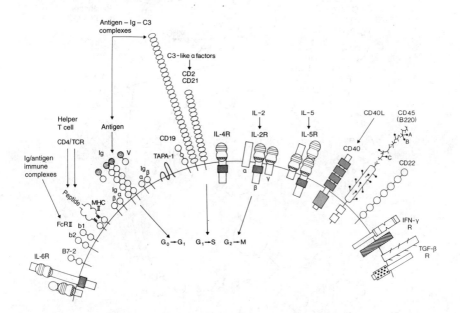

Fig. 3. Molecules expressed on the surface of B cells that are involved in the control of proliferation and maturation to Ig-secreting and memory B cells. The functions of these molecules are reviewed in the following articles: IL-6R (Taga and Kishimoto, 1993), B7-2 (June et al., 1994), FCR (Fridman, 1993), MHC class II (Wade et al., 1993), Ig-α and β (Reth, 1994), TAPA-1, CD19, CR1, CD21 (CR2) (Fearon, 1993), IL-2R, Il-4R, IL-5R (Kishimoto et al., 1994; Paul and Seder, 1994), CD40 (Noelle et al., 1992; Armitage, 1994), CD45 (Kishihara et al., 1993), CD22, IFN-γR (Aguet et al., 1988; Bazan, 1990), TGF-βR (Miyazono et al., 1994).

Since CR2 is associated with CR1, it is also likely that C3b bound to antigen–Ig complexes is enzymatically converted by CR1 to C3bi and to C3d/g which, in turn, can bind to CR2, stimulating B cells to enter into S phase. Hence, CR2 might be used by two alternative forms of C3 in two phases of B-cell response.

Once B cells have passed this control point in the cycle they reach yet another restriction point, in the G_2 phase. It is controlled by helper T-cell-derived lymphokines such as IL-2 and IL-5. Signalling through the corresponding lymphokine receptors allows the cell to enter mitosis. It has yet to be determined where in the

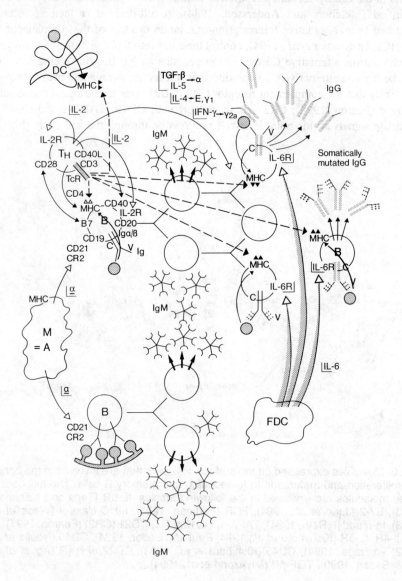

cycle CD40 acts when it is bound by CD40 ligand in T-dependent responses. LPS, as the prototype type I T-independent antigen (Mosier and Subbarao, 1982), has the apparent capacity to replace helper T-cell-derived cytokines. Type II T-independent antigens, on the other hand, still require these cytokines as co-stimulators, and it remains unclear how this is achieved in apparently T-independent ways.

There are many more receptors on the surface of B cells that influence proliferation, maturation to Ig secretion, class switching, hypermutation of V regions of Ig genes, migration and survival. Some, but not all of them, are shown in Fig. 3. Some of their possible functions are discussed below, others in articles that are quoted in the caption of Fig. 3 and in the text.

The unique role of surface Ig is to safeguard the resting state of a mature B lymphocyte against the possible proliferation-inducing actions of antigen-non-specific, polyclonally acting cell–cell contacts and soluble lymphokines (Andersson et al., 1980). Ig does this in multiple ways. First, it controls the expression of some of the contact molecules and receptors on the surface. Second, it connects them to the intracellular signal-transmitting pathways that induce cell cycle progression. Third, by taking up, processing and presenting antigen on MHC molecules to helper T cells it prepares the B cells for cognate interactions. Thereby, it induces T cells to upregulate surface expression of cell contact molecules (such as CD40 ligand) and synthesis and secretion of lymphokines (such as IL-2, etc.). As a result, clonal selection by antigen will only activate specific B cells to proliferation and Ig secretion.

Once B cells have divided, i.e. completed mitosis, they are likely to fall back into a resting, G_0 state unless antigen, acting through Ig, keeps them excited. In limiting concentrations of antigen, only those B cells with the most avid and the most specific Ig receptors will continue clonal expansion.

Fig. 4. T-dependent (top) and T-independent (bottom) B-cell responses. In T-dependent responses antigen (●) is taken up by dendritic cells (DC), processed and presented as peptides (▲▲) on MHC class II molecules. Helpers T cells (T_H) recognize these MHC–peptide complexes with their T cell receptors (TCR) anchored in the surface via the complex of CD3 molecules. B cells bind antigen via Ig variable (V) regions with their constant (C) region Ig is in the membrane via the Igα/β complex. Details of the excitation cascade are described in Fig. 2 and in the text. In response B cells divide, secrete first IgM (✲), then switch to other classes of Ig (Y) with the help of cytokines produced by T cells. B cells also somatically hypermutate the V regions of the Ig H and L chain genes making, amongst other things, better fitting receptors that are continuously selected by even lower concentrations of antigen, and stimulated with the help of T cells, in cognate TCR–MHC–peptide interactions. The responses occur in follicles structured by a net of follicular dendritic cells (FDC), and develop to germinal centres in which somatically hypermutated, Ig-class-switched cells develop as memory cells. T-independent responses develop by cross-linking of Ig with polyvalent antigens (✐✐✐) which may not be taken up by the cells. Macrophages producing α factors, interacting with complement receptor 2 (CD21) on B cells, co-stimulate these responses, which only lead to IgM secretion, but not to switching and hypermutation. No memory cells develop.

MEMORY B-CELL DEVELOPMENT AND RESPONSES

Memory does not usually develop to T-independent antigens such as polysaccharides. They stimulate B cells, by Ig cross-linking without the help of T cells, to proliferate and to develop into IgM-secreting plasma cells (Fig. 4, bottom). Switching of Ig genes to other classes and somatic hypermutation of V regions of Ig H and L chain genes is infrequent. The IgM-secreting plasma cells are short-lived. A second encounter with a T-independent antigen appears to recruit newly made B cells, and not cells that might have developed from a previous encounter with the T-independent antigen.

Memory usually develops to T-cell-dependent antigens. They stimulate B cells, with the help of T cells (outlined above), to proliferate and to develop into cells that initially secrete IgM (Fig. 4, top left). This part of the response appears to be dependent on the production of IL-2 by helper T cells, and on complement C3-like α factors either produced by A cells or generated in enzymatic processing of C3b bound to antigen-Ig-complexes. In the later phase of the response, lymphokines (produced by helper T cells and possibly also by other cells) induce Ig class switching (Snapper and Finkelman, 1993). IL-4 induces switches to IgG_1 and IgE, interferon γ (IFN-γ) to IgG_{2a} and IL-5, or transforming growth factor β (TGF-β) to IgA. Cognate interactions of TCR on T cells with MHC class II plus peptide continue to be required.

A primary B-cell clone responding to antigen also hypermutates the rearranged V-gene segments of the expressed Ig H and L chain genes so that approximately one point mutation is introduced per division (Berek *et al.*, 1991; Jacob *et al.*, 1991; Kelsoe and Zheng, 1993). Among the many possible mutations displayed in Ig molecules on the surface of the B cells in the dividing clone, those better fitting the antigen are selected by a more efficient binding and uptake of the antigen at lower concentrations, which results in longer presentation and thus stimulation to proliferation by helper T cells specific for the processed antigen presented on self-MHC class II molecules (Fig. 4, top right).

T-dependent responses are crucially dependent on the interaction of the CD40 ligand on T cells with CD40 on B cells. Targeted disruption of the CD40 gene in mice results in the inability to respond to T-dependent antigens with switched, somatically hypermutated antibody responses, while T-independent responses are not affected (Kawabe *et al.*, 1994). Mutations in the CD40 ligand gene in humans are known to cause X-linked hyper-IgM syndrome, an immunodeficiency that is characterized by the absence of serum IgG, IgA and IgE (Callard *et al.*, 1993; Kinnon *et al.*, 1993).

Conversely, T-independent responses are impaired in CBA/N mice, while T-dependent responses remain normal (Sher, 1982). CBA/N mice carry a mutation in the *btk* gene on the X-chromosome that encodes a cytoplasmic tyrosine kinase. The homologous gene can also be defective in humans, leading to X-linked immunodeficiency disease, called XLA (Tsukada *et al.*, 1993; Vetrie *et al.*, 1993).

The development of T-dependent B-cell responses appears to require also the co-stimulatory action of IL-6, since targeted disruption of the IL-6 gene impairs these responses (Kopf et al., 1994). In lymphoid organs, such as spleen and lymph nodes, T-dependent B-cell responses occur in follicles that develop to germinal centres during the response (Liu et al., 1992). Before the initiation of a B-cell response, follicles consist of a network of follicular dendritic cells (FDC), in which recirculating mature sIgM$^+$ sIgD$^+$ B cells are found. FDC may well produce the IL-6 required in the development of switched, hypermutated Ig responses of B cells. Foreign antigen, possibly in the form of immune complexes with complement, is taken up by FDC, and can be stored for many months in undegraded form. B cells take up the antigen from FDC, process it and present it on MHC to T cells that have previously been activated by the antigen on (non-follicular) dendritic cells (Fig. 4, top left), possibly outside the follicles. Proliferative expansion, switching and hypermutation of responding B cells then is propagated in germinal centres which develop into a dark zone, a basal light zone, and apical light zone, an outer zone and a follicular mantle zone. Bcl-2, a gene involved in the inhibition of programmed cell death (apoptosis) (Tsujimoto et al., 1985) is not yet expressed in the central parts of the germinal centre, but is upregulated in B cells that are found in the follicular mantle zone, and in extrafollicular resting B cells after they have left the area of the antigen-driven responses.

Germinal centres appear to be IgV-region mutant breeding sites of the immune system. The monovalent interaction of a hapten on a monomeric antigen in T-dependent responses monitors the affinity of interaction between antigen and surface Ig probably by simple mass law. In the polyvalent interactions of repetitive determinants on a T-independent antigen, the apparent avidities of antigen binding to surface Ig on the B cells might be quite high, while the individual interactions of single haptenic determinants of the T-independent antigen with individual Ig molecules on the B cell surface might often be quite low. It is evident that the monovalent interactions of T-dependent antigens are much more affinity dependent than the polyvalent interactions of T-independent antigens. Mutations to higher affinities are, therefore, likely to be selected and cells expressing these high-affinity, mutated Ig molecules will still be stimulated even at lower concentrations of antigen and, thereby, enter more rounds of division. This, in turn, will increase the chances for more mutations and switches.

Finally, even a B cell with better-fitting Ig receptors and switched classes of H chains would vanish quickly from the system if that B cell did not change its lifestyle. The experience with antigen, i.e. the selection for better-fitting Ig, is preserved by longevity. While primary B cells live less than a week, secondary 'memory' B cells may remain in the system as long as the antigen persists, and maybe for life (MacLennan and Gray, 1986; Osmond, 1986). They may be able to migrate to and reside in areas of the immune system that are reserved for such antigen-experienced cells.

SELF-REACTIVE B-CELL MUTANTS

The pool of mature, virgin, antigen-reactive B cells contains high frequencies of self-antigen-binding B cells. Most of them, however, have receptors with low avidities.

Elimination of these low-avidity self-reactive B cells from this pool of virgin cells is at least not visible (Dighiero et al., 1983; Prabhakar et al., 1983; Vakil and Kearney 1986; Rolink et al., 1987). High-avidity self-reactive B cells undergo arrest of differentiation deletion or anergy, as discussed later.

Self-reactive B cells could become a threat if they mutated to cells with better-fitting surface Ig. High-rate somatic mutations are likely to occur in proliferating B cells but is much less likely, if at all, in resting cells. Therefore, mutations to better-fitting self-reactive Ig molecules are only a threat in responding cells. Self-reactive B cells cannot be induced to a T-dependent response, since the system of mature T cells is purged of self-reactive helper T cells in the thymus. It is possible that a T-independent response is produced in self-reactive B cells by self-T-independent antigens or by foreign T-independent antigens that cross-react with self. The polyvalent interactions of multiple haptens on one B cell, however, do not select for high-affinity interactions of potentially mutated, better-fitting Ig with self-antigens, unlike a T-dependent response to a monovalent antigen. A primary B cell clone will die after 5–15 divisions, even if it contains better-fitting mutants.

It is also conceivable that self-reactive B cells of high avidity arise as somatic mutants of a B-cell clone that has been induced in a helper T-cell-dependent response to a foreign antigen. The surface Ig of this mutant B cell would no longer bind the foreign antigen; it could instead take up and process a self-antigen. The processed self-carrier presented on self-MHC would now no longer find the help of the T cells activated to the foreign carrier, as it would not find any self-reactive helper T cell, since they have been eliminated in the thymus. Consequently, this mutant self-reactive B cell would not grow, and hence no autoimmune response would develop. Autoimmune danger remains in somatically mutated B cells that happen to recognize foreign as well as self-haptens with high-affinity surface Ig, since they could bind, take up and present both foreign and self-carriers and, therefore, be helped to a response by the foreign carrier-specific, self-MHC-restricted helper T cells. How often this happens in somatically acquired autoimmune diseases remains to be seen.

In summary, the dual signalling of hapten binding to surface Ig, and of the TCR of a helper T cell binding to carrier plus self-MHC on a B cell, constitutes a safety device against reactions to self-antigens.

SUPPRESSIVE MECHANISMS INVOLVING THE IG MOLECULE

Two mechanisms of suppression of B-cell responses are known to involve Ig molecules on B cells. One is the suppression of maturation to Ig secretion of stimulated B cells by Ig-specific antibodies *in vivo* and *in vitro* (Kincade et al., 1970; Andersson et al., 1974; Kearney et al., 1976; see also articles in Möller, 1980). *In vitro* studies have shown that proliferation is unaffected. Since mitogens, such as

LPS, were used to stimulate B cells polyclonally and under the apparent circumvention of the binding to surface Ig, the inhibition of LPS-driven B-cell maturation has led to the hypothesis that a complex of Ig with other receptors in the surface membrane could regulate B-cell proliferation versus maturation.

The second suppressive mechanism is the binding of the Fc portion of secreted Ig into an Fc receptor (Fridman, 1993) (Fig. 3). If the FcγRIIb$_1$ receptor is expressed on B cells, binding of Fc will inhibit proliferation and maturation. Sometimes, however, and particularly with antibodies to Fc receptor, stimulation of B cells can also be observed. In this case, FcγRIIb$_2$, FcγRIII or FcεRII receptors appear to function. Antigen binding to the secreted Ig in an immune complex can act as a concentrating device that binds with a second determinant to surface Ig of an antigen-specific B cell and, thereby, locally concentrates enough Fc to inhibit the antigen-specific B cell. This is thought of as a feedback mechanism in which the secreted antigen-specific product (Ig) inhibits the precursor cells (the antigen-sensitive B cell) to produce more production sites for Ig (the Ig-secreting plasma cells).

Although we know that sIg is anchored in the surface membrane with the *mb*-1 and *B29* gene-encoded Igα and Igβ membrane proteins (Reth, 1993), and although we can begin to dissect the signal-transducing pathway inside a B cell (DeFranco, 1993; Cantrell, 1994; Sefton and Taddie, 1994), we still do not understand how sIg, MHC with peptide, CD40, lymphokine receptors and other interaction molecules on B cells (Fig. 3) cooperate in positive and negative signalling to produce different responses of B cells.

It is not known at present how immature and mature B cells are deleted or anergized when they are specific for autoantigens (see below), and whether the Fc receptor-mediated suppressive mechanisms play any role in this establishment of B-cell tolerance.

ROLES OF IG AND IG-LIKE MOLECULES IN EARLY PHASES OF B-CELL DEVELOPMENT

B-lymphocyte development from the earliest progenitor before the onset of Ig gene rearrangements to the mature Ig-secreting plasma cells has traditionally been divided into two phases. The two phases are separated at the point of the development of the antigen-sensitive, mature, resting, virgin B cell. All steps from this mature B cell to Ig-secreting plasma cells and memory cells are thought to be dependent on stimulation by foreign antigen, while the earlier phases are thought to be either antigen independent, or dependent on some unknown self-antigens.

Ig-like molecules expressed on early precursors and Ig molecules on immature B cells play important roles in the development of B cells, i.e. in the early phases of the B lineage from stem cells to virgin, resting, antigen-sensitive, mature B cells.

For an understanding of the roles of Ig and Ig-like molecules in this early development, the following facts and observations should be remembered.

46 F. Melchers

Nomenclature

This paper	Pro-B	Pre-B-I	Large Pre-B-II	Small Pre-B-II	Immature B	Mature B
Osmond	Early Pro B	Intermediate Pro B	Late Pro B/ Large Pre B	Small Pre B	Immature B	Mature B
Nishikawa	B-Pro I PA6 only	B-Pro II PA6-IL7	CFU-IL-7 only			
Hardy	–	A, B, C Pre Pro B, Pro B, Pro B	C'	D Pre B	Immature B	Mature B

COMPARTMENT SIZE

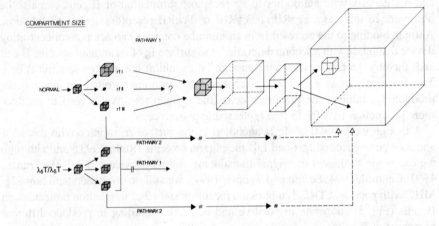

Status of Ig gene rearrangements

H	G/G	[DJ–DJ]/[DJ–DJ] DJ/DJ		VDJp	VDJp / VDJnp or DJ	VDJp / VDJnp or DJ	VDJp / VDJnp or DJ
L	G/G	G/G			G→VJ / G	VJp→VJ / G or VDJnp	VJp

Ig-LIKE COMPLEXES

p130 — p50/65 — SL — rfI/III   DJCμ — SL — rfII   μH — SL   μH L   μH L   μH L δH L

Markers

B220 (CD45)	–	⊕		⊕	⊕	⊕	⊕
c-kit	⊕	⊕		–	–	–	–
CD43	?	⊕		⊕	–	–	–
IL-7R	⊕	⊕		⊕	?	–	–
TdT	⊕	⊕		–	–	–	–
RAG-1/RAG-2	⊕	⊕		⊕	⊕	⊕	–
Cμ	–	– fII DJCμ		⊕ >90%	⊕ >90%	⊕ >90%	⊕ >90%
IL-2Rα	–	–		⊕	⊕	–	⊕↓

Functional properties

Clonability on stroma IL-7 in vitro	⊕	⊕ long term		+ short term	–	–	–
Population of SCID in vitro	⊕	⊕					

1. The immune system of the mouse contains some 5×10^8 resting, mature B cells that are built up by proliferative expansion and differentiation from pluripotent (Weissman, 1994) and later B-cell-lineage-committed (Rolink et al., 1991) progenitors and precursors within 5 weeks, one week before and four weeks after birth. A steady state of 5×10^7 immature B cells are generated daily in young adult mouse bone marrow (Osmond, 1986). Over 98% of all B-lineage cells on bone marrow have a half-life of between 2 and 4 days. From these, $2-3 \times 10^6$ cells are selected each day to enter the pool of mature, antigen-sensitive B cells. The half-life of mature B cells in the periphery is at least 4–6 weeks, and often longer (Schitteck and Rajewsky, 1990). Repertoires of developing, immature B cells in bone marrow are screened to eliminate or anergize high-affinity autoreactive B cells (Goodnow et al., 1989; Nemazee and Bürki, 1989a,b; Gay et al., 1993; Hartley et al., 1993; Radic et al., 1993; Tiegs et al., 1993).

2. Progenitor (pro) B cells (with all Ig genes in germline configuration) and precursors (pre) B cells (with Ig gene loci at different stages of gene rearrangements) can be subdivided and characterized by the expression of surface markers, intracytoplasmic markers, by growth properties in vitro and in vivo (cycling vs. resting cells, requirements for contacts with stromal cells and for molecules expressed on them, requirements for cytokines) and by capacities to repopulate immunodeficient hosts (Rolink and Melchers, 1993). Some of these distinguishing properties are illustrated in Fig. 5.

Fig. 5. Compartments of precursors of the B-lymphocyte lineage of differentiation in bone marrow of mice. The nomenclature of the various compartments used by different laboratories (Osmond, 1986; Hardy, 1991; Tsubata and Nishikawa, 1991; Rolink and Melchers, 1993) is compared for classification. Two pathways of B-lineage differentiation are indicated with large, cycling cells and small, resting cells. G denotes an Ig gene locus in germline configuration. Two configurations of one locus indicate the possible configurations of the two alleles. D_H to J_H rearrangements can occur in three reading frames (rf I, II, III). V_H to $D_H J_H$ and V_L to J_L rearrangements can occur in-frame (p, productive) or out-of-frame (np, non-productive), giving rise to cells that produce neither H nor L chains, only H chains, only L chains or full Ig molecules. Pre-B-I cells with H-chain loci rearranged in rf I or III express a p130/p50–65–surrogate L chain (SL) complex on the surface, while those rearranged in rf II might express a $D_H J_H C_\mu$ protein–SL complex. A $D_H J_H C_\mu$–SL complex on the surface of these cells might not only suppress the expansion into the pre-B-I cell pool, but also alter their chances to enter the pathway of differentiation. Pre-B-I cells from λ5T/λ5T mice, as well as such cells from RAG-2T/RAG-2T and μMT/μMT mice cannot enter pathway 1, but only pathway 2. The immature cells generated from such pre-B-I cells from λ5T/λ5T mice express sIg and are selectable into the periphery as mature sIg+ B cells, while cells from RAG-2T/RAG-2T and μMT/μMT mice cannot express sIg, and therefore cannot be selected and die. Large pre-B-II cells express μH chains together with SL. Small pre-B-II cells express μH chain only intracytoplasmically. The sizes of the compartments vary with age (Rolink et al., 1993b) and might be different in different inbred strains of mice. The figure depicts bone marrow of a 4–6-week-old mouse. For a comparison with human B cell development, see Chapter 1. (Modified from Figure 1 in Rolink et al., 1994.)

Ig gene rearrangements occur in a stepwise fashion. They begin with D_H to J_H rearrangements on both H chain gene alleles, continue with V_H to $D_H J_H$-rearrangements, then with V_κ to J_κ-rearrangements, and finally with V_λ to J_λ-rearrangements in the κL and then λL chain gene loci (Tonegawa, 1983). Different stages of pre-B cell development, described in Fig. 5, accumulate cells that have been produced at different points in this successive, stepwise pathway of Ig gene rearrangements.

3. Two populations of pre-B cells, pro/pre-B-I and large pre-B-II cells express two genes, V_{pre-B} and λ5, which are not expressed in more mature B cells and are, in fact, not expressed in any other cells of the body so far tested (Melchers et al., 1993). The proteins encoded by these two genes associate with each other to form a light (L) chain-like structure, the surrogate L chain (SL). It can form Ig-like complexes with three partners on the surface of three types of pre-B cells. It forms a μH chain–surrogate L chain complex on large pre-B-II cells, a $D_H J_H C_\mu$–surrogate L chain complex on cells at the transition of pre-B-I to pre-B-II, and a gp 130/gp35–65–surrogate L chain complex on pro/pre-B-I cells (Figs 5 and 6; Karasuyama et al., 1993). The Ig-anchoring molecules Igα and Igβ are expressed and function in pre-B-I and pre-B-II cells. IgH, IgL and IgH/L transgenic mice, the latter also with autoantibody specificities, and mice with targeted genetic deletions, particularly of the transmembrane portion of the μH chain (Kitamura et al., 1991; Gu et al., 1991), the J_H region of the H chain gene locus (Ehlich et al., 1993), the λ5 gene (Kitamura et al., 1992), and the *RAG*-1 and *RAG*-2 genes (Mombaerts et al., 1992; Shinkai et al., 1992) have helped to clarify the roles of the Ig-like and the Ig receptors and pro-B and pre-B cells, respectively on immature B cells.

The possible functions of surrogate L chain complexes with gp130/gp35–65 are not known. The complex of $D_H J_H C_\mu$ protein with surrogate L chain appears to mediate suppression of pre-B cells with H chain gene loci rearranged in reading frame II (Haasner et al., 1994). Best understood are functions of the μH chain–surrogate L chain pre-B receptor.

FUNCTIONS OF THE μH CHAIN–SURROGATE L CHAIN PRE-B RECEPTOR

In the $V_H D_H J_H$-rearranged pre-B-II compartment (Figs 5 and 6) over 95% of all cells express cytoplasmic μH chains. Therefore, they appear positively selected for productively over non-productively $V_H D_H J_H$-rearranged cells (Fig 5 and 6, pathway 1). Approximately half of the large, cycling pre-B-II cells, i.e. approximately 10% of all pre-B-II cells, express a pre-B-cell receptor of μH chains with the surrogate L chain, which appears on the surface when cells prepared *ex vivo* are allowed to incubate for 1 hour at 37 °C in tissue culture (Winkler, T., et al., 1995). The rest of the large and all small pre-B-II cells do not express the surrogate L chain, but only μH chain—and only in the cytoplasm.

In the bone marrow of λ5-deficient mice (λ5T), generated by targeted gene

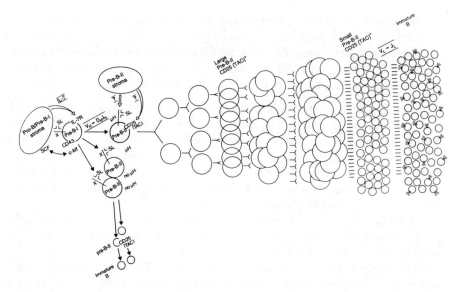

Fig. 6. Two pathways of pre-B-cell development in bone marrow. Pre-B-I cells expressing c-*kit*, the IL-7 receptor and a complex of surrogate L chain with SP130/SP35–65 interacts with stromal cells which provide the c-*kit* expand; stem cell factor (SCF), and IL-7. When all sites are occupied on the stroma, a subsequent division will be asymmetric, so that one of the daughter cells remains attached while the other becomes detached. Removal of contact and lack of supply of IL-7 induces differentiation and concomitant $V_H \rightarrow D_H J_H$ rearrangements. Two are out-of-frame (lower two) and differentiate without divisions to pre-B-II cells and to CD25 (TAC)$^+$ immature sIg cells. One is in-frame, hence productive, and allows expression of surrogate L chain together with μH chain as pre-B receptor on the surface. This pre-B receptor may recognize a stromal cell determinant (?), and the contact may also induce signals leading to cytokine production by the stromal cells (?). The pre-B receptor signals the large, CD25 (TAC)$^+$ cell to proliferate for five to six divisions, then to fall into a resting state, in which L chain gene rearrangements are likely to be induced. One of three rearrangements will be in-frame, leading to the expression of IgM molecules on the surface of immature B cells. Differentiation to this stage from the resting pre-B-II stage occurs without division. The immature B-cell pool is then subjected to negative and possibly also positive selective processes which delete or anergize the autoreactive B cells, and select some cells as mature, long-lived cells into the peripheral pool.

disruption in embryonic stem cells (Kitamura *et al.*, 1992), at least 98% of all pre-B-II cells and of the immature B cells are missing, while the pro/pre-B-I compartments are only slightly enlarged. The same defect in B-cell development is seen in *RAG*-1T, *RAG*-2T, J_HT and μMT mice (Ehlich *et al.*, 1993; Rolink *et al.*, 1993a).

These findings have suggested that the μH chain–surrogate L chain pre-B-cell receptor has several roles to play.

1. It signals large pre-B-II cells to expand 100 fold by proliferation. Thereby, μH chain-expressing cells outnumber the non-productively $V_H D_H J_H$-rearranged, μH chain-negative pre-B-II cells (Fig. 6).

2. The pre-B-cell receptor signals the downregulation of expression of the surrogate L chain-encoding V_{pre-B} and $\lambda 5$ gene.
3. It might also signal these cells to turn off TdT, RAG1 and RAG2 and to turn off V_H to $D_H J_H$ rearrangements on the other H chain allele, thus ensuring allelic exclusion, i.e. the production of only one H chain in a B cell.

However, it is unlikely that the pre-B receptor signals the induction of κL and λL gene rearrangements. Evidence against such an inductive role comes from experiments with $J_H T$ mice (Ehlich et al., 1993) and with selected pre-B-I cell lines (Grawunder et al., 1993) both of which are unable to express μH chains but can rearrange κL chain genes and express κL chains at rates and in quantities comparable to those in normal mice and in pre-B-I cell lines that are normally capable of expressing μH chains upon differentiation.

Little is known of the co-stimulatory requirements of pre-B-II cells expressing the pre-B receptor. It can be expected that special adhesion molecules mediate the contacts between pre-B-II and stromal cells, and that soluble cytokines and cell–cell contacts cooperate in signal transduction. For pre-B-I cells, some of these cytokines and contacts are known (Rolink and Melchers, 1991) but they could be different for the pre-B-II cell. Nevertheless, it might well be that the Ig-like pre-B receptor has a similarly unique and dominating role in the control of signal transduction, as Ig has in mature B cells. Besides all these co-stimulating interactions it remains to be seen whether pre-B-cell receptors recognize (a) ligand(s) in the environment of bone marrow, or (b) whether mere surface deposition suffices to establish its functions.

Finally, pre-B-I cells from λ5T mice can develop in vitro to sIg^+ B cells at normal rates and in normal numbers (Rolink et al., 1993a), and even pro/pre-B-I cells from Severe Combined Immune Deficiency (SCID) and RAG-2T mice can differentiate to $CD25(TAC)^+$ pre-B-II/immature B-like cells in which sterile transcription of the κL chain gene locus is induced (Grawunder, U., Melchers, F. and Rolink, A., submitted for publication). These results suggest that a cellular differentiation programme of B cell development is controlled by cytokines and cell contacts, and can occur without expression of surrogate L chain, without deposition of μH chains into membranes, without productive H chain gene rearrangements, and even without any Ig gene rearrangements at all. Hence, at the transition from pre-B-I cell to pre-B-II cell the role of the pre-B receptor is not to induce L chain gene rearrangements and differentiation programmes controlling CD25(TAC) expression and activation of sterile transcription of the L chain loci, but rather to expand the number of cells in which these rearrangements and this differentiation occurs.

THE ROLES OF IG ON IMMATURE B CELLS

Immature B cells express IgM but not IgD on their surface. They are unable to respond to polyclonal activators such as LPS by proliferation and maturation to IgM-secreting plasma cells.

Immature sIg⁺ B cells emerging from *in vitro* differentiation of normal mouse pre-B-I cells die by apoptosis within 2–4 days. When animals are given bromodeoxyuridine (BrdU) in the drinking water the same rapid apoptosis is observed *in vivo* in bone marrow where BrdU is incorporated into the DNA of pro-B, pre-B-I, pre-B-II and immature B cells; the BrdU label is lost within 2–4 days when BrdU is withdrawn. This indicates that a mere surface deposition of Ig molecules does not allow sIg⁺ B cells to enter the long-lived pool of peripheral, mature B cells.

Immature sIg⁺ and sIg⁻ B cells of *bcl*-2 transgenic mice do not die by apoptosis. This fact has allowed an analysis of the state of differentiation of immature B cells when kept in culture for prolonged periods of time (Rolink *et al.*, 1993a). Immature sIg⁺ and sIg⁻ B cells keep the expression of *RAG*-1 and *RAG*-2 upregulated. The cells continue to rearrange L chain gene loci even in the absence of proliferation in an apparently ordered fashion, so that κL⁺ sIg⁺ cells can become λL⁺ sIg⁺, or sIg⁻ cells, while λL⁺ sIg⁺ can become sIg⁻ but not κL⁺ sIg⁺ cells. Thus, deposition of Ig on the surface of an immature B cell, by itself, does not stop the rearrangement machinery.

Immature B cells of *bcl*-2 transgenic mice, as those of normal mice, are unresponsive to polyclonal activators. Hence, *bcl*-2 expression in immature, now long-lived sIg⁺ B cells, by itself, does not render these cells mature, mitogen reactive.

All of these observations make it likely that additional molecular changes have to be induced to render immature B cells mature. The situation is reminiscent of that in the thymus where immature α/β TCR-expressing, CD4/CD8 double-positive thymocytes are induced by self-MHC plus self-peptides in concert with co-stimulatory cell–cell contacts and cytokines to differentiate to CD4⁺ or CD8⁺ single-positive helper and killer T cells (positive selection), or are actively arrested in differentiation and induced to die by apoptosis when their interaction with MHC plus peptide is very strong (negative selection) (von Boehmer, 1994). Immature thymocytes with TCR molecules that do not fit MHC plus peptides at all are ignored and die within 2–4 days in the thymus. From 5×10^7 immature thymocytes, $2–3 \times 10^6$ mature, single-positive T cells are selected each day for release into the periphery.

The similarity between the above figures and the number of cells that are selected to enter the peripheral pool of mature B cells each day ($2–3 \times 10^6$) from immature sIg⁺ B cells (5×10^7) is remarkable. The repertoire of these immature B cells is screened against autoreactive B cells by negative selection, which is effected by deletion or anergy of the strongly autoreactive cells. However, if positive selection renders $2–3 \times 10^6$ B cells mature each day, it remains to be investigated which antigens do the selecting.

Ig-transgenic mice have helped to illuminate some of the mechanisms by which autoreactive B cells are selected, and where in the pathway of B-cell development this may occur. (Melchers *et al.*, 1995.)

DELETION VERSUS ANERGY OF B CELLS

Three types of transgenic mice have been analysed in detail. The first carries an H and L chain of an antibody with specificity for MHC class I (K^k or K^b) molecules

(Nemazee and Bürki, 1989a,b). The expression of the MHC self-antigen can be controlled in two ways: either its total presence or absence by breeding the transgenic Ig onto the right (K^k or K^b) or wrong (other K) MHC haplotype, or its central or only peripheral presence, the latter in special MHC-K^b transgenics which express the transgenic K^b only on plasma membranes in the liver and kidney (Tiegs et al., 1993). Central expression of the autoantigen resulted in the arrest of differentiation and deletion of transgenic, autoreactive B cells at the stage of an immature B cell (Fig. 5).

However, deletion is not complete when differentiation is arrested at the immature B-cell stage. The rearrangement machinery of RAG-1 and RAG-2 expression is upregulated. sIg^+ B cells are rescued into the periphery, in which the transgenic L chain has been replaced with an endogenously encoded and rearranged L chain and in which the κ/λ ratio is lower than normal. This indicates that continuous rearrangements of κL chain gene loci and increasingly of λL chain gene loci occur that allow the immature B cells to express a new, no longer autoreactive Ig and thus evade induction of apoptosis (Tiegs et al., 1993).

In this so-called receptor editing, occupancy of sIg on immature B cells is expected to signal the cell to keep its immature, RAG-expressing state. On the other hand, cross-linking of sIg on immature B cell lines from N-myc transgenic mice that express sIg and RAG-1 and RAG-2 simultaneously induces rapid and reversible downregulation of RAG-1 and RAG-2 (Ma et al., 1992). This latter scenario would be in line with the expectation that positive selection by some unknown antigens *down*regulate the rearrangement machinery, in line with the observation that this machinery is inactive in mature B cells of the periphery. It remains to be seen how the mode of Ig-receptor occupancy, additional cell–cell contacts with an antigen-presenting environment, and cytokine production and reception influence the reactions of an immature B cell.

When the MHC-K^b autoantigen is presented only in the liver and kidney, B-cell development in bone marrow is normal, but the mouse has virtually no B cells in spleen and lymph nodes. In this case, it is likely that deletion of the B cells occurs either at the transition from immature to mature B cells or in mature B cells, i.e. later than the deletion centrally in bone marrow, and there is no longer any receptor editing (Fig. 5).

The second transgenic mouse expresses an H and L chain of an antibody with specificity for DNA (Gay et al., 1993; Radic et al., 1993). The autoantigen is present ubiquitously, i.e. centrally. As expected, and in agreement with the results obtained with the MHC-$K^{k/b}$ specific mouse in which the autoantigen is centrally expressed, B-cell development is impaired at the transition of pre-B-II to immature B cells, i.e. immature B cells are depleted in bone marrow. Again, receptor editing is observed, so that B cells appear in the periphery that have replaced the transgenic L chain with an endogenously rearranged L chain. Again, the endogenous L chains are more frequently of λ-type.

The third transgenic mouse expresses an H and L chain gene encoding an antibody specific for hen egg lysozyme (HEL), which can be expressed as IgM and IgD (Goodnow et al., 1989). It can be confronted with two forms of HEL, one membrane bound (Hartley et al., 1993), the other soluble (Goodnow et al., 1989). Confrontation with the membrane-bound form leaves B-cell development in bone marrow intact but deletes B cells in the peripheral lymphoid organs (spleen, lymph nodes). This deletion is comparable to the peripherally induced deletion of MHC-$K^{k/b}$-specific B cells at the transition from immature to mature B cells, with no receptor editing (Fig. 5). Exposure to the soluble form of HEL, however, does not delete B cells in bone marrow, nor in the periphery, but renders peripheral transgenic B cells anergic. These anergic cells die within 2–4 days. Transfer of the anergic, i.e. antigen-unreactive, cells into a normal host can rescue these cells to become antigen reactive. This indicates that continued presence of the autoantigen is required to maintain the anergic state of the self-reactive cells in the periphery.

In conclusion, exposure of ligands (antigens) to sIg at the point of transition from an sIg^- pre-B-II cell to an sIg^+ immature B cell in the B cell-generating organ (i.e. in bone marrow, 'centrally') can signal the immature B cells to upregulate the rearrangement machinery and edit receptors. Once these immature B cells have traversed this state of immaturity and are ready to be released into the mature B-cell compartments, high-avidity interactions of Igs with their ligands will delete the cells without the chance of further editing. The ligands (antigens) obviously have to be presented in specific (membrane bound?) fashion. Interactions of lower avidity might well allow the cells to pass into the mature compartment, but Ig transgenic mice that allow the generation of mature, transgenic B cells have not been investigated for their possible lower avidity interactions with potential self-antigens. Ligands (antigens) presented in soluble form with high avidity at the transition of an immature to a mature B cell anergize these cells. It is unclear what could happen when the avidity of interactions is lowered. Positive and negative selection of emerging B cell repertoires have been discussed in detail in a recent review (Melchers et al., 1995).

CONCLUSIONS

The roles of Ig and Ig-like molecules on the surface of B-lymphocyte lineage cells change with the development of the cell lineage.

1. The first role of the Ig H chain in association with the surrogate L chain as pre-B receptor is to signal the proliferative expansion of precursor cells of the lineage, to turn off surrogate L chain, RAG1, RAG2 and TdT expression and, possibly, to turn off rearrangements at the other H chain allele, securing allelic exclusion.
2. The second role, now of Ig, on immature B cells is to signal upregulation of the rearrangement machinery so that secondary L chain gene rearrangement can replace an original by a new L chain, thus avoiding autoreactivity.
3. The third role of Ig, at the transition from an immature to a mature B cell, is to signal arrest of differentiation and induction of apoptosis, if the interaction is of high avidity and properly presented. If the avidity of interactions is lower the signal may change so that differentiation from a short-lived immature to a long-lived mature B cell is now induced.

4. On a mature resting cell, Ig binds antigen and thus initiates reactions that excite this resting, mature cell to become susceptible to the action of proliferation-propagating lymphokines. On the one hand, surface Ig can be cross-linked by T-independent antigen. On the other hand, it can bind T-dependent antigen, take up the antigen inside the cell and therefore allow presentation of processed antigen with MHC class II molecules to helper T cells. Binding of TCR to self-MHC plus processed antigen continues the signalling to the B cell. The cell cycle is completed by the synergistic action of lymphokines (C3-like, IL-2, IL-4, IL-5, etc.) and cell–cell contacts (CD40 ligand/CD40).
5. After completion of each cell cycle of an activated B cell, i.e. after mitosis early in the G_1 phase, Ig will repeat its interaction with T-independent antigens and initiate the next cell cycle. In T-dependent responses it is not clear whether Ig has to take up, process and present new antigen at every cell cycle.

REFERENCES

It should be clear to the reader that the list of references cannot be complete nor can justly balance the quoting of all the important, ground-breaking discoveries of the fields of immunological reseach, where Ig is a player. It should, however, suffice for a more intense secondary search into original publications relevant to the issues of this chapter.

Aguet, M., Dembic, A. and Merlin, G. (1988). *Cell* **55**, 273.
Allen, R.M. (1987). *Immunol. Today* **8**, 270.
Andersson, J., Sjöberg, O. and Möller, G. (1972). *Transplant Rev.* **11**, 131.
Andersson, J., Bullock, W.W. and Melchers, F. (1974). *Eur. J. Immunol.* **4**, 715.
Andersson, J., Coutinho, A. and Melchers, F. (1979). *J. Exp. Med.* **149**, 553.
Andersson, J., Schreier, M.H. and Melchers, F. (1980). *Proc. Natl Acad. Sci. USA* **77**, 1612–1616.
Armitage, R.J. (1994). *Curr. Opin. Immunol.* **6**, 407.
Bazan, J.F. (1990). *Cell* **61**, 753.
Berek, C., Berger, A. and Apel, M. (1991). *Cell* **67**, 1121.
Bretscher, M. and Cohn, M. (1970). *Science* **169**, 1042.
Burnet, F.M. (1959). 'The Clonal Selection Theory of Acquired Immunity'. Vanderbilt University Press, Nashville.
Cahen-Kramer, Y., Mårtensson, L. and Melchers, F. (1994). *J. Exp. Med.*, in press.
Callard, R.E., Armitage, R.J., Fanslow. W.C. and Spriggs, M.K. (1993). *Immunol. Today* **14**, 559.
Cammisuli, S. and Henry, C. (1978). *Eur. J. Immunol.* **8**, 663.
Cammisuli, S., Henry, C. and Wofsy, L. (1978). *Eur. J. Immunol.* **8**, 656.
Cantrell, D. (1994). *Curr. Opin Immunol.* **6**, 380.
Chesnut, R.W. and Grey, H.M. (1981). *J. Immunol.* **126**, 1075.
Chesnut, R.W. and Grey, H.M. (1986). *Adv. Immunol.* **39**, 51.
Chused, T.M., Kassn, S.S. and Mosier, D.E. (1976). *J. Immunol.* **116**, 1579.
Claman, H.N., Chaperon, E.A. and Triplett, R.F. (1966). *Proc. Soc. Exp. Biol. Med.* **122**, 1167.
Corbel, C. and Melchers, F. (1984). *Immunol. Rev.* **78**, 51.
Coutinho, A. and Möller, G. (1975). *Adv. Immunol.* **21**, 114.
DeFranco, A.L. (1993). *Curr. Opin. Immunol.* **6**, 364.
Dighiero, G., Lymberi, P., Mazié, J., Rouyse, S., Butler-Browne, G.S., Whalen, R.G. and Avrameas, S. (1983). *J. Immunol.* **131**, 2267.
Ehlich, A., Schaal, S., Gu, H., Kitamura, D., Müller, W. and Rajewsky, K. (1993). *Cell* **72**, 695.
Ehrlich, P. (1904). In 'Gesammelte Arbeiten zur Immunitäts-forschung'. Verlag von August Hirschwald, Berlin.

Fearon, D.T. (1993). *Curr. Opin. Immunol.* **5**, 341.
Fridman, W.H. (1993). *Curr. Opin. Immunol.* **5**, 355.
Gay, D., Saunders, T., Camper, S. and Weidger, M. (1993). *J. Exp. Med.* **177**, 999.
Goodnow, C.C., Crosbie, J., Jorgensen, H., Brink, R.A. and Basten, A. (1989). *Nature* **342**, 385.
Gowans, J.L. (1957). *Br. J. Exp. Pathol.* **38**, 67.
Grawunder, U., Haasner, D., Melchers, F. and Rolink, A. (1993). *Int. Immunol.* **5**, 1609.
Gu, H., Kitamura, D. and Rajewsky, K. (1991). *Cell* **65**, 47.
Haasner, D., Rolink, A. and Melchers, F. (1994). *Int. Immunol.* **6**, 21.
Hardy, R.R. (1991). *Curr. Opin. Immunol.* **4**, 181.
Hartley, S.B., Cooke, M.P., Fulcher, D.A., Harris, A.W., Cory, S., Basten, A. and Goodnow, C.C. (1993). *Cell* **72**, 325.
Hayakawa, K., Hardy, R.R. and Herzenberg, L.A. (1986). *Eur. J. Immunol.* **16**, 450.
Howard, M. and Paul, W.E. (1983). *Annu. Rev. Immunol.* **1**, 307.
Humphrey, J.H. and Porter, R.R. (1957). *Lancet* **i**, 196.
Jacob, J., Kelsoe, G., Rajewsky, K. and Weiss, U. (1991). *Nature* **354**, 389.
Jerne, N.K. (1955). *Proc. Natl Acad. Sci. USA* **41**, 849.
June, C.H., Bluestone, J.A., Nadler, L.M. and Thompson, C.B. (1994). *Immunol. Today* **15**, 321.
Kabat, E.A. (1956). *J. Immunol.* **77**, 377.
Karasuyama, H., Rolink, A. and Melchers, F. (1993). *J. Exp. Med.* **178**, 469.
Katz, D.H., Hamoaoka, T., Dorf, M.E. and Benacerraf, B. (1973). *Proc. Natl Acad. Sci. USA* **70**, 2624.
Kawabe, T., Naka, T., Yoshida, K., Tanaka, T., Fujiwara, H., Suematsu, S., Yoshida, N., Kishimoto, T. and Kikutani, H. (1994). *Immunity* **1**, 167.
Kearney, J.E.M., Cooper, M.D. and Lawton, R.R. (1976). *J. Immunol.* **116**, 1664.
Kelsoe, G. and Zheng B. (1993). *Curr. Opin. Immunol.* **5**, 418.
Kincade, P.W., Lawton, A.R., Brockman, D.E. and Cooper, M.D. (1970). *Proc. Natl Acad. Sci. USA* **67**, 1918.
Kinnon, C., Hinshelwood, S., Levinsky, R.J. and Lovering, R.C. (1993). *Immunol. Today* **14**, 554.
Kishihara, K., Penninger, J., Wallace, V.A., Kündig, T.M., Kawai, K., Wakeham, A., Timms, E., Pfeffer, K., Ohashi, P.S., Thomas, M.L., Furlonger, C., Paige, C.J. and Mak, T.W. (1993). *Cell* **74**, 143.
Kishimoto, T. (1985). *Annu. Rev. Immunol.* **3**, 133.
Kishimoto, T. and Ishizaka, K. (1971). *J. Immunol.* **107**, 1567.
Kishimoto, T., Miyake, T., Nishizawa, Y., Watanabe, T. and Yamamura, Y. (1975). *J. Immunol.* **115**, 1179.
Kishimoto, T., Taga, T. and Akira, S. (1994). *Cell* **76**, 253.
Kitamura, D., Roes, J., Kühn, R. and Rajewsky, K. (1991). *Nature* **350**, 423.
Kitamura, D., Kudo, A., Schaal, S., Müller, W., Melchers, F. and Rajewsky, K. (1992). *Cell* **69**, 823.
Kopf, M., Baumann, H., Freer, G., Freudenberg, M., Lamers, M., Kishimoto, T., Zinkernagel, R., Bluethmann, H. and Köhler, G. (1994). *Nature* **368**, 339.
Lanzavecchia, A. (1985). *Nature* **314**, 537.
Lernhardt, W., Karasuyama, H., Rolink, A. and Melchers, F. (1987). *Immunol. Rev.* **99**, 241.
Liu, Y.J., Johnson, G.D., Gordon, J. and MacLennan, I.C.M. (1992). *Immunol. Today* **13**, 17.
Ma, A., Fisher, P., Dildrop, R., Olta, E., Rathbun, G., Achacoso, P., Stall, A. and Alt, F.W. (1992). *EMBO J.* **11**, 2727.
MacLennan, I.C.M. and Gray, D. (1986). *Immunol. Rev.* **91**, 62.
Melchers, F. and Andersson, J. (1984). *Cell* **37**, 715.
Melchers, F., Andersson, J., Lernhardt, W. and Schreier, M.H. (1980). *Eur. J. Immunol.* **10**, 679.
Melchers, F., Karasuyama, H., Haasner, D., Bauer, S., Kudo, A., Sakaguchi, N., Jameson, B. and Rolink, A. (1993). *Immunol. Today* **14**, 60.
Melchers, F., Rolink, A., Grawunder, U., Winkler, T.H., Karasuyama, H., Ghia, P. and Andersson, J. (1995). *Curr. Opin. Immunol.* **7**, 214–227.
Mitchison, N.A. (1971). *Eur. J. Immunol.* **1**, 18.
Miyazono, K., Ten Dijke, P., Ichojo, H. and Heldin, C.H. (1994). *Adv. Immunol.* **55**, 181.
Möller, G. (ed.) (1980). *Immunol. Rev.* **52**, 1.
Mombaerts, P., Iacomini, J., Johnson, R.S., Herrup, K., Tonegawa, S. and Papaionannou, V.E. (1992). *Cell* **68**, 869.

Mosier, D.E. (1967). *Science* **158**, 1573.
Mosier, D.E. and Subbarao, B. (1982). *Immunol. Today* **3**, 217.
Nemazee, D. and Bürki, K. (1989a). *Nature* **337**, 562.
Nemazee, D. and Bürki, K. (1989b). *Proc. Natl Acad. Sci. USA.* **86**, 8039.
Noelle, R.J., Ledbetter, J.A. and Aruffo, A. (1992). *Immunol. Today* **13**, 431.
Okumura, K., Hayakawa, K. and Tada, T. (1982). *J. Exp. Med.* **156**, 443.
Osmond, D.G. (1986). *Immunol. Rev.* **93**, 105.
Paul, W.E. and Seder, R.A. (1994). *Cell* **76**, 241.
Prabhakar, B.S., Saeguza, J., Onodera, T. and Notkins, A.B. (1983). *J. Immunol.* **133**, 2815.
Radic, M.Z., Erikson, J., Litwin, S. and Weigert, M. (1993). *J. Exp. Med.* **177**, 1165.
Rajewsky, K., Schirrmacher, V., Nase, S. and Jerne, N.K. (1969). *J. Exp. Med.* **129**, 1131.
Reth, M. (1993). *Curr. Opin. Immunol.* **6**, 3.
Rolink, A. and Melchers, F. (1991). *Cell* **66**, 1081.
Rolink, A. and Melchers, F. (1993). *Curr. Opin. Immunol.* **5**, 207.
Rolink, A.G., Radaszkiewcz, T. and Melchers, F. (1987). *J. Exp. Med.* **165**, 1675.
Rolink, A., Kudo, A., Karasuyama, H., Kukuchi, Y. and Melchers, F. (1991). *EMBO J.* **10**, 327.
Rolink, A., Grawunder, U., Haasner, D., Strasser, A. and Melchers, F. (1993a). *J. Exp. Med.* **178**, 1263.
Rolink, A., Karasuyama, H., Grawunder, U., Haasner, D., Kudo, A. and Melchers, F. (1993b). *Eur. J. Immunol.* **23**, 1284.
Rolink, A., Karasuyama, H., Haasner, D., Grawunder, U., Mårtensson, I.-L., Kudo, A. and Melchers, F. (1994). *Immunol. Rev.* **137**, 185.
Rosenthal, A.S. and Shevach, E.M. (1973). *J. Exp. Med.* **138**.
Schimpl, A. and Wecker, E. (1972). *Nature New Biol.* **237**, 15.
Schittek, B. and Rajewsky, K. (1990). *Nature* **463**, 749.
Schreier, M.H., Andersson, H., Lernhardt, W. and Melchers, F. (1980). *J. Exp. Med.* **151**, 194.
Sefton, B.M. and Taddie, J.A. (1994). *Curr. Opin. Immunol.* **6**, 372.
Sell, S. and Gell, P.G.H. (1965). *J. Exp. Med.* **122**, 423.
Sher, I. (1982). *Immunol. Rev.* **64**, 117.
Shinkai, Y., Ratbun, G., Lam, K.P., Oltz, E.M., Stewart, W., Mendelson, M., Charron, J., Datta, M., Young, F., Stall, A.M. and Alt, F.M. (1992). *Cell* **68**, 855.
Simonsen, M. (1957). *Acta Pathol. Microbiol. Scand.* **11**, 480.
Snapper, C.M. and Finkelman, F.D. (1993). 'Fundamental Immunology', 3rd edn, vol. 22, p. 837.
Sprent, J. (1973). *Cell. Immunol.* **7**, 10.
Sprent, J. (1978). *Immunol. Rev.* **42**.
Taga, T. and Kishimoto, T. (1993). *FASEB J.* **7**, 3387.
Tiegs, S.L., Russell, D.M. and Nemazee, D. (1993). *J. Exp. Med.* **177**, 1009.
Tonegawa, S. (1983). *Nature* **302**, 575.
Tsubata, T. and Nishikawa, S.-I. (1991). *Curr. Opin. Immunol.* **3**, 186.
Tsujimoto, Y., Cossman, J., Jaffe, E. and Croce, C. (1985). *Science* **228**, 1440.
Tsukada, S., Saffron, D.C., Rawlings, D.J., Parolini, O., Allen, R.C., Klisak, I., Sparkes, R.S., Kubagawa, H., Mohanda, T., Quan Quan, S., Belmont, J.W., Cooper, M.D., Conley, M.E. and Witte, O.N. (1993). *Cell* **72**, 279.
Unanue, E.R. and Cerottini, J.C. (1970). *J. Exp. Med.* **131**, 711.
Vakil, M. and Kearney, J.F. (1986). *Eur. J. Immunol.* **16**, 1151.
Vetrie, D., Vorechosvsky, I., Sideras, P., Holland, J., Davies, A., Flinter, F., Hammarström, L., Kinnon, Cl., Levinsky, R., Bobrow, M., Smith, C.I.E. and Bentley, D.R. (1993). *Nature* **361**, 226.
von Boehmer, H. (1994). *Cell* **76**, 219.
Wade, W.F., Davoust, J., Salamero, J., André, P., Watts, T.H. and Cambier, J.C. (1993). *Immunol. Today* **14**, 539.
Waldmann, H. and Munro, A. (1973). *Nature* **243**, 356.
Weissman, I.L. (1994). *Cell* **76**, 207.
Winkler, T.H., Rolink, A., Melchers, F. and Karasuyama, H. (1995). *Eur. J. Immunol.* **25**, 446.

3

The maturation of the antibody response

César Milstein and Cristina Rada

MRC Laboratory of Molecular Biology, Hills Road, Cambridge, UK

Introduction 57
Methodological approaches 58
The primary response and the germinal centre reaction............................62
Somatic hypermutation..67
Selection and affinity maturation 74

INTRODUCTION

The most remarkable property of the immune system is its ability to 'learn' molecular recognition. This occurs in a stepwise evolutionary manner. First there is initial recognition which, imperfect as it may be, is able to trigger a complex series of events, generally referred to as the maturation of the response. It is during this maturation process that the improvement of the molecular recognition (antibody affinity maturation) takes place.

The idea that the quality as well as the quantity of antibody present in an antiserum improves during successive immunizations is almost a century old (Kraus, 1903). However, convincing quantitative evidence came only with the development of methods for measuring affinity constants around the middle of the century (Jerne, 1951; Eisen and Siskind, 1964). Nevertheless, the understanding of the molecular changes that underlie this process is much more recent. Indeed no such explanation could be advanced before the elucidation of the genetic origin of antibody diversity. We now know that in mice and humans this arises by the combinatorial assembly V-(D)-J fragments of immunoglobulin genes, the junctional diversity created by imprecise joining of the fragments, and by additional somatic point mutations. Even so, the elucidation of the precise nature of the learning process had to await the analysis of the changes in the antibodies produced following repeated immunizations. This was made possible by hybridoma technology coupled to emerging methods of mRNA sequencing and reliable measurements of antigen–antibody dissociation constants (Velick *et al.*, 1960).

In this chapter we describe developments in the study of the maturation of the response in the mouse. There is evidence that similar events take place in humans. However, this is not necessarily the case in other species. The driving force of affinity maturation in mice is somatic hypermutation; the molecular bases of this process remain unidentified, but the features described below strongly suggest a highly specialized function which may have appeared very recently in evolution. There is no compelling evidence to indicate that in birds (McCormack *et al.*, 1991) or even in rabbits (Knight, 1992) a similar process of affinity maturation takes place. It is possible that the molecular basis of affinity maturation in these or other species is of a different nature. However, even in birds (where gene conversion plays a major role in the somatic generation of the primary repertoire in the bursa of Fabricius) it has been suggested that point mutations play a role in maturation (Parvari *et al.*, 1990). Mice and humans do not have a bursa and, in spite of proposals to the contrary (Maizels, 1989; David *et al.*, 1992), the available evidence indicates that gene conversion plays no significant role in either the generation of the primary repertoire or the maturation of the response (Wysocki and Gefter, 1989; Lebecque and Gearhart, 1990; Wysocki *et al.*, 1990; Milstein *et al.*, 1992).

METHODOLOGICAL APPROACHES

Idiotypic responses give rise to a preponderant expression of antibodies with highly related and sometimes identical sequences. Thus, they usually derive from a single V_L and V_H gene combination, and often differ from each other in the boundary residues between V, D and J. These variations are attributable to differences in the primary B-cell clones from which they derive. Further diversification arises by point mutation following antigen-induced proliferation during the maturation stage. For the analysis of the diversification by somatic mutation, it is essential to ensure that the mutated sequence derives from a given germline gene and not instead from a closely related one. This is not always easy, but it is becoming increasingly reliable as the germline V-gene loci are being extensively characterized.

The hybridoma approach

The onset and maturation of the antibody response to a variety of antigens was initially studied by analysing hybridomas at different times after immunization (Griffiths *et al.*, 1984; Rudikoff *et al.*, 1984; Sablitzky *et al.*, 1985; Cumano and Rajewsky, 1986; Wysocki *et al.*, 1986; Claflin *et al.*, 1987; Malipiero *et al.*, 1987; Manser *et al.*, 1987). This approach became only practical when coupled to fast sequencing methods (Hamlyn *et al.*, 1981). More recently, direct mRNA sequencing has been replaced by procedures involving a polymerase chain reaction (PCR) amplification step followed by cloning and sequencing of multiple clones.

The hybridoma approach still remains the best way in which the affinity and the amino acid sequence changes of the derived antibodies can be readily and reliably correlated. It has provided clear evidence for the emergence of antibodies with improved binding affinity as a result of single point mutations (Berek and Milstein, 1987; Allen et al., 1988). For example, a mutation of His for Asn at position 34 of the $V_\kappa Ox1$ germline gene results in an affinity increase of 10 fold for the hapten phOx (Berek and Milstein, 1987). In the case of antibodies against p-azophenylarsonate, three point mutations were identified in the heavy chain that together increase the affinity for the hapten by a factor of 200 (Sharon, 1990). However the retrospective analysis of the effect of each amino acid substitution on the affinity required antibody engineering methods.

However, hybridomas do not immortalize subpopulations of B-cells at random. Various lines of evidence (Claflin and Williams, 1978; Apel and Berek, 1990; Brown and Willcox, 1991) suggest that blast germinal centre B-cells are preferentially immortalized. This is advantageous in terms of the isolation of antigen-specific hybridomas but probably restricts the range of mutations found to those introduced in the early stages of the traffic through the germinal centre. Therefore, the mutations detected by hybridomas at any given time may be less than those present in the memory cell population.

Hybridomas provide clear evidence of a common clonal origin of many of the shared mutations found in different monoclonal antibodies. Such clonal relationship can be firmly established by analysing the rearrangements of the expressed heavy and light chains as well as the configuration of the allelically excluded loci (Clarke et al., 1985). Not only can mutations of both the light and heavy chains expressed by a single B-cell be analysed, but other genes or gene fragments present in the same clone can also be looked at, for instance the non-expressed immunoglobulin (Ig) genes (Weiss and Wu, 1987; Roes et al., 1989), non-allelically excluded transgenes (Sharpe et al., 1991) or multiple copies of identical transgenes (Rogerson et al., 1991). Hybridomas also make it possible to study the hypermutation in DNA segments that are too long to be amplified by conventional PCR procedures.

The PCR approach

Amplified gene fragments from populations of B-cells can be directly cloned into M13 vectors and sequenced (Orlandi et al., 1989). A major advantage of this approach is its speed and simplicity. Usually, productively rearranged V genes are amplified using primers specific for DNA or mRNA (via cDNA). Attempts to systematically clone both heavy and light chains from the same cell in mixed cell populations (in-cell PCR) have only been partially successful (Embleton et al., 1992), and therefore the mutations of light and heavy chains are usually analysed as independent events. This problem does not arise when single B-cells or B-cell clones are analysed by PCR but this is laborious and technically more demanding.

The maximum size of the amplified fragments is restricted and the method does not provide information on antibody affinity. Artefacts and errors of the procedure must be carefully considered. The most dangerous one is the occurrence of crossover-like events between the products of closely related genes (Shuldiner et al., 1989; Paabo et al., 1990; Marton et al., 1991). The frequency of such artefactual crossovers depends on PCR conditions, and it increases with the length of the amplified fragments (Rada, 1993). PCR amplification also gives rise to misincorporations, which cannot be distinguished from authentic point mutations. Misincorporations are critically dependent on PCR amplification conditions and their frequency must be taken into consideration when analysing data. This can be a serious drawback when the frequency of mutants is low. That is the case in early primary responses or when the rate of mutation is low, for instance when modifications in regulatory elements are introduced. In addition some V genes seem to mutate at a lower rate than others (Cumano and Rajewsky, 1986; Berek et al., 1987; Blier and Bothwell, 1987; Rickert and Clarke, 1993). The more recent use of polymerases with proof-reading properties in the PCR procedure is likely to decrease considerably the background misincorporation error. Their effects on artificial crossovers are not yet clear.

In spite of these difficulties, the method is becoming increasingly popular because it yields large databases and is fast. One major advantage is that it is not restricted to any particular cell subpopulation. Indeed when DNA is amplified, all cells seem to have equal probability of being represented, regardless of their state of differentiation. It is a powerful approach when used judiciously to answer specific questions. It has been used to analyse genes from antigen, idiotype or class-specific B-cells and other subpopulations of B-cells (Weiss and Rajewsky, 1990; Gu et al., 1991; Rada et al., 1991; Schittek and Rajewsky, 1992), cells taken from histologically defined areas of germinal centres (Jacob et al., 1993; Ziegner et al., 1994), and also from single B-cells or B-cell clones (McHeyzer et al., 1991; Küppers et al., 1993). The advantages and disadvantages of each of these methods are summarized in Table I.

The transgenic approach

Transgenic animals are being extensively used to analyse hypermutation. Transgenes containing relevant immunoglobulin sequences mutate (O'Brien et al., 1987), and when the relevant flanking elements are included, the rate of hypermutation is of the same order of magnitude as endogenous counterparts (Sharpe et al., 1991). So far this is the assay system that has been mostly used to analyse the regulatory elements controlling hypermutation. An important bonus is that there is no question about the affiliation of the germline gene from which the mutants arise. Transgene constructs containing light or heavy chains and also reporter genes have been used (Sharpe et al., 1990; Carmack et al., 1991; Umar et al., 1991; Azuma et al., 1993; Giusti and Manser, 1993; Sohn et al., 1993; Taylor et al., 1994). However, constructs

Table I Methods of analysis of somatic hypermutation

Hybridomas	PCR of selected subpopulations	PCR of single cells	PCR Peyer's patches
Hyperimmunization (2 months)	Immunization (14 days to 2 months)	Hyperimmunization optional	FACS sorting PNA^{high} cells (hours)
Hybridomas (3 months) Each hybridoma separately analysed	Selection of cells (hours) Cloning and sequencing	Selection of cells by sorting or micromanipulation	Cloning and sequencing
Advantages	*Advantages*	*Advantages*	*Advantages*
No PCR error	No hybridomas needed	No hybridomas needed	Yields large databases
All the components of a mutated cell can be analysed and compared (e.g. transgene vs. endogenous)	Yields large databases	H and L segments from the same antibody can be studied	No hyperimmunization needed
	Small subpopulations can be studied	Small subpopulations can be studied	Primers can select the V genes analysed
Correlation between structure and function	Primers can select the V genes analysed	Primers can select the V genes analysed	
Disadvantages	*Disadvantages*	*Disadvantages*	*Disadvantages*
Time and labour consuming	PCR artefacts	Small databases	Not antigen specific
Small databases	H and L segments do not derive from the same clone	Higher PCR artefacts	PCR artefacts
Restricted to B cells giving hybridomas		Time and labour consuming	H and L segments do not derive from the same clone

containing reporter genes may not have included all the necessary flanking sequences required for full hypermutation.

The analysis of hypermutation of transgenes has been mostly based on the derivation of hybridomas, but more recently alternative faster methods have been used, in particular direct amplification and cloning of the transgene from Peyer's patches germinal centre cells (González-Fernández and Milstein, 1993; Betz et al., 1994). The possibility of manipulating the germline of mice by homologous recombination in embryonic stem (ES) cells offers new avenues for the analysis of both the phenomenon itself and the factors involved.

THE PRIMARY RESPONSE AND THE GERMINAL CENTRE REACTION

After stimulation by antigen, the first step in the development of a humoral response is the rapid proliferation of antigen-binding clones from the primary antibody repertoire. This has been best studied in idiotypic responses to simple antigens (haptens) in which one or very few canonical $V_L V_H$ combinations predominate (Moller, G, (ed.) 1987; Rajewsky et al., 1987; Berek and Milstein, 1988; Blier and Bothwell, 1988; French et al., 1989; Kocks and Rajewsky, 1989).

The naive repertoire of B-cell receptors is the result of the non-random combinatorial joining of fragments: V and J for light chains and V, D and J for heavy chains. The number of active V genes for each of the chains present in the germline is unlikely to exceed 200–300 while the number of J and D fragments is much smaller. This gives rise to considerable diversity, which is vastly increased by additions and deletions during the joining of genetic fragments. The potential diversity thus generated is huge. Theoretical calculations suggest values $> 10^9$ (Milstein, 1987; Berek and Milstein, 1988), much higher than the number of virgin clones in a mouse. Virgin B-cells are short lived and constantly regenerated (MacLennan and Gray, 1986). It is therefore reasonable to assume that the repertoire expressed by a mouse (available repertoire) on first encounter with antigen will differ from animal to animal, and even at different times in the same animal (Fig. 1). However, as stated above, the combinatorial recombination is not random. Some gene combinations and junctional diversity variants are likely to arise with much higher probability than others for a variety of genetic reasons (Tuaillon et al., 1994). For instance, V segments closer to D or J segments seem to have higher probability of genetic recombination (Jeong and Teale, 1989; Freitas et al., 1990; Malynn et al., 1990), random recombination between individual V_κ and J_κ genes is not observed, and even the nature of deletions and additions of junctional residues is not a random event (Lafaille et al., 1989; Milstein et al., 1992; Victor et al., 1994). Furthermore, the available repertoire is likely to be subjected to selective pressures from self components and from environmental antigens (Gu et al., 1991; Grandien et al., 1994). Thus, the repertoire is far from a random representation of potential clones. Some of them are so frequent that they are likely to be repeatedly present in several

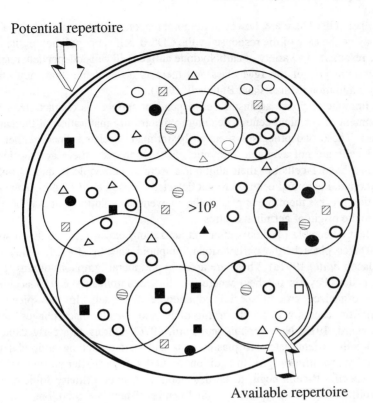

Fig. 1. The potential repertoire of antigen receptors is considerably larger than the number of B-cell clones expressing them. Hence, the available repertoire at any single moment (inner circles), particularly in a small animal like a mouse, is only a fraction of the total potential (outer circle). However different receptors (distinct elements within circles) occur at different frequencies, so that at any given time identical or closely related receptors are most likely to be expressed in several independent clones while others are very rarely present. The primary response to an immunogen draws, from this available repertoire, cells recognizing antigen above a certain affinity threshold. The primary response is thus dominated not only by initial affinity but also by clonal frequency at the time of immunization.

clones in all animals. Others are rarely expressed. Most if not all idiotypic responses probably derive from the first category. Lower affinity clones have a head start, when present at high frequency. Indeed potentially better antibodies derived from germline gene combinations may be so rare as to be absent at the time of the primary immunization (Berek and Milstein, 1988). Therefore, the selective proliferation in the primary responses is related not only to the affinity for antigen but also to the precursor frequency of B-cells that produce suitable antibody.

Antigen induced B-cell activation is a complex process. It is likely to involve an initial stage of antigen binding prior to its internalization and eventual presentation to T-helper cells. There are, however, humoral responses that do not require T-cells. This may be the case of the response by the CD5 B-cell subpopulation in the mouse and the response to a variety of carbohydrate antigens. T-cell-independent responses are predominantly of the IgM class and there is no evidence that they undergo affinity maturation (Maizels and Bothwell, 1985).

The immune response against T-cell-dependent antigens is characterized by the development of germinal centres, leading to the affinity maturation of the antibody response. These two processes seem to be intimately related since neither occur when T-cells are not available (athymic mice). Moreover, there are species (e.g. frogs) that have T-cells but their antibodies seem to be unable to undergo affinity maturation and germinal centres are not detected (Wilson et al., 1992). The affinity maturation process therefore appears to be an event in evolution that may be closely linked to the origin of germinal centres.

The first detectable (48 hours after immunization) expansion of antigen-specific B-cells takes place in the periarteriolar lymphoid sheaths (PALS) (Gray et al., 1986; Jacob et al., 1991a). These are areas in peripheral lymphoid organs (spleen, lymph nodes, Peyer's patches) where T-cells (predominantly $CD4^+$) accumulate and are organized around specialized endothelial vessels. In the spleen, such PALs are the access gate for the traffic of B- and T-cells, macrophages, antigen, etc. (Howard, 1972; Nieuwenhuis and Ford, 1976). During this early expansion, some B-cells differentiate into plasma cells and start producing antibody, which is generally low-affinity IgM. In addition, around 4–5 days after an immunization, antigen-specific B-cells begin to be detectable in splenic primary follicles where they proliferate to form typical germinal centres. There are therefore two proliferating populations of activated B-cells: one leading to primary antibody production, the other to maturation and memory. Most likely both develop from the same lineage (Jacob et al., 1992), but alternative possibilities have been put forward (Linton et al., 1989). Germinal centres are highly organized structures that develop around the follicular dendritic cell (FDC) network. The formation of the germinal centres critically depends on the presence of antigen–antibody complexes trapped in the interdigitating processes of FDCs (Klaus et al., 1980). This is the key to their function since it is believed they provide a regulated microenvironment for the antigen-dependent maturation of the response and for the generation of memory (Moller, G. (ed.) 1992). Moreover, the evidence that hypermutation takes place in the germinal centres is very persuasive (Berek et al., 1991; Jacob and Kelsoe, 1992; González-Fernández and Milstein, 1993). In germinal centres, several cellular compartments have been defined histologically: dark, light and follicular mantle zones and a few differentiation markers of the subpopulations of B-cells within germinal centres are known in humans. In particular CD38 is a marker for germinal centre cells (Hardie et al., 1993). In the mouse, germinal centre B-cells are brightly stained by the lectin peanut agglutinin (PNA^{high}) (Rose et al., 1980) and this has been the marker most commonly used to fractionate them. There are a number of studies that suggest functional differences between the different germinal centre compartments (Fig. 2).

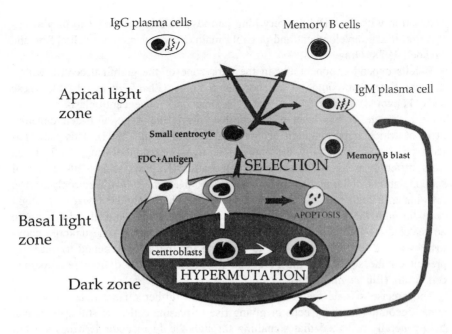

Fig. 2. The histologically defined areas in the germinal centre may correspond to functionally distinct compartments. The dark zone, an area of intense B-cell proliferation, is assumed to be the microenvironment where hypermutation takes place. The Ig⁻centroblasts undergo several rounds of mutation and then migrate to the light zone through the network of follicular dendritic cells where antigen is deposited. Interaction through the newly expressed mutated Ig receptor delivers a signal to rescue the cell from apoptosis. The rescued centrocytes can receive a differentiating signal to become a plasma cell. Alternatively they might re-enter the dark zone in a new round of expansion, mutation and selection, or can directly migrate out of the germinal centre to become a recirculating memory cell.

The kinetics of the expansion of antigen-specific cells, the appearance of B-cell clones bearing somatically mutated antibodies and the development of germinal centres are very tightly correlated (Jacob *et al.*, 1991a; Liu *et al.*, 1991a; Rada *et al.*, 1991; Hollowood and Macartney, 1992; Weiss *et al.*, 1992). The whole process starts at around day 7 after primary immunization, it peaks one week later and then decays over the following couple of weeks. This timing is for spleen responses to intraperitoneal stimulation but it can be different depending on the route and type of immunization. For example, it is faster in lymph nodes (Kallberg *et al.*, 1993) or with carrier pre-immunization (Levy *et al.*, 1989; Liu *et al.*, 1991a). It is likely that the decay in the response is directly correlated with the elimination of the circulating antigen. When this is cleared, the germinal centres gradually decrease in size and almost disappear so that very few antigen-specific B-cells can be detected. However, small amounts of antigen persist in association with the FDC for very long periods and this is thought to play a critical role in the development and possibly maintenance of memory cells (Tew and Mandel, 1978; Gray, 1993).

B-cells expand exponentially in the dark zone of the germinal centre, with a division time that could be as short as 6 hours (Zhang et al., 1988). These cells, known as primary blasts, later become dark zone centroblasts and continue to divide. It has been proposed that this is the stage of B-cell development at which hypermutation occurs (MacLennan et al., 1992). This view has received strong support by the analysis of hypermutation of single cells taken from different areas of human germinal centres (Küppers et al., 1993) and of subpopulations of human tonsil B-cells (Pascual et al., 1994). Centroblasts stop dividing at a high rate and become centrocytes which migrate towards the light zone through the FDC network. This is when the hypermutated cells seem to be selected by antigen. Many of the centrocytes in the light zone undergo apoptosis, while a few might be able to compete for the antigen bound to antibody present on the surface of the FDC and therefore be rescued from programmed cell death (Liu et al., 1989).

The precise signals that allow centrocytes to further differentiate, eventually either becoming memory cells or giving rise to plasma cells, are still speculative. It is generally accepted that signalling through the Ig receptor is necessary but probably not sufficient at this stage of B-cell differentiation. Centrocytes can be prevented from entering apoptosis if they are activated both through their receptors for antigen and CD40 (Liu et al., 1989). CD40 is a surface molecule present on B-cells while its ligand is found on T-cells (Armitage et al., 1992; Lane et al., 1992; Noelle et al., 1992). It is therefore possible that B-cells with improved antibodies on their surface might be able to internalize antigen, process it and present it via class II MHC to the $CD4^+$ T-cells, which in turn might deliver the required signals leading to further differentiation. A recombinant soluble fragment of CD23 (a molecule on the surface of the FDCs) plus IL-1a can also rescue germinal centre B-cells from apoptosis (Liu et al., 1991b). However, rescue from apoptosis in either case seems to induce the germinal centre cells to differentiate to either resting B-cells or to plasmablasts respectively. In both cases recovery from apoptosis is accompanied by expression of the oncogene bcl-2 (Liu et al., 1991c). Centroblasts in the dark zone do not express bcl-2, low expression is detected in the apical light zone in centrocytes presumably after antigenic selection, while all B-cells in the follicular mantle express high levels of bcl-2 (Korsmeyer, 1992). While these data and other evidence (McDonnell et al., 1989; Nuñez et al., 1991) suggest that bcl-2 is essential to prevent apoptosis after the cell stops dividing, other factor(s) have also been implicated in the prevention of apoptosis in centrocytes after hypermutation (Holder et al., 1993).

The memory cells specific for a given antigen constitute a very small recirculating population (Hayakawa et al., 1987; Weiss and Rajewsky, 1990). Restimulation with the same antigen leads to a rapid expansion of the memory cells, eliciting a high-affinity response. Secondary responses usually yield higher titres of antibody in the serum that persist for a long time, although the germinal centres are significantly smaller than in primary responses (Liu *et al.*, 1991a; Hollowood and Macartney, 1992). Cells that have undergone somatic hypermutation and class switch are characteristically found in the memory repertoire. Thus, most of the cells responsible for secondary responses are of the IgG or IgA class, have mutated antibody genes and show evidence of antigen selection for improved binding.

SOMATIC HYPERMUTATION

Hypermutation occurs following a change in the differentiation state of antigen-stimulated cells. Early proposals linking two events that occur after antigenic stimulation, namely class switch and somatic hypermutation, were abandoned when bona fide somatic mutations could be detected in IgM antibodies. At the same time it was shown that very early responses express non-hypermutated IgG, indicating that there is at first proliferation with no mutation (Griffiths *et al.*, 1984). Therefore class switch occurs independently from hypermutation, although both occur in cells committed to the same differentiation pathway. The hypermutation rate has been estimated in a few instances. Although there are preliminary indications suggesting that different V genes mutate at different rates (Cumano and Rajewsky, 1986; Berek *et al.*, 1987; Blier and Bothwell, 1987; Rickert and Clarke, 1993), the differences are not likely to be in the order of magnitude range. Early estimates of the rate of hypermutation did not take into consideration the possible distortions introduced by antigenic selection and gave estimates of around 10^{-3} per base pair per generation (McKean *et al.*, 1984). Values of around $3-4 \times 10^{-4}$ are more realistic estimates assuming a cell division time of 6–10 hours (Berek and Milstein, 1988; Levy *et al.*, 1989; Rada *et al.*, 1991; Weiss *et al.*, 1992). The difference between the rate of accumulation of silent versus total mutations suggests that, on average, for every mutation that improves antigen binding the cell acquires three or four which are neutral (Berek and Milstein, 1988). Mutations that are deleterious also occur but are rarely detected in ordinary situations. In the case of transgenic animals carrying multiple copies of identical genes they occur very commonly, and are in some cases selected by antigen to favour expression of a single copy (Lozano *et al.*, 1993). Indeed, mutations to stop codons in transgenes are found at higher than randomly expected frequency.

The rate of hypermutation is orders of magnitude higher than the mutation rate of antibody genes in myeloma cells in culture (Adetugbo *et al.*, 1977), and indeed consistent with theoretical estimates of the optimum rates for generating maximum diversity without catastrophic degeneracy (Allen *et al.*, 1987). However, the rate of hypermutation may not be constant during B-cell differentiation. Indeed it has been argued that the maximum rate could be even higher than the values quoted (Weigert, 1986).

The hypermutation target area

Hypermutation is restricted to a segment of around 1000 bases that includes the whole of the VJ or VDJ segments (Kim et al., 1981; Gearhart and Bogenhagen, 1983). Thus, C regions are unmutated although C_λ genes have been found with occasional mutations (Motoyama et al., 1991). This is probably because the intron separating C and J is much shorter in the λ than in the κ or μ loci. The upstream hypermutation boundary is well defined in κ chains. It is in the leader intron, and thus hypermutation does not occur in most of the secretory signal peptide (Rada et al., 1994; Rogerson, 1994). A pyrimidine (mostly T)-rich motif occurs at the boundary position, but its significance is unknown. There seems to be a hypermutation boundary in the 5' end of V_H and $V_\lambda 1$ genes but there is some uncertainty as to its precise location (Rothenfluh et al., 1993; Motoyama et al., 1994; Rogerson, 1994).

It should be noted, however, that upstream of this hypermutation boundary some mutations are also detected, at a frequency that is at least one order of magnitude lower than in the V segment. These mutations extend into the 5' flanking region, upstream of the promoter. They could originate as leaky events resulting from abnormal recognition of putative sequence motifs defining the 5' boundary. However there is no evidence that these mutations are related to the hypermutation machinery itself.

The mutations at the 3' end extend well beyond the J segment. No clear-cut boundary is found at this end, and the mutation rate seems to decrease gradually over a range of 500 bases or longer (Lebecque and Gearhart, 1990; Weber et al., 1991; Steele et al., 1992; Rickert and Clarke, 1993).

Passenger transgenes hypermutate

Although hypermutation requires antigen stimulation, the hypermutated gene does not need to contribute to the antigen-specific antibody. There are examples of non-allelically excluded B-cells from transgenic animals where the light chain of the antibody is of endogenous origin and the co-expressed light chain transgene does not contribute to antigen binding (Sharpe et al., 1991; Betz et al., 1993b). For that reason, they are considered passengers subjected to the same hypermutation 'bombardment' as the endogenous gene that has driven the B-cell into the germinal centre. Passenger transgenes are useful for dissociating the hypermutation itself from the biases introduced by antigen selection. However, they may not be totally irrelevant if they can act as competitors of the endogenous counterpart.

Somatic mutants are not randomly distributed

As stated above, the mutation process is targeted to the variable region segment, with a sharp boundary at the 5' end and a gradual decay at the 3' end. However, within the coding segment, there are large variations in the frequency at which mutants are found. In particular the CDR1 has repeatedly been found more mutated than other regions (Fig. 3) (Malipiero et al., 1987; Betz et al., 1993b). The relative frequency of hypermutations in CDR3 is more difficult to assess as many variants originate from diversity created during the recombination of fragments. However, this problem does not arise in the analysis of transgene mutations, and the preliminary evidence does not suggest a high mutation level in CDR3 (Fig. 3). It has been suggested that the preferential accumulation of mutations is the result of a polarity intrinsic to the mutation mechanism (Steele et al., 1992). An alternative view is that, regardless of such putative polarity, the accumulation of mutations in CDR1 is largely a reflection of the accumulation of mutational hotspots (Betz et al., 1993b).

The frequency at which individual nucleotides mutate varies considerably within the V segment (Fig. 3). Although reasonably large databases are available in only a few examples, mutational hotspots are found in all cases. Hotspots can originate in two ways, namely by antigenic selection and by an intrinsic mutational bias that is independent of selection. It is not always possible to discriminate between both possibilities (Betz et al., 1993a), and indeed some hotspots may have a combined origin. Certain mutations must be antigen selected not only because they are known to increase antigen affinity, but more important because they only occur with one antigen but not with others. For instance, His_{34} and Tyr_{36} of the gene V_kOx1 mutate with extreme frequency to Asn or Gln and to Phe in responses to the hapten 2-phenyloxazolone. The same mutations are rarely if ever found in the heterogeneous population of Peyer's patches B-cells, when the same gene is a passenger in anti-NP responses (Fig. 3) or in responses to phycoerythrin (Betz et al., 1993b). Therefore neither position is an 'intrinsic' mutational hotspot.

The idea that hotspots could be an intrinsic feature of the mechanism of somatic hypermutation is quite old (Secher et al., 1977). Intrinsic hotspots were first identified by statistical analysis of the frequency of repeated silent mutations (Berek and Milstein, 1987). However the extreme examples of hotspots (Ser_{31} in Fig. 3) are not silent. They are not the result of antigenic selection because they are repeatedly found in the response to different antigens as well as in passenger transgenes. The existence of intrinsic hotspots has been attributed to peculiarities of DNA structure, including palindromes, repeats (Milstein et al., 1986; Golding et al., 1987; Kolchanov et al., 1987) and primary sequence targeted motifs. The latter category includes the motifs TAA, RGYW (R = A or G, Y = C or T, W = A or T) (Rogozin and Kolchanov, 1992), CAGCT/A and AAGTT (Betz et al., 1993a).

It is possible that there is an evolutionary advantage in concentrating hotspots in the CDR1 region of some genes. The diversity in CDR2 and CDR3 segments pre-exists before antigen stimulation. Diversity in CDR3 is largely generated by the combinatorial and junctional diversity, while CDR2 is the most diverse segment among members of the same gene family (e.g. Fig. 3) (Chothia et al., 1992; Tomlinson et al., 1992; González-Fernández and Milstein, 1993). On the other hand, in the naive repertoire, the diversity of CDR1 is not very extensive, at least in some families (e.g. $V_\kappa Ox$ family; see Fig. 3). The variations of CDR1, largely created by somatic mutation, are therefore more likely to include new forms that improve antigen affinity. In turn, germline evolution could favour intrinsic hotspot motifs in CDR1 to increase the chances of somatic mutations in that region. In the case of

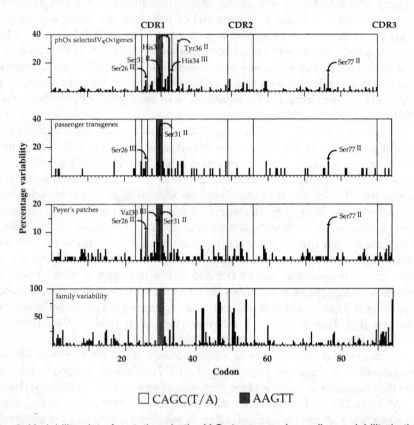

Fig. 3. Variability plot of mutations in the $V_\kappa Ox1$ gene and germline variability in the $V_\kappa Ox$ family genes. Variability is given as the percentage of sequences that carry a mutation at each particular residue. Individual hotspots are marked by their codon designation with the particular nucleotide in the codon indicated by roman numerals. The first panel shows the variability observed in an extensive compilation of $V_\kappa Ox1$ sequences from anti-phOx cells (hybridomas and antigen-selected cells (Griffiths et al., 1984; Berek et al., 1987; Rada et al., 1991)). In the second panel the variability found in passenger transgenes is shown (Sharpe et al., 1991; Betz et al., 1993b). The third panel shows the mutations found in the transgenic $V_\kappa Ox1$ V region in cells isolated from germinal centres from Peyer's patches (González-Fernández and Milstein, 1993). Finally, the germline variability found in the $V_\kappa Ox$ family is shown. The shared hotspots are likely to be due to the intrinsic hypermutation mechanism, including Val_{30} (since it is always silent). Grey shaded bars indicate the location of two consensus sequences shown at the bottom of the figure. The comparison of the somatic variability and the germline shows that while the hypermutation hotspots cluster in the CDR1, the germline variability is maximum in the CDR2.

mouse λ1 chains, where the 'family' is one or at most three genes and there is no combinatorial rearrangement of J segments, mutations in CDR2 and CDR3 are at least as frequent as in CDR1 (González-Fernández, A., et al (1994).

The nucleotide substitution bias

There are two types of substitution biases, namely the one introduced by antigen selection and the one intrinsic to the hypermutation mechanism. Antigen selection favours amino acid substitutions that improve binding, while those which decrease or destroy binding result in the elimination of the corresponding B-cell. As a consequence, the surviving cells have a higher than randomly expected ratio of expressed over silent mutations (Weigert, 1986). This ratio (opposite to the conservative changes in evolution) is sometimes used as evidence for unknown antigenic selection (e.g. environmental antigens).

The intrinsic base substitution bias introduced by the hypermutation process may give important clues as to its mechanism. The analysis of such bias should avoid potential distortions which may arise by antigenic selection. Silent mutations should be free of this problem, but databases are much smaller if so restricted. Even then they are not totally free of distortions, since in many codons transitions are silent while transversions are not. Non-coding segments are not free of selective pressures since they contain controlling elements and their base composition is far from random.

Probably the best estimates of mutational biases come from combining databases of several responses to defined antigen (excluding events suspected to be selected by antigen), in addition to databases of passenger transgenes, and of V regions from Peyer's patches of transgenic animals. The data in Table II confirm and expand certain trends already noticed with earlier analysis (Golding et al., 1987; Levy et al., 1988; Both et al., 1990; Manser, 1991; Rogerson et al., 1991). Transversions, which

Table II Substitution preferences of somatic hypermutation

To From	T	C	A	G	Total	Corrected
T	—	0.52	0.31	0.17	119	0.13
C	0.74	—	0.09	0.17	246	0.21
A	0.29	0.19	—	0.52	304	0.29
G	0.09	0.32	0.59	—	406	0.37

The distribution of base substitutions is given as the proportion of the total for each base. The table shows a compilation of mutations in the mouse $V_\kappa Ox1$ gene, the mouse $V_H 1$ 86.2 gene, the mouse $V_H Ox1$ and the human $V_H 5$ gene. Data taken from Betz (1994). The corrected proportions show the mutations occuring in each nucleotide after correction for base composition. Mutations are scored from the coding strand. A total of 1075 mutations are analysed. The intrinsic bias against mutations in pyrimidines (specially T) is clear, and it implies a polarity in the hypermutation machinery that suggests mutations are introduced in only one of the strands.

on a random bases should account for two-thirds of the events, occur at lower frequency than transitions. This is commonly observed in DNA changes during evolution (Li *et al.*, 1984). On the other hand, the clear prevalence of mutations involving A vs. T and G vs. C residues indicates that the mutation process affects only one strand, a selectivity which is referred to as strand polarity. Such polarity is observed in certain types of repair mechanism (transcription-linked DNA repair in mammalian cells (Hanawalt, 1991; Mullender *et al.*, 1991)). It is also noticeable that purines mutate more frequently than pyrimidines. This could reflect the specificity of the putative error-prone repair/replication event.

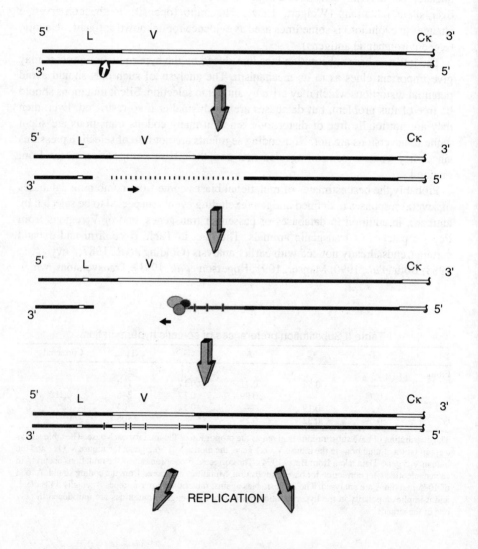

How does it work?

There are several mechanisms that have been proposed to explain the hypermutation process. The ones we believe best fit the data derive from the basic idea that mutations are introduced by an error-prone DNA polymerase that is specifically targeted to one of the two strands of the V(D)J locus by specific recognition elements. Two mechanisms have been proposed to achieve this. One (Brenner and Milstein, 1966; Lebecque and Gearhart, 1990) is by localized DNA nicks leading to error-prone single-strand DNA repair (Fig. 4). This proposal finds considerable support in the emerging similarities between hypermutation and excision repair of damaged DNA (e.g. Seidman *et al.*, 1987). The other mechanism (Manser, 1990) is by unidirectional localized error-prone DNA replication of the lagging strand (Okazaki fragments). Both mechanisms invoke an error-prone polymerase (such as polymerase β, which has an error rate in the order of 3×10^{-4} (Kunkel and Alexander, 1986)) that acts independently of chromosomal replication.

The important difference between the two is that the second includes a step of synthesis and expression of a mutant receptor required for cell division. Resumption of division is contingent upon antigenic recognition of the mutated receptor. The model thus requires the cyclic expression of mutated receptor Ig. Furthermore, hypermutation and selection seem to occur on two different cell types (centroblasts and centrocytes respectively). In addition, the newly synthesized receptor must be destroyed or internalized before the next mutated form is expressed to avoid competition between mutant chains.

A very different model has been proposed whereby mutations occur following reverse transcription and via cDNA template which becomes reintegrated into the

---

Fig. 4. A version of the error-prone repair model of somatic hypermutation in κ light chains. A specific single-strand nick is introduced in the transcribed strand. This is followed by exonucleolytic trimming of the nicked strand with subsequent error-prone repair. Mutations are introduced in one strand and are fixed during DNA replication. This accounts for the observed strand polarity, which is deduced from the biases intrinsic to hypermutation (see Table II). The preferential repair of the transcribed DNA strand is a recognized feature of nucleotide excision repair (Hanawalt and Mellon, 1993). In this version of the model, the preferred cleavage site is located upstream of the V-region coding segment and determines a boundary of hypermutation. In the case of mouse κ chains, this is in the leader intron. The exonucleolytic activity 'falls off' but only after an average of about 500–1000 bp. In this version of the model, hypermutation and transcription could be in competition. Alternative versions that are not incompatible with the strand polarity, the sharp boundary at the 5′ end and the decay at the 3′ end, while preserving the nuclease/polymerase polarity, are possible. For instance, there could be a large number of cleavage sites at the J–C intron of the non-transcribed strand and a sharp stop of the nucleolytic cleavage in the leader intron. (From Brenner and Milstein, 1966.)

correct position in the chromosome (Steele and Pollard, 1987). The distribution of mutations, particularly the sharp boundary about 150 bases downstream of the cap site, that emerges from the more recently extended databases does not support the original predictions of the model. Neither does it support the suggestion put forward by Rogerson *et al.* (1991) that hypermutation is initiated by a factor binding a single-stranded DNA segment located upstream of the V gene transcriptional promoters. Furthermore, the upstream segment containing the promoter can be substituted by a non-Ig promoter without significant changes in the hypermutation rate (Betz *et al.*, 1994).

Indeed, the information that is emerging from the analysis of transgenic animals, while short of definitive answers, is most illuminating. Hypermutation happens over a very short span of B-cell development, probably due to the induction of factors (DNA-binding proteins and enzymes) that act in *trans*. It is very likely that there are not one but several *cis* DNA motifs, binding *trans* factor(s) that are required to various degrees to achieve hypermutation. There is also convincing evidence that hypermutation is not necessarily a 'yes' or 'no' event, and that certain changes can drastically reduce without totally silencing the somatic mutation process. For instance, low frequency of mutation is observed in transgenic heavy chains (Sohn *et al.*, 1993), except when the transgene becomes occasionally translocated to its homologous chromosomal location (Giusti and Manser, 1993). Similarly, removal of the κ 3′ enhancer drastically impairs, but may not totally inhibit, the hypermutation process (Betz *et al.*, 1994).

In κ chains, at least two DNA fragments are required for full hypermutation (Betz *et al.*, 1994). One is included in the matrix attachment region (MAR) and intron enhancer fragment (located upstream of C_κ), the other in the 3′ enhancer located about 9 kb downstream of the C_κ exon. This and the comments in the previous paragraph show striking parallels between regulation of somatic hypermutation and transcription. It is possible that there are some factors which are unique and others shared and required to regulate both events. This is also thought to be the case of the nucleotide excision repair in eukaryotes (Bootsma and Hoeijmakers, 1993; Hanawalt and Mellon, 1993; Friedberg, 1994). Indeed a better understanding of the correlation between transcription and hypermutation will throw considerable light on the process.

SELECTION AND AFFINITY MATURATION

The affinity maturation of the humoral antibodies arises by two complementary strategies. One is the drift of the primary amino acid sequence of primary response antibodies fuelled by somatic hypermutation and selection of improvements in antigenic recognition. The other, the shift of the canonical repertoire, is the result of the proliferation (and somatic mutation) of clones that were very poorly represented at, or appeared after, the time of immunization (Berek and Milstein, 1987). The maturation is the result of the competition for antigen by the B-cells with new improved surface Ig including those generated after somatic hypermutation. The maturation of the response by hypermutation and selection is basically a Darwinian evolutionary process (Milstein, 1990) because it involves generation of variants that are heritable and under selective pressure. The selective pressure is driven by competition for an increasingly limited supply of antigen. The selected variants are heritable but only at the somatic level: the micro-evolutionary experience is unique to each individual and dies with that individual. The variation generated by point mutations is, as discussed above, not totally random due to biases in the hypermutation event. These biases presumably are dependent on DNA sequences of the germline and therefore are under germline evolutionary selection.

The process of somatic hypermutation of antibody genes in the germinal centres gives rise to all sorts of mutations, some deleterious (e.g. stop codons) or impairing antigen binding, others neutral and a few that improve antigenic recognition. Indeed it has been estimated that 25% of all mutations are selected against (Weigert, 1986) and that, as in the case of the hapten phOx, only one out of four mutations are selected due to affinity improvement (Berek and Milstein, 1988). There is also a certain probability that cross-reactive anti-self specificities appear by chance. This may indeed be the origin of certain autoimmune conditions (Shlomchik *et al.*, 1987). An increase in cross-reactivity of the antiserum in a secondary response has also been reported (Sperling *et al.*, 1983). This may arise not only by the drift of primary sequence but also by shift of the repertoire. It has been suggested that such clones are eliminated before they are rescued from apoptosis by interaction with T-cells (Linton *et al.*, 1989, 1991, 1992).

The efficiency of the selection of variants with improved affinity for antigen is remarkable. For example, within a period of 7 days, the majority of the hybridomas against phOx have drifted from the unmutated sequence to one carrying the changes in residues 34 and 36 of the $V_\kappa Ox1$ light chain (Griffiths *et al.*, 1984), even though those changes are not in intrinsic hotspots nor do they represent an example of biased base substitution.

A critical stage of selection occurs in germinal centres where, following hypermutation, cells die by apoptosis unless rescued by antigen. This is clearly a most efficient selective method but it is insufficient. For instance, if the mutations referred to above were to arise each with a frequency of about 4×10^{-4}/generation (the estimated mutation rate), the double mutation would occur in one cell when, on average, the population exceeds 10^9 cells. Multiple cycles of mutations and selection are therefore required. Mathematical models of the most efficient way to achieve this process involve separating the stages of proliferation and accumulation of mutants from the negative selection by apoptosis (Agur et al., 1991; Kepler and Perelson, 1993). The derivation of hypermutated memory cells in the primary response is therefore likely to involve multiple cycles of hypermutation, negative selection, proliferation (perhaps with positive antigenic stimulation) and new rounds of hypermutation and selection. The process presumably drags to a halt when circulating antigen is exhausted. This could be the reason why the accumulation of mutants stops at 3–4 weeks after primary immunization (Rada et al., 1991; Weiss et al., 1992). Affinity maturation is thus achieved by cyclic somatic hypermutation of antigen-selected B-cells. Both steps, hypermutation and antigen selection, take place in germinal centres.

The secondary response may follow a rather similar pattern, except that in this case the process may be shorter and more synchronized (Fig. 5). The proliferating population arises from the memory pool and by day 3 after immunization hypermutation starts, followed, around day 5, by survival and proliferation of cells selected for their improved binding to antigen (Rada et al., 1991). Since the secondary response is much more efficient than the primary, antigen depletion is much faster and the whole process is likely to be much shorter. However the evidence for this scheme is indirect. It is based on the fact that silent mutations accumulate even in tertiary responses (Berek and Milstein, 1988) and supported by the increase in the replacement vs. silent mutations at the end of the secondary

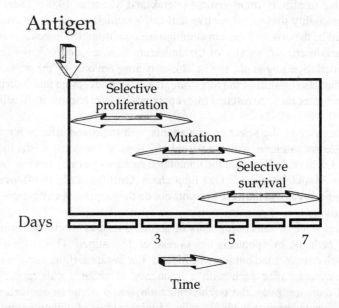

Fig. 5. Mutation and selection in secondary responses. The two-headed arrows span the temporal predominance of proliferation, mutation and selection proposed, based on the mutation patterns and frequencies in antigen-specific clones (Rada et al., 1991). The initial antigenic stimulation leads to the very rapid and synchronized expansion of cells from the memory compartment. Some of these cells are drawn to the germinal centres and undergo hypermutation. Finally after several rounds of mutation strong competition determines the selective survival of cells that might form the new memory compartment.

response, and the appearance of deleterious mutations in clones that have been antigen selected (Rada *et al.*, 1991). However, clones taken from a primary response that have been expanded in an adopted transfer experiment failed to show evidence of further mutation (Siekevitz *et al.*, 1987). It is possible that lack of suitable germinal centres in the adoptive transfer experiments accounts for this apparent discrepancy.

Competition for free antigen is thought to be one of the selective forces in affinity maturation. Probably both thermodynamic and kinetic parameters are involved. The kinetic parameters, however, are likely to play a more important role in the repertoire shift (Foote and Milstein, 1991). Indeed, the kinetic ON rate of antigen–antibody interaction is likely to be limited by stereochemical considerations. For instance, if the antigen fits in a deep but narrow cleft (as is the case with one canonical structure recognizing the hapten phOx (Alzari *et al.*, 1990)) collision theory predicts that the on rate will be limited by the low probability of a correctly oriented interaction. On the other hand, other gene combinations may provide a different stereochemical interaction (e.g. involving a fully surface-exposed binding site (McManus and Riechmann, 1991)) with excellent on rates. If these are not common in the naive repertoire, the kinetic advantage is likely eventually to overcome the deficit. In either case, however, as maturation proceeds the scope for improving binding interactions by somatic mutations will be determined by other factors, such as structural constraints and the ease and speed at which a favourable mutation(s) occurs in a particular V gene.

The efficiency of antigen selection seems also to be dependent on the density (or efficiency of synthesis) of the receptor antibody (George *et al.*, 1993). This is evident in mice expressing multiple copies of a transgene. The maturation of the response favours deleterious mutations in copies which otherwise compete with the copy that has acquired mutations improving binding (Lozano *et al.*, 1993). Antigenic selection of cells based on such competition has implications in terms of the process of antigen selection of high-affinity mutants. Not only thermodynamic parameters (affinity) but also the number of receptor molecules seem to play a critical role in B-cell selection leading to affinity maturation.

Thus the 'success' of initially dominant responses is likely to dependent on its structural potential (Berek *et al.*, 1985; Manser *et al.*, 1987; Milstein, 1990). For instance, somatic mutations of the canonical (V_H186.2-V1) dominant anti-NP responses do not seem to improve its binding beyond a factor of about 10 (Allen *et al.*, 1988). Improved responses arise not by further mutations but by shifting the repertoire to other V_H/V_L combinations. Even when the initially dominant response is capable of much larger improvements by somatic hypermutation (as is the case with the response to phOx), the repertoire shift plays a critical role in the emergence of mature antibodies of vastly improved affinity. It seems, therefore, that although antigen selection is capable of extremely efficient evolution of improved variants, the selection also allows scope for the survival and emergence of antibodies with lower affinity, which in some cases provide the raw material for the best antibodies in later responses.

Acknowledgements

We thank M. Neuberger and J. Foote for helpful comments. We gratefully acknowledge the generous support from the National Foundation for Cancer Research.

REFERENCES

Adetugbo, K., Milstein, C. and Secher, D.S. (1977). *Nature* **265**, 299–304.
Agur, Z., Mazor, A. and Meilijson, I. (1991). *Proc. R. Soc. Lond. B.* **245**, 147–150.
Allen, D., Cumano, A., Dildrop, R., Kocks, C., Rajewsky, K., Rajewsky, N., Roes, J., Sablitzky, F. and Siekevitz, M. (1987). *Immunol. Rev.* **96**, 5–22.
Allen, D., Simon, T., Sablitzky, F., Rajewsky, K. and Cumano, A. (1988). *EMBO J.* **7**, 1995–2001.
Alzari, P.M., Spinelli, S., Mariuzza, R.A., Boulot, G., Poljak, R.J., Jarvis, J.M. and Milstein, C. (1990). *EMBO J.* **9**, 3807–3814.
Apel, M. and Berek, C. (1990). *Int. Immunol.* **2**, 813–819.
Armitage, R.J., Fanslow, W.C., Strockbine, L., Sato, T.A., Clifford, K.N., Macduff, B.M., Anderson, D.M., Gimpel, S.D., Davis, S.T., Maliszewski, C.R., Clark, E.A., Smith, C.A., Grabstein, K.H., Cosman, D. and Spriggs, M.K. (1992). *Nature* **357**, 80–82.
Azuma, T., Motoyama, N., Fields, L.E. and Loh, D.Y. (1993). *Int. Immunol.* **5**, 121–130.
Berek, C. and Milstein, C. (1987). *Immunol. Rev.* **96**, 23–41.
Berek, C. and Milstein, C. (1988). *Immunol. Rev.* **105**, 5–26.
Berek, C., Griffiths, G.M., and Milstein, C. (1985). *Nature* **316**, 412–418.
Berek, C., Jarvis, J.M. and Milstein, C. (1987). *Eur. J. Immunol.* **17**, 1121–1129.
Berek, C., Berger, A. and Apel, M. (1991). *Cell* **67**, 1121–1129.
Betz, A.G. (1994). 'Somatic hypermutation of immunoglobulin κ light chain genes.' PhD Thesis, Cambridge.
Betz, A.G., Neuberger, M.S. and Milstein, C. (1993a). *Immunol. Today* **14**, 405–411.
Betz, A.G., Rada, C., Pannell, R., Milstein, C. and Neuberger, M.S. (1993b). *Proc. Natl Acad. Sci. USA* **90**, 2385–2388.
Betz, A.G., Milstein, C., González-Fernández, A., Pannell, R., Larson, T. and Neuberger, M.S. (1994). *Cell* **77**, 239–248.
Blier, P.R. and Bothwell, A. (1987). *J. Immunol.* **139**, 3996–4006.
Blier, P.R. and Bothwell, A.L.M. (1988). *Immunol. Rev.* **105**, 27–43.
Bootsma, D. and Hoeijmakers, J.H.J. (1993). *Nature* **363**, 114–115.
Both, G.W., Taylor, L., Pollard, J.W. and Steele, E.J. (1990). *Mol. Cell. Biol.* **10**, 5187–5196.
Brenner, S. and Milstein, C. (1966). *Nature* **211**, 242–246.
Brown, A.N. and Willcox, H.N. (1991). *Immunology* **74**, 600–605.
Carmack, C.E., Camper, S.A., Mackle, J.J., Gerhard, W.U. and Weigert, M.G. (1991). *J. Immunol.* **147**, 2024–2033.
Chothia, C., Lesk, A.M., Gherardi, E., Tomlinson, I.M., Walter, G., Marks, J.D., Llewelyn, M.B. and Winter, G. (1992). *J. Mol. Biol.* **227**, 799–817.
Claflin, J.L., Berry, J., Flaherty, D. and Dunnick, W. (1987). *J. Immunol.* **138**, 3060–3068.
Claflin, L. and Williams, K. (1978). *Curr. Top. Microbiol. Immunol.* **81**, 107–109.
Clarke, S.H., Huppi, K., Ruezinsky, D., Staudt, L., Gerhard, W. and Weigert M. (1985). *J. Exp. Med.* **161**, 687–704.
Cumano, A. and Rajewsky, K. (1986). *EMBO J.* **5**, 2459–2468.
David, V., Folk, N.L. and Maizels, N. (1992). *Genetics* **132**, 799–811.
Eisen, H.N. and Siskind, G.W. (1964). *Biochemistry* **3**, 996–1014.
Embleton, M.J., Gorochov, G., Jones, P.T. and Winter, G. (1992). *Nucleic Acids Res.* **20**, 3831–3837.
Foote, J. and Milstein, C. (1991). *Nature* **352**, 530–532.

Freitas, A.A., Andrade, L., Lembezat, M.P. and Coutinho, A. (1990). *Int. Immunol.* **2**, 15–23.
French, D.L., Laskov, R. and Scharff, M.D. (1989). *Science* **244**, 1152–1157.
Friedberg, E.C., Bardwell, A.J., Bardwell, L., Weaver, W.J., Kornberg, R.D., Svejstrop, J.Q., Tomkinson, A.E. and Wang, Z. (1995). *Phil. Trans. R. Soc. Lond. B.* **347**, 63–68.
Gearhart, P.J. and Bogenhagen, D.F. (1983). *Proc. Natl Acad. Sci. USA* **80**, 3439–3443.
George, J., Penner, S.J., Weber, J., Berry, J. and Claflin, J.L. (1993). *J. Immunol.* **151**, 5955–5965.
Giusti, A.M. and Manser, T. (1993). *J. Exp. Med.* **177**, 797–809.
Golding, G.B., Gearhart, P.J. and Glickman, B.W. (1987). *Genetics* **115**, 169–176.
González-Fernández, A. and Milstein, C. (1993). *Proc. Natl Acad. Sci. USA* **90**, 9862–9866.
González-Fernández, A., Gupta, S.H., Pannell, R., Neuberger, M.S., and Milstein, C. (1994). *Proc. Natl Acad. Sci. USA.* **91**, 12614–12618.
Grandien, A., Modigliani, Y., Freitas, A., Andersson, J. and Coutinho, A. (1994). *Immunol. Rev.* **137**, 53–89.
Gray, D. (1993). *Annu. Rev. Immunol.* **11**, 49–77.
Gray, D., MacLennan, I.C. and Lane, P.J. (1986). *Eur. J. Immunol.* **16**, 641–648.
Griffiths, G.M., Berek, C., Kaartinen, M. and Milstein, C. (1984). *Nature* **312**, 271–275.
Gu, H., Tarlinton, D., Muller, W., Rajewsky, K. and Forster, I. (1991). *J. Exp. Med.* **173**, 1357–1371.
Hamlyn, P.H., Gait, M.J. and Milstein, C. (1981). *Nucleic Acids Res.* **9**, 4485–4494.
Hanawalt, P.C. (1991). *Mutat. Res.* **247**, 203–211.
Hanawalt, P.C. and Mellon, I. (1993). *Curr. Biol.* **3**, 67–69.
Hardie, D.L., Johnson, G.D., Khan, M. and MacLennan, I.C. (1993). *Eur. J. Immunol.* **23**, 997–1004.
Hayakawa, K., Ishii, R., Yamasaki, K., Kishimoto T. and Hardy R.R. (1987). *Proc. Natl Acad. Sci. USA* **84**, 1379–1383.
Holder, M.J., Wang, H., Milner, A.E., Casamayor, M., Armitage, R., Spriggs, M.K., Fanslow, W.C., MacLenan, I.C., Gregory, C.D. and Gordon, J. (1993). *Eur. J. Immunol.* **22**, 2368–2371.
Hollowood, K. and Macartney, J. (1992). *Eur. J. Immunol.* **22**, 261–266.
Howard, J.C. (1972). *J. Exp. Med.* **135**, 185–199.
Jacob, J. and Kelsoe, G. (1992). *J. Exp. Med.* **176**, 679–687.
Jacob, J., Kassir, R. and Kelsoe, G. (1991a). *J. Exp. Med.* **173**, 1165–1175.
Jacob, J., Kelsoe, G., Rajewsky, K. and Weiss U. (1991b). *Nature* **354**, 389–392.
Jacob, J., Przylepa, J., Miller, C., Kelsoe, G. (1993). *J. Exp. Med.* **178**, 1293–1307.
Jeong, H.D. and Teale, J.M. (1989). *J. Immunol.* **143**, 2752–2760.
Jerne, N.K. (1951). *Acta Pathol. Microbiol. Scand. Suppl.* **87**, 2.
Kallberg, E., Gray, D. and Leanderson, T. (1993). *Int. Immunol.* **5**, 573–581.
Kepler, T.B. and Perelson, A.S. (1993). *Immunol. Today* **14**, 412–415.
Kim, S., Davis, M., Sinn, E., Patten, P. and Hood, L. (1981). *Cell* **27**, 573–581.
Klaus, G.G., Humphrey, J.H., Kunkl, A. and Dongworth, D.W. (1980). *Immunol. Rev.* **53**, 3–28.
Knight, K.L. (1992). *Annu. Rev. Immunol.* **10**, 593–616.
Kocks, C. and Rajewsky, K. (1989). *Annu. Rev. Immunol.* **7**, 537–559.
Kolchanov, N.A., Solovyov, V.V. and Rogozin, I.B. (1987). *FEBS Lett.*, **214**, 87–91.
Korsmeyer, S.J. (1992). *Immunol. Today* **13**, 285–288.
Kraus, R. (1903). *Zentralbl. Bakt.* **34**, 488–496.
Kunkel, T.A. and Alexander, P.S. (1986). *J. Biol. Chem.* **261**, 160–166.
Küppers, R., Zhao, M., Hansmann, M.L. and Rajewsky, K. (1993). *EMBO J.* **12**, 4955–4967.
Lafaille, J.J., DeCloux, A., Bonneville, M., Takagaki, Y. and Tonegawa, S. (1989). *Cell* **59**, 859–870.
Lane, P., Traunecker, A., Hubele, S., Inui, S., Lanzavecchia, A. and Gray, D. (1992). *Eur. J. Immunol.* **22**, 2573–2578.
Lebecque, S.G. and Gearhart, P.J. (1990). *J. Exp. Med.* **172**, 1717–1727.
Levy, N.S., Malipiero, U.V., Lebecque, S.G. and Gearhart, P.J. (1989). *J. Exp. Med.* **169**, 2007–2019.
Levy, S., Mendel, E., Kon, S., Avnur, Z. and Levy, R. (1988). *J. Exp. Med.* **168**, 475–489.
Li, W.-H., Wu, C.-I. and Luo, C.-C. (1984). *J. Mol. Evol.* **21**, 58–71.
Linton, P.L., Decker, D.J. and Klinman, N.R. (1989). *Cell* **59**, 1049–1059.
Linton, P.J., Rudie, A. and Klinman, N.R. (1991). *J. Immunol.* **146**, 4099–4104.

Linton, P.J., Lo, D., Lai, L., Thorbecke, G.J. and Klinman, N.R. (1992). *Eur. J. Immunol.* **22**, 1293–1297.
Liu, Y.J., Joshua, D.E., Williams, G.T., Smith, C.A., Gordon, J. and MacLennan, I.C. (1989). *Nature* **342**, 929–931.
Liu, Y.J., Zhang, J., Lane, P.J., Chan, E.Y. and MacLennan, I.C. (1991a). *Eur. J. Immunol.* **21**, 2951–2962.
Liu, Y.J., Cairns, J.A, Holder, MJ., Abbot, S.D., Jansen, K.U., Bonnefoy, J.Y., Gordon, J. and MacLennan, I.C. (1991b). *Eur. J. Immunol.* **21**, 1107–1114.
Liu, Y.J., Mason, D.Y., Johnson, G.D., Abbot, S., Gregory, C.D., Hardie, D.L., Gordon J. and MacLennan I.C. (1991c). *Eur. J. Immunol.* **21**, 1905–1910.
Lozano, F., Rada, C., Jarvis, J.M. and Milstein, C. (1993). *Nature* **363**, 271–273.
McCormack, W.T., Tjoelker, L.W. and Thompson, C.B. (1991). *Ann. Rev. Immunol.* **9**, 219–241.
McDonnell, T.J., Deane, N., Platt, F.M., Nunez, G., Jaeger, U., McKearn, J.P. and Korsmeyer, S.J. (1989). *Cell* **57**, 79–88.
McHeyzer, W.M., Nossal, G.J. and Lalor, P.A. (1991). *Nature* **350**, 502–505.
McKean, D., Huppi, K., Bell, M., Staudt, L., Gerhard, W. and Weigert, M. (1984). *Proc. Natl Acad. Sci. USA* **81**, 3180–3184.
MacLennan, I.C. and Gray, D. (1986). *Immunol. Rev.* **91**, 61–85.
MacLennan, I.C., Liu, Y.J. and Johnson, G.D. (1992). *Immunol. Rev.* **126**, 143–161.
McManus, S. and Riechmann, L. (1991). *Biochemistry* **30**, 5851–5857.
Maizels, N. (1989). *Trends Genet.* **5**, 4–8.
Maizels, N. and Bothwell, A. (1985). *Cell* **43**, 715–720.
Malipiero, U.V., Levy, N.S. and Gearhart, P.J. (1987). *Immunol. Rev.* **96**, 59–74.
Malynn, B.A., Yancopoulos, G.D., Barth, J.E., Bona, C.A. and Alt, F.W. (1990). *J. Exp. Med.* **171**, 843–859.
Manser, T. (1990). *Immunol. Today* **11**, 305–308.
Manser, T. (1991). Regulation, timing and mechanism of antibody V gene somatic hypermutation: lessons from the arsonate system. In 'Somatic Hypermutation in V-regions' (E.J. Steel, ed.). CRC Press, Boca Raton.
Manser, T., Wysocki, L.J., Margolies, M.N. and Gefter, M.L. (1987). *Immunol. Rev.* **96**, 141–162.
Marton, A., Delbecchi, L. and Bourgaux, P. (1991). *Nucleic Acids Res.* **19**, 2423–2426.
Milstein, C. (1987). *Biochem. Soc. Trans.* **15**, 779–787.
Milstein, C. (1990). *Proc. R. Soc. Lond. B.* **239**, 1–16.
Milstein, C., Even, J. and Berek, C. (1986). *Biochem. Soc. Symp.* **51**, 173–182.
Milstein, C., Even, J., Jarvis, J.M., González-Fernández, F.A. and Gherardi, E. (1992). *Eur. J. Immunol.* **22**, 1627–1634.
Moller, G. (ed.) (1987). *Immunol. Rev.* **96**, 23–41.
Moller, G. (ed.). (1992). *Immunol. Rev.* **126**.
Motoyama, N., Okada, H. and Azuma, T. (1991). *Proc. Natl Acad. Sci. USA* **88**, 7933–7937.
Motoyama, N., Miwa, T., Suzuki, Y., Okada, H. and Azuma, T. (1994). *J. Exp. Med.* **179**, 395–403.
Mullender, L.H., Vireling, H. and Venema, J. (1991). *Mutat. Res.* **247**, 203–211.
Nieuwenhuis, P. and Ford, W.L. (1976). *Cell. Immunol.* **23**, 254–267.
Noelle, R.J., Ledbetter, J.A. and Aruffo, A. (1992). *Immunol. Today* **13**, 431–433.
Nuñez, G., Hockenbery, D., McDonnell, T.J., Sorensen, C.M. and Korsmeyer, S.J. (1991). *Nature* **353**, 71–73.
O'Brien, R.L., Brinster, R.L. and Storb, U. (1987). *Nature* **326**, 405–409.
Orlandi, R., Gussow, D.H., Jones, P.T. and Winter, G. (1989). *Proc. Natl Acad. Sci. USA* **86**, 3833–3837.
Paabo, S., Irwin, D.M. and Wilson, A.C. (1990). *J. Biol. Chem.* **265**, 4718–4721.
Parvari, R., Ziv, E., Lantner, F., Heller, D. and Schechter, I. (1990). *Proc. Natl Acad. Sci. USA* **87**, 3072–3076.
Pascual, V., Liu, Y.-J., Magalski, MA., de Bouteiller, O., Banchereau, J. and Capra, J.D. (1994). *J. Exp. Med.* **180**, 329–339.
Rada, C. (1993). 'Somatic hypermutation of antibodies.' PhD Thesis, Cambridge.
Rada, C., Gupta, S.K., Gherardi, E. and Milstein, C. (1991). *Proc. Natl Acad. Sci. USA* **88**, 5508–5512.

Rada, C., González-Fernández, A., Jarvis, J.M. and Milstein, C. (1994). *Eur. J. Immunol.* **24**, 1453–1457.
Rajewsky, K., Forster, I. and Cumano, A. (1987). *Science* **238**, 1088–1094.
Rickert, R. and Clarke, S. (1993). *Int. Immunol.* **5**, 255–263.
Roes, J., Huppi, K., Rajewsky, K. and Sablitzky, F. (1989). *J. Immunol.* **142**, 1022–1026.
Rogerson, B.J. (1994). *Mol. Immunol.* **31**, 83–98.
Rogerson, B., Hackett, J.J., Peters, A., Haasch, D. and Storb, U. (1991). *EMBO J.* **10**, 4331–4341.
Rogozin, I.B. and Kolchanov, N.A. (1992). *Biochim. Biophys. Acta* **1171**, 11–18.
Rose, M.L., Birbeck, M.S., Wallis, V.J., Forrester, J.A. and Davies, A.J. (1980). *Nature* **284**, 364–366.
Rothenfluh, H.S., Taylor, L., Bothwell, A.L., Both, G.W. and Steele, E.J. (1993). *Eur. J. Immunol.* **23**, 2152–2159.
Rudikoff, S., Pawlita, M., Pumphrey, J. and Heller, M. (1984). *Proc. Natl Acad. Sci. USA* **81**, 2162–2166.
Sablitzky, F., Wildner, G. and Rajewsky, K. (1985). *EMBO J.* **4**, 345–350.
Schittek, B. and Rajewsky, K. (1992). *J. Exp. Med.* **176**, 427–438.
Secher, D.S., Milstein, C. and Adetugbo, K. (1977). *Immunol. Rev.* **36**, 51–72.
Seidman, M.M., Bredberg, A., Seetharam, S. and Kraemer, K.H. (1987). *Proc. Natl Acad. Sci. USA* **84**, 4944–4948.
Sharon, J. (1990). *Proc. Natl Acad. Sci. USA* **87**, 4814–4817.
Sharpe, M.J., Neuberger, M., Pannell, R., Surani, M.A. and Milstein, C. (1990). *Eur. J. Immunol.* **20**, 1379–1385.
Sharpe, M.J., Milstein, C., Jarvis, J.M. and Neuberger, M.S. (1991). *EMBO J.* **10**, 2139–2145.
Shlomchik, M.J., Marshak, R.A., Wolfowicz, C.B., Rothstein, T.L. and Weigert, M.G. (1987). *Nature* **328**, 805–811.
Shuldiner, A.R., Nirula, A. and Roth, J. (1989). *Nucleic Acids Res.* **17**, 4409.
Siekevitz, M., Kocks, C., Rajewsky, K. and Dildrop, R. (1987). *Cell*, **48**, 757–770.
Sohn, J., Gerstein, R.M., Hsieh, C.L., Lerner, M. and Selsing, E. (1993). *J. Exp. Med.* **177**, 493–504.
Sperling, R., Francus, T. and Siskind, G.W. (1983). *J. Immunol.* **131**, 882–885.
Steele, E.J. and Pollard, J.W. (1987). *Mol. Immunol.* **24**, 667–673.
Steele, E.J., Rothenfluh, H.S. and Both, G.W. (1992). *Immunol. Cell Biol.* **70**, 129–141.
Taylor, L.D., Carmack, C.E., Huszar, D., Higgins, K.M., Mashayekh, R., Kay, R.M., Woodhouse, C.S. and Lonberg, N. (1994). *Int. Immunol.* **6**, 579–591.
Tew, J.G. and Mandel, T. (1978). *J. Immunol.* **120**, 1063–1069.
Tomlinson, I.M., Walter, G., Marks, J.D., Llewelyn, M.B. and Winter, G. (1992). *J. Mol. Biol.* **227**, 776–798.
Tuaillon, N., Miller, A.B., Lonberg, N., Tucker, P.W. and Capra, J.D. (1994). *J. Immunol.* **152**, 2912–2920.
Umar, A., Schweitzer, P.A., Levy, N.S., Gearhart, J.D. and Gearhart, P.J. (1991). *Proc. Natl Acad. Sci. USA* **88**, 4902–4906.
Velick, S.F., Parker, C.W. and Eisen, H.N. (1960). *Prod. Natl Acad. Sci. USA* **46**, 1470–1475.
Victor, K.D., Vu, K., Feeney, A.J. (1994). *J. Immunol.* **152**, 3467–3475.
Weber, J.S., Berry, J., Manser, T. and Claflin, J.L. (1991). *J. Immunol.* **146**, 3652–3655.
Weigert, M. (1986). *Prog. Immunol.* **6**, 139–144.
Weiss, S. and Wu, G.E. (1987). *EMBO J.* **6**, 927–932.
Weiss, U. and Rajewsky, K. (1990). *J. Exp. Med.* **172**, 1681–1689.
Weiss, U., Zoebelein, R. and Rajewsky, K. (1992). *Eur. J. Immunol.* **22**, 511–517.
Wilson, M., Hsu, E., Marcuz, A., Courtet, M., Du, P.L. and Steinberg, C. (1992). *EMBO J.* **11**, 4337–4347.
Wysocki, L.J. and Gefter, M.L. (1989). *Annu. Rev. Biochem.* **58**, 509–531.
Wysocki, L., Manser, T. and Gefter, M.L. (1986). *Proc. Natl Acad. Sci. USA* **83**, 1847–1851.
Wysocki, L.J., Gefter, M.L. and Margolies, M.N. (1990). *J. Exp. Med.* **172**, 315–323.
Zhang, J., MacLennan, I.C., Liu, Y.J. and Lane, P.J. (1988). *Immunol. Lett.* **18**, 297–299.
Ziegner, M., Steinhauser, G. and Berek, C. (1994). *Eur. J. Immunol.* **24**, 2393–2400.

4

Development of B-cell subsets

Sandra M. Wells*, Alan M. Stall*, Aaron B. Kantor
and Leonore A. Herzenberg†[1]

* Department of Microbiology, College of Physicians & Surgeons, Columbia
University, New York, NY, USA
† Beckman Center, Department of Genetics, Stanford University Medical
Center, Stanford, CA, USA

Introduction 83
Features of B-cell development 84
B-cell subsets and lineages..86
Distribution and phenotypic characteristics of B cells in peripheral lymphoid
  organs 87
Self-replenishment of B-cell subsets 91
Differentiation of B cells in response to antigenic stimulation 92
Differences in repertoire between B-cell subsets 94
Origins of the B-cell lineages 95
Additional B-cell lineages 98
The layered evolution of the immune system...............................99

INTRODUCTION

Antibody production is the key factor in the humoral defense against invading pathogens and other potentially harmful substances that may be introduced into the body. Antibody-forming cells (AFC) are derived from specialized lymphocytes named B cells, which are individually differentiated to produce antibody molecules that have distinctive antigen-combining sites (specificities) and functional components (isotypes).

The enormous diversity of antibody reactivities required to protect against the wide variety of antigens that an animal may encounter throughout life is generated by a complex differentiation process consisting of two basic stages. The variable region, which defines the antigen-combining site of the unique immunoglobulin molecule

[1] Address correspondence to Dr Leonore A. Herzenberg

that each B cell produces, is initially determined by the developmentally-controlled juxtaposition of V_H, D and J_H elements of the heavy chain and V_L and J_L elements of the light chain. Variable-region rearrangement occurs during the early stages of B cell differentiation in the bone marrow and fetal liver. The first constant region domains (μ and δ) used in B cell development are determined by differential processing of the RNA transcripts. Later, as mature B cells in the periphery encounter specific antigens, the antibody molecules they produce may be further diversified by mechanisms that enable 'switching' of constant region (isotype) concomitant with somatic mutation of variable-region genes. This latter process, which can ultimately yield antibodies with higher affinities for the antigen, matures from a series of cell clones producing distinct but related immunoglobulin (Ig) molecules that often have markedly different binding affinities for the antigen.

The differentiation stages through which B cells and their progenitors pass on their way to becoming AFC initially appeared to follow a single developmental pathway. However, phenotypic and cell transfer studies begun in the early 1980s identified at least two independent B cell development pathways (lineages) whose progenitors are now distinguishable at an early developmental stage (pro-B cell), when the first Ig ($D-J_H$) rearrangement begins (Hayakawa et al., 1984, 1985). T cells have now also been shown to differentiate along similarly separable developmental lineages, suggesting that the commitment to give rise to particular T and B lineages reflects the activity of distinctive lymphoid stem cells (Ikuta et al., 1990).

In this chapter, we focus primarily on the features that distinguish the developmental pathways of the two most well-established B cell lineages, commonly referred to as the B-1a (earlier designated Ly-1 or CD5⁺ B) and conventional (or B-2) lineages (Kantor et al., 1991). We begin with a relatively brief description of the phenotypic and functional characteristics of the developmental stages of B cells. Then, starting with the most mature B cells and working back towards the early stages of B cell development, we detail the differences between the cells in each lineage at these various stages of development. Finally, we discuss the features that potentially distinguish additional B cell developmental lineages and discuss the evolutionary and functional consequences of these multiple layers of functionally related cells in the immune system.

FEATURES OF B-CELL DEVELOPMENT

The progressive Ig gene rearrangements that occur as cells move along the B cell developmental pathway are accompanied by differentiation events that alter the size and surface phenotype of the developing cells (summarized in Table I). This process has been well studied for conventional B cell development in adult bone marrow (Hardy et al., 1991; Rolink and Melchers, 1991). In essence, the initial Ig heavy chain rearrangement, which joins a D and a J_H element is accompanied by a shift from the pro-B cell to the early pre-B phenotype. The rearrangement that joins this

Table I Phenotypic changes during B-cell development

	Pro-B cells	Pre-B cells	Immature B cells	Mature B cells	Memory B cells	AFC plasma cells
Surface Phenotype						
IgM	Negative	Negative	High	Intermediate	Negative	High*
IgD	Negative	Negative	Neg/low	High	Negative	Low*
B220	Low	Intermediate	High	High	?	Low*
HSA	Negative	Intermediate	High	Low	?	High*†
CD43	High	Low	Low	Negative	?	High
CD23	Negative	Negative	Negative	High	?	Negative*
BP-1	Negative	High	Negative	Negative	?	?
Size (FSC)	Large	Large/small	Small	Small	Small	Large*†
Ig gene rearrangements	Heavy chain D to J	Heavy chain V to DJ light chain	Rearranged	Rearranged	Rearranged/isotype switching	Rearranged/isotype switching
Functional	Antigen unresponsive	Antigen unresponsive	Selection/tolerance?	Antigen responsive	Late stages of response	Late stages of response
Role of antigen	Antigen independent	Antigen independent	Antigen independent	Antigen dependent	Antigen dependent	Antigen dependent
Anatomic sites	Bone marrow	Bone marrow	Bone marrow and periphery	Bone marrow and periphery	Periphery	Periphery

* Antigen-specific activation of B-1 cells *in vivo*.
† Activation of spleen cells *in vitro*.
These data are based upon the following references: Allman, *et al.*, (1993); Hardy *et al.*, (1984, 1991); Hathcock *et al.* (1992); Metcalf and Klinman, 1977; Nishimura *et al.* (1992); Nossal, (1983); Rolink and Melchers (1991); Waldschmidt *et al.* (1992).

D–J_H segment to a variable-region (V_H) gene is accompanied by a shift to the late pre-B stage. Transcription and splicing of the RNA creates an expressible μ heavy chain that is found in the cytoplasm and probably on the surface in association with the pseudo-light chain proteins, λ5 and V_{pre-B} (Rolink and Melchers, 1991). The rearrangement of the light chain and the expression as surface IgM move the cell to the immature B cell stage. Finally, B cells migrate to the periphery where they are selected into the long-lived pool of recirculating B cells. This is marked by an increase in IgD expression and additional phenotypic changes that define the cell as a mature conventional B cell.

Hardy and Hayakawa have defined distinctive FACS (Fluorescence Activated Cell Sorter) phenotypes of the pro-B and pre-B cell developmental stages (see Table I) (Hardy et al., 1991). These differences primarily involve coordinated quantitative changes in the expression of several surface molecules, notably leukosialin (CD43/S7), B220/6B2, and heat stable antigen (HSA; 30-F11), as well as the IgM molecule itself. As conventional B cells are selected into the mature pool, they can be distinguished by surface expression of IgM, IgD, CD23 and low levels of HSA. Following antigen-specific activation, the mature B cells can undergo additional differentiation steps to become AFC/plasma cells or memory B cells. Plasma cells are the differentiated effector cells responsible for producing large amounts of secreted Ig, while memory B cells, which persist for long periods and can later give rise to plasma cells, are crucial for the immune system's ability to mount an accelerated and enhanced response upon its second encounter with a particular antigen (Gray, 1993; Parker, 1993).

B-CELL SUBSETS AND LINEAGES

Early studies of lymphocyte subsets focused largely on the T cell compartment. Antibody-mediated cytotoxicity and FACS studies showed that expression of the Lyt-2 (CD8) surface molecule distinguishes the suppressor/cytotoxic subset from the helper/inducer subset (Cantor and Boyse, 1975, 1977). The expression of Lyt-1 (CD5) was initially used in antibody-mediated cytotoxicity studies to identify and deplete the helper/inducer subset (Herzenberg et al., 1976); however, subsequent FACS studies with monoclonal anti-CD5 antibodies showed that the difference in CD5 expression between the subsets was quantitative rather than qualitative (Ledbetter et al., 1980). The use of anti-CD5 in T cell subset studies terminated with discovery of the CD4 surface antigen, whose expression on the helper/inducer subset definitively distinguishes this subset. Ironically, the expression of CD5 later proved to distinguish the B cell subsets/lineages (Lanier et al., 1981; Hardy et al., 1982; Manohar et al., 1982; Herzenberg et al., 1986).

B cells were initially thought to be much more homogeneous than T cells, differing principally with respect to the isotype and specificity of the Ig molecules they produced. Phenotypically distinct subsets of B cells were recognized; however, these subsets tended to be viewed as consisting of B cells at different stages of differentiation (e.g. naive and memory B cells) that have linearly descended or alternatively differentiated from the same progenitors to provide specialized capabilities related to antigen recognition or antibody production.

This paradigm was replaced/extended by the recognition that B cell progenitors are committed to differentiate along developmental pathways that culminate in the development of different functional subsets. By now, a wide variety of phenotypic, anatomical, developmental and functional differences between these major B cell lineages have been verified. Two major lineages are generally recognized: conventional B cells (also called B-2 cells), which develop predominantly from progenitors in the adult bone marrow; and B-1a cells, which develop predominantly from fetal progenitors (Kantor and Herzenberg, 1993). The genetically defined developmental program of these lineages is crucial to defining their overall capabilities and has an important, possibly decisive, influence in defining their basic antibody repertoires and the specific roles they play in immune functions.

Phenotypic, locational and functional differences between the two major B cell lineages can be found at every stage of development; however, considerably more is known about these differences amongst the mature B cells. Thus, we begin here by defining the developmental and functional subsets of mature B cells and their progeny within each lineage and contrasting the lineage to lineage variation of these mature subsets.

DISTRIBUTION AND PHENOTYPIC CHARACTERISTICS OF B CELLS IN PERIPHERAL LYMPHOID ORGANS

B cells are small lymphocytes that express surface Ig molecules. In general, they do not secrete Ig but can be stimulated to differentiate into phenotypically distinct AFC responsible for performing this function. In adult animals, the majority of all B cells located in the peripheral lymphoid organs (i.e. spleen, lymph nodes and Peyer's patches) belong to the conventional B cell lineage. Small numbers of B-1 lineage cells can be found in these organs, most notably in spleen; however, most studies characterizing B-1 cells draw these cells from the peritoneal cavity, where they represent the predominant B cell population (Hayakawa et al., 1985; Herzenberg et al., 1986; Kantor and Herzenberg, 1993).

Conventional B cells

The major conventional B cell subset (follicular B cells) contains cells that are relatively small and found predominantly in lymphoid follicles, where they express low levels of IgM, high levels of IgD and characteristic levels of other surface molecules (Table II, Figs 1 and 2). These B cells constitute virtually the entire B cell population in lymph nodes; however, in the spleen, the overall B cell population also

Table II Phenotype of peripheral B-cell subsets

	Follicular	Immature	Marginal zone	B-1
IgM	Intermediate	High	High	High
IgD	High	Negative/low	Low	Low
CD45	High	High	High	High
B220 (6B2)	High	High	Intermediate	Low
B220 (2C2)	High	High	High	High
HSA (J11d)	Intermediate	?	High	Intermediate
HSA (M1/69)	Low	High	Intermediate	Intermediate
HSA (53–10)	Negative	?	Intermediate	Intermediate
CD23	High	Negative	Negative	Negative
CD22	High	Low	High	High
L-selectin (MEL-14)	High	Low ?	Low	Negative
CTLA-4Ig-binding	Negative	Negative	Negative	Low
CD43 (S7)	Negative	Low	Negative	Intermediate
MAC-1 (CD11b)	Negative	Negative	Negative	Spleen: negative PerC: low
CD5	Negative	Negative	Negative	B-1a: low B-1b: negative
Size	Small	Small	Small	Large

The values given here represent the relative intensity of the major population within each subset. For any given antigen there can be considerable heterogeneity within the subset. For more detailed analysis see the FACS plots in Fig. 2. The data are based upon the following references: Manohar et al. (1982); Hayakawa et al. (1983); Hardy et al. (1984); Herzenberg et al. (1986); Kantor et al. (1992); Nishimura et al. (1992); Waldschmidt et al. (1992); Allman et al. (1993); Wells et al. (1994).

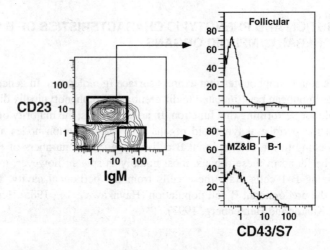

Fig. 1. Phenotypic identification of splenic B cell subpopulations. Spleen cells from 1–2-month-old BAB/25 mice were stained for CD23 (TexasRed [TR]), IgM (Fluorescein [Fl]) and CD43/S7 (Phycoerythin [PE]). The CD23, IgM phenotype of total spleen cells is shown in the left-hand panel. Both marginal zone and immature cells (MZ&IB) and B-1 cells are found within the IgM^{bright}, CD23$^-$ gate, but these two populations can be distinguished by their CD43/S7 phenotype. The MZ&IB, which make up the majority of this population, are IgM^{bright}, CD23$^-$, CD43/S7$^-$. The B-1 cells, however, are IgM^{bright}, CD23$^-$, CD43/S7$^+$.

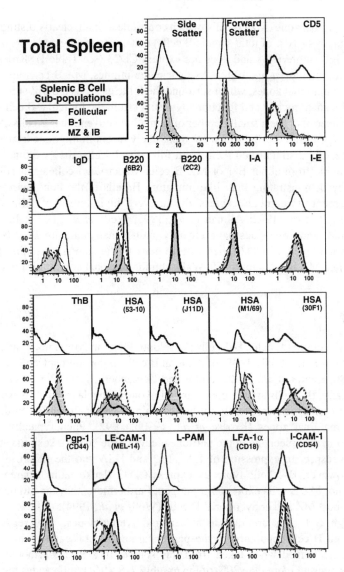

Fig. 2. Comparative phenotypes of splenic B-1 and marginal zone and immature B cells (MZ&IB). Spleen cells from 1–2-month-old BAB/25 mice were stained for CD23 (biotin [bi]), IgM (Fluorescein [Fl]), CD43/S7 (Phycoerythin [PE]) and CD5 (allo phycocyanin [APC]). In all other samples, cells were stained for CD23 (TexasRed [TR]), IgM (APC), CD43/S7 (PE) and Fl conjugates of antibodies to the antigens specified. The expression of each antigen on total spleen (upper histograms) and on follicular, B-1 cells and MZ&IB cells (lower histograms) is shown. The B cell subpopulations were gated as shown in Fig. 1. The y-axis is a normalized scale for each histogram. Follicular, MZ&IB and B-1 cells comprise approximately 85%, 12% and 3% of the total B cells respectively. All antibodies were obtained from Pharmingen (San Diego) with the exception of anti-IgM (331) and anti-ThB (49H4) which were prepared in our laboratory.

contains marginal zone (MZ) B cells, which constitute a small, clearly distinguishable subset (roughly 5–10% of total splenic B cells) that expresses higher levels of surface IgM, less IgD, more HSA and does not express CD23 (see Table I) (Kroese et al., 1990a; Waldschmidt et al., 1992). As their name implies, MZ B cells are found in areas called marginal zones, which surround the follicles. Thus the MZ subset is phenotypically, anatomically and functionally distinct from the follicular B cell subset.

Adult spleen also has a small number of immature conventional B cells. Although the phenotype of these cells, which are most likely recent immigrants from the bone marrow, readily distinguished them from follicular B cells, they were not initially distinguishable from either B-1 or MZ B cells. Cancro and colleagues solved this problem by demonstrating that, like immature B cells in the bone marrow, these putative recent immigrants express distinctively high levels of HSA (M1/69) (Waldschmidt et al., 1992; Allman et al., 1993; Wells et al., 1994). In addition, Waldschmidt and colleagues have recently shown that immature B cells express lower surface levels of CD22 than mature cells in the spleen, including both the follicular and MZ B cell subsets (T. Waldschmidt, personal communication).

B-1 cells

Unlike conventional B cells, B-1 cells are seldom found in lymph nodes and Peyer's patches and represent only 1–3% of the total B cells in the adult spleen (Herzenberg et al., 1986; Kantor et al., 1992; Kantor and Herzenberg, 1993; Wells et al., 1994). Anatomically, B-1 cells are not concentrated in one particular area of the spleen and tend to be found most frequently in the red pulp, occasionally in the follicles, and rarely in marginal zones (Kroese et al., 1990b, 1992). Phenotypically, the B-1 population in the spleen, as in the peritoneal cavity (see below), is similar to MZ B cells with respect to expression of IgM, IgD and HSA and the failure to express CD23. However, B-1 cells typically express CD43 (leukosialin) whereas neither follicular nor MZ B cells express this antigen. Furthermore, they tend to be larger in size than most MZ and conventional B cells (Wells et al., 1994).

Although B-1 cells are rare in spleen and lymph nodes, they represent the predominant B cell population in the peritoneal and pleural cavities and are most readily characterized when taken from this source. In a BALB/c mouse, B cells typically represent more than 80% of the roughly $7-8 \times 10^6$ lymphocytes recoverable from the peritoneal cavity. B-1 cells typically comprise 60–80% of these B cells (depending upon strain). Conventional B cells that are phenotypically identical to the follicular B cells found in the spleen and lymph nodes comprise the remainder (Herzenberg et al., 1986; Kantor and Herzenberg, 1993).

Peritoneal and splenic B-1 cells are phenotypically identical with one major exception: splenic B-1 cells do not express MAC-1 (CD11b), whereas peritoneal and pleural B-1 cells express detectable levels of this surface antigen. Curiously, this difference in MAC-1 expression appears either to be environmentally determined or to determine the migratory behavior of the B-1 cells on which it is expressed: FACS-sorted splenic B-1 cells, which lack MAC-1, reconstitute peritoneal B-1 cells that express MAC-1 in transfer recipients; and, conversely, peritoneal B-1 cells that express MAC-1 reconstitute splenic B-1 cells that do not express this antigen (Kantor et al., 1995).

B-1 cells were originally called Ly-1 B cells. This earlier name reflected their initial recognition as a subset of B cells that expresses CD5, a pan T cell surface glycoprotein formerly called Lyt-1 or Ly-1 (see above). The current name (B-1) was provisionally selected to allow inclusion of another subset/lineage of cells that do not express CD5 but are otherwise phenotypically and locationally indistinguishable from the CD5$^+$ B-1 cells (Kantor et al., 1991). In this terminology, which is now widely used, B-1 cells that express CD5 are referred to as B-1a cells and the B-1 cells that do not express CD5 are referred to as B-1b cells. Other differences between B-1a and B-1b cells include differences in antibody repertoire (Kantor et al., 1994), in frequency of occurrence in different mouse strains (Herzenberg et al., 1986; Stall et al., 1992; Wells et al., 1994), and in the time at which they develop (Kantor et al., 1992; Stall et al., 1992). These differences, particularly at the developmental level, suggest that B-1a and B-1b cells belong to distinct, albeit closely related lineages (see below).

The surface phenotypes of the adult splenic B cell subsets described are shown in details in Figs 1 and 2 and summarized in Table II. Follicular B cells from both spleen and lymph node are most easily identified as small lymphocytes that express CD23, low surface levels of IgM, high levels of IgD, and high levels of B220/6B2. MZ, immature and B-1 cells, in contrast, all express high levels of IgM, low levels of IgD and no CD23. These latter three populations, which are found in the spleen, are distinguished by their differential expression of HSA, B220/6B2, CD22 and CD43: immature B cells express high levels of HSA, intermediate B220/6B2, low CD22, and do not express CD43; MZ B cells express intermediate levels of HSA and B220/6B2, high CD22, and also lack CD43 expression; B-1a (CD5$^+$) and B-1b (CD5$^-$) cells in both the spleen and peritoneal cavity express intermediate levels of HSA, low B220/6B2, high CD22 and, in contrast to follicular, MZ and immature B cells, low levels of CD43. Finally, splenic B-1 cells do not express MAC-1 whereas peritoneal B-1 cells express low levels of MAC-1 (Kantor et al., 1992; Waldschmidt et al., 1992; Allman et al., 1993; Wells et al., 1994).

SELF-REPLENISHMENT OF B-CELL SUBSETS

In addition to the unique phenotypic and anatomical characteristics of B-1 and conventional B cells, distinct functional differences further define these two lineages. A feedback mechanism regulates development of the B-1 population from immature progenitors. Decreasing B-1a progenitor activity plus this feedback regulation results in the inhibition of newly emerging B-1 cells into the periphery after weaning, while still allowing conventional B cells to continue to develop from Ig$^-$ progenitors in the adult bone marrow (Lalor et al., 1989a,b; Watanabe, L.A. Herzenberg and A.B.

Kantor, in preparation). Data demonstrating that B-1 and conventional B cells turn over at similar rates (Förster and Rajewsky, 1990; Deenen and Kroese, 1993), and that adoptively transferred peritoneal B-1 cells completely replenish the peripheral B-1 population (Hayakawa et al., 1986) demonstrate that the B-1 lineage is mainly self-replenished from Ig⁺ cells. In contrast, adoptive transfers of splenic or lymph node conventional B cells indicate that although these cells may persist for long periods, they have little or no capacity for expansion or self-replenishment (Sprent et al., 1991; Kantor and Herzenberg, 1993; Kantor, 1995).

A number of experiments suggest that the preferential expansion/self-replenishment of the B-1 cells may be the result of ongoing receptor stimulation, possibly due to the low-affinity self-reactive specificities common within the population. Analysis of B-1 cell hyperplasias in NZB/W mice show that individual clones within the same mouse utilized the same V_H, D_H and V_L genes, indicating that the expansion was associated with a given specificity or idiotope (Tarlinton et al., 1988). Similarly, the expansion of anti-phosphatidylcholine reactive B-1 cells, which account for 10–15% of peritoneal B-1, has been attributed to antigenic selection and expansion (Pennell et al., 1988, 1989, 1990). This is supported by the fact that a large fraction of B-1 cells specific for phosphatidylcholine (bromelain-treated mouse erythrocytes) appear to have their membrane immunoglobulin (mIg) occupied by antigen (Carmack et al., 1990). However, these data, which indicate antigenic stimulation, do not undermine the lineage distinction between B-1 and conventional B cells, which is based on developmental differences that are manifest very early in the B cell developmental pathway.

DIFFERENTIATION OF B CELLS IN RESPONSE TO ANTIGENIC STIMULATION

Primary immune responses to both T-independent (TI) and T-dependent (TD) antigen are characterized by a relatively weak, short-lived burst of antibody (generally IgM) production. TD antigens are also able to elicit secondary or anamnestic responses resulting in a more rapid rise in antibody level and a significantly higher steady-state level that persists long after first contact with the antigen. The cells responsible for the antibody production in the secondary response have undergone affinity maturation in order to produce antibodies that are more effective in clearing the invading antigen.

Secondary responses are also characterized by a switch to the production of isotypes other than IgM. Each Ig heavy chain isotype exhibits a unique pattern of effector functions that participate in these host-defense responses. B-1 cells expressing isotypes other than IgM have been found *in vivo* (Herzenberg et al., 1986; Förster and Rajewsky, 1987; Kroese et al., 1989; Solvason et al., 1991). Furthermore *in vitro* studies (Braun and King, 1989; Waldschmidt et al., 1992; Tarlinton et al., 1994; Whitmore et al., 1992) have shown that isotype switching can occur in B-1 cells.

However, conventional B cells participating in antibody responses tend to switch to certain isotypes more frequently than do B-1 cells, resulting in quite distinct isotype profiles in antibody responses produced by these two lineages (Herzenberg et al., 1986). The significance of this difference in isotype profiles between B-1 and conventional B cells is not yet known, but understanding these differences will be important in determining the role of B-1 cells in immune responses.

In addition to isotype switching, antigen-stimulated cells undergo affinity maturation, which is characterized by production of Ig whose combining site structure has changed due to somatic hypermutation. This somatic hypermutation process, which characteristically occurs during the generation of memory B cells, appears to occur substantially less frequently in B-1 cells than in conventional B cells. Analysis by Hardy and colleagues of B-1 cells that produce anti-phosphatidylcholine antibodies demonstrated that both the variable heavy chain (V_H) and variable light chain (V_L) remain totally unmutated (Hardy, 1989). Shirai et al. (1991) in contrast, have found evidence of somatic mutation in neoplastic B-1 cells. These findings suggest that B-1 and conventional B cells differ with respect to somatic mutation; however, data are too sparse at present to determine whether these differences reflect inherent differences in the ability to undergo somatic mutation or whether they stem secondarily from other differences between the two types of B cells, e.g. the local environment in which the antibody response develops; the basic antibody repertoire from which the response is drawn; or the nature of the T cell help evoked by the antigen and/or the responding B cell.

Generally, B cells give rise to two distinct cell types in primary responses to TD antigens: AFC or plasma cells that secrete predominantly low-affinity IgM antibodies and are located in the periphery of the periarteriolar lymphoid sheaths (PALs) of the spleen and lymph nodes (van Rooijen et al., 1986); and memory B cells, which (according to current understanding) are generated in the germinal centers of peripheral lymphatic organs and upon later stimulation yield the plasma cells responsible for the high-affinity, predominantly non-IgM antibodies produced during secondary responses (Coico et al., 1983; for review see Gray, 1993.

Several theories have been proposed to account for the simultaneous generation of these two populations. The most widely accepted explanation holds that an unequal division or differentiation of primary B cells following antigen-specific activation results in the formation of both AFC and memory B cells (Williamson et al., 1976). In support of this, Jacob and Kelsoe (1992) have shown that the cells responsible for early antibody production in the primary response to (4-hydroxy-3-nitrophenyl)acetyl (NP) share a common clonal origin to germinal center B cells arising in the same response. Other studies by Klinman and colleagues, however, strongly suggest that there are two distinct populations of B cells responsible for the primary response (J11d^{low}) and secondary responses (J11d^{hi}) (Linton et al., 1989, 1992; Linton and Klinman, 1992) These 'separate populations' and 'unequal division' theories, however, are not mutually exclusive. Properties of the antigenic response being studied (dose, route of injection, etc.) as well as the B cell subsets involved may favor one model over the other in a given situation.

DIFFERENCES IN REPERTOIRE BETWEEN B-CELL SUBSETS

Unlike conventional B cells, the B-1 population has a repertoire that seems to be skewed toward the production of antibodies reactive with self antigens such as bromelain-treated mouse red blood cells and DNA (Hayakawa et al., 1984; Mercolino et al., 1986, 1988). B-1 cells also respond well to many TI antigens, particularly those associated with microorganismal coat antigens such as lipopolysaccharide (Su et al., 1991), phosphorylcholine (PC) (Masmoudi et al., 1990; Taki et al., 1992), $\alpha(1\rightarrow3)$ dextran (Förster and Rajewsky, 1987) and $\alpha(1\rightarrow6)$ dextran (Wang et al., 1994); however, they do not respond to all TI antigens and they do respond to certain T-dependent (TD) antigens (see below). Similarly, although conventional B cells usually produce the high-affinity (affinity-matured) memory responses to TD antigens, they also can produce low-affinity primary responses to certain TI antigens. Thus, although some general trends exist, the responses of B-1 and conventional B cell responses cannot be definitively categorized according to the nature of the responses in which they participate.

The unique repertoire and responsiveness of the B-1 population has generated considerable speculation on the idea that B-1 and conventional B cells can be distinguished by their ability to respond to TI and TD antigens, respectively. Indeed, the acquisition of the B-1 phenotype itself has been proposed to result from the interaction of a resting conventional B cell with a TI antigen (Rabin et al., 1992; Haughton et al., 1993). This theory, however, is inconsistent with data, alluded to above, showing that the responses of B-1 and conventional B cells cannot be categorized solely as TD or TI. For example, B-1 cells do not respond well to many TD antigens, including sheep erythrocytes, Trinitrophenol (TNP) and NP haptens, presented in TD forms (Förster and Rajewsky, 1987; Hayakawa and Hayakawa et al., 1984; Hardy, 1988); however, they produce virtually the entire primary TD response to phosphorylcholine-keyhole limpet hemocyanin (PC-KLH) (which is composed of T15id$^+$ antibodies) (Masmoudi et al., 1990; Taki et al., 1992). Furthermore, although they do respond to certain TI antigens, B-1 cells do not respond at detectable levels to antigens such as TNP or NP presented in a TI form, i.e. coupled to ficoll (Hayakawa et al., 1984; Förster and Rajewsky, 1987; Hayakawa and Hardy, 1988).

Data from studies such as those outlined above suggest that response to a given antigen will often be produced by either the B-1 or the conventional B cell population, but not both. However, Hayakawa et al. (1984) showed many years ago that presenting Dinitrophenol (DNP) on *Brucella abortus* elicited antibody responses from both B-1 and conventional B cells. In a more extensive series of experiments, Wells et al. have recently shown that the immune response to $\alpha(1\rightarrow6)$ dextran, unlike the vast majority of the antibody responses studied thus far, is elicited by both B-1 and conventional B cells (S.M. Wells, D. Wang, E.A. Kabat and A.M. Stall, submitted). These latter data demonstrate conclusively that the B-1 and conventional B cell populations can respond to the same antigen and that, following challenge with a defined antigen, both are capable of eliciting an antibody response that is similar with respect to kinetics and isotype profile.

Structural studies that characterize the repertoires of B-1 and conventional B cell populations at various locations in adult and neonatal animals are in their early stages. Differences in V_H gene usage have been reported; however, although the overall picture from these studies is most likely correct, the methodology used leaves some room for question concerning the conclusions reached. Current technology, which allows definitive sequencing of Ig heavy and light chains from individual B cells, will provide a clearer view of these repertoire difference. For example, Ig heavy chain sequence data from individually sorted B cells from the adult peritoneal cavity indicate that the fraction of cells that produce antibodies lacking N-region additions is significantly higher in the B-1a subset that in the conventional B cell subset. The B-1b subset is more similar to conventional B cells. The differences found are consistent with the idea that B-1 cells present in adults tend to have been generated early in development, when pro-B cells do not express terminal deoxynucleotidyl transferase (TdT) and thus are not able to increase antibody diversity by the addition of N-region nucleotides (Kantor et al., 1994; A.B Kantor, C.E. Merrill, L.A. Herzenberg and J.L. Hillson, in preparation).

ORIGINS OF THE B-CELL LINEAGES

The initial separation of B cells into two lineages was essentially based on differences in how the lineages maintain their number in adult animals. Conventional B cells, being the predominant B cell population in spleen and lymph nodes, are continually replenished by *de novo* differentiation of progenitors in the bone marrow (Hayakawa et al., 1985, 1986). In addition, they can be readily reconstituted in irradiated recipients by transfers of relatively undifferentiated (B220⁻) adult bone marrow cells but not by transfers of mature conventional B cells in bone marrow or in the periphery (A.B. Kantor, A.M. Stall, S. Adams, K. Watanabe and L.A. Herzenberg, submitted).

B-1a cells, in contrast, are well reconstituted by transfers of fetal and neonatal sources of lymphoid progenitors but not by transfers of adult bone marrow (Hayakawa et al., 1985; Solvason et al., 1991; Kantor et al., 1992). Furthermore, they can be readily reconstituted by transfers of mature Ig⁺ B-1 cells from the peritoneal cavity (Hayakawa et al., 1986). Thus, unlike conventional B cells, the B-1 population has a substantial capacity for self-replenishment (see above) and therefore is able to maintain its numbers in adults in the absence of continued *de novo* differentiation from early progenitors. The selective lack of B-1a progenitor activity in adult bone marrow indicates that the progenitors for B-1a cells are distinct from progenitors for conventional B cells and hence that these cells belong to separate developmental lineages.

The expression of several 'activation markers', such as CD43, on B-1 cells, the restriction of a particular antibody response to either B-1 or conventional B cell populations, and the apparent skewing of B-1 cell antibody production towards auto- and anti-bacterial specificities have led others (Rabin *et al.*, 1992; Haughton *et al.*, 1993) to propose that B-1 cells are conventional B cells that have differentiated to the B-1 phenotype in response to activation with a particular type of antigen (i.e. TI type II). This idea has been strongly advocated by Wortis and colleagues, who demonstrated that stimulating conventional B cells *in vitro* with anti-IgM and IL-6 induces expression of CD5 and certain other markers associated with B cell activation (Ying-zi *et al.*, 1991). The authors view this anti-IgM stimulation as a model of TI II responses and interpret the phenotypic shift towards the B-1 phenotype *in vitro* as evidence that the B-1 population seen *in vivo* is generated by stimulating conventional B cells with TI II antigens.

The demonstration that B cell progenitors in bone marrow fail to give rise to more than a few B-1a cells while progenitors from fetal and neonatal animals readily generate these cells would appear to negate this hypothesis. However, proponents of the T type II hypothesis rationalize these developmental differences by postulating that, unlike B-1 cells, bone marrow-derived B cells arise relatively late in life and do not tend to make antibodies that react with TI type II antigens. This selective non-reactivity is ascribed to the N-region insertions that the TdT enzyme introduces into the Ig heavy chains produced by the bone marrow-derived cells (Haughton *et al.*, 1993). These insertions would be rare in B cells generated in young animals since TdT is not expressed in fetal pro-B cells (Li *et al.*, 1993). Thus, following the logic of this hypothesis, conventional B cells generated from progenitors in neonates would be more likely to produce antibodies reactive with TI type II antigens and hence would be more likely to give rise to B-1 cells than conventional B cells generated from (TdT-expressing) progenitors in adult bone marrow.

The antibody responses of B-1 cells and conventional B cells, however, cannot be readily defined in terms of either the form in which the antigen is presented or the characteristics of the epitope to which the response is produced. Nor can the antibodies themselves be defined in terms of particular structural characteristics. For example, as indicated above, some TI and some TD antigens readily stimulate antibody production in B-1 cells; however, others only stimulate antibody production in conventional B cells. Furthermore, as we have recently shown (S.M. Wells, D. Wang, E.A. Kabat and A.M. Stall, submitted) $\alpha(1\rightarrow6)$ dextran, a well-studied TI type II antigen, stimulates antibody secretion by both B-1 and conventional B cells. Finally, also as indicated above, we have recently shown (Kantor *et al.*, 1994; A.B. Kantor, C.E. Merrill, L.A. Herzenberg and J.L Hillson, in preparation) that although B-1a cells tend to have few N-region insertions more frequently than conventional cells, many B-1a cell V_H transcripts have substantial numbers of these insertions. Thus, the weight of the antibody specificity and structure data argues strongly against either of these factors being involved in determining the subset/lineage fate of developing B cells.

While B-1 cells share some phenotypic characteristics with B cells activated following lipopolysaccharide or anti-IgM stimulation, such as lower expression of B220/6B2 and IgD and the absence of CD23, it is too simplistic to characterize them merely as activated B cells. Our recent data indicates that the phenotype of the peritoneal and splenic B-1 cells is distinct from that typically associated with B cells activated with particular protocols. Studies with *in vivo* antigen-activated B-1 cells reveal that the morphology and surface phenotype of B-1 cells is significantly altered following activation (S.M. Wells, A.B. Kantor and A.M. Stall, in preparation). The B-1 cells found in the spleen and peritoneal cavity may have previously interacted with antigen and be in a primed (semi-activated) state ready for further antigenic stimulation.

The two lineage model

The idea that there are two distinct lymphocyte lineages in the murine immune system, which was initially posited to account for the developmental differences between B-1a and conventional B cells, is supported both by studies of T cell and B cell development. In B cell development, B-1a cells are generated from fetal and neonatal progenitors but not from progenitors in adult bone marrow; in T cell development, the first waves of γ/δ T cells to develop in the neonatal thymus have similarly been shown to derive from progenitors that are present in fetal lymphoid progenitor sources but not in adult bone marrow (Ikuta *et al.*, 1992a,b). Furthermore, both in T cell and B cell development, bone marrow progenitors have been shown to give rise to lymphocytes that are not normally found in large numbers in fetal or neonatal tissues. These developmental patterns suggest that the lymphocytes that develop primarily during fetal life and those that develop primarily in adults may belong to separate lineages derived from independent lymphoid stem cells that have diverged prior to the emergence of progenitors that are committed to differentiate to B or T cells.

Little is known about differences amongst lymphoid stem cells; however, adoptive transfers of various sources of lymphoid progenitors into irradiated recipients have shown a clear locational separation between progenitors for B-1a and conventional B cells. Progenitors that give rise to conventional B cells are present in both the adult bone marrow and the fetal liver. In contrast, B-1a progenitors are rarely present in adult bone marrow, but are readily detectable in the fetal and neonatal liver (Hayakawa *et al.*, 1985; Solvason *et al.*, 1991; Kantor *et al.*, 1992). Furthermore, other early sources of progenitors, such as the fetal omentum and day 9 para-aortic splanchnopleura, give rise to B-1a cells but fail to reconstitute conventional B cells (Solvason *et al.* 1991; Godin *et al.*, 1993); These studies provide strong evidence that the progenitors for B-1a cells are not only distinct from, but also arise earlier in ontogeny than, progenitors for conventional B cells.

The conclusions from these adoptive transfer studies are further supported by the *in vitro* experiments of Hardy and colleagues. These studies show that FACS-sorted fetal pro-B cells cultured on appropriate stromal layers give rise to B-1a cells; however, cultures of similarly sorted adult bone marrow pro-B cells give rise to conventional B cells but not to B-1a cells (Hardy and Hayakawa, 1991). Thus, pro-B cells, the earliest identifiable cells in the B cell developmental pathway, are already committed to develop into B cells that belong to one or the other lineage.

Until recently, phenotypic distinction between any of the B cell precursors from fetal sources such as fetal and neonatal liver and from adult sources such as adult bone marrow was not possible. However, recent studies by Lam and Stall of MHC class II expression on pre-B cells clearly demonstrate phenotypic differences between fetal and adult B cell development. They show that in contrast to pre-B cells in the adult bone marrow, which express I-A and I-E (Tarlinton, 1994), these class II molecules are absent on the surface of pre-B and newly generated sIgM$^+$ cells in the fetal liver (Lam and Stall, 1994). They further show that the while the 'adult-type' B cell developmental pathway is predominantly found in bone marrow, 'fetal-type' B cell developmental pathway is present in many neonatal lymphoid organs, providing additional evidence of two distinct B cell developmental pathways during ontogeny.

Taken together, the adoptive transfer and *in vitro* studies strongly argue for the existence of two B cell developmental lineages. While these studies cannot rule out the possibility that a conventional B cell could be stimulated to give rise to a cell with a B-1 phenotype, the body of adoptive transfer (Hayakawa *et al.*, 1985; Hardy and Hayakawa, 1992; Kantor *et al.*, 1992) and anti-allotype depletion (Lalor *et al.*, 1989b; K. Watanabe, L.A. Herzenberg and A.B. Kantor, in preparation) experiments indicate that in a normal developing animal B-1 cells are derived predominantly, if not exclusively, from the fetal/neonatal lineage.

ADDITIONAL B-CELL LINEAGES

While the evidence for separate lineages is strongest for B-1a and conventional B cells other subpopulations of B cells may also represent distinct developmental pathways. As discussed above, stem cell progenitors for B-1a and conventional B cells have not been phenotypically distinguished; however, their temporal and spatial isolation in fetal versus adult lymphoid progenitor sources have allowed the distinction of these lineages. These distinctions are not as clear cut for the B-1b populations. All tissue sources studied thus far have progenitor activity for B-1b cells (Solvason *et al.*, 1991), Godin *et al.*, 1993), although adult bone marrow is deficient when compared to fetal liver (Kantor *et al.*, 1992). Thus, although phenotypic and developmental evidence strongly suggest that B-1b cells constitute a separate lineage derived from independent progenitors present both in fetal lymphoid progenitor sources and in adult bone marrow, the current evidence can also be interpreted as indicating that the progenitors for B-1a cells *and* the progenitors for conventional B cells both give rise to B-1b cells, i.e. that B-1b cells belong to both lineages rather than constituting a lineage of their own.

MZ and follicular B cells have also been proposed as separate B cell lineages

(MacLennan et al., 1982). The functional and anatomical differences between these two B cell subsets argue strongly for this distinction (Kroese et al., 1992; Waldschmidt et al., 1992). However, progenitors from both fetal liver and adult bone marrow equally reconstitute both populations and no source has yet been found to preferentially provide progenitors for one and not the other. Thus the lineage origins of the MZ and follicular B cells are still unclear. Similarly, although Klinman and colleagues have demonstrated functional differences between populations of B cells that give rise to primary and secondary responses (Linton et al., 1989), it is not yet clear whether these represent distinct lineages that can be distinguished within B cell precursor populations.

THE LAYERED EVOLUTION OF THE IMMUNE SYSTEM

The importance of antibody production to mammalian survival provides a clear rationale for the evolution of the complex processes involved in Ig gene rearrangement, isotype switching and B cell development. Similarly, the evolutionary value of the various functions that T cells perform clearly provides a strong impetus for the evolution of these cells. But while we, as students of the immune system, can readily offer explanations for 'why' it evolved, we have little concept of how.

We have recently pointed out that the existence of distinct B cell and T cell lineages that develop successively during ontogeny may reflect an evolutionary process in which lymphocytes capable of more advanced functions were acquired in layers as mammals evolved from more primitive organisms (Herzenberg and Herzenberg, 1989; Kantor and Herzenberg, 1993). In essence, this model proposes that evolution has created a series of at least two mammalian hematopoietic stem cells that begin functioning sequentially during ontogeny and give rise to lymphocytes (and erythroid and myeloid cells) with progressively more advanced capabilities. Thus, B-1a cells and the first wave of T cells to enter the thymus would be generated from the earliest stem cell, which begins functioning in the mouse during the second trimester of fetal life, whereas follicular and MZ cells would be generated from later stem cells, which also give rise to evolutionarily more mature T cells. The validity of this model is clearly open to question; however, its value in stimulating experiments aimed at distinguishing different types of stem cells and the lymphocyte lineages they engender, as reviewed here and elsewhere (Ikuta et al., 1990, 1992b), is well established.

Acknowledgements

This work is supported by NIH grants HD-01287 and LM-04836 to Leonore Herzenberg, and NRSA awards AI-07937 and AI-07290 to A.B. Kantor. A.M. Stall is a Irma T. Hirschl Scholar and is supported by ACS JFRA-338.

REFERENCES

Allman, D.M., Ferguson, S.E., Lentz, V.M. and Cancro, M.P. (1993). *J. Immunol.* **151**, 4431–4444.
Braun, J. and King, L. (1989). *J. Mol. Cell. Immunol.* **4**, 121–127.
Cantor, H. and Boyse, E.A. (1975). *J. Exp. Med.* **141**, 1376–1389.
Cantor, H. and Boyse, E.A. (1977). *Immunol. Rev.* **33**, 105–124.
Carmack, C.E., Shinton, S.A., Hayakawa, K. and Hardy, R.R. (1990). *J. Exp. Med.* **172**, 371–374.
Coico, R.F., Bhogal, B.S. and Thorbecke, G.J. (1983). *J. Immunol.* **131**, 2254–2257.
Deenen, G.J. and Kroese, F.G. (1993). *Eur. J. Immunol.* **23**, 12–16.
Förster, I. and Rajewsky, K. (1987). *Eur. J. Immunol.* **17**, 521–528.
Förster, I. and Rajewsky, K. (1990). *Proc. Natl Acad. Sci. USA* **87**, 4781–4784.
Godin, I.E., Garcia, P.J., Coutinho, A., Dieterlen, L.F. and Marcos, M.A. (1993). *Nature* **364**, 67–70.
Gray, D. (1993). *Annu. Rev. Immunol.* **11**, 49–77.
Hardy, R.R., (1989). *Curr. Opin. Immunol.* **2**, 189–198.
Hardy, R.R. and Hayakawa, K. (1991). *Proc. Natl Acad. Sci. USA* **88**, 11550–11554.
Hardy, R.R. and Hayakawa, K. (1992). In 'CD5 B Cells in Development and Disease' L.A. Herzenberg, K. Rajewsky and G. Haughton, eds), Vol. 651 pp. 84–98. *Ann. N.Y. Acad. Sci.*
Hardy, R., Hayakawa, K., Haaijaman, J. and Herzenberg, L.A. (1982). *Ann. N.Y. Acad. Sci.* **399**, 112–121.
Hardy, R.R., Hayakawa, K., Parks, D.R., Herzenberg, L.A. and Herzenberg, L.A. (1984). *J. Exp. Med.* **159**, 1169–1188.
Hardy, R.R., Carmack, C.E., Shinton, S.A., Kemp, J.D. and Hayakawa, K. (1991). *J. Exp. Med.* **173**, 1213–1225.
Hathcock, K.S., Hirano, H., Murakami, S. and Hodes, R.J. (1992). *J. Immunol.* **149**, 2286–2294.
Haughton, G., Arnold, L.W., Whitmore, A.C. and Clarke, S.H. (1993). *Immunol. Today* **14**, 84–87.
Hayakawa, K. and Hardy, R.R. (1988). *Annu. Rev. Immunol.* **6**, 197–218.
Hayakawa, K., Hardy, R.R., Honda, M., Herzenberg, L.A., Steinberg, A.D. and Herzenberg, L.A. (1984). *Proc. Natl Acad. Sci. USA* **81**, 2494–2498.
Hayakawa, K., Hardy, R.R., Herzenberg, L.A. and Herzenberg, L.A. (1985). *J. Exp. Med.* **161**, 1554–1568.
Hayakawa, K., Hardy, R.R., Stall, A.M., Herzenberg, L.A. and Herzenberg, L.A. (1986). *Eur. J. Immunol.* **16**, 1313–1316.
Herzenberg, L.A. and Herzenberg, L.A. (1989). *Cell* **59**, 953–954.
Herzenberg, L.A., Okumura, I., Cantor, H., Sato, V.L., Fung-Win, S., Boyse, E.A. and Herzenberg, L.A. (1976). *J. Exp. Med.* **144**, 330–334.
Herzenberg, L.A., Stall, A.M., Lalor, P.A., Sidman, C., Moore, W.A., Parks, D.R. and Herzenberg, L.A. (1986). *Immunol. Rev.* **93**, 81–102.
Ikuta, K., Kina, T., Macneil, I., Uchida, N., Peault, B., Chien, Y. and Weissman, I.L. (1990). *Cell* **62**, 863–874.
Ikuta, K., Kina, T., MacNeil, I., Uchida, N., Peault, B., Chien, Y.H. and Weissman, I.L. (1992a). *Ann. N.Y. Acad. Sci.* **651**, 21–32.
Ikuta, K., Uchida, N., Freedman, J. and Weissman, I.L. (1992b). *Annu. Rev. Immunol.* **10**, 759–783.
Jacob, J. and Kelsoe, G. (1992). *J. Exp. Med.* **176**, 679–687.
Kantor, A.B. and Herzenberg, L.A. (1993). *Annu. Rev. Immunol.* **11**, 501–538.
Kantor, A.B. et al. (1991). *Immunol. Today* **12**, 338.
Kantor, A.B., Stall, A.M., Adams, S., Herzenberg, L.A. and Herzenberg, L.A. (1992). *Proc. Natl Acad. Sci. USA* **89**, 3320–3324.
Kantor, A.B., Stall, A.M., Adams, S., Watanabe, K. and Herzenberg, L.A. (1995). *Intl. Immunol.* **7**, 55–68.
Kantor, A.B., Merrill, C.E., Herzenberg, L.A. and Hillson, J.L. (1994). *J. Cell. Biochem.* **18A**, 437.
Kroese, F.G., Butcher, E.C., Stall, A.M., Lalor, P.A., Adams, S. and Herzenberg, L.A. (1989). *Int. Immunol.* **1**, 75–84.
Kroese, F.G.M., Butcher, E.C., Lalor, P.A., Stall, A.M. and Herzenberg, L.A. (1990a). *Eur. J. Immunol.* **20**, 1527–1534.

Kroese, F.G.M., Timens, W. and Nieuwenhuis, P. (1990b). *Curr. Top. Pathol.* **84**, 103–148.
Kroese, F.G.M., Ammerlaan, W.A.M. and Deenen, G.J. (1992). In 'CD5 B Cells in Development and Disease' (Herzenberg, L.A. Rajewsky, K. and Haughton, G, eds), Vol. 651 pp.44–58. *Ann. N.Y. Acad. Sci.*
Lalor, P.A., Herzenberg, L.A., Adams, S. and Stall, A.M. (1989a). *Eur. J. Immunol.* **19**, 507–513.
Lalor, P.A., Stall, A.M., Adams, S. and Herzenberg, L.A. (1989b). *Eur. J. Immunol.*, **19**, 501–506.
Lam, K.P. and Stall, A.M. (1994). *J. Exp. Med.* **180**, 507–516.
Lanier, L.L., Warner, N.L., Ledbetter, J.A. and Herzenberg, L.A. (1981). *J. Exp. Med.* **153**, 998–1003.
Ledbetter, J.A., Rouse, R.V., Micklem, H.S. and Herzenberg, L.A. (1980). *J. Exp. Med.* **152**, 280–295.
Li, Y.-S., Hayakawa, K. and Hardy, R.R. (1993). *J. Exp. Med.* **178**, 951–960.
Linton, P.J. and Klinman, N.R. (1992). *Semin. Immunol.* **4**, 3–9.
Linton, P.J., Decker, D.J. and Klinman, N.R. (1989). *Cell* **59**, 1049–1056.
Linton, P.J., Lo, D., Lai, L., Thorbecke, G.J. and Klinman, N.R. (1992). *Eur. J. Immunol.* **22**, 1293–1297.
MacLennan, I.C.M., Gray, D., Kumararatne, D.S. and Bazin, H. (1982). *Immunol. Today* **3**, 921–928.
Manohar, V., Brown, E., Leiserson, W.M. and Chused, T.M. (1982). *J. Immunol.* **129**, 532–538.
Masmoudi, H., Mota-Santos, T., Huetz, F., Coutinho, A. and Cazenve, P.A. (1990). *Int. Immunol.* **2**, 515–520.
Mercolino, T.J., Arnold, L.W. and Haughton, G. (1986). *J. Exp. Med.* **163**, 155–165.
Mercolino, T.J., Arnold, L.W., Hawkins, L.A. and Haughton, G. (1988). *J. Exp. Med.* **168**, 687–698.
Metcalf, E.S. and Klinman, N.R. (1977). *J. Immunol.* **118**, 2111–2116.
Nishimura, H., Hattori, S., Abe, M., Hirose, S. and Shirai T (1992). *Cell Immunol.* **140**, 432–443.
Nossal, G.J.V. (1983). *Ann. Rev. Immunol.* **1**, 33–62.
Parker, D.C. (1993). *Annu. Rev. Immunol.* **11**, 331–360.
Pennell, C.A., Arnold, L.W., Haughton, G. and Clarke, S.H. (1988). *J. Immunol.* **141**, 2788–2796.
Pennell, C.A., Mercolino, T.J., Grdina, T.A., Arnold, L.W., Haughton, G. and Clarke, S.H. (1989). *Eur. J. Immunol.* **19**, 1289–1295.
Pennell, C.A., Maynard, E., Arnold, L.W., Haughton, G. and Clarke, S.H. (1990). *J. Immunol.* **145**, 1592–1597.
Rabin, E., Ying-Zi, C. and Wortis, H. (1992). *Ann. N.Y. Acad. Sci.* **651**, 130–142.
Rolink, A. and Melchers, F. (1991). *Cell* **66**, 1081–1094.
Shirai, T., Hirose, S., Okada, T. and Nishimura, H. (1991). *Clin. Immunol. Immunopathol.* **59**, 173–186.
Solvason, N., Lehuen, A. and Kearney, J.F. (1991). *Int. Immunol.* **3**, 543–550.
Sprent, J., Schaefer, M., Hurd, N., Surh, C.D. and Ron, Y. (1991). *J. Exp. Med.* **174**, 717–728.
Stall, A.M., Adams, S., Herzenberg, L.A. and Kantor, A.B. (1992). In 'CD5 B Cells in Development and Disease' (Herzenberg L.A., Rajewsky K. and Haughton G. eds), Vol. 651 pp. 33–43. *Ann. N.Y. Acad. Sciences.*
Su, S.D., Ward, M.M., Apicella, M.A. and Ward, R.E. (1991). *J. Immunol.* **146**, 327–331.
Taki, S., Schmitt, M., Tarlinton, D., Forster, I. and Rajewsky, K. (1992). In 'CD5 B Cells in Development and Disease' (Herzenberg, L.A. Rajewsky, K. and Haughton, G. eds), Vol. 651 pp. 328–335. *Ann. N.Y. Acad. Sciences.*
Tarlinton, D. (1994). *Int. Immunol.* **5**, 1629–1635.
Tarlinton, D., Stall, A.M. and Herzenberg, L.A. (1988). *EMBO J.* **7**, 3705–3710.
Tarlinton, D.M., McLean, M. and Nossal, G.J.V. (1994). *J. Biol. Chem.* **18D**, 440.
van Rooijen, N., Claassen, E. and Eikelenboom, P. (1986). *Immunol. Today* **7**, 193–196.
Waldschmidt, T., Snapp, K., Foy, T., Tygrett, L. and Carpenter, C. (1992). In 'CD5 B Cells in Development and Disease' (Herzenberg, L.A., Rajewsky, K. and Haughton G., eds), Vol. 651 pp. 84–98. *Ann. N.Y. Acad. Science.*
Wang, D., Wells, S.M., Stall, A.M. and Kabat, E.A. (1994). *Proc. Natl Acad. Sci. USA*, **91**, 2502–2506.
Wells, S.M., Kantor, A.B. and Stall, A.M. (1994). *J. Immunol.*, **153**, 5503–5515.
Whitmore, A.C., Haughton, G. and Arnold, L.W. (1992). In 'CD5 B Cells in Development and Disease' (Herzenberg L.A., Rajewsky K. and Haughton G., eds) Vol. 651 pp. 143–151. *Ann. N.Y. Acad. Sci.*
Williamson, A.R., Zitron, I.M. and McMichael, A.J. (1976). *Fed. Proc.* **35**, 2195–2201.
Ying-Zi, C., Rabin, E. and Wortis, H.H. (1991). *Int. Immunol.* **3**, 467–476.

5

Regulation of B-cell differentiation by stromal cells

Elizabeth A. Faust, Douglas C. Saffran and Owen N. Witte

Department of Microbiology and Molecular Genetics, Howard Hughes Medical Institute, University of California, Los Angeles, California, USA

Introduction 103
What are stromal cells? 104
Phenotypic characteristics of stromal cells that support B lymphopoiesis 106
Experimental approaches that revealed the interactions between B cells and
  stroma 110
Cytokine signals: complex, multifunctional, redundant, synergistic 114
Concluding remarks..121

INTRODUCTION

This chapter will focus on early stages of development in the B lymphocyte pathway that require bone marrow stromal elements (either contact or stromal-derived factors) and are antigen independent (for detailed review see Rolink and Melchers, 1991). The earliest event in B cell development involves commitment of the pluripotent hematopoietic stem cell to the B cell pathway. Stem cells are found in the fetal liver and bone marrow of adults (Metcalf and Moore, 1971; Till and McCulloch, 1980; Visser *et al.*, 1990). They are defined by their capacity to repopulate all the hematopoietic lineages in marrow-ablated animals and function continuously for a large fraction of the lifespan of an animal. The progenitor (pro-) B cell is the most primitive cell of the B lineage and is most closely related to the pluripotential stem cell. Pro-B cells can be defined as B lineage restricted cells that have not rearranged immunoglobulin (Ig) genes, require undefined signals derived from stroma for growth, and express the pan B cell antigen B220 (Davidson *et al.*, 1984; Muller-Sieburg *et al.*, 1986; Rolink and Melchers, 1991; Faust *et al.*, 1993). Bone marrow sorted on the basis of antigen expression indicates that pro-B cells express low levels of Thy-1 antigen (Tidmarsh *et al.*, 1989; Hardy *et al.*, 1991; Melchers and Potter, 1991).

The next stage of B cell development is the immature pre-B cell (review by Rolink and Melchers, 1991). These cells have rearranged diversity (D) and joining (J) region segments on at least one allele of the Ig heavy (H) chain and express B220. The surrogate light (L) chains λ and V_{pre-B} and the *mb-1/B29* genes whose products are essential for export of IgM to the surface are expressed at this stage (Bauer *et al.*, 1988; Sakaguchi *et al.*, 1988; Hombach *et al.*, 1990). Immature pre-B cells require signals derived from stroma in addition to interleukin (IL)-7 for growth (Hayashi *et al.*, 1990; Hardy *et al.*, 1991). Immature pre-B cells differentiate into pre-B cells upon productive H chain variable (V)–DJ region joining and express the resultant IgH chain in the cytoplasm. IL-7 stimulates pre-B cells to divide in the absence of stroma (Namen *et al.*, 1988a; Lee *et al.*, 1989; Sudo *et al.*, 1989) and IL-7 can synergize with steel locus factor (SLF) to provide a potent growth stimulus for these cells (Palacios *et al.*, 1987; McNiece *et al.*, 1991). Newly described stromal cell-derived factors, which are discussed later, also influence the growth of these cells.

Production of functional Ig L chain and surface expression of IgM, IgD, and Ia molecules correlates with the accumulation of B cells in the periphery and loss of responsiveness to IL-7 (Hardy *et al.*, 1991). Cells capable of responding to antigen migrate and accumulate in peripheral lymphoid tissues where they can be activated by antigen in concert with secondary factors, some of which are derived from stromal-like elements (Osmond 1990; Rolink and Melchers, 1991). These selected and activated cells expand in number and secrete large amounts of Ig into the lymphatic and vascular spaces.

In this chapter we address key issues of the B cell differentiation pathway, some controversial, to illuminate potential mechanisms by which B cells develop. First, what are stromal cells and what is their origin? Second, does the stromal microenvironment direct B cell development, or is there a stochastic mechanism for which stroma provides a support function? Third, what are the signals derived from stroma that direct or influence B cell growth and development? This will include a discussion of cytokines that either positively or negatively influence B cell development. Finally, we discuss natural and artificial mutations that have given us insight into the regulation of B cell development.

WHAT ARE STROMAL CELLS?

Origin of stromal cells

It is important to clarify that 'stromal cells' does not refer to a single cell type but rather a collection of cells presumably of non-hematopoietic origin. A good working definition of stromal cells for this review is a cell type that supports the growth and differentiation of hematopoietic cells. The formation of a stromal cell matrix is necessary during development before other cell lineages, such as hematopoietic cells, can begin their developmental pathways (Weiss, 1976; Weiss and Sakai, 1984; Dorshkind, 1990). The events leading to the establishment of the stromal cell matrix *in vivo* are unclear but it is believed that stromal cells are derived from mesenchymal cells (Metcalf and Moore, 1971; Kincade *et al.*, 1989; Dorshkind, 1990). It has been shown *in vitro* and *in vivo* that bone marrow stromal cells can differentiate into several non-hematopoietic cell types, suggesting that stroma contains multipotential mesenchymal cells (Kincade *et al.*, 1989).

Several examples exist of stroma that possess hematopoietic support capabilities and differentiation potential. The fibroblast lines Swiss 3T3 and CH3/10T½ can each support myelopoiesis under Dexter type culture conditions (Roberts *et al.*, 1987; Kincade *et al.*, 1989). However after exposure to azacytidine they can differentiate into adipocytes, chondrocytes, or myocytes. Stromal cells in human long-term bone marrow cultures (LTBMC) which support granulopoiesis express α-sphingomyelin (α-SM) microfilaments characteristic of smooth muscle (Galmiche *et al.*, 1993). This phenotype is similar to immature fetal smooth muscle cells and subendothelial intimal smooth muscle cells. The potential *in vivo* counterpart for these cells are myoid cells, which are smooth muscle cells in origin and line the sinuses in the bone marrow compartment. Myoid cells also support granulopoiesis in the bone marrow. Additionally expression of nerve growth factor receptor (NGFR) has been demonstrated *in vitro* in human LTBMC on adventitial reticular cells (Cattoretti *et al.*, 1993). These NGFR-positive adventitial reticular cells (ARCs) are found *in vivo* in fetal bone marrow prior to hematopoiesis and later in adult bone marrow.

In other cases, stromal cells are osteogenic precursors that have the ability to support hematopoiesis or develop into bone (Owen, 1985; Friedenstein, 1990). Bone marrow-derived stroma have been shown to exhibit bone formation *in vivo* after implantation in diffusion chambers or when placed under the kidney capsule (Friedenstein *et al.*, 1974; Ashton *et al.*, 1980; Hirano and Urist, 1981). Stromal cells derived from primary culture of rat bone marrow have also been shown to form bone *in vitro* in the presence of β-glycerophosphate and dexamethasone (Maniatopoulos *et al.*, 1988).

Do stromal cells and hematopoietic cells share a common progenitor?

Since stromal cells and hematopoietic cells share similar environments it has been suggested that these cell types are related. There are recent and previous claims that stromal cells and hematopoietic cells come from a common precursor. Evidence to suggest this came from findings that adherent cells derived from LTBMC of bone marrow transplant patients were of donor origin (Keating *et al.*, 1982; Singer *et al.*, 1984). However it cannot be ruled out that stroma was also transplanted along with hematopoietic cells in the patients (Dorshkind, 1990). Huang and Terstappen (1992) recently reported that a common $CD34^+$, $HLA-DR^-$, $CD38^-$, low forward light scatter cells in human fetal bone marrow could form both hematopoietic and stromal cell types with high frequency (1 : 20 to 1 : 100 cells). None of these precursors gave rise to only one lineage, while CD34$^+$, HLA-DR$^+$, CD38$^-$ cells only gave rise to hematopoietic cells. They proposed an interesting model that a common stem cell first gives rise to stromal cells, which then provide a signal for new CD34$^+$ precursors to develop into the hematopoietic lineage.

However there is a large body of evidence that stem cells in humans and mice can only give rise to hematopoietic cells and not stromal elements (Till and McCulloch, 1980; Spangrude et al., 1988; Visser et al., 1990; Smith et al., 1991; Dexter and Allen, 1992; Ogawa, 1993). In addition there is evidence that donor-derived stroma does not develop in the host after bone marrow transplantation (Simmons et al., 1987), which refutes the findings of Keating et al. (1982) and Singer et al. (1984). This, in addition to the recent retraction of findings by Huang and Terstappen (1994), suggests that stromal cells come from a precursor distinct from hematopoietic cells.

To clearly define the role of stroma in supporting hematopoiesis requires formal demonstration of a functional association between stromal and hematopoietic cells. One approach has been to utilize *in vitro* culture systems that support hematopoietic cell growth to examine these interactions. In many cases the *in vitro* systems have been informative with respect to the true *in vivo* situation (see p. 110). Two widely used culture systems include a myeloid culture system developed by Dexter et al. (1977) and a lymphoid culture system developed by Whitlock and Witte (1982). Both systems rely on the initial establishment of an adherent layer, which is required for the subsequent growth of hematopoietic lineage cells. This adherent layer has been especially useful to examine interactions with hematopoietic cells and assess the type of cells that support growth, which are extensively described in the following sections.

PHENOTYPIC CHARACTERISTICS OF STROMAL CELLS THAT SUPPORT B LYMPHOPOIESIS

Antigen expression

Morphological features of stromal cell lines are not well correlated with function. Attempts to associate antigen expression and extracellular matrix (ECM) components have been unsuccessful, but are ongoing. Whitlock et al. (1987) isolated the AC series of stromal cell lines (Table I) from Whitlock–Witte cultures. The series contains lines that support the growth and differentiation of B cells and those that do not. Within the AC lines, expression of the 6C3/BP-1 antigen is correlated with the capacity to support B lymphopoiesis (Whitlock et al., 1987). However, lines from other groups, notably S10 and S17, that can support B lineage growth lack expression of 6C3/BP-1 (Collins and Dorshkind, 1987; Henderson et al., 1990). In addition, heterogeneous stromal layers from Whitlock–Witte cultures do not express this antigen (Witte et al., 1987). A description of surface antigens expressed by stromal cells is listed in Table I.

Extracellular matrix

Stromal cell lines produce an ECM consisting of a heterogeneous mixture of molecules including proteoglycans, fibronectin, laminin, and collagen. It is clear that ECM plays a critical role in the binding and proliferation of hematopoietic cells (Zuckerman and Wicha, 1983). The ECM can bind cytokines, which correlates with the necessity of cellular contact between stromal cells and hematopoietic precursors (Gordon et al., 1987; Roberts et al., 1988). For example the ECM contains heparan sulfate proteoglycans, which can mediate interaction of growth factors with their specific receptors. One molecule called N-syndecan (syndecan 3) can bind basic fibroblast growth factor (bFGF) and pleiotrophin to promote nervous tissue development (Chernousov and Carey, 1993; Raulo et al., 1994). The hematopoietic cytokines IL-3, GM-CSF, and IFN-γ all have been shown to associate with heparin and heparan sulfate (Roberts et al., 1988; Lortat-Jacob et al., 1991). However no specific heparan sulfated proteoglycans involved in B cell development have been described.

Most stromal cell lines that have been examined produce some type of collagen (Table I). Different types of collagen have been associated with different cell lineages. For example, collagen types I and III are produced by fibroblasts (Gay et al., 1976), whereas collagen type IV is characteristic of endothelial cells (Howard et al., 1976; Jaffe et al., 1976). There is little correlation between the ability to support hematopoiesis and the type of collagen produced by stromal cells. This lack of consistency may be due to culture adaptation of the lines or may have significance yet to be identified.

Stromal cells also express adhesion molecules that are mediators of direct cell–cell interactions with developing blood cells. One example is the VLA-4/VCAM-1 interaction. VLA-4 is an integrin expressed on lymphoid cells and binds to VCAM-1, which is expressed by stromal cells. Antibodies to VLA-4 block binding of lymphoid cells to stroma (Simmons et al., 1992) and inhibit lymphopoiesis in Whitlock–Witte cultures and myelopoiesis in Dexter cultures (Miyake et al., 1991).

Developing pre-B cells have been shown to interact with ECM proteins secreted by stromal cells. Stromal cells secrete fibronectin, which is recognized by pre-B cells via a fibronectin receptor termed α5β1 integrin (Kincade et al., 1989; Ruoslahti, 1991). This interaction is essential since antibodies or peptides that block this interaction inhibit proliferation of pre-B cells (Lemoine et al., 1990). Stromal cells also secrete hyaluronate, which pre-B cells bind to via CD44 (Miyake et al., 1990a). This interaction is also essential in vitro, since antibodies that recognize CD44 have been shown to inhibit stromal cell-dependent pre-B cell growth (Miyake et al., 1990b).

Table I Characteristics of stromal cell lines

Stroma	Origin	Extracellular matrix	Surface antigen	Cytokine production	Support of B lymphopoiesis	References
Primary bone marrow	Murine bone marrow	Actin+, collagen IV+	No Lgp 100, MHC-II, Mac-1, B220, BP-1, Mac-2, Mac-3, Thy-1	M-CSF, GM-CSF, IL-7, SLF, TGF-β (others)	Proliferation and differentiation of B lineage cells	Kincade et al. (1989)
PA6	Culture of calvaria cells from newborn mice	Collagens+	ND	SLF; no G-CSF, GM-CSF, IL-3 or IL-7	Maintenance of early B progenitors (without differentiation) from bone marrow in Whitlock–Witte conditions (B220− sIgM− progenitors transferred onto ST2 stromal cells line mature into pre-B cells)	Nishikawa et al. (1988)
BMS 1, BMS 2	Adherent cell layer from short-term, 5-fluorouracil-treated marrow cultures	Actins+, collagen types I+, IV+, III− (BMS1) III+ (BMS2)	KM16+, Thy-1, Mac-3, N-cadherin; variable expression of HSA and 6C3; weak expression of Mac-2; no Ig, CD45, TCRβ, LFA-1, MHC-II, Lgp100, F4/80, B220, AA4.1, Mac-1	M-CSF, IL-6, TGF-β, IL-7; no IL-1, IL-3, IL-4, IL-5 or GM-CSF	Proliferation of B lineage cells	Kincade et al. (1989)
TC-1	Adherent cell layer from Dexter culture	ND	ND	M-CSF, GM-CSF, IL-4, G-CSF and SLF; no IL-1, IL-2, IL-3, IL-5, IL-6, IL-7	Supports lymphoid cells growth, proliferation of activated B blasts and low-density spleen B cells, and maturation of B220− B cell progenitors into pre-B cells	Quesenberry et al. (1987a)

Name	Source		Markers	Factors	Function	Reference
S10	Adherent cell layer from MPA-treated Dexter cultures	ND	MHC-I+, Thy 1−, Mac 1−, CD5−, 6C3−, Ig−, MHC-II−	IL-7; signals required from maturation of pre-B cells to the sIgM+ stage, no IL-4	Maturation of pre-B to the stage of sIgM+ cells	Collins and Dorshkind (1987)
S17	Adherent cell layer from MPA-treated Dexter cultures	ND	MHC-I+, Thy-1−, Mac-1−, CD5−, 6C3−, Ig−, MHC-II−	IGF-1, SLF, GM-CSF; no IL-1, IL-2, IL-3, IL-4, IL-7	Growth of sIgM− B220+ pre-B cells; differentiation of B lineage precursors up to the pre-B (cytoplasmic μ+) cells transferred onto S10 mature into sIg+ B cells	Collins and Dorshkind (1987)
AC series	Adherent cell layer from Whitlock–Witte cultures	ND	MHC-I+, Thy-1−(AC-3 and AC-11 are Thy-1+/−); Mac-1+(AC-8); 6C3+(AC-3, 4, 6, 6.4), 6C3+/−(AC-8, AC-11); 3% 6C3+ cells in AC-10 line	All lines produce M-CSF; IL-7 (AC-3, 6); G-CSF and GM-CSF (AC-6); no IL-3 or IL-4	Myelopoiesis (AC-11, 6); support of cells from Whitlock–Witte: long-term support by AC-3, 4, 6 lines, no support by AC-10, 11 lines; no differentiation of B cell precursors into sIg+ cells (transfer onto mixed stromal layer is required for maturation up to the sIg+ stage)	Whitlock et al. (1987)
ALC	Adherent cell layer from Whitlock–Witte cultures	Collagen type I+	6C3+	M-CSF, G-CSF, IL-7; no GM-CSF, IL-1, IL-2, IL-3, IL-4	Pre-B cell growth from Whitlock–Witte	Hunt et al. (1987)
ST2	Adherent cell layer from Whitlock–Witte	Collagen+	6C3+, R25+	Induction of IL-7 by IL-1, hematopoietic cells or B cells; no IL-3, G-CSF, GM-CSF	Differentiation of B lineage cells up to the stage of B220+ cells, giving rise to sIgM+ cells in the presence of IL-7	Nishikawa et al. (1988)

G-CSF, granulocyte-colony stimulating factor; GM-CSF, granulocyte macrophage colony stimulating factor; HSA, heat shock antigen; IGF-1, insulin-like growth factor-1; IL, interleukin; M-CSF, macrophage-colony stimulating factor; MHC-I, class I histocompatibility antigen; MHC-II, class II histocompatibility antigen; MPA, mycophenolic acid; ND, not determined; sIgM, surface immunoglobulin class M; SLF, steel locus factor; TCRβ, T cell receptor β chain; TGF-β, transforming growth factor-β; Whitlock–Witte, long-term bone marrow cultures described by Whitlock and Witte (1982).

EXPERIMENTAL APPROACHES THAT REVEALED THE INTERACTIONS BETWEEN B CELLS AND STROMA

In vivo approaches to study B cell development

The association between developing B cells in the bone marrow with stromal cells has been difficult to study. The most extensive studies have been performed by Osmond and colleagues and are summarized in a review (Osmond, 1986). The bone marrow consists of up to 25% small lymphocytes, of which the majority are 14.8 or B220 positive cells (Park and Osmond, 1987). Infusion of ^{3}H-thymidine into the bone marrow to monitor turnover of small lymphocytes revealed two populations: rapidly renewing cells, which make up 75–95% of the total; and a slowly renewing population, which consists of mature B cells, including memory cells, that recirculate from the periphery into the bone marrow. Double-labeling studies of ^{3}H-thymidine and anti-sIgM staining demonstrate that 50% of marrow small lymphocytes are sIg$^+$ and are renewing rapidly (Osmond, 1986). Using similar technology, the remaining B lineage cells in the bone marrow comprise small C_μ^+ su$^-$ pre-B cells (Rahal and Osmond, 1984). These cells are divided into two subsets: dividing cells and non-dividing cells (75%), which are small and mature into the sIg$^+$ population.

Histological examination in the bone marrow demonstrates that ^{3}H-labeled lymphocytes are scattered in the extravascular compartment (Osmond, 1986). It appears that there is an increase in the labeling index from peripheral to central marrow zones, and eventually many rapidly renewing cells in the intravascular region which are primed to exit the bone marrow. Perfusion of radiolabeled anti-IgM into mice demonstrated that sIgM$^+$ B cells are concentrated towards the center of the marrow in the sinusoidal region, contiguous with ARCs (Osmond, 1986). Jacobsen et al. (1992) have recently demonstrated that a monoclonal antibody (KM16), which recognizes a determinant on a bone marrow stromal cell line (BMS2), could label stromal cells after perfusion into the bone marrow. The labeled stromal cells were associated with cells of lymphoid morphology, possibly representing pluripotential stem cells or early B cell precursors.

In vitro approaches to study B cell development

Culture systems have been established in vitro as a result of the need to study development of hematopoiesis under controlled conditions. The key features of these systems is the obligate interaction of hematopoietic precursors with an established stromal cell layer. Dexter et al. (1977) originally demonstrated that culture of bone marrow cells at 33 °C in the presence of hydrocortisone results in the maintenance of myeloid cells. The non-adherent cells in these cultures included mature myeloid cells as well as cells able to form colony forming unit-spleen (CFU-S) after transfer to irradiated mice (Schrader and Schrader, 1978; Jones-Villeneuve and Phillips, 1980).

Although these cultures did not contain any cells of lymphoid morphology, they contained cells that could reconstitute B and T cells after transfer to lethally irradiated hosts (Schrader and Schrader, 1978; Jones-Villeneuve and Phillips, 1980).

Subsequent to this, Whitlock and Witte (1982) developed a culture system to grow B cells *in vitro*. Similar to Dexter cultures the successful establishment of B lymphopoiesis first required establishment of an adherent layer after 2 weeks, followed by the appearance of foci representing lymphoid outgrowths. Cultures contained cells that represent the entire B cell lineage, from early pro-B cells, to pre-B cells that express cytoplasmic IgH chain, and sIg$^+$ B cells (Denis and Witte, 1986). In addition cells from these cultures derived from normal animals were transferred to immunodeficient CBA/N or SCID mice and could reconstitute normal B cell function (Kurland *et al.*, 1984; Dorshkind *et al.*, 1986).

In order to study the interaction that occurs between developing hematopoietic cells and stroma, several groups have utilized cloned stromal cells *in vitro* to support myelopoiesis or lymphopoiesis. This approach provides an opportunity to dissect specific growth requirements that are provided by the cloned stromal cell line. Several laboratories have utilized this strategy to clone stromal cell lines to study B cell development (Table I). Extensive characterization of surface molecules expressed and growth factors secreted by these cells has revealed some of the requirements for B cell growth (see pp. 106 and 114).

Dorshkind and colleagues cloned several stromal cell lines from the adherent layer of Dexter cultures that could support hematopoiesis (Collins and Dorshkind, 1987). One of the lines, S17, was shown to support both myelopoiesis and lymphopoiesis, while another, S10, supported only myelopoiesis. Further characterization of these lines demonstrated that both S17 and S10 could support the growth of cells that could form CFU-B colonies (Henderson *et al.*, 1990). Interestingly, culture of Ig- and adherent cell-depleted bone marrow on S17 gave rise to cells expressing B220 but no IgM, whereas culture or transfer to S10 gave rise to B220$^+$ cells of which a small but significant proportion expressed IgM. This suggested that S17 allowed maturation to the pre-B cell stage, and that culture on S10 allowed further differentiation to sIg$^+$B cells.

Our laboratory has also utilized the S17 line to grow early cells in the B cell lineage. Scherle *et al.* (1990) were able to grow on S17 clonal populations of early B cell progenitors that expressed the chimeric *bcr/abl* P210 oncoprotein. These clonal outgrowths expressed B220, were negative for surface or cytoplasmic IgH chain expression, and retained IgH chain genes in the germline configuration. Additionally, these cells were able to differentiate into mature, Ig-producing B cells after transfer to SCID mice. Similarly we have also been able to grow enriched populations of pro-B cells, although not likely clonal, without expression of the *bcr/abl* oncogene (Saffran *et al.*, 1992; Faust *et al.*, 1993). The cells in these cultures were highly enriched for B220, had Ig germline genes, and could differentiate into mature B cells in SCID mice.

Similar to the findings of Henderson *et al.* (1990), Nishikawa *et al.* (1988) had shown that successive stages of B cell development could be supported by different stromal cell clones. One clone, PA6, was able to support growth of a population of

Table II Factors that regulate B-cell growth and development

Factor	Activities on B cells and their precursors	Isoforms	References
M-CSF	*Fms* receptor on normal, malignant B cells; synergizes with IL-3 and GM-CSF on stem cell proliferation	Soluble, membrane, matrix bound	McNiece (1988); Baker (1993)
GM-CSF	Negative regulator of B cells in LTBMC	Soluble	Dorshkind (1991)
G-CSF	Synergizes with IL-3 to shorten G_0 of stem cells; negative regulator	Soluble	Ikebuchi (1988)
IL-1β	Acts as cofactor in B cell growth and differentiation; potentiates surface Ig expression; negative regulator of B cells in LTBMC	Soluble, membrane bound	Dorshkind (1988)
IL-4	Proliferation and differentiation; increases expression of MHC-II; inhibits mature B in Whitlock–Witte; S17 treated with IL-4 does not support B cells	Soluble	Rabin (1986); Boothby et al. (1988); Hofman et al. (1988); Peschel et al. (1989); Billips et al. (1990)
IL-6	Induces Ig secretion; synergizes with IL-3 to shorten G_0 of stem cells	Soluble	Leary (1988); Kishimoto (1989)
IL-7	Proliferation of pre-B alone and in synergy with other factors; no differentiation	Soluble	Cumano (1990); Namen et al. (1988); Lee et al. (1989); Morrissey et al. (1991); Damia et al. (1992); Rich et al. (1993)
IL-10	Increases expression of MHC-II; enhances viability of small dense splenocytes	Soluble	Go et al. (1990)
IL-11	Proliferation of IL-6 dependent plasmacytoma; differentiation of T cell dependent Ig-producing B cells; synergy with other factors on stem cells	Soluble	Paul et al. (1990); Du and Williams (1994)

SLF	Synergizes with IL-7 on pre-B cells; synergizes with IL-6, G-CSF, IL-3, IL-1α, GM-CSF on stem cells	Soluble, membrane bound	de Vries (1991); Matsui (1991); McNeice (1991); Metcalf (1991); Migloccio (1991); Ogawa et al. (1991); Okada (1991); Tsuji (1991); Billips et al. (1992); Ikuta and Weissman (1992); Williams (1992); Funk et al. (1993); Bodine (1993); Orlic (1993)
LIF	Enhances effect of IL-3 on stem cells	Soluble, matrix bound	Hilton (1992)
TGF-B	Blocks κ-chain expression and response to IL-7	Soluble, matrix bound	Lee et al. (1987); Sing et al. (1988); Border (1992)
IGF-1	Growth and differentiation of IL-7 dependent cells	Soluble	Landreth et al. (1992); Gibson et al. (1993)
INF-γ	Induces Ig secretion	Soluble, matrix bound	Leibson (1984); Sidman (1984); Brunswick (1985)
PBSF	Synergizes with IL-7 on pre-B cell growth	Soluble	Tashiro et al. (1993); Nagasawa et al. (1994)
FL	Enhances response of stem cells to IL-3, IL-6, GM-CSF	Soluble	Samal et al. (1994)
BST-1	Supports growth of pre-B cell line DW34	Membrane bound	Hannum et al. (1994)
PBEF	Synergizes with IL-7 and SLF on pre-B cell growth	Soluble	Kaisho et al. (1994); Kincade (1994)

BST-1, bone marrow stromal antigen-1; FL, fetal liver kinase ligand; Ig, immunoglobulin; INF-γ, interferon-γ; LIF, leukemia inhibitory factor; LTBMC, long-term bone marrow culture; PBEF, pre-B cell colony enhancing factor; PBSF, pre-B cell growth stimulating factor. For other abbreviations see footnote to Table I.

cells that was devoid of lymphocytes. However after transfer of those cells to a secondary stromal cell clone, ST2, B lymphopoiesis progressed to both pre-B and mature B cell stages. Thus PA6 was able to maintain an early B cell progenitor that could differentiate under conditions lacking in PA6. The AC series of cell lines, derived from Whitlock–Witte cultures by Whitlock et al. (1987), and ALC line similarly derived by Hunt et al. (1987) have also been shown to support pre-B cells from Whitlock–Witte cultures. However in no case was there differentiation to sIg$^+$ cells in culture on these lines. This event required transfer onto a heterogeneous feeder layer derived from bone marrow (Whitlock et al., 1987). Although both the AC and ST2 lines were developed from Whitlock–Witte cultures, the differences in ability to induce differentiation reflect heterogeneity of adherent layers in long-term culture.

In addition to cloning stromal cells, pro-B and pre-B cell clones have also been derived to examine development in vitro. Pro-B and pre-B cell clones from either fetal liver or bone marrow have been established in the presence of lymphokines. Palacios and others have derived clones after culture in IL-3. The clones from IL-3 dependent cultures are said to retain Ig genes in the germline configuration and can differentiate into mature Ig-secreting B cells both in vitro and in vivo (McKearn et al., 1985; Palacios et al., 1987; Kinashi et al., 1988). In contrast Rolink et al. (1991) have utilized stromal cell clones and IL-7 to derive pre-B cell clones that have D–J rearrangements. These clones can differentiate into mature B cells in vitro after removal from stroma and IL-7, and in vivo when injected into SCID mice. In addition, pre-B cell lines have been established after culture in IL-4 and IL-5 (Ogawa et al., 1988; Preschel et al., 1989). These lines are D–J rearranged and exhibit limited maturation in vitro. However, while IL-7 is well recognized as having a positive influence on pre-B cell growth, the role of IL-3, IL-4, and IL-5 on early B cell development remains questionable (see p. 121).

CYTOKINE SIGNALS: COMPLEX, MULTIFUNCTIONAL, REDUNDANT, SYNERGISTIC

Positive regulators of B cell growth

Stromal cell-derived cytokines, both membrane bound and secreted, play an important role in regulating the growth and differentiation of B lymphocytes. The majority of cytokines function pleiotropically, exhibiting a wide range of biological effects on a variety of tissues. We will limit our discussion to the effects on B cells.

Table II lists factors that are secreted by stromal cells and have effects on the growth and differentiation of B lineage cells and their precursors. The most striking feature of B cell growth factors is the redundancy of their activities. This brings up the question of why there are numerous factors that promote B cell growth and whether or not the activities of these factors is misrepresented by in vitro studies. For example, both IL-1 and IL-4 increase expression of class II antigens, potentiate surface Ig expression by maturing pre-B cells, and increase B cell viability (Boothby et al., 1988; Hofman et al., 1988). The activities of IL-4 and IL-10 on small dense splenic B cells are indistinguishable (Go et al., 1990). IL-6 and IL-11 have many of the same functions. IL-11 was identified by the ability to support the growth of an IL-6 responsive plasmacytoma line (Paul et al., 1990).

Extensive redundancy is seen in factors that target pre-B cells. IL-7 was the first factor described that promotes the growth of pre-B cells. Subsequently SLF, insulin-like growth factor-1 (IGF-1), pre-B cell colony enhancing factor (PBEF), pre-B cell growth stimulating factor (PBSF) and bone marrow stromal antigen-1 (BST-1) have been shown to either have similar activities or synergize with IL-7. Most of these factors acts as cofactors with IL-7 on B lineage cells.

IL-7 was isolated from an SV40-transformed stromal cell line and identified by the ability to stimulate the proliferation of pre-B cells from Whitlock–Witte cultures (Namen et al., 1988b). Proliferation of B220$^+$ cells is supported by IL-7, but this factor does not support the maturation of B cells to mitogen responsiveness (Lee et al., 1989; Cumano et al., 1990; Morrissey et al., 1991; Billips et al., 1992; Damia et al., 1992). *In vivo* administration of IL-7 to mice results in an increase in the number of immature B cells (B220$^+$ C$_\mu^+$ sIgM$^-$) and IL-7 transgenic mice have an increased number of B220$^+$ sIgM$^+$ cells in the bone marrow, lymph node, and thymus (Morrissey et al., 1991; Damia et al., 1992). These mice also suffer B and T cell lymphomas (Rich et al., 1993).

Two other factors, SLF and IGF-1, were not originally defined as B cell growth factors. However, both synergize with IL-7 to promote pre-B cell growth. SLF was cloned by several groups and originally defined as a mast cell or stem cell growth factor (Anderson et al., 1990; Copeland et al., 1990; Flanagan and Leder, 1990; Huang et al., 1990; Martin et al., 1990; Williams et al., 1990; Witte, 1990; Zsebo et al., 1990). To determine if SLF plays a role in B lymphopoiesis, a monoclonal antibody (ACK2) directed against the SLF receptor, c-*kit*, was added to cultures of pre-B cell clones (Rolink et al., 1991). The proliferation of these clones was inhibited by ACK2. However, the ability to differentiate into sIgM$^+$ B cells was unaffected. The stimulation of mature B cells by mitogens was unimpaired by this same monoclonal antibody (Rolink et al., 1991). These data indicate that SLF is required for the growth of pre-B cells, but not for their differentiation. However, ACK2 injected into mice does not affect the number of B220$^+$, IL-7 responsive pre-B cells (Ogawa et al., 1991). The inconsistency in these results may reflect the use of clones as opposed to primary B cells in the *in vitro* experiments. Other experiments in which pre-B cells isolated from bone marrow were examined suggest that SLF is not necessary for pre-B cell growth, but acts as a cofactor to enhance the activities of IL-7. White spotting mice (see p. 118), which have a mutation of the c-*kit* gene, have normal numbers of B220$^+$ pre-B cells in the bone marrow (Landreth et al., 1984). Several groups have shown that SLF alone cannot support the growth of pre-B cells, but that it exhibits potent synergy with IL-7 to support proliferation of pre-B cells isolated from bone marrow and pre-B cell lines (McNeice et al., 1991; Billips et al., 1992; Funk et al., 1993).

IGF-1 was originally defined as a mitogenic polypeptide acting on mesenchymal cells, but subsequent studies have shown that IGF-1 is highly pleiotropic with activities on myoblasts, osteoblasts, neurons and other cell lineages (Schmid et al., 1983; Froesch et al., 1985; Xue et al., 1988; Sara and Hall, 1990). The effects of IGF-1 on B cells was discovered when it was identified as the differentiation factor secreted by the S17 stromal line that drives B220⁻ C_μ^- cells to C_μ^+ cells (Landreth et al., 1992). IGF-1 also synergizes with IL-7 (much like SLF) to promote the growth of pre-B cells (Gibson et al., 1993).

Despite the identification of IL-7, SLF and IGF-1, additional factors that influence the earliest events in B cell development remain to be identified. The proliferation and differentiation of B220⁺ C_μ^- B cell progenitors is dependent upon the S17 stromal layer (see Table I) and cannot be stimulated with either IL-7, SLF, or a combination of the two (Billips et al., 1992; Faust et al., 1993). Other studies suggest that additional stromal cell-derived cytokines control the developmental step in which the maturation of immature B cell precursors to the IL-7 and SLF responsive state occurs (Landreth et al., 1992). Recently, several groups have taken different approaches to identifying novel cytokines with influences on B cells. These efforts have resulted in the discovery of PBEF, PBSF, BST-1 and fetal liver kinase-2 ligand (FL).

Rolink et al., (1991) showed that the stromal cell line PA6 potentiated the proliferative effect of IL-7 on early pre-B cell clones. Recent experiments have shown that this synergistic activity is not due to SLF (Ogawa et al., 1991). To identify this novel factor, Nagasawa et al., (1994) constructed an expression library from PA6. Clones were screened for the ability to support the growth of the pre-B cell clone DW34. PBSF, an 89 amino acid polypeptide was identified. The same factor was cloned by Honjo and colleagues and termed SDF-1α (Tashiro et al., 1993). Alone PBSF has little effect on bone marrow-derived pre-B cells, but it does support the growth of the pre-B cell clone DW34. Similar to SLF and IGF-1, PBSF synergizes with IL-7 to support the growth of bone marrow-derived pre-B cells (Nagasawa et al., 1994). Thus far there has been no distinction between the activities of these three factors in synergy with IL-7.

Samal and colleagues were also searching for novel hematopoietic factors. Their strategy was to use a degenerate oligonucleotide probe that was designed on the basis of similarity in nucleotide sequences surrounding and coding for the signal peptidase cleavage sites of a number of cytokines, i.e. GM-CSF, IL-1β, IL-2, IL-3, IL-6. This probe was used to screen a human peripheral blood lymphocyte cDNA library and PBEF, a 52 kDa secreted protein, was identified. Conditioned medium from COS 7 cells transiently expressing PBEF revealed that PBEF alone has no activity on pre-B cell colony formation, but synergized with SLF and IL-7 to enhance colony formation (Samal et al., 1994). Synergy with IL-7 alone has not been reported.

Hematopoietic growth factors function as soluble, membrane-bound and matrix-bound molecules (see Table II). The factors discussed thus far were cloned as soluble molecules. Kaisho et al. (1994) attempted to isolate molecules bound to the surface of stromal layers that support B lymphopoiesis. This group prepared numerous transformed bone marrow stromal cell lines from normal individuals and patients with rheumatoid arthritis. Some of the lines supported proliferation of the DW34 pre-B cell line. To identify molecules that were required for B lymphopoiesis, a monoclonal antibody that preferentially recognized stromal cells with high support capability was used to isolate the BST-1 gene encoding this antigen. BST-1 was introduced into non-supporting fibroblasts, which could then support B lymphopoiesis (Kaisho et al., 1994; Kincade, 1994).

Fetal liver kinase-2 (FLK-2) encodes a tyrosine kinase receptor that is expressed on stem cells and pro-B cells (Matthews et al., 1991a,b; E.A. Faust and O.N. Witte, unpublished observation). Recently the human homologue of the FLK-2 receptor, termed STK-1, was cloned from enriched human hematopoietic stem cells (Small et al., 1994). This protein shares 85% identity and 92% similarity to the murine FLK-2 receptor. The FLK-2/STK-1 receptors are closely related to c-*kit* (the SLF receptor) and c-*fms* (the M-CSF receptor) suggesting that the FLK-2 ligand may be a related growth factor. The FLK-2 ligand (FL), a 65 kDa homodimer, was recently cloned in both membrane and soluble forms (Lyman et al., 1993; Hannum et al., 1994). The gene was most highly expressed in the spleen and lung. Bone marrow was not analyzed for expression. FL enhances the response of murine stem cells to IL-3, IL-6 and GM-CSF. It also synergizes with IL-7 to promote the growth of thymocytes (Hannum et al., 1994). Synergy with IL-7 on B cells has not been reported.

Controversies over function of B cell growth factors

The controversy over whether or not SLF is essential for the proliferation of pre-B cells has been discussed above. Another controversy surrounds the effects of IL-7 and SLF on pro-B cells. Initial reports (McNeice et al., 1991) indicated that bone marrow depleted of B220+ cells differentiated in response to IL-7 and SLF. Two other groups (Billips et al., 1992; Funk et al., 1993) observed no response of the same cellular population. This discrepancy may be due to different degrees of pre-B cell contamination in the target population. Using a culture system to selectively grow pro-B cells (Saffran et al., 1992), we concluded that pro-B cells do not respond to IL-7 or SLF either alone or in combination (Faust et al., 1993). This is supported further by analysis of IL-7 receptor (IL-7R) knockout mice, which have reduced numbers of pre-B cells undergoing gene rearrangement, but have normal development to the pro-B cell stage of development (Peschon et al., 1994). Pro-B cells are likely dependent on alternative growth signals derived from bone marrow stroma. It is possible that one or a combination of the recently identified factors described above could be the elusive factor(s).

The role of IL-3 in B lymphopoiesis is also controversial. Two different groups have described pro-B cell lines that are dependent on IL-3 for growth (McKearn et al., 1985; Palacios et al., 1987). Using a culture system developed in our laboratory we could detect no effect of IL-3 on pro-B cells (Scherle et al., 1990; Faust et al., 1993). In fact, one report claims that IL-3 acts as a negative regulator of B lymphopoiesis (Hirayama et al., 1994). This evidence is discussed further on p. 121. There is also a lack of evidence to demonstrate production of IL-3 by bone marrow stroma. For example, none of the stromal cell lines listed in Table I have been shown to produce detectable levels of IL-3. However, there is one report of IL-3 expression by stromal cells from Dexter cultures using a reverse transcriptase–polymerase chain reaction (RT–PCR) with oligonucleotide primers specific for IL-3 cDNA (Kittler et al., 1992). IL-3 expression was not observed by Northern blot analysis, which is less sensitive, either before or after stimulation of stroma with pokeweed mitogen. This report suggests that minute amounts of IL-3 may be produced by bone marrow stroma that could regulate lymphohematopoietic growth.

What have we learned from natural and artificial mutations that affect hematopoietic cell/stromal interaction?

Valuable information on the interactions between B cells and the stromal microenvironment can be obtained from *in vitro* studies and from the administration, to whole animals, of cytokines or antibodies directed against stromal molecules. Another worthwhile approach is the study of naturally occurring mutations that lead to hematopoietic defects of mice and humans. The number of naturally occurring mutations is limited. However, technology allowing the germline disruption of genes is a powerful tool that has led to a better understanding of the bone marrow microenvironment and how cytokines function. We will discuss mice in which the genes encoding cytokines, cytokine receptors and signaling molecules have been disrupted by gene knockout techniques as well as naturally occurring mutations of growth factors and their receptors.

The study of the white spotting (*W*) and steel (*Sl*) mice led to the discovery of the interaction between c-*kit* and SLF. *W* and *Sl* were independently identified by mutations that affect hematopoiesis, gametogenesis and melanogenesis (Bennett, 1956; Russell, 1979; Silvers, 1979). Numerous alleles of both *Sl* and *W* have been identified; they are semidominant mutations, and the different alleles vary in their effects on the different cell lineages (Russell, 1979; Silvers, 1979). *In vivo* transplantation experiments and coculture experiments between *W* and *Sl* mutants showed that the defect in *W* is intrinsic to the hematopoietic stem cell, while the defect in *Sl* is in the marrow stromal cell environment. Bone marrow from *Sl* mice transplanted into *W* recipients resulted in a hematologically normal phenotype (Bernstein et al., 1968). Dexter cultures initiated with bone marrow from *Sl* or *W* bone marrow gave rise to defective cultures, consistent with the *in vivo* phenotype. The addition of *Sl* bone marrow to preexisting *W* stromal layers resulted in long-term hematopoiesis (Dexter and Moore, 1977; Boswell et al., 1987). The *W* locus encodes the c-*kit* tyrosine kinase receptor (Chabot et al., 1988; Geissler et al., 1988). Based on the phenotypic similarities and complementary nature of the defects produced by *W* and *Sl* mutations, several groups hypothesized that the ligand for c-*kit* represents the product of the *Sl* locus (Chabot et al., 1988; Geissler et al., 1988).

This hypothesis was proven true when SLF was identified and shown to bind selectively to cells that express c-*kit* (Huang *et al.*, 1990; Williams *et al.*, 1990). As discussed above, *W* mice have been shown to have normal numbers of B220⁺ pre-B cells in the bone marrow (Landreth *et al.*, 1984). This indicates that SLF is not essential for the development of pre-B cells and reinforces the idea of redundancy among hematopoietic growth factors.

Osteopetrotic (*op/op*) mice are characterized by an autosomal recessive inactivating mutation in another hematopoietic growth factor gene, M-CSF (Yoshida *et al.*, 1990). As a consequence young *op/op* mice have impaired mononuclear phagocyte development characterized by a deficiency of both macrophage and osteoclasts (Stanley and Yuspa, 1983; Felix *et al.*, 1990a,b). M-CSF treatment of these mice results in repair of the bone resorption and the appearance of macrophages (Felix *et al.*, 1990a). Studies of *op/op* mice from 2 weeks to 5 months of age demonstrate a progressive change in the structure of the marrow and a normalization in the macrophage content as animals age (Begg *et al.*, 1993). These results suggest that hematopoietic systems have the capacity to use alternative mechanisms to compensate for the multifunctional cytokine M-CSF. This may explain the functional redundancy among cytokines.

A mouse in which the leukemia inhibitory factor (LIF) gene was disrupted by gene knockout technology exemplifies the extensive redundancy of growth factors. LIF is a highly pleiotropic growth factor with activities on embryonic stem cells (Williams *et al.*, 1988; Nichols *et al.*, 1990) and hematopoietic stem cells (Escary *et al.*, 1993), as well as neuronal cells, osteoblasts, adipocytes, hepatocytes and endothelial cells (Heath, 1992). A mouse lacking LIF would be predicted to have multiple defects, but in fact it appears normal. The one defect in these mice is that females are infertile as a result of a failure of the embryo to implant (Stewart *et al.*, 1992). Perhaps a strategy to study redundant growth factors is to create knockout mice, then cross the mice to determine if the offspring are rescued. Such a strategy could be useful in distinguishing the activity of SLF and PBEF on B cells.

Disruption of cytokine receptor genes as well as cytokine genes has revealed the extent to which certain cytokines are required for the normal development and function of B cells. A mouse in which the IL-7R gene is disrupted has allowed analysis of the specific stage at which IL-7 affects B cell development. These mice have small spleens and thymuses compared to normal litter mates. Analysis of bone marrow by flow cytometry revealed a paucity of early and late pre-B cells undergoing Ig gene rearrangements. As discussed above, *in vitro* experiments on sorted pro-B cells and long-term cultures of pro-B cells have indicated that pro-B cells do not respond to IL-7 (Billips *et al.*, 1992; Faust *et al.*, 1993; Funk *et al.*, 1993). This is supported in the IL-7R knockout mice where there are normal numbers of pro-B cells, and the stages following this are diminished. In addition, a limited number of mature B cells were detected in the periphery of these mice (Peschon *et al.*, 1994). The functional capacity of these cells has not been tested. The mature B cells that are present indicate that, to a limited extent, lymphoid development can proceed in the absence of an IL-7R-mediated phase of expansion. Perhaps growth factors with overlapping function or those that augment the function of IL-7 such as SLF, IGF-1, PBEF, PBSF and BST-1 contribute to this alternative pathway. It would be interesting to determine the B cell profile of animals that lacked the IL-7R and the receptor for these other factors.

The human disease X-linked severe combined immunodeficiency (XSCID) is characterized by profound defects of early T cell development and normal numbers of B cells. However, these B cells are non-functional, and mature B cells of XSCID female carriers have non-random X-chromosome inactivation, indicating that there is an intrinsic defect of the B cells (Conley, 1992). Genetic linkage analysis indicates that the IL-2 receptor (IL-2R) γ chain gene and the locus for XSCID are the same (Noguchi *et al.*, 1993a). Receptor studies have shown that many cytokines consist of two polypeptide chains, a ligand-binding receptor and a non-binding signal transducer. The same β chain is shared by receptors for IL-3, IL-5, and GM-CSF (Miyajima *et al.*, 1992). Another molecule, gp130, serves as a signaling molecule by the receptors for IL-6, LIF, oncostatin M and ciliary neurotrophic factor receptors (Gearing *et al.*, 1992). This sharing of receptor subunits may explain the redundancy of growth factor activities. Since IL-2 is important for T cell proliferation (Leonard, 1992), the disruption of a portion of the IL-2R explains the paucity of T cells, but not the lack of functional B cells in XSCID patients. In addition, IL-2-deficient SCID patients (Weinberg and Parkman, 1990) and mice (Schorle *et al.*, 1991) have normal numbers of T cells. Therefore, the hypothesis that the IL-2R γ chain is a receptor component of other cytokines that contribute to T and B cell development seems plausible. Two teams have shown that IL-2R γ chain can form part of two other cytokine receptors: those for IL-4 and IL-7. The group headed by Warren Leonard, in two separate experiments (Noguchi *et al.*, 1993b; Russell *et al.*, 1993), transfected COS 7 cells with either the IL-4 or IL-7 receptors either alone or with IL-2R γ. They demonstrated that ^{125}I-labeled IL-4 or IL-7 respectively bound with low affinity to cells that did not express the IL-2R γ chain, but with high affinity when the IL-2R γ chain was expressed. They were also able to co-immunoprecipitate labeled factor, the IL-2R γ chain and the other receptor component. Another group was able to show that antibodies raised against the IL-2R γ chain inhibited IL-4-induced cell growth and high-affinity binding of IL-4 to the murine T cell line CTLL-2 (Kondo *et al.*, 1993). These data suggest that the IL-2R γ chain, now referred to as the common γ chain, is shared by IL-2, IL-4, and IL-7. This arrangement may explain the redundant effects of cytokines because different ligand-binding molecules can share the same signal transducer.

Internal and external signals are important for the development of the stromal microenvironment which is then required for B lymphopoiesis. Mice in which the *c-fos* gene is disrupted demonstrate this. The *fos* protein is a member of a large multigene family and is a major component of the activator protein 1 (AP-1) transcription factor complex (see Distel and Spiegelman, 1990; Angel and Karin, 1991 for reviews). *In vitro* experiments suggest that c-*fos* plays an important role in signal transduction, cell proliferation, and cell differentiation. Mice lacking c-*fos* develop severe osteopetrosis with deficiencies in bone remodeling and exhibit extramedullary hematopoiesis, thymic atrophy, and altered B cell development (Johnson et al., 1992; Wang et al., 1992). The spleens of these mice have an increase in the number of myeloid (Mac-1⁺) cells and a drastic reduction in the number of B220⁺ B cells (Wang et al., 1992). *In vitro* differentiation and bone marrow reconstitution experiments demonstrated that hematopoietic stem cells lacking c-*fos* can give rise to all mature myeloid as well as lymphoid cells, suggesting that the observed B lymphopenia in the mutant mice is due to an altered environment. Whitlock–Witte bone marrow stromal layers derived from the mutant mice do not support B lymphopoiesis (Okada et al., 1994).

Cytokines as negative regulators of B lymphopoiesis

Factors that inhibit hematopoiesis are not as well characterized as those that stimulate proliferation and differentiation. Some factors provide both positive and negative stimuli for different types of hematopoietic cells. GM-CSF is a positive regulator of myelopoiesis, but interferes with B lymphopoiesis in long-term B cell cultures. Systemic treatment of mice with GM-CSF resulted in reduced numbers of pre-B cells in the bone marrow and recovery occurred only after cessation of administration (Dorshkind, 1991). Other studies have shown that treatment of stromal cells with G-CSF or GM-CSF abrogates their ability to support B cell differentiation and suggested that this negative effect may be mediated by alterations of stromal function (Dorshkind and Landreth, 1992). IL-1 increases the production of GM-CSF and G-CSF from stroma, which stimulate the growth of myeloid cells at the expense of lymphoid cells in culture (Dorshkind, 1988). IL-4 included at the initiation of LTBMC causes the lymphoid cells to retain a more immature phenotype consisting of B220⁻ cells (Rennick et al., 1987; Peschel et al., 1989). When the S17 stromal line is treated with either IL-1 or IL-4, it loses the capacity to support B cell growth because of uncharacterized negative signals produced by the stroma (Billips et al., 1990). Antibodies against transforming growth factor β (TGF-β) enhance B lymphopoiesis in long-term culture. TGF-β also blocks κ expression on maturing pre-B cells (Lee et al., 1987) and the growth responses of pre-B cells to IL-7 (Sing et al., 1988). IL-3 added to lymphoid progenitor colonies in a two-step methylcellulose culture system inhibited B lymphopoiesis. *In vivo* transfer of these B lymphoid colonies to SCID mice results in the production of serum IgM. However, in the presence of IL-3 no serum IgM was detected in reconstituted mice (Hirayama et al., 1994) indicating that IL-3 exhibits a negative influence on the maturation of B cells.

CONCLUDING REMARKS

Long-term lymphoid cultures have facilitated the study of B cell differentiation. Stromal cell lines cloned from these cultures and directly from tissue have led to the discovery of numerous growth factors. It is evident from the number of new factors isolated recently that much effort is being focused on identifying factors involved in B cell development. These factors may be useful in the treatment of human disorders. For example, following bone marrow transplantation or during transient hypogammaglobulinemia in infancy, administration of pre-B cell growth factors may be advantageous to enhance regeneration of the B cell lineage.

The extensive redundancy among pre-B cell growth factors is an area that can be addressed by exploiting new technology. Creating mice deficient in various redundant cytokines will elucidate the necessity of these factors. In addition, crossing various cytokine knockout animals to look for rescue of defects will be a powerful tool. Other technology such as *in situ* hybridization can be used to explore the production of cytokines *in vivo*. Many cytokines are bound to the stroma via the ECM or other mechanisms. This sequestering of cytokines may play a role in creating specialized niches for the growth and differentiation of hematopoietic cells within the bone marrow. These types of studies may indicate the extent to which clonal stromal cell lines reflect the *in vivo* environment and lead to the development of better *in vitro* systems.

Research to identify the particular surface determinants and ECM components on stromal cells required for the support of hematopoiesis is important as well. Several protocols for retroviral gene transfer into human cells in gene therapy currently include transduction on allogeneic stroma (Andrews *et al.*, 1990; Moore *et al.*, 1992). However this presents potential problems associated with transfer into humans of stromal cells predisposed to grow *in vitro*, and possibly *in vivo*, in a malignant fashion. A desired goal would be to develop cell-free systems utilizing ECM components and growth factors for the expansion of human hematopoietic cells specifically for gene transfer.

Understanding the mechanisms that regulate B cell development will also facilitate therapy during hyperactive conditions, such as leukemia or autoimmune disorders. This may be achieved by controlling the production or function of growth factors and cell surface molecules on stroma essential for growth. Manipulation of cells *ex vivo* may also be one way to alter the disease process. For example, culture of bone marrow cells from chronic myelogenous leukemia patients under Dexter conditions results in death of affected leukemic cells while sparing normal hematopoietic progenitors (Udomsakdi *et al.*, 1992). The utilization of similar techniques as well as the development of new strategies may be useful to control B cell disorders in the future.

Acknowledgements

E.A.F. is funded by an NIH Biotechnology Training Grant (1 T32 GM08375), D.C.S. is a Leukemia Society of America Fellow and O.N.W. is an Investigator with the Howard Hughes Medical Institute. This work was supported by NIH grant R35 CA 53867 and the Howard Hughes Medical Institute.

REFERENCES

Anderson, D.M., Lyman, S.D., Baird, A., Wignall, J.M., Eisenman, J., Rauch, C., March, C.J., Boswell, H.S., Gimpel, S.D. and Cosman, D. (1990). *Cell* **63**, 235–243.
Andrews, R.G., Singer, J.W. and Bernstein, I.D. (1990). *J. Exp. Med.* **172**, 355–358.
Angel, P. and Karin, M. (1991). *Biochim. Biophys. Acta* **1072**, 129–157.
Ashton, B.A., Allen, T.D., Howlett, C.R., Eagleson, C.C., Hattori, A. and Owen, M. (1980). *Clin. Orthop.* **151**, 294–306.
Baker, A.H., Ridge, S.A., Hoy, T., Cachia, P.G., Culligan, D., Baines, P., Whittaker, J.A., Jacobs, A. and Padua, R.A. (1993). *Oncogene* **8**, 371–378.
Bauer, S.R., Kudo, A. and Melchers, F. (1988). *EMBO J.* **7**, 111.
Begg, S.K., Radley, J.M., Pollard, J.W., Chisholm, O.T., Stanley, E.R. and Bertoncello, I. (1993). *J. Exp. Med.* **177**, 237–242.
Bennett, D. (1956). *J. Morphol.* **98**, 199–234.
Bernstein, S.E., Russell, E.S. and Keighley, G.H. (1968). *Ann. N.Y. Acad. Sci.* **149**, 475–485.
Billips, L.G., Petitte, D. and Landreth, K.S. (1990). *Blood* **75**, 611–619.
Billips, L.G., Petitte, D., Dorshkind, K., Narayanan, R., Chiu, C.-P. and Landreth, K.S. (1992). *Blood* **79**, 1185–1192.
Bodine, D.M., Seidel, N.E., Zsebo, K.M. and Orlic, D. (1993). *Blood* **82**, 445–455.
Boothby, M., Gravallese, E., Liou, H.-C. and Glimcher, L.H. (1988). *Science* **242**, 1559–1562.
Border, W.A., Noble, N.A., Yamamoto, T., Harper, J.R., Yamaguchi, Y., Pierschbacher, M.D. and Ruoslahti, E. (1992). *Nature* **360**, 361–364.
Boswell, H.S., Albrecht, P.R., Shupe, R.E., Williams, D.E. and Burgess, J. (1987). *Exp. Hematol.* **15**, 46–53.
Brunswick, M. and Lake, P. (1985). *J. Exp. Med.* **161**, 953–971.
Cattoretti, G., Schiró, R., Orazi, A., Soligo, D. and Colombo, M.P. (1993). *Blood* **81**, 1726–1738.
Chabot, B., Stephenson, D.A., Chapman, V.M., Besmer, P. and Bernstein, A. (1988). *Nature* **335**, 88–89.
Chernousov, M.A. and Carey, D.J. (1993). *J. Biol. Chem.* **268**, 16810–16814.
Collins, L.S. and Dorshkind, K. (1987). *J. Immunol.* **138**, 1082–1087.
Conley, M.E. (1992). *Annu. Rev. Immunol.* **10**, 215–238.
Copeland, N.G., Gilbert, D.J., Cho, B.C., Donovan, P.J., Jenkins, N.A., Cosman, D., Anderson, D., Lyman, S.D. and Williams, D.E. (1990). *Cell* **63**, 175–183.
Cumano, A., Dorshkind, K., Gillis, S. and Paige, C.J. (1990). *Eur. J. Immunol.* **20**, 2183–2189.
Damia, G., Komschlies, K.L., Faltynek, C.R., Ruscetti, F.W. and Wiltrout, R.H. (1992). *Blood* **79**, 1121–1129.
Davidson, W.F., Fredrickson, T.N., Rudikoff, E.K., Coffman, R.L., Hartley, J.W. and Morse, H.C., III (1984). *J. Immunol.* **133**, 744–753.
Denis, K.A. and Witte, O.N. (1986). *Proc. Natl Acad. Sci. USA* **83**, 441–445.
Dexter, M. and Allen, T. (1992). *Nature* **360**, 709–710.
Dexter, T.M. and Moore, M.A.S. (1977). *Nature* **269**, 412–414.
Dexter, T.M., Allen, T.D. and Lajtha, L.G. (1977). *J. Cell. Physiol.* **91**, 335–344.
Distel, R.J. and Spiegelman, B.M. (1990). *Adv. Cancer Res.* **55**, 37–55.
Dorshkind, K. (1988). *J. Immunol.* **141**, 531–538.
Dorshkind, K. (1990). *Annu. Rev. Immunol.* **8**, 111–137.
Dorshkind, K. (1991). *J. Immunol.* **146**, 4204–4208.
Dorshkind, K. and Landreth, K.S. (1992). *Int. J. Cell Cloning* **10**, 12–17.
Dorshkind, K., Denis, K.A. and Witte, O.N. (1986). *J. Immunol.* **137**, 3457–3463.
Du, X.X. and Williams, D.A. (1994). *Blood* **83**, 2023–2030.
Escary, J.-L., Perreau, J., Duménil, D., Ezine, S. and Brûlet, P. (1993). *Nature* **363**, 361–364.
Faust, E.A., Saffran, D.C., Toksoz, D., Williams, D.A. and Witte, O.N. (1993). *J. Exp. Med.* **177**, 915–923.
Felix, R., Cecchini, M.G. and Fleisch, H. (1990a). *Endocrinology* **127**, 2592–2594.

Felix, R., Cecchini, M.G., Hofstetter, W., Elford, P.R., Stutzer, A. and Fleisch, H. (1990b). *J. Bone Miner. Res.* **5**, 781–787.
Flanagan, J.G. and Leder, P. (1990). *Cell* **63**, 185–194.
Friedenstein, A.J. (1990). *Bone Miner. Res.* **7**, 243–272.
Friedenstein, A.J., Chailakhyan, R.K., Latsinik, N.B., Panasyuk, A.F. and Keiliss-Borok, Z.V. (1974). *Transplantation* **17**, 331–340.
Froesch, E.R., Schmid, C., Schwander, J. and Zapf, J. (1985). *Annu. Rev. Physiol.* **47**, 443–467.
Funk, P.E., Varas, A. and Witte, P.L. (1993). *J. Immunol.* **150**, 748–752.
Galmiche, M.C., Koteliansky, V.E., Brière, J., Hervé, P. and Charbord, P. (1993). Blood **82**, 66–76.
Gay, S., Martin, G.R., Müller, P.K., Timpl, R. and Kühn, K. (1976). *Proc. Natl Acad. Sci. USA* **73**, 4037–4040.
Gearing, D.P., Comeau, M.R., Friend, D.J., Gimpel, S.D., Thut, C.J., McGourty, J., Brasher, K.K., King, J.A., Gillis, S., Mosley, B., Ziegler, S.F. and Cosman, D. (1992). *Science* **255**, 1434–1437.
Geissler, E.N., Ryan, M.A. and Housman, D.E. (1988). *Cell* **55**, 185–192.
Gibson, L.F., Piktel, D. and Landreth, K.S. (1993). *Blood* **82**, 3005–3011.
Go, N.F., Castle, B.E., Barret, R., Kastelein, R., Dang, W., Mosmann, T.R., Moore, K.W. and Howard, M. (1990). *J. Exp. Med.* **172**, 1625–1631.
Gordon, M.Y., Riley, G.P., Watt, S.M. and Greaves, M.F. (1987). *Nature* **326**, 403–405.
Hannum, C., Culpepper, J., Campbell, D., McClanahan, T., Zurawski, S., Bazan J.F., Kastelein, R., Hudak, S., Wagner, J., Mattson, J., Luh, J., Duda, G., Martina, N., Peterson, D., Menon, S., Shanafelt, A., Muench, M., Kelner, G., Namikawa, R., Rennick, D., Roncarolo, M.-G., Zlotnik, A., Rosnet, O., Dubreuil, P., Birnbaum, D. and Lee, F. (1994). *Nature* **368**, 643–648.
Hardy, R.R., Carmack, C.E., Shinton, S.A., Kemp, J.D. and Hayakawa, K. (1991). *J. Exp. Med.* **173**, 1213–1225.
Hayashi, S.-L., Kunisada, T., Ogawa, M., Sudo, T., Kodama, H., Suda, T., Nishikawa, S. and Nishikawa, S.-I. (1990). *J. Exp. Med.* **171**, 1683–1695.
Health, J.K. (1992). *Nature* **359**, 17.
Henderson, A.J., Johnson, A. and Dorshkind, K. (1990). *J. Immunol.* **145**, 423–428.
Hilton, D.J. (1992). *TIBS* **17**, 72–76.
Hirano, H. and Urist, M.R. (1981). *Clin. Orthop.* **154**, 234–248.
Hirayama, F., Clark, S.C. and Ogawa, M. (1994). *Proc. Natl Acad. Sci. USA* **91**, 469–473.
Hofman, F.M., Brock, M., Taylor, C.R. and Lyons, B. (1988). *J. Immunol.* **141**, 1185–1190.
Hombach, J., Tsubata, T., Leclercq, L., Stappert, H. and Reth, M. (1990). *Nature* **343**, 760–762.
Howard, B.V., Macarak, E.J., Gunson, D. and Kefalides, N.A. (1976). *Proc. Natl Acad. Sci. USA* **73**, 2361–2364.
Huang, E., Nocka, K., Beier, D.R,. Chu, T.-Y., Buck, J., Lahm, H.-W., Wellner, D., Leder, P. and Besmer, P. (1990). *Cell* **63**, 225–233.
Huang, S. and Terstappen, L.W.M.M. (1992). *Nature* **360**, 745–749.
Huang, S. and Terstappen, L.W.M.M. (1994). *Blood*, **83**, 1515–1526.
Hunt, P., Robertson, D., Weiss, D., Rennick, D., Lee, F. and Witte, O.N. (1987). *Cell* **48**, 997–1007.
Ikebuchi, K., Clark, S.C., Ihle, J.N., Souza, L.M. and Ogawa, M. (1988). *Proc. Natl Acad. Sci. USA* **85**, 3445–3449.
Ikuta, K. and Weissman, I.L. (1992). *Proc. Natl. Acad. Sci. USA* **89**, 1502–1506.
Jacobsen, K., Miyake, K., Kincade, P.W. and Osmond, D.G. (1992). *J. Exp. Med.* **176**, 927–935.
Jaffe, E.A., Minick, C.R., Adelman, B., Becker, C.G. and Nachman, R. (1976). *J. Exp. Med.* **144**, 209–225.
Johnson, R.S., Spiegelman, B.M. and Papaioannou, V. (1992). *Cell* **71**, 577–586.
Jones-Villeneuve, E. and Phillips, R.A. (1980). *Exp. Hematol.* **8**, 65–76.
Kaisho, T., Ishikawa, J., Oritani, K., Inazawa, J., Tomizawa, H., Muraoka, O., Ochi, T. and Hirano, T. (1994). *Proc. Natl Acad. Sci. USA* **91**, 5325–5329.
Keating, A., Singer, J.W., Killen, P.D., Striker, G.E., Salo, A.C., Sanders, J., Thomas, E.D., Thorning, D. and Fialkow, P.J. (1982). *Nature* **298**, 280–283.
Kinashi, T., Inaba, K., Tsubata, T., Tashiro, K., Palacios, R. and Honjo, T. (1988). *Proc. Natl Acad. Sci. USA* **85**, 4473–4477.

Kincade, P.W. (1994). *Proc. Natl Acad. Sci. USA* **91**, 2888–2889.
Kincade, P.W., Lee, G., Pietrangeli, C.E., Hayashi, S.-I. and Gimble, J.M. (1989). *Annu. Rev. Immunol.* **7**, 111–143.
Kishimoto, T. (1989). *Blood* **74**, 1–10.
Kittler, E.L.W., McGrath, H., Temeles, D., Crittenden, R.B., Kister, V.K. and Quesenberry, P.J. (1992). *Blood*, **79**, 3168–3178.
Kondo, M., Takeshita, T., Ishii, N., Nakamura, M., Watanabe, S., Arai, K. and Sugamura, K. (1993). *Science* **262**, 1874–1877.
Kurland, J.I., Ziegler, S.F. and Witte, O.N. (1984). *Proc. Natl Acad. Sci. USA* **81**, 7554–7558.
Landreth, K.S., Kincade, P.W., Lee, G. and Harrison, D.E. (1984). *J. Immunol.* **132**, 2724–2729.
Landreth, K.S., Narayanan, R. and Dorshkind, K. (1992). *Blood* **80**, 1207–1212.
Leary, A.G., Ikebuchi, K., Hirai, Y., Wong, G.G., Yang, Y-C., Clark, S.C., and Ogawa, M. (1988). *Blood* **71**, 1759–1763.
Lee, G., Ellingsworth, L.R., Gillis, S., Wall, R. and Kincade, P.W. (1987). *J. Exp. Med.* **166**, 1290–1299.
Lee, G., Namen, A.E., Gillis, S., Ellingsworth, L.R. and Kincade, P.W. (1989). *J. Immunol.* **142**, 3875–3883.
Leibson, H.J., Gefter, M., Zlotnik, A., Marrack, P. and Kappler, J.W. (1984). *Nature* **309**, 799–801.
Lemoine, F., Dedhar, S., Lima, G.M. and Eaves, C.J. (1990). *Blood* **76**, 2311–2320.
Leonard, W.J. (1992). 'Interleukin-2: Frontiers in Pharmacology and Therapeutics', vol. 1, pp. 29–46. Blackwell Scientific Publications, Boston.
Lortat-Jacob, H., Kleinman, H.K. and Grimaud, J.A. (1991). *J. Clin. Invest.* **87**, 878–883.
Lyman, S.D., James, L., Bos, T.V., de Vries, P., Brasel, K., Gliniak, B., Hollingsworth, L.T., Picha, K.S., McKenna, H.J., Splett, R.R., Fletcher, F.A., Maraskovsky, E., Farrah, T., Foxworthe, D., Williams, D.E. and Beckmann, M.P. (1993). *Cell* **75**, 1157–1167.
McKearn, J.P., McCubrey, J. and Fagg, B. (1985). *Proc. Natl Acad. Sci. USA* **82**, 7414–7418.
McNiece, I.K., Robinson, B.E. and Quesenberry, P.J. (1988). *Blood* **72**, 191–195.
McNiece, I.K., Langley, K.E. and Zsebo, K.M. (1991). *J. Immunol.* **146**, 3785–3790.
Maniatopoulos, C., Sodek, J. and Melcher, A.H. (1988). *Cell Tissue Res.* **254**, 317–330.
Martin, F.H., Suggs, S.V., Langley, K.E., Lu, H.S., Ting, J., Okino, K.H., Morris, C.F., McNiece, I.K., Jacobsen, F.W., Mendiaz, E.A., Birkett, N.C., Smith, K.A., Johnson, M.J., Parker, V.P., Flores, J.C., Patel, A.C., Fisher, E.F., Erjavec, H.O., Herrera, C.J., Wypych, J., Sachdev, R.K., Pope, J.A., Leslie, I., Wen, D., Lin, C.-H., Cupples, R.L. and Zsebo, K.M. (1990). *Cell* **63**, 203–211.
Matsui, Y., Toksoz, D., Nishikawa, S., Nishikawa, Shin-I., Williams, D., Zsebo, K. and Hogan, B.L.M. (1991). *Nature* **353**, 750–752.
Matthews, W., Jordan, C.T., Gavin, M., Jenkins, N.A., Copeland, N.G. and Lemischka, I.R. (1991a). *Proc. Natl Acad. Sci. USA* **88**, 9026–9030.
Matthews, W., Jordan, C.T., Wiegand, G.W., Pardoll, D. and Lemischka, I.R. (1991b). *Cell* **65**, 1143–1152.
Melchers, F. and Potter, M. (1991). Mechanisms of cell neoplasia. Workshop at the Basel Institute for Immunology, 14–16 April, pp. 1–19.
Metcalf, D. (1991). *Proc. Natl Acad. Sci. USA* **88**, 11310–11314.
Metcalf, D. and Moore, M.A.S. (1971). In 'Haemopoietic Cells', vol. 1, pp. 220–221. North-Holland, Amsterdam.
Migliaccio, G., Migliaccio, A.R., Valinsky, J., Langley, K., Zsebo, K., Visser, J.W.M. and Adamson, J.W. (1991). *Proc. Natl Acad. Sci. USA* **88**, 7420–7424.
Miyajima, A., Kitamura, T., Harada, N., Yokota, T. and Arai, K. (1992). *Annu. Rev. Immunol.* **10**, 295–331.
Miyake, K., Underhill, C.B., Lesley, J. and Kincade, P.W. (1990a). *J. Exp. Med.* **172**, 69–75.
Miyake, K., Medina, K.L., Hayashi, S.-I., Ono, S., Hamaoka, T. and Kincade, P.W. (1990b). *J. Exp. Med.* **171**, 477–488.
Miyake, K., Weissman, I.L., Greenberger, J.S. and Kincade, P.W. (1991). *J. Exp. Med.* **173**, 599–607.
Moore, K.A., Deisseroth, A.B., Reading, C.L., Williams, D.E. and Belmont, J.W. (1992). *Blood* **79**, 1393–1399.
Morrissey, P.J., Conlon, P., Charrier, K., Braddy, S., Alpert, A., Williams, D., Namen, A.E. and Mochizuki, D. (1991). *J. Immunol.* **147**, 561–566.

Müller-Sieburg, C.E., Whitlock, C.A. and Weissman, I.L. (1986). *Cell* **44**, 653–662.
Nagasawa, T., Kikutani, H. and Kishimoto, T. (1994). *Proc. Natl Acad. Sci. USA* **91**, 2305–2309.
Namen, A.E., Lupton, S., Hjerrild, K., Wignall, J., Mochizuki, D.Y., Schmierer, A., Mosley, B., March, C.J., Urdal, D., Gillis, S., Cosman, D. and Goodman, R.G. (1988a). *Nature* **333**, 571–573.
Namen, A.E., Schimierer, A.E., March, C.J., Overell, R.W., Park, L.S., Urdal, D.L. and Mochizuki, D.Y. (1988b). *J. Exp. Med.* **167**, 988–1002.
Nichols, J., Evans, E.P. and Smith, A.G. (1990). *Development* **110**, 1341–1348.
Nishikawa, S.-I., Ogawa, M., Nishikawa, S., Kunisada, T. and Kodama, H. (1988). *Eur. J. Immunol.* **18**, 1767–1771.
Noguchi, M., Yi, H., Rosenblatt, H.M., Filipovich, A.H., Adelstein, S., Modi, W.S., McBride, O.W. and Leonard, W.J. (1993a). *Cell* **73**, 145–157.
Noguchi, M., Nakamura, Y., Russell, S.M., Ziegler, S.F., Tsang, M., Cao, X. and Leonard, W.J. (1993b). *Science* **262**, 1877–1880.
Ogawa, M. (1993). *Blood* **81**, 2844–2853.
Ogawa, M., Nishikawa, S., Kohichi, I., Yamamura, F., Naito, M., Takahashi, K. and Nishikawa, S.-I. (1988). *EMBO J.* **7**, 1337–1343.
Ogawa, M., Matsuzaki, Y., Nishikawa, S., Hayashi, S.-I., Kunisada, T., Sudo, T., Kina, T., Nakauchi, H. and Nishikawa, S.-I. (1991). *J. Exp. Med.* **174**, 63–71.
Okada, S., Nakauchi, H., Nagayoshi, K., Nishikawa, S., Nishikawa, S., Miura, Y. and Suda, T. (1991). *Blood* **78**, 1706–1712.
Okada, S., Wang, Z.-Q., Grigoriadis, A.E., Wagner, E.F. and von Rüden, T. (1994). *Mol. Cell. Biol.* **14**, 382–390.
Orlic, D., Fischer, R., Nishikawa, Shin-I., Nienhuis, A.W. and Bodine, D.M., (1993). *Blood* **82**, 762–770.
Osmond, D.G. (1986). *Immunol. Rev.* **93**, 103–124.
Osmond, D.G. (1990). *Immunology* **2**, 173–180.
Owen, M. (1985). In 'Bone and Mineral Research' (W.A. Peck, ed.), vol. 3, pp. 1–25. Elsevier, Amsterdam.
Palacios, R., Karasuyama, H. and Rolink, A. (1987). *EMBO J.* **6**, 3687–3693.
Park, Y.-H. and Osmond, D.G. (1987). *J. Exp. Med.* **165**, 444–458.
Paul, S.R., Bennett, F., Calvetti, J.A., Kelleher, K., Wood, C.R., O'Hara, R.M., Leary, A.C., Sibley, B., Clark, S.C., Williams, D.A. and Yang, Y.-C. (1990). *Proc. Natl Acad. Sci. USA* **87**, 7512–7516.
Peschel, C., Green, I. and Paul, W.E. (1989). *J. Immunol.* **142**, 1558–1568.
Peschon, J.J., Morrissey, P.J., Grabstein, K.H., Ramsdell, F.J., Maraskovsky, E., Gliniak, B.C., Park, L.S., Ziegler, S.F., Williams, D.E., Ware, C.B., Meyer, J.D. and Davison, B.L. (1994). *J. Exp. Med.*, **180**, 1955–1960.
Rabin, E.M., Mond, J.J., Ohara, J. and Paul, W.E. (1986). *J. Exp. Med.* **164**, 517–531.
Rahal, M.D. and Osmond, D.G. (1984). *Cell. Immunol.* **87**, 379–388.
Raulo, E., Chernousov, M.A., Carey, D.A.J., Nolo, R. and Rauvala, H. (1994). *J. Biol. Chem.* **269**, 12999–13004.
Rennick, D., Yang, G., Gemmell, L. and Lee, F. (1987). *Blood* **69**, 682–691.
Rich, B.E., Campos-Torres, J., Tepper, R.I., Moreadith, R.W. and Leder, P. (1993). *J. Exp. Med.* **177**, 305–316.
Roberts, R.A., Spooncer, E., Parkinson, E.K., Lord, B.I., Allen, T.D. and Dexter, T.M. (1987). *J. Cell. Physiol.* **132**, 203–214.
Roberts, R., Gallagher, J., Spooncer, E., Allen, T.D., Bloomfield, F. and Dexter, T.M. (1988). *Nature* **332**, 376–378.
Rolink, A. and Melchers, F. (1991). *Cell* **66**, 1081–1094.
Rolink, A., Kudo, A., Karasuyama, H., Kikuchi, Y. and Melchers, F. (1991). *EMBO J.* **10**, 327–336.
Ruoslahti, E. (1991). *J. Clin. Invest.* **87**, 1–5.
Russell, E.S. (1979). *Adv. Genet.* **20**, 357–459.
Russell, S.M., Keegan, A.D., Harada, N., Nakamura, Y., Noguchi, M., Leland, P., Friedmann, M.C., Miyajima, A., Puri, R.K., Paul, W.E. and Leonard, W.J. (1993). *Science* **262**, 1880–1883.
Saffran, D.C., Faust, E.A. and Witte, O.N. (1992). *Curr. Top. Microbiol. Immunol.* **182**, 34–44.

Sakaguchi, N., Kashiwamura, S.I., Kimoto, M., Thaimann, P. and Melchers, F. (1988). *EMBO J.* **7**, 3457–3464.
Samal, B., Sun, Y., Stearns, G., Xie, C., Suggs, S. and McNiece, I. (1994). *Mol. Cell. Biol.* **14**, 1431–1437.
Sara, V.R. and Hall, K. (1990). *Physiol. Rev.* **70**, 591–614.
Scherle, P.A., Dorshkind, K. and Witte, O.N. (1990). *Proc. Natl Acad. Sci. USA* **87**, 1908–1912.
Schmid, C., Steiner, T. and Froesch, E.R. (1983). *FEBS Lett.* **161**, 117–121.
Schorle, H., Holtschke, T., Hunig, T., Schimpl, A. and Horak, I. (1991). *Nature* **352**, 621–624.
Schrader, J.W. and Schrader, S. (1978). *J. Exp. Med.* **148**, 832–928.
Sidman, C.L., Marshall, J.D., Shultz, L.D., Gray, P.W. and Johnson, H.M. (1984). *Nature* **309**, 801–804.
Silvers, W.K (1979). In 'The Coat Colors of Mice: A Model for Gene Action and Interaction', vol. 1, pp. 206–241. Springer-Verlag, New York.
Simmons, P.J., Przepiorka, D., Thomas, E.D. and Torok-Storb, B. (1987). *Nature* **320**, 429–432.
Simmons, P.J., Masinovsky, B., Longenecker, B.M., Berenson, R., Torok-Storb, B. and Gallatin, W.M. (1992). *Blood* **80**, 388–395.
Sing, G.K., Keller, J.R., Ellingsworth, L.R. and Ruscetti, F.W. (1988). *Blood* **72**, 1504–1511.
Singer, J.W., Keating, A., Cuttner, J., Gown, A.M., Jacobson, R., Killen, P.D., Moohr, J.W., Najfeld, V., Powell, J., Sanders, J., Striker, G.E. and Fialkow, P.J. (1984). *Leuk. Res.* **8**, 535–545.
Small, D., Levenstein, M., Kim, E., Carow, C., Amin, S., Rockwell, P., Witte, L., Burrow, C., Ratajczak, M.Z., Gewirtz, A.M. and Civin, C.I. (1994). *Proc. Natl Acad. Sci. USA* **91**, 459–463.
Smith, C., Gasparetto, C., Collins, N., Gillio, A., Muench, M.O., O'Reilly, R.J. and Moore, M.A.S. (1991). *Blood* **77**, 2122–2128.
Spangrude, G.J., Heimfeld, S. and Weissman, I.L. (1988). *Science* **241**, 58–62.
Stanley, J.R. and Yuspa, S.H. (1983). *J. Cell Biol.* **96**, 1809–1814.
Stewart, C.L., Kaspar, P., Brunet, L.J., Bhatt, H., Gadi, I., Kontgen, F. and Abbondanzo, S.J. (1992). *Nature* **359**, 76–79.
Sudo, T., Ito, M., Ogawa, Y., IIzuka, M., Kodama, T., Kunisada, T., Hayashi, S.-I., Ogawa, M., Sakai, K., Nishikawa, S. and Nishikawa, S.-I. (1989). *J. Exp. Med.* **170**, 333–338.
Tashiro, K., Tada, H., Heiker, R., Shirozu, M., Nakano, T. and Honjo, T. (1993). *Science* **261**, 600.
Tidmarsh, G.F., Heimfeld, S., Whitlock, C.A., Weissman, I.L. and Müller-Seiburg, C.E. (1989). *Mol. Cell. Biol.* **9**, 2665–2671.
Till, J.E. and McCulloch, E.A. (1980). *Biochim. Biophys. Acta* **605**, 431–459.
Tsuji, K., Zsebo, K.M. and Ogawa, M. (1991). *Blood* **78**, 1223–1229.
Udomsakdi, C., Eaves, C.J., Swolin, B., Reid, D.S., Barnett, M.J. and Eaves, A.C. (1992). *Proc. Natl Acad. Sci. USA* **89**, 6192–6196.
Visser, J.W.M., Hogeweg-Platenberg, M.G.C., de Vries, P., Bayer, J.A. and Ploemacher, R.E. (1990). In 'The Biology of Hematopoiesis' N. Dainiak, E.P. Cronkite, R. McCaffrey and R.K. Shaduck, eds), vol. 1, pp. 1–8. Wiley-Liss, New York.
de Vries, P., Brasel, K.A., Eisenman, J.R., Alpert, A.R. and Williams, D.E. (1991). *J. Exp. Med.* **173**, 1205–1211.
Wang, Z.-Q., Ovitt, C., Grigoriadis, A.E., Möhle-Steinlein, U., Ruther, U. and Wagner, E.F. (1992). *Nature* **360**, 741–745.
Weinberg, K. and Parkman, R. (1990). *N. Engl. J. Med.* **322**, 1718–1723.
Weiss, L. (1976). *Anat. Rec.* **186**, 161–184.
Weiss, L. and Sakai, H. (1984). *Am. J. Anat.* **170**, 447–463.
Whitlock, C.A. and Witte, O.N. (1982). *Proc. Natl Acad. Sci. USA* **79**, 3608–3612.
Whitlock, C.A., Tidmarsh, G.F., Müller-Sieburg, C. and Weissman, I.L. (1987). *Cell* **48**, 1009–1021.
Williams, D.E., Eisenman, J., Baird, A., Rauch, C., Van Ness, K., March, C.J., Park, L.S., Martin, U., Mochizuki, D.Y., Boswell, H.S., Burgess, G.S., Cosman, D. and Lyman, S.D. (1990). *Cell* **63**, 167–174.
Williams, N., Bertoncello, I., Kavnoudias, H., Zsebo, K. and McNiece, I. (1992). *Blood* **79**, 58–64.
Williams, R.L., Hilton, D.J., Pease, S., Willson, T.A., Stewart, C.L., Gearing, D.P., Wagner, E.F., Metcalf, D., Nicola, N.A. and Gough, N.M. (1988). *Nature* **336**, 684–687.
Witte, O.N. (1990). *Cell* **63**, 5–6.

Witte, P.L., Burrows, P.D., Kincade, P.W. and Cooper, M.D. (1987). *J. Immunol.* **138**, 2698–2705.
Xue, Z.G., Le Douarin, N.M. and Smith, J. (1988). *Cell Differ. Dev.* **25**, 1–10.
Yoshida, H., Hayashi, S.I., Kunisada, T., Ogawa, M., Nishikawa, S., Okamura, H., Sudo, T., Shultz, L.D. and Nishikawa, S.I. (1990). *Nature* **345**, 442–444.
Zsebo, K.M., Wypych, J., McNiece, I.K., Lu, H.S., Smith, K.A., Karkare, S.B., Sachdev, R.K., Yuschenkoff, V.N., Birkett, N.C., Williams, L.R., Satyagal, V.N., Tung, W., Bosselman, R.A., Mendiaz, E.A. and Langley, K.E. (1990). *Cell* **63**, 195–201.
Zuckerman, K.S. and Wicha, M.S. (1983). *Blood* **61**, 540–547.

6

Antigen receptors on B lymphocytes

Michael Reth

Max-Planck Institut für Immunbiologie, Freiburg, Germany

Introduction 125
Membrane-bound immunoglobulins 126
The Igα/Igβ heterodimer 129
Interaction of the Igα/β heterodimer with PTK 133
Internalization function of the receptor 135
Conclusion 136

INTRODUCTION

Immunoglobulin can exist as membrane-bound molecules (mIg) on the cell surface or as secreted antibody molecules (sIg) in the serum. The mIg molecule is part of the antigen receptor (BCR) on mature B lymphocytes. As postulated in Niels Jerne's and Sir MacFarlane Burnet's selection theories the antigen receptor plays a central role in the specific activation of B cells. A mature B lymphocyte carries about 10^5 identical antigen receptors on the cell surface. Due to the enormous variability of the V domains of the mIg molecules most B cells differ from each other in the antigen-binding specificity of their antigen receptors. A particular antigen selects only those possessing a complementary antigen receptor among the approximately 10^8 B cells in the mouse. Upon aggregation by antigen the antigen receptor generates activation signals inside the B cells. Activated B cells multiply and under the influence of activated T cells and T cell-derived factors finally differentiate into either memory B cells or plasma cells secreting large amounts of antibody.

The differences between sIgM and mIgM reside only in the C terminus of the Ig heavy chain. The secretory form has a specific glycosylated C terminus of 28 amino acids, whereas the C-terminus part of the membrane-bound form consists of 48 amino acids: an extracellular linker, a transmembrane (Tm) part and a 3 amino acid cytoplasmic part. These differences at the C terminus determine the fate of the two

Ig forms and their association with different molecules. Through a cysteine at its C terminus sIgM can covalently bind to the J chain whereas the membrane-proximal C domain and the Tm sequence are required for a non-covalent association of mIg with the Igα/Igβ heterodimer. Igα and Igβ are glycosylated Tm proteins encoded by the B cell-specific genes *mb*-1 and *B29*, respectively. These proteins are similar in their gene and protein structure to the components of the CD3 complex associated with the T cell receptor (TCR). A signalling-competent BCR is a complex between the Igα/Igβ heterodimer and the ligand-binding mIg molecule. While mIgM is the ligand-binding subunit, the Igα/Igβ heterodimer seems to be the signalling subunit of the receptor complex. The cytoplasmic part of Igα and Igβ carries a conserved amino acid motif that couples the receptor to protein tyrosine kinases (PTKs). The structure and function of the BCR has been also described in several recent reviews (Reth, 1992; DeFranco, 1993; Gold and DeFranco, 1994).

MEMBRANE-BOUND IMMUNOGLOBULINS

Each C gene of the Ig heavy chain locus has two Tm exons situated 5′ of the last C-domain exon (Alt *et al.*, 1980; Kehry *et al.*, 1980). These Tm exons are used in an alternative splicing process to generate transcripts for the mIg heavy chain (Fig. 1).

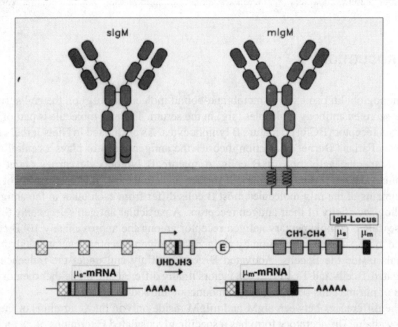

Fig. 1. Generation of two types of mRNA and proteins from the Ig heavy chain (H) locus by alternative splicing. The V-gene segments and μ exons of the IgH locus are shown as boxes. For simplicity the two short μ_m-specific (exon) is given as one box. The domain structure of sIgM and mIgM is shown above the IgH locus.

The Tm exons are only used when primary transcripts proceed through the part 5' of the last CH exon. Indeed, in plasma cells the transcription is often terminated before the Tm exons are reached and thus only transcripts for the sIg heavy chain are produced (Yuan et al., 1990). The molecular details of this termination control are not known. In mature B cells both transcripts are found and here an alternative splicing factor is thought to control the amount of the two transcripts (Tanaka et al., 1994). The expression of the two forms of Ig is not only under a transcriptional but also a post-transcriptional control. Mature B lymphocytes produce sIgM molecules intracellularly but do not secrete them. This transport control is due to the binding of the chaperon heavy chain *binding protein* (BIP) to the μ_s-specific C terminus and the retention of the sIgM molecule in the endoplasmic reticulum (ER) (Alberini et al., 1990; Sitia et al., 1990). In plasma cells the sIgM monomers are assembled into the sIgM pentamer, which is not retained by BIP in the ER and is secreted in large amounts. The mIgM molecules, however, are not transported on the cell surface of plasma cells (Sitia et al., 1987; Hombach et al., 1988a). This is due to the fact that murine plasma cells do not express the *mb*-1 gene coding for Igα (Hombach et al., 1988b). In the absence of the Igα/Igβ heterodimers all mIgM molecules are arrested in the ER.

Two Tm exons encode the Tm region of the mIg molecule. This sequence consists of a negatively charged extracellular linker, the Tm part crossing the membrane and a cytoplasmic tail. The Tm region sequences for all murine and most of the human mIg classes have been determined (Fig. 2). The extracellular linker is quite variable in sequence and length between the different mIg classes. However, 13 of the 25 amino acids of the Tm part are conserved in most mIg classes. In the cytoplasmic

```
             --------transmembrane-----  ------cytoplasmic----------
                                         1........10........20........30
                                         ,         ,         ,         ,
             ** **   *   * ***** *
EVNAEEEGFEN  LWTTASTFIVLFLLSLFYSTTVTLF   KVK                                     mouse IgM
--S-D------  --A----------------------   ---                                     human IgM

DSYMDL-EENG  --P-MC--VA----T-L--GF--FI   ---                                     mouse IgD
DDYTTFDDVGS  ----L---VA--I-T-L--GI--FI   ---                                     human IgD

CAE-QDGELDG  ----ITI--S-----VC--AS----   ---WIFSSVVELKQTISPDYRNMIGQGA            mouse IgG₂ₐ
CAE-KDGELDG  ----ITI--S-----VC--AS----   --------------K-------------            mouse IgG₂ᵦ
CAE-QDGELDG  ----ITI--S-----VC--AA----   ----------LV-E-K-----AP                 mouse IgG₁
CAE-QDGELDG  ----ITI--S-----VC--AS----   ----------QV---AI----------             mouse IgG₃

CAE-QDGELDG  ----ITIL-T-----VC--A---F-   ----------D-------I---------            human IgG₃

DILE--APGAS  --P-TV--LT----------AL-VT   T-RGP-G-KEVPQY                          mouse IgA
ETLE--TPGA-  --P-TI--LT----------AL-VT   S-RGPSGNREGPQY                          human IgA₁
ETLE--TPGA-  --P-TI--LT----------AL-VT   S-RGPSGKREGPQY                          human IgA₂

IEEV-G-EL-E  ---SICV--T-----VS-GA---VL   ----VL-TPMQDTPQTFQ--A-ILQTR-            mouse IgE
VEE--G-APW   T--GLCI-AA-----VS--AAL--L   M-QRFL-ATRQGRPQL-L---T-VLQPH-           human IgE
             ^^ ^    ^^^^^   ^^  ^^
```

* conserved polar amino acids > 50 %
^ conserved amino acids > 85 %
bold: complete conservation

Fig. 2. Comparison of the mouse and human transmembrane region of the different mIgM heavy chains. The transmembrane region of mIgM molecules consists of a linker, a transmembrane (Tm) and a cytoplasmic sequence. For simplicity only the last 11 amino acids of the linker are shown. A dash indicates identity to the upper sequence. The position of polar and conserved (>85%) amino acids in the Tm sequence are indicated by a star and an arrowhead, respectively. (Modified from Reth, 1992.)

part there is a striking difference between the mIg classes predominantly expressed on mature B cells (mIgM and mIgD) and the mIg classes expressed on memory B cells (mIgG, mIgA and mIgE). While mIgM and mIgD both have only a short cytoplasmic tail of 3 amino acids (Lys, Val, Lys), mIgG and mIgE have a cytoplasmic sequence of 28 and mIgA a cytoplasmic sequence of 14 amino acids.

The Tm part of the mIg molecule has several interesting features. First, it is not exclusively composed of hydrophobic residues as it contains several polar amino acids. In the Tm part of mIgM for example, 9 of the 25 amino acids are serines or threonines. Second, the Tm sequence is highly conserved throughout evolution. The Tm sequences of human and mouse mIgM display 96% identity. Conservation of this sequence is seen even in mIgM molecules of the shark, which show 58% homology with human mIgM. These features suggest a more diverse function of the Tm sequence than just serving as a membrane anchor of the mIgM molecule. Indeed recent data suggest that the Tm sequence mediates the contact between the mIgM molecule and other transmembrane proteins. For example, the binding of mIgM to the Igα/Igβ heterodimer is disrupted by a mutation that changes four conserved polar amino acids in the Tm sequence into alanine (Williams et al., 1990). The 25 amino acids of the Tm part of mIg are thought to cross the lipid bilayer as an α-helix. Interestingly most of the 13 amino acids conserved in all mIg classes are lying on one side of the α-helix (Fig. 3). As all mIg classes are associated with the same Igα/Igβ heterodimer (Venkitaraman et al., 1991) this conserved side is presumably the interaction surface between these molecules (Reth, 1992). The side opposite to this conserved surface is mostly composed of amino acids specific for the various mIg classes and could form an interaction surface for Tm proteins binding to mIg in a class-specific manner. Such protein specifically interacting with the Tm sequences of either mIgM or mIgD have recently been discovered (Kim et al., 1994).

Although exceptions have been found (see below) most mIg molecules are not transported on the cell surface unless they assemble with the Igα/Igβ heterodimer. The polar amino acids in the Tm sequence seem to be responsible for this transport control as mutation of these amino acids results in mIg surface expression without

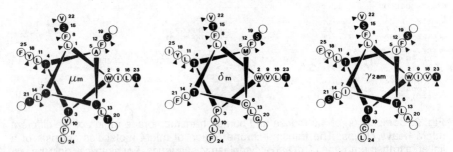

Fig. 3. Comparison of the transmembrane sequence of μ_m, δ_m and $\gamma 2a_m$ given as an α-helical scheme. Polar amino acids are given as white letters on a black background and amino acids conserved in all mIgH classes are marked by black arrowheads. Numbers indicate the positions of the amino acids in the sequence. (Modified from Reth, 1992.)

Igα/Igβ heterodimer (Williams et al., 1990). Apparently a 'detector' in the ER membrane can recognize these unpaired or unsheathed polar amino acids and this results in the retention and degradation of the mIg molecule (Klausner and Sitia, 1990). The molecules and the mechanism involved in this retention of unpaired Tm proteins in the ER are presently unknown. In contrast to mIgM, the mIgD and mIgG molecules are not arrested in the ER and can appear on the cell surface also without the Igα/Igβ heterodimer (Williams et al., 1993; Weiser et al., 1994; Williams et al., 1994). How these mIg classes are shielding their polar Tm amino acids and are escaping the retention control is not known. This could be due to mIg aggregation or to the binding to membrane molecules different from the Igα/Igβ heterodimer.

THE IGα/IGβ HETERODIMER

The Igα and Igβ proteins on mouse B cells are classical type I Tm proteins of 34 kDa and 39 kDa consisting of 220 and 228 amino acids, respectively (Table I). According to the CD nomenclature these proteins are also referred to as CD79a and CD79b. Each of them carry a leader sequence that is cleaved off upon entry of these proteins into the ER (Fig. 4). Murine Igα and Igβ have an extracellular sequence of 109 and 129 amino acids, respectively. These sequences show all the hallmarks of proteins belonging to the Ig superfamily (Williams and Barclay, 1988). Both sequences contain two conserved cysteine residues forming the intra-domain disulphide bond and the conserved tryptophan residue, which is important in filling the hydrophobic space inside an Ig domain (see bold-type amino acids in Fig. 4). Murine Igα and Igβ are glycoproteins and their extracellular domains contain two and three N-linked glycosylation sites respectively. Most of these N-linked glycosylation sites are lying on the loops connecting the β-strands of the Ig domain. Human Igα (Flaswinkel and Reth, 1992; Ha et al., 1992; Leduc et al., 1992; Yu and Chang, 1992) has four

Table I Characteristics of Igα and Igβ

Protein	kDa	Amino acids	N-linked CHO	Accession No.	Isoelectric point	Chromosome*
Murine Igα	34	220 (192)	2	MMMB1.Em–Ro (X13450)	4.71	9?
Human Igα	47	226 (194)	6	HSIGMAMB.Em–Pr (M74721)	4.78	19q13
Murine Igβ	39	228 (198)	3	MMB29.Em–Ro (J03857)	5.78	ND
Human Igβ	37	229 (199)	3	HSB29.Em–Pr (M80461)	6.05	17q23

* Chromosomal assignments are taken from Wood et al. (1993) and Hashimoto et al. (1993). The chromosomal assignment of the murine Igα is based on the isogenic organization of the mouse and human genome, and is therefore preliminary.
ND, not determined.

(a)
```
         <----------- leader ------------>1                                  .
    -32  MPGGPGVLQALPATIFLLFLLSAVYLGPGCQALWMHKVPASLMVSLGEDA                  human
         ||||  :.|.|||      ||::||  .:|||||||||::.  .|:||  |.|||:|
    -28  MPGGLEALRALP....LLLFLSYACLGPGCQALRVEGGPPSLTVNLGEEA                  mouse

                       . Δ              .              .            .
     19  HFQCPHNSSNNANVTWWRILHGNYTWPPEFLGPGEDPNGTLIIQNVNKSH                  human
         ::  |  .|.:.|:|:|||    |::|.||||  `||||::..|  |:::.:|||.|
     19  RLTC.ENNGRNPNITWWFSLQSNITWPPVPLGPGQGTTGQLFFPEVNKNH                  mouse

                   Δ     .     ◊       .              .      <------
     69  GGIYVCRVQEGNESYQQSCGTYLRVRQPPPRPFLDMGEGTKNRIITAEGI                  human
         |:|.|.|  |  |:  ...|||||||||.|.||||||||||||||||||||
     68  RGLYWCQVIE.NNILKRSCGTYLRVRNPVPRPFLDMGEGTKNRIITAEGI                  mouse

         --- tm (22) -->          .              .              .
    119  ILLFCAVVPGTLLLFRKRWQNEKLGLDAGDEYEDENLYEGLNLDDCSMYEDI                human
         ||||||||||||||||||||||||||:|:|  .|:||||||||||||||||||||
    117  ILLFCAVVPGTLLLFRKRWQNEKFGVDMPDDYEDENLYEGLNLDDCSMYEDI                mouse
                                          YxxL          YxxI

                      .              .
    171  SRGLQGTYQDVGSLNIGDVQLEKP   194                                     human
         ||||||||||||||.|:|||.|||
    169  SRGLQGTYQDVGNLHIGDAQLEKP   192                                     mouse
```

(b)
```
         <----------- leader ---------->1     . Δ             .
    -30  MARLALSPVPSHWMVALLLLLSAEPVPAARSEDRYRNPKGSACSRIWQSP                  human
         ||  |.||.:|:||::  ||||:|:|||||  |.|   |  .||:||.|||  |
    -30  MATLVLSSMPCHWLLFLLLLFSGEPVPAMTSSDLPLNFQGSPCSQIWQHP                  mouse

                      . Δ              .              .
     21  RFIARKRRFTVKMHCYMNSASGNVSWLWKQEMDENPQQLKLEKGRMEESQ                  human
         ||  |:||.    ||:|||  |    ||.:.|::|.:  .:.||:|  |.||  :.|
     21  RFAAKKRSSMVKFHCYTNH.SGALTWFRKRGSQQ.PQELVSEEGRIVQTQ                  mouse

                          . Δ  Δ   .        ◊        .
     71  NESLATLTIQGIRFEDNGIYFCQQKCNNTS.EVYQGCGTELRVMGFSTLA                  human
         |:|:  |||||.|..:|||||||.|||:...  :|  ::|||||  |:|||||.
     69  NGSVYTLTIQNIQYEDNGIYFCKQKCDSANHNVTDSCGTELLVLGFSTLD                  mouse

              <------ tm (22) ------->           .
    120  QLKQRNTLKDGIIMIQTLLIILFIIVPIFLLLDKDDSKAGMEEDHTYEGL                  human
         |||.||||||||| ||||||||||||||||||||||:|||||||||||||
    119  QLKRRNTLKDGIILIQTLLIILFIIVPIFLLLDKDDGKAGMEEDHTYEGL                  mouse
                                                          YxxL

                       .              .
    170  DIDQTATYEDIVTLRTGEVKWSVGEHPGQE   199                               human
         :||||||||||||||||||||||||||||||
    169  NIDQTATYEDIVTLRTGEVKWSVGEHPGQE   198                               mouse
             YxxI
```

Fig. 4. Comparison of the human and mouse Igα (a) and Igβ (b) protein sequence. Between the sequences a dash indicates identity and a dot similarity of the amino acids. The position of the leader and the Tm sequence is indicated by arrows. Intra-domain cysteines are indicated by a triangle and inter-domain cysteines by a diamond. N-linked glycosylation sites are underlined in the sequence. The position of the TAM is indicated as YxxL or YxxI below the sequence. The mouse sequences are taken from Hermanson et al. (1988) and Sakaguchi et al. (1988) and the human sequences from Flaswinkel and Reth (1992), Ha et al. (1992), Leduc et al. (1992), Müller et al. (1992), Yu and Chang (1992), Hashimoto et al. (1993) and Wood et al. (1993).

additional N-linked glycosylation sites and is thus much larger (47 kDa) than its murine counterpart (Van Noesel et al., 1992). The Ig domains of Igα and Igβ are followed by an extracellular spacer containing a cysteine, which presumably mediates the disulphide bond between the two proteins (marked as diamonds in Fig. 4). The α/β heterodimerization may not only be mediated by the Ig domains but also by the extracellular linker sequences of Igα and Igβ, which contain complementary charged amino acids.

The 22 amino acid Tm sequences of Igα and Igβ show some similarity to each other (32% identity). Most remarkable is a negatively charged amino acid at position 5 of the Tm sequence of Igα. The Tm sequence of the ζ chain of the TCR carries a negative charge at the same position. This amino acid position may play a role in the dimerization of the ζ homodimer and the Igα and Igβ heterodimer.

The cytoplasmic parts of Igα and Igβ are 61 and 48 amino acids, respectively. These sequences are highly conserved between the mouse and human proteins (> 90% identity) and are dominated by negatively charged amino acids. Igα and Igβ are therefore acidic proteins with isoelectric points of 4.7 and 3.7, respectively. Four tyrosines are found in the cytoplasmic Igα sequence and two tyrosines in the cytoplasmic Igβ sequence. Interestingly two of the tyrosines of Igα and both tyrosines of Igβ belong to a conserved sequence motif (Reth, 1989), which is found not only in the BCR complex but also in components of the TCR and the Fc receptors (Fig. 5).

	Consensus:	xxDxx E	YxxL I	xxxxxxx	YxxL I
1	CD3-γ	QNEQL	YQPL	KDREYDQ	YSHL
2	CD3-δ	KNEQL	YQPL	RDREDTQ	YSSL
3	CD3-ε	VPNPD	YEPI	RKGQRDL	YSGL
4	ζ-a	DPNQL	YNEL	NLGRREE	YDVL
5	ζ-b	PQEGV	YNAL	QKDKMAEA	YSEI
6	ζ-c	GHDGL	YQGL	STATKDT	YDAL
7	FcεRI-γ	KADAV	YTGL	NTRSQET	YETL
8	FcεRI-β	PDDRL	YEEL	NVYSPI.	YSAL
9	BCR/Ig-α	EDENL	YEGL	NLDDCSM	YEDI
10	BCR/Ig-β	EEDHT	YEGL	NIDQTAT	YEDI
11	BLV-gp30-1	KPDSD	YQAL	LPSAPEI	YSHL
12	BLV-gp30-2	SAPEI	YSHL	SPTKPD.	YINL
13	EBV-LMP2A	DRHSD	YQPL	GTQDQSL	YLGL
14	γ2a	TISPD	YRNM	IGQGA	

Fig. 5. Occurrence of the tyrosine activation motif (TAM) in components of the T cell receptor, FcεRI receptor and the B cell receptor as well as in two viral transmembrane proteins (gp30 of BLV and LMP2A of EBV). The conserved tyrosine and leucine/isoleucine are boxed.

This sequence motif consists of two tyrosines and two leucines in precise spacing to each other (YxxLxxxxxxxYxxL) and often is accompanied by two N-terminal negatively charged amino acids. This motif is known under several different names, such as TAM (tyrosine-based activation motif; Klausner and Samelson, 1991), ARAM (antigen receptor activation motif; Weiss, 1993), ARH1 (antigen receptor homology 1; Clark *et al.*, 1992) and YxxL motif (Wegener *et al.*, 1992). It is also found in two viral Tm proteins (Alber *et al.*, 1993; Beaufils *et al.*, 1993). A mutational analysis of this motif shows that the tyrosines as well as the leucines are required for signal transduction via the different receptors (Letourneur and Klausner, 1992; Flaswinkel and Reth, 1994). All classes of mIg are associated with the same Igα/Igβ heterodimer (Venkitaraman *et al.*, 1991). A model of the BCR complex (Reth, 1992) suggests that due to its bilateral symmetry each mIg molecule is bound by two Igα/Igβ molecules (Fig. 6). The binding between mIg and the Igα/Igβ heterodimer requires the presence of the last (membrane-proximal) C domain and the Tm part of the mIg molecule (Hombach *et al.*, 1990).

Igα and Igβ are encoded by the B cell-specific genes *mb*-1 (Sakaguchi *et al.*, 1988) and *B29* (Hermanson *et al.*, 1988), respectively. The promotors of these genes have no TATA boxes but an initiation region that directs initiation of transcription from multiple sites (Hagman *et al.*, 1991; Travis *et al.*, 1991; Omori and Wall, 1993). However, these promotors carry such elements as an octamer, an *ets* box and E3-like-box also known from other lymphocyte-specific genes. The *mb*-1 promotor carries

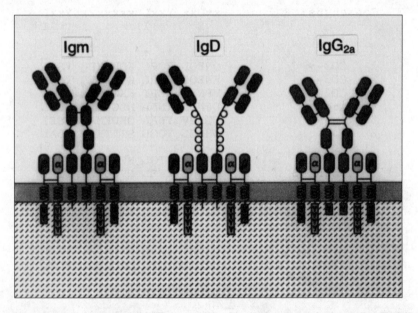

Fig. 6. Structural model of the B cell antigen receptor of class IgM, IgD and IgG$_{2a}$. Note that all classes of immunoglobulin are associated with the same Igα/Igβ heterodimer. Each Ig domain is indicated by an oval and the tyrosine in the cytoplasmic sequence is indicated by a Y.

the DNA element 5'-CAAGGGAAT-3' called BlyF, which is necessary and sufficient to direct its tissue-specific expression (Hagman et al., 1991; Feldhaus et al., 1992). This element is bound by a novel nuclear factor, termed early B cell factor (EBF), that is specifically expressed in pre-B and B lymphocytes (Hagman et al., 1993). A mutational analysis of the cloned EBF cDNA identified the amino-terminal cysteine-rich domain of EBF as a novel DNA-binding site. This factor may be responsible for the tissue-specific expression of the mb-1 gene (Verschuren et al., 1993).

INTERACTION OF THE IGα/β HETERODIMER WITH PTK

The Igα and Igβ heterodimer is competent to transduce signals and thus seems to be the signalling subunit of the BCR. Indeed the cytoplasmic sequences of Igα or Igβ alone can already transduce signals. This can be demonstrated in experiments with chimeric receptors (Irving and Weiss, 1991; Romeo and Seed, 1991) carrying the sequence of CD8 extracellularly and in the membrane and either Igα or Igβ intracellularly (Kim et al., 1993a). Expression of these chimeric receptors on the B cell surface and cross-linking by anti-CD8 antibodies results in PTK activation and calcium mobilization. Similarly, chimeric mIgM molecules in which the cytoplasmic part of either Igα or Igβ is covalently attached to the C terminus of the μ_m heavy chain can signal in B cells (Sanchez et al., 1993, Law, 1993). In some of these chimeric mIgM molecules the hydrophobic amino acids of the Tm part have been mutated to allow their surface expression without association with the endogenous Igα/Igβ heterodimer.

The activation of PTKs is the first event seen after the cross-linking of the complete BCR (Campbell and Sefton, 1990). A kinetic study of B cell activation demonstrates a rapid increase in substrate phosphorylation 15–30 s after stimulation of the BCR (Kim et al., 1993b). Two different types of PTK become phosphorylated and presumably activated via the BCR (Fig. 7). These are the src-related PTKs lyn, blk, fyn and lck (Rudd et al., 1993) as well as the cytoplasmic PTK syk (Taniguchi et al., 1991; Hutchcroft et al., 1992). All B cells express syk but the expression of the four src-related PTKs can vary depending on the developmental stage of the B cells (Law et al., 1992; Kim et al., 1993b). The src-related PTK blk for example is not expressed in myeloma cells whereas most B-cells express lyn. Most substrate proteins of these kinases are unknown at present, but tyrosines in the cytoplasmic sequence of Igα and Igβ are targets of phosphorylation (Gold, et al., 1991).

How and in what order these PTKs become activated via the BCR is still a matter of debate. A threefold interaction between the BCR and the PTKs can be envisioned: first, an interaction before or during receptor aggregation; second, a classical kinase–substrate relationship resulting in the phosphorylation of the Igα and Igβ heterodimer; and third, the binding of the PTKs to the phosphorylated receptor via their SH2 domains. From these possible interactions the first one is the less well defined. If the BCR is purified from lysates of B cells, src-related PTKs are coprecipated although the stoichiometry of this coprecipitation (2–3%) is rather low. Thus either the

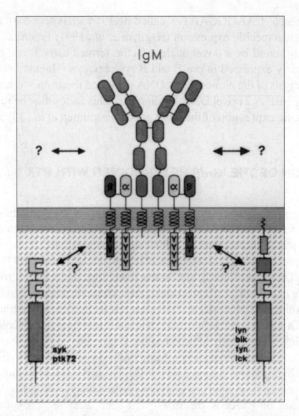

Fig. 7. Possible interaction of the B cell receptor with the *src*-related tyrosine kinases *lyn*, *blk*, *fyn*, *lck*, and the cytoplasmic tyrosine kinase *syk*. Boxes indicate the position of SH3, SH2 and the catalytic domain of the kinases. Arrows with question marks indicate as yet unknown relationships of the BCR to co-receptors on the cell surface or to the intracellular kinase.

BCR–PTK interaction may be of low affinity and not stable in detergent lysates or only a few receptors are in the active conformation allowing binding to PTKs (Clark *et al.*, 1992). Alternatively only PTKs in a pre-activated form may be able to interact with the receptor and the amount of these activated PTKs may be limited in the B cell. Which part of the *src*-related PTKs interact with the BCR and whether the *src*-related PTKs are indeed the first ones to interact with the BCR remain to be established.

On the receptor site, however, it is clear that it is the TAM of Igα, and to a lesser extent that of Igβ, which are involved in PTK activation. Mutations in either of the two tyrosines of the TAM abolish the PTK activation detected after cross-linking chimeric CD8/Igα receptors (Sanchez *et al.*, 1993; Flaswinkel and Reth, 1994). However, mutation of a tyrosine residue outside the TAM does not result in a signalling defect. Mutations of the tyrosines of TAM also hamper signal transduction from the complete BCR.

The cytoplasmic tail of Igα carries four tyrosines that are potential phosphate acceptors. In an *in vitro* kinase assay with bacterial fusion proteins (GST/Igα) and the baculovirus-expressed *src*-related PTKs, *lyn* and *fyn* only the first tyrosine of TAM becomes phosphorylated (Flaswinkel and Reth, 1994). Mutation of this tyrosine also prevents any phosphorylation of Igα normally seen after BCR cross-linking in the myeloma line. These results reveal a very high specificity of *lyn* and *fyn* for certain tyrosines of a substrate protein. Apparently amino acids around the tyrosines must play an important role in determining the efficiency of the substrate–kinase interaction. Preliminary data show that in contrast to the myeloma the second tyrosine of the TAM also becomes phosphorylated in B lymphoma cells. This would suggest that in these cells a second kinase with a different phosphorylation specificity must be active.

Phosphorylated Igα and Igβ are potential binding targets for SH2 domains. Many proteins involved in signal transduction carry one or two SH2 domains (Songyang *et al.*, 1993). Among these are also the *src*-related PTKs with one SH2 domain and *syk* with two SH2 domains. The three-dimensional structure of several SH2 domains has been determined (Booker *et al.*, 1992; Waksman *et al.*, 1992). These structures show that the binding of tyrosine-phosphorylated peptides involves two binding pockets in the SH2 domain, one for the phosphotyrosine and one for the Y+3 amino acid residue. Binding studies with SH2 domains and a library of phosphopeptides show indeed a preference of certain amino acids at the Y+3 position. The SH2 domains of *lyn* and *syk* for example bind preferentially phosphopeptides with a YxxL/I motif (Waksman *et al.*, 1992). As these sequences are part of the TAM the phosphorylated TAM could be a target for SH2 domains of *lyn* and *syk*. Preliminary biochemical data show that GST/*lyn*SH2 and GST/*syk*SH2 fusion proteins can precipitate the phosphorylated Igα/Igβ heterodimer (M. Barner and M. Reth, unpublished data). These fusion proteins, however, also bind other PTK substrate proteins in activated B cells. In particular GST/*lyn*SH2 can bind many different tyrosine-phosphorylated proteins. Other SH2-carrying signalling proteins like SHC may also be able to bind the phosphorylated TAM (Pelicci *et al.*, 1992; Ravichandran *et al.*, 1993). The *in vivo* association of these SH2-carrying molecules with the BCR component remains to be demonstrated.

INTERNALIZATION FUNCTION OF THE RECEPTOR

A second role of the BCR is the specific uptake of bound antigens, which are processed intracellularly and presented as major histocompatibility complex (MHC)-bound peptides on the B cell surface (Lanzavecchia, 1990). The molecular mechanisms of BCR internalization are poorly defined. Bound monovalent antigen is rapidly endocytosed via clathrin-coated pits (Taylor *et al.*, 1971; Lanzavecchia, 1985; Watts *et al.*, 1989), and even the antigen-free BCR seems to be internalized at a high rate (Patel and Neuberger, 1993). After cross-linking of the BCR by antigen, capping (Taylor et al., 1971) occurs, a process that requires energy and is dependent on the association of the mIg with various components of the cytoskeleton (Rosenspire and Choi, 1982; Bourguignon and Bourguignon, 1985). The capped receptor–ligand complexes are also rapidly endocytosed (Braun et al., 1978).

The treatment of B cells with PTK inhibitors like tyrphostin or genistein prevents both signalling and internalization of the BCR (Puré and Tardelli, 1992; Shimo et al., 1993; Shuler and Owen, 1993). This would suggest that the molecular requirements for signalling and internalization are similar. The internalization of the BCR, however, does not require the activation of PTKs. mIgD and mIgG molecules expressed on the B cell surface without an Igα/Igβ heterodimer are efficiently internalized although their aggregation does not result in PTK activation (Weiser et al., 1994).

It has been proposed that the TAM motif is involved in the internalization of the TCR or BCR (Amigorena et al., 1992). Mutant IgM molecules (Patel and Neuberger, 1993), with an exchange of hydrophilic Tm amino acids to hydrophobic amino acids, are not associated with the Igα/Igβ heterodimer and are defective in internalization unless covalently linked to the cytoplasmic part of Igβ. Mutations of tyrosines of the TAM of Igβ, however, do not alter the internalization of the chimeric mIgM/Igβ molecules, demonstrating that the cytoplasmic part of Igβ carries an internalization signal independent of the TAM.

CONCLUSION

Although we have learned a great deal in recent years about the structure of the BCR and about its various interactions with intercellular components, the molecular details of these interactions still remain to be revealed. In particular it will be interesting to determine the different requirements for the signalling and the internalization functions of the BCR. It has been known for a long time that after cross-linking the BCR becomes insoluble because of its connections to cytoskeleton components. How this connection is made, and whether it plays an important role in the internalization function of the BCR, is bound to be an active field of research in the coming years.

REFERENCES

Alber, G., Kim, K.-M., Weiser, P., Riesterer, C., Carsetti, R. and Reth, M. (1993). *Curr. Biol.* **3**, 333–339.
Alberini, C.M., Bet, P., Milstein, C. and Sitia, R. (1990). *Nature* **347**, 485–487.
Alt, F.W., Bothwell, A., Knapp, M., Siden, E., Mather, E., Koshland, M. and Baltimore, D. (1980). *Cell* **20**, 293–301.
Amigorena, S., Salamero, J., Davoust, J., Fridman, W.H. and Bonnerot, C. (1992). *Nature* **358**, 337–341.
Beaufils, P., Choquet, D., Mamoun, R.Z. and Malissen, B. (1993). *EMBO J.* **12**, 5105–5112.
Booker, G.W., Breeze, A.L., Downing, A.K., Panayotou, G., Gout, I., Waterfield, M.D. and Campbell, I.D. (1992). *Nature* **358**, 684–687.
Bourguignon, L.A.W. and Bourguignon, G.J. (1985). *Int. Rev. Cytol.* **87**, 195–208.

Braun, J., Fujiwara, K., Pollard, T.D. and Unanue, E.R. (1978). *J. Cell Biol.* **79**, 409–414.
Campbell, M.A. and Sefton, B.M. (1990). *EMBO J.* **9**, 2125–2132.
Clark, M.R., Campbell, K.S., Kazlauskas, A., Johnson, S.A., Hertz, M., Potter, T.A., Pleiman, C. and Cambier, J.C. (1992). *Science* **258**, 123–126.
DeFranco, A.L. (1993). *Annu. Rev. Cell Biol.* **9**, 377–410.
Feldhaus, A.L., Mbangkollo, D., Arvin, K.L., Klug, C.A. and Singh, H. (1992). *Mol. Cell. Biol.* **12**, 1126–1133.
Flaswinkel, H. and Reth, M. (1992). *Immunogenetics* **36**, 266–269.
Flaswinkel, H. and Reth, M. (1994). *EMBO J.* **13**, 83–89.
Gold, M.R. and DeFranco, A.L. (1994). *Adv. Immunol.* **55**, 221–295.
Gold, M.R., Matsuuchi, L., Kelly, R.B. and DeFranco, A.L. (1991). *Proc. Natl Acad. Sci. USA* **88**, 3634–3638.
Ha, H.J., Kubagawa, H. and Burrows, P.D. (1992). *J. Immunol.* **148**, 1526–1531.
Hagman, J., Travis, A. and Grosschedl, R. (1991). *EMBO J.* **10**, 3409–3417.
Hagman, J., Belanger, C., Travis, A., Turck, C.W. and Grosschedl, R. (1993). *Genes Dev.* **7**, 760–773.
Hashimoto, S., Gregersen, P.K. and Chiorazzi, N. (1993). *J. Immunol.* **150**, 491–498.
Hermanson, G.G., Eisenberg, D., Kincade, P.W. and Wall, R. (1988). *Proc. Natl Acad. Sci. USA* **85**, 6890–6894.
Hombach, J., Leclercq, L., Radbruch, A., Rajewsky, K. and Reth, M. (1988b). *EMBO J.* **7**, 3451–3456.
Hombach, J., Sablitzky, F., Rajewsky, K. and Reth, M. (1988a). *J. Exp. Med.* **167**, 652–657.
Hombach, J., Tsubata, T., Leclercq, L., Stappert, H. and Reth, M. (1990). *Nature* **343**, 760–762.
Hutchcroft, J.E., Harrison, M.L. and Geahlen, R.L. (1992). *J. Biol. Chem.* **267**, 8613–8619.
Irving, B.A. and Weiss, A. (1991). *Cell* **64**, 891–901.
Kehry, M.S., Ewald, S., Douglas, R., Silbey, C., Raschke, W., Frambrough, D. and Hood, I. (1980). *Cell* **21**, 393–400.
Kim, K.-M., Alber, G., Weiser, P. and Reth, M. (1993b). *Immunol. Rev.* **132**, 125–146.
Kim, K.M., Alber, G., Weiser, P. and Reth, M. (1993a). *Eur. J. Immunol.* **23**, 911–916.
Kim, K.-M., Adachi, T., Nielsen, P.J., Terashima, M., Lamers, M.C., Köhler, G. and Michael, M. (1994). *EMBO J.*, **13**, 3793–3800.
Klausner, R.D. and Samelson, L.E. (1991). *Cell* **64**, 875–878.
Klausner, R.D. and Sitia, R. (1990). *Cell* **62**, 611–614.
Lanzavecchia, A. (1985). *Nature* **314**, 537–541.
Lanzavecchia, A. (1990). *Annu. Rev. Immunol.* **8**, 773–793.
Law, D.A., Chan, V.W.F., Datta, S.K. and DeFranco, A. (1993). *Current Biology* **3**, 645–657.
Law, D.A., Gold, M.R. and DeFranco, A.L. (1992). *Mol. Immunol.* **29**, 917–926.
Leduc, I., Preud'homme, J.L. and Cogné, M. (1992). *Clin. Exp. Immunol.* **90**, 141–146.
Letourneur, F. and Klausner, R.D. (1992). *Science* **255**, 79–82.
Müller, B., Cooper, L. and Terhorst, C. (1992). *Eur. J. Immunol.* **22**, 1621–1625.
Omori, S.A. and Wall, R. (1993). *Proc. Natl Acad. Sci. USA* **90**, 11723–11727.
Patel, K.J. and Neuberger, M.S. (1993). *Cell* **74**, 939–946.
Pelicci, G., Lanfrancone, L., Grignani, F., McGlade, J., Cavallo, F., Forni, G., Nicoletti, I., Grignani, F., Pawson, T. and Pelicci, P.G. (1992). *Cell* **70**, 93–104.
Puré, E. and Tardelli, L. (1992). *Proc. Natl Acad. Sci. USA* **89**, 114–117.
Ravichandran, K.S., Lee, K.K., Songyang, Z., Cantley, L.C., Burn, P. and Burakoff, S.J. (1993). *Science* **262**, 902–905.
Reth, M. (1989). *Nature* **338**, 383.
Reth, M. (1992). *Annu. Rev. Immunol.* **10**, 97–121.
Romeo, C. and Seed, B. (1991). *Cell* **64**, 1037–1046.
Rosenspire, A.L. and Choi, Y.S. (1982). *Mol. Immunol.* **19**, 1515–1521.
Rudd, C.E., Janssen, O., Prasad, K.V., Raab, M., da, S.A., Telfer, J.C. and Yamamoto, M. (1993). *Biochim. Biophys. Acta* **1155**, 239–266.
Sakaguchi, N., Kashiwamura, S., Kimoto, M., Thalmann, P. and Melchers, F. (1988). *EMBO J.* **7**, 3457–3464.

Sanchez, M., Misulovin, Z., Burkhardt, A.L., Mahajan, S., Costa, T., Franke, R., Bolen, J.B. and Nussenzweig, M. (1993). *J. Exp. Med.* **178**, 1049–1055.
Shimo, K., Gyotoku, Y., Arimitsu, Y., Kakiuchi, T. and Mizuguchi, J. (1993). *FEBS Lett.* **323**, 171–174.
Shuler, R.L. and Owen, C.S. (1993). *Immunol. Cell Biol.* **71**, 1–11.
Sitia, R., Neuberger, M.S. and Milstein, C. (1987). *EMBO J.* **6**, 3969–3977.
Sitia, R., Neuberger, M., Alberini, C., Bet, P., Fra, A., Valetti, C., Williams, G. and Milstein, C. (1990). *Cell* **60**, 781–790.
Songyang, Z., Shoelson, S.E., McGlade, J., Olivier, P., Pawson, T., Bustelo, X.R., Barbacid, M., Sabe, H., Hanafusa, H., Yi, T. *et al.* (1993). *Cell* **72**, 767–778.
Tanaka, K., Watakabe, A. and Shimura, Y. (1994). *Mol. Cel. Biol.* **14**, 1347–1354.
Taniguchi, T., Kobayashi, T., Kondo, J., Takahashi, K., Nakamura, H., Suzuki, J., Nagai, K., Yamada, T., Nakamura, S. and Yamamura, H. (1991). *J. Biol. Chem.* **266**, 15790–15796.
Taylor, R.B., Dufus, W.P.H., Raff, M.C. and De Petris, S. (1971). *Nature* **233**, 225–229.
Travis, A., Hagman, J. and Grosschedl, R. (1991). *Mol. Cell. Biol.* **11**, 5756–5766.
Van Noesel, C., Brouns, G.S., van, S.G., Bende, R.J., Mason, D.Y., Borst, J. and Van Lier, R.A. (1992). *J. Exp. Med.* **175**, 1511–1519.
Venkitaraman, A.R., Williams, G.T., Dariavach, P. and Neuberger, M.S. (1991). *Nature* **352**, 777–781.
Verschuren, M.C., Comans, B.W., Kapteijn, C.A., Mason, D.Y., Brouns, G.S., Borst, J., Drexler, H.G. and van Dohgen, J.J. (1993). *Leukemia* **7**, 1939–1947.
Waksman, G., Kominos, D., Robertson, S.C., Pant, N., Baltimore, D., Birge, R.B., Cowburn, D., Hanafusa, H., Mayer, B.J., Overduin, M. and *et al.* (1992). *Nature* **358**, 646–653.
Watts, C., West, M.A., Reid, P.A. and Davidson, H.W. (1989). *Cold Spring Harbor Symp. Quant. Biol.* **1**, 345–352.
Wegener, A.M., Letourneur, F., Hoeveler, A., Brocker, T., Luton, F. and Malissen, B. (1992). *Cell* **68**, 83–95.
Weiser, P., Riesterer, C. and Reth, M. (1994). *Eur. J. Immunol.* **24**, 665–671.
Weiss, A. (1993). *Cell* **73**, 209–212.
Williams, A. and Barclay, A.N. (1988). *Annu. Rev. Immunol.* **6**, 381–405.
Williams, G.T., Venkitaraman, A.R., Gilmore, D.J. and Neuberger, M.S. (1990). *J. Exp. Med.* **171**, 947–952.
Williams, G.T., Dariavach, P., Venkitaraman, A.R., Gilmore, D.J. and Neuberger, M.S. (1993). *Mol. Immunol.* **30**, 1427–1432.
Williams, G.T., Peaker, C.J., Patel, K.J. and Neuberger, M.S. (1994). *Proc. Natl Acad. Sci. USA* **91**, 474–478.
Wood, W.J., Thompson, A.A., Korenberg, J., Chen, X.N., May, W., Wall, R. and Denny, C.T. (1993). *Genomics* **16**, 187–192.
Yu, L.M. and Chang, T.W. (1992). *J. Immunol.* **148**, 633–637.
Yuan, D., Dang, T. and Sanderson, C. (1990). *J. Immunol.* **145**, 3491–3496.

PART II: ORGANIZATION AND REARRANGEMENT OF IMMUNOGLOBULIN GENES

7

Immunoglobulin heavy chain loci of mouse and human

Tasuku Honjo and Fumihiko Matsuda

Department of Medical Chemistry, Faculty of Medicine and Center for Molecular Biology and Genetics, Kyoto University, Sakyo-ku, Kyoto 606, Japan

Introduction 145
Human V_H locus 146
Mouse V_H locus 156
C_H locus 164

INTRODUCTION

The immunoglobulin (Ig) molecule is composed of the heavy (H) and light (L) chains, both of which consist of the variable (V) and constant (C) regions. The V region is responsible for antigen binding whereas the C_H region specifies the isotype of Ig. Genes encoding IgH V regions are split into V_H, D_H and J_H segments, each of which is comprised of multiple copies. One each of the three segments is generally assembled into a functional V_H gene by a somatic genetic event called VDJ recombination. The V_H locus that contains these gene segments is located on human chromosome 14q32.33 (Croce *et al.*, 1979; Kirsch *et al.*, 1982) or mouse chromosome 12 (D'eustachio *et al.*, 1980). There are nine and eight C_H genes in the human and murine C_H loci, respectively. The cluster of C_H genes is referred to as the C_H locus. The V_H and C_H loci are tightly linked on the chromosome. The distance between the V_H locus segment nearest the 3' end (J_H) and the C_H gene nearest the 5' end (C_μ) is less than 8 kb in both mouse and human (Liu *et al.*, 1980; Ravetch *et al.*, 1981). Recent studies have shown that the human V_H and C_H loci span at least 1.1 megabases (Mb) (Matsuda *et al.*, 1993; Cook *et al.*, 1994) and 0.3 Mb (Bottaro *et al.*, 1989; Hofker *et al.*, 1989), respectively. Thus the IgH locus combining the V_H and C_H loci constitutes a huge multigene family. The aim of this chapter is to summarize the current knowledge of the organization and structure of the mouse and human IgH loci and to discuss its biological significance and implications.

There are a number of reasons why the complete elucidation of the IgH locus is important to immunologists as well as geneticists. From a geneticist's point of view it is fascinating to elucidate how the multigene family has evolved and how a large number of V_H segments are maintained. These questions may be answered more clearly by direct comparison of the IgH loci between related species. Complete knowledge of the organization and structure of the germline V_H segments may provide answers to a number of questions essential to Ig repertoire formation and Ig expression. Obviously, the total number of V_H segments determines the upper limit of the germline Ig repertoire, although somatic genetic events including VDJ recombination and hypermutation amplify the expressed repertoire tremendously. Immunologists are interested in polymorphic variation of the number and repertoire of germline V_H segments and C_H genes and in the association of such polymorphisms with disease susceptibility. It is known that some V_H segments are overrepresented, suggesting that V_H usage may not be random. It is important to know whether the germline organization or structure of V_H segments have anything to do with such a biased usage. Isolation of the total human V_H segments has played important roles for generation of human Ig in J segment-disrupted mice carrying human Ig mini loci (Green et al., 1994; Taylor et al., 1994). Needless to say, studies on the complete organization of the C_H locus are the basis for understanding the molecular mechanism of class switching (see Chapter 11).

Since the publication of the first edition of this book (1989), a major breakthrough was made in elucidation of the human IgH locus by completion of the physical mapping of human V_H segments using yeast artificial chromosome (YAC) clones. Organization and structure of human V_H segments have been extensively studied, providing a tremendously useful reference to map expressed V_H genes, and their polymorphisms. Unfortunately, however, little progress has been made in the study of the mouse V_H locus, though comparison of V_H organization between mouse and human would be extremely interesting.

HUMAN V_H LOCUS

V_H subgroups and families

Human V_H regions were divided into three subgroups based on amino acid sequences (reviewed in Kabat et al., 1991). These protein subgroups have been further subdivided into six distinct V_H families defined by nucleotide sequence homology; V_H segments that show 80% or greater similarity are considered to be in the same family while V_H segments that have less than 70% similarity to one another form different V_H families (Kodaira et al., 1986; Lee et al., 1987; Shen et al., 1987; Berman et al., 1988). Such criteria have been supported by construction of the phylogenetic tree of 33 functional V_H segments located in the 3' 0.8-Mb region of the V_H locus (Fig. 1) (Haino et al., 1994). According to this phylogenetic tree human V_H segments first

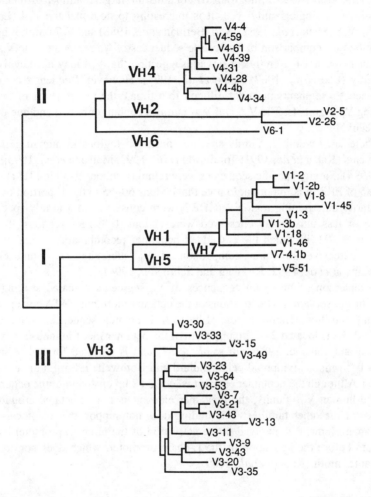

Fig. 1. Phylogenetic tree of human germline functional V_H segments. (Modified from Haino et al., 1994.)

diverged into two groups: subgroup II and ancestor of subgroup I and III. Subgroup II was then divided into V_H2, V_H4 and V_H6 families. Subgroup I was split into V_H1 and V_H5 families. A unique set of V_H segments, which share high homology (78–82%) with V_H1 but differ from V_H1 at a clustered region between framework 2 (FR2) and FR3, has been proposed to be classified as V_H7 family (Schroeder et al., 1990). According to the above definition of the V_H family, V_H7 should be a subfamily of V_H1 or a family captured in transition from V_H1 to independence (Kirkham and

Schroeder, 1994). Nonetheless, the classification of V_H7 is useful as at least six members of V_H7 have been found and mapped at dispersed positions in the V_H locus (van Dijk *et al.*, 1993). Subgroup III contains the largest number of members, yet constitutes a single family, V_H3. It is interesting to note that the V_H4 (Lee *et al.*, 1987), V_H5 (Shen *et al.*, 1987), V_H6 (Berman *et al.*, 1988) and V_H7 families have been identified by comparison of nucleotide sequences of V_H segments. The V_H4 family members are most strongly conserved, suggesting that V_H4 may have evolved most recently (Lee *et al.*, 1987; Haino *et al.*, 1994). However, frequent recombination between V_H segments makes it difficult to estimate the precise time of divergence among V_H segments. The V_H5 and V_H6 families contain only two and one members, respectively.

There are several V_H family-specific conserved regions in human germline V_H segments (Kabat *et al.*, 1991; Tomlinson *et al.*, 1992; Matsuda *et al.*, 1993; Haino *et al.*, 1994). Family-specific sequences were found in codons 9–30 in FR1 and codons 60–85 of FR3. It is important to note that codons 60–65 in the 3′ portion of complementarity determining region 2 (CDR2) were conserved in a family-specific way. More or less universally conserved were codons 1–8, FR2 (codons 38–47) and codons 86–92, in which the embedded heptamer recombination signal is located. More extensive structural comparison of V_H subregions is found elsewhere (Tomlinson *et al.*, 1992; Kirkham and Schroeder, 1994).

Comparison of upstream sequences of V_H segments revealed striking family-specific conservation. The locations of the octamer motif and TATA box are different among families (Haino *et al.*, 1994). A heptamer sequence with consensus CTCATGA is located 2–22 bp upstream of the octamer motif in mouse V_H segments (Eaton and Calame, 1987; Siu *et al.*, 1987) and is required for full V_H promoter activity in mouse lymphoid cells (Ballard and Bothwell, 1986; Eaton and Calame, 1987). Although the heptamer element is located 2 bp upstream of the octamer motif of the human V_H1 family, the heptamer element is not detectable around similar places of the other families. This finding does not support the hypothesis that the heptamer element is involved in the activation of the H-chain promoter by the *oct* protein before the activation of the L-chain promoter, which does not contain the heptamer motif (Kemler *et al.*, 1989).

Physical mapping of human V_H segments

Studies on physical mapping of the human V_H locus were initiated by cosmid cloning (Kodaira *et al.*, 1986). Distribution of V_H families on 23 cosmid clones with average size 40 kb has shown that members of different V_H families are interspersed, in contrast to the finding that the same family members tend to cluster in the mouse V_H locus (Kemp *et al.*, 1981; Rechavi *et al.*, 1982). Another important conclusion from early studies on physical mapping using phage and cosmid vectors is the presence of abundant pseudogenes (about 40%), many of which are highly conserved with only a few point mutations (Givol *et al.*, 1981; Kodaira *et al.*, 1986). A similar type of analysis mapped the D3 segment only 22 kb upstream of the J_H cluster (Buluwela et al., 1988; Matsuda et al., 1988) and the V_H6 segment 20 kb upstream to the D4 segment (Buluwela and Rabbitts, 1988; Sato et al., 1988; Schroeder et al., 1988).

A more general overview of the whole human V_H locus has been provided by studies using pulsed field gel electrophoresis (PFG). Human DNA digested with rare restriction site enzymes was separated by PFG and hybridized with various V_H family-specific probes. Such an analysis allowed the V_H content to be examined on DNA fragments of a few hundred to one thousand kilobases. The total size of the human V_H locus was estimated to be about 2.5–3.0 Mb (Berman et al., 1988; Matsuda et al., 1988) including the D5-hybridizing fragments that later mapped to chromosome 15. Such studies also confirmed the previous conclusion that human V_H families are intermingled. PFG analysis using two-dimensional electrophoresis has provided a more precise determination of the total V_H locus of about 1.2 Mb, on which 76 human V_H segments were mapped using SfiI, BssHII and NotI digests (Walter et al., 1990). Although the precise location of each V_H segment cannot be determined and some V_H segments were inevitably missed, this study has made a great contribution to defining the overall organization of the human V_H locus. The same group further refined the mapping using the deletion profile of V_H segments associated with VDJ recombination in human B cell lines (Walter et al., 1991a).

Introduction of the YAC vector has been essential to complete the physical mapping of the human V_H locus (Fig. 2). The first report using YAC cloning identified and located five V_H segments proximal to the D_H segments (Shin et al., 1991). These authors proposed to rename all the V_H segments by the family number and the order from the 3' end of the V_H locus. For example, V3-36P indicates a V_H3 family member located thirty-sixth from the V_H segment nearest the 3' end, i.e. V6-1. P indicates the pseudogene. An insertional polymorphic V_H segment is indicated by a number with decimal point. This nomenclature of V_H segments was controversial not only because the investigators named V_H segments idiosyncratically but also because many expressed V_H sequences containing somatic mutations could not be easily assigned as different V_H segments. The newly proposed nomenclature defined V_H segments only when they were mapped on the chromosome, which had been expected to be completed in a few years. The same group in Kyoto has completed mapping of about 70% (0.8 Mb) of the human V_H locus by analysing more than seven overlapping YAC clones. All the YAC clones were subcloned into either cosmids, phages or plasmids, and the nucleotide sequences of 64 V_H segments were determined (Matsuda et al., 1993). V_H segments at around 770–740, 710, 555, 430, 360 and 200–100 kb upstream of the J_H segments have the same transcriptional orientation as the J_H segments. The results indicate that there is no gross inversion in the human V_H locus, and that the majority of V_H segments rearrange to associate with the D_H and J_H segments by looping-out but not by inversion, in contrast to the human V_κ locus (see Chapter 8).

Subsequently, Cook et al. (1994) identified the 5' end of the human V_H locus using human telomere activity in yeast and a chromosome translocation that places telomere-proximal V_H segments onto chromosome 8. A 200-kb clone (yIgH6) was isolated and subcloned into cosmids. V_H family-specific primers were used to

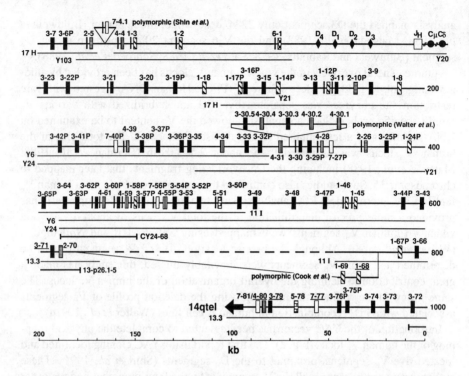

Fig. 2. Organization of the human IgH locus. The 1.1-Mb DNA is shown by six thick horizontal lines with the 3′ end at the top right corner. YAC and cosmid clones covering the region are shown with their names by horizontal lines below. V_H segments belonging to V_H1, 2, 3, 4, 5, 6 and 7 families are indicated by the symbols ◨, ☰, ■, ⊞, ⊟, ☱ and □. Five V_H segments whose sequences have not been determined are underlined.

amplify V_H coding regions and to determine 19 V_H sequences identified on yIgH6. The two V_H segments nearest the 3′ end on yIgH6 were virtually identical to V3-63P and V3-64, which are the V_H segments nearest the 5′ end previously identified by the Kyoto group (Matsuda *et al.*, 1993), thus linking the entire V_H locus physically by YAC clones. The Kyoto group also isolated an independent clone containing the V_H segments nearest the 5′ end and the telomere (F. Matsuda *et al.*, unpublished data). Interestingly, the distance between the telomere sequence and the 5′-end V_H segment (V7-81) is only a few kilobases. The telomere-containing YAC clone (13.3) of the Kyoto group completely overlaps with yIgH6 down to V2-70, confirming the telomeric end of the physical map by the Cambridge group. However, another YAC clone (11I) extending toward the telomeric end from V3-50P contains a striking difference starting from V1-67P. The Kyoto group carried out two-colour *in situ* hybridization of interphase nuclei using cosmid clones as probes and estimated that the gap between CY24-68 and 13-p26.1-5 (Fig. 2) is either 194 ± 27 kb or 109 kb depending on haplotypes. The distance between the same clones is about 90 kb

Table I Summary of the human V_H segments*

Chromosomes (length)	V_H families							Total
	1	2	3	4	5	6	7	
14q32 (1100 kb)	14	4	48	12	2	1	6	87
15q11 (>250 kb)	6	0	1	1	0	0	0	8
16p11 (>700 kb)	4	1	11	0	0	0	0	16
Total (>2 Mb)	24	5	60	13	2	1	6	111

* The number of V_H segments on chromosome 14 is calculated by the results from Matsuda *et al.* (1993) and Cook *et al.* (1994). Seven polymorphic V_H segments reported (Shin *et al.*, 1991; Walter *et al.*, 1993) are included. Information of V_H segments on chromosomes 15 and 16 is taken from Nagaoka *et al.* (1994) and Tomlinson *et al.* (1994).

according to the map of yIgH6. It is clear that there is a large insertion polymorphism (about 90 kb) between V1-67P and V2-70. yIgH6 is likely to represent a haplotype with a large deletion as described (Walter *et al.*, 1990). The physical mapping of the human V_H locus completed by the linkage between yIgH6 and Y6 (and Y24) shows that the total number of V_H segments is 87, although there are several polymorphic insertions or deletions (Table I). The only caveat to the above conclusion is cloning artefacts by YAC cloning. The best way to exclude such artefacts is to isolate independent overlapping clones, which has been done for most of the V_H regions so far analysed.

V_H segment number and polymorphisms

One of the most important goals in the study of the human V_H locus is to determine the total number of V_H segments. Given the almost complete physical map of the human V_H locus, the total number of V_H segments in the smallest haplotype is calculated to be 81 with an additional six or more polymorphic V_H segments (Table I), which is in general agreement with the previous estimation using two-dimensional PFG (Walter *et al.*, 1990) or specific amplification of V_H segments by polymerase chain reaction (PCR) (Tomlinson *et al.*, 1992). Polymorphic insertion (or deletion) appears to cluster at three regions: V4-4/V2-5, V3-30/V4-31 and V1-67P/V2-70. It is interesting to know whether any particular DNA sequences are responsible for these clustered polymorphic deletions (or insertions). A project to determine the entire nucleotide sequence of the human V_H locus is under way by the Kyoto group and will provide an answer to the above question. As previously analysed there are abundant pseudogenes. At least 32 out of 87 mapped V_H segments are pseudogenes; four V_H segments were not found in the expressed V_H database; 46 V_H segments were shown to be used; and the remaining five V_H segments were unsequenced (Matsuda *et al.*, 1993; Cook *et al.*, 1994; Haino *et al.*, 1994). The 46 V_H segments used are classified into 8 V_H1, 3 V_H2, 22 V_H3, 10 V_H4, 1 V_H5, 1 V_H6 and 1 V_H7 family segments.

Restriction fragment length polymorphism (RFLP) and DNA sequencing have shown that there are a number of polymorphic V_H alleles. One of the most polymorphic V_H segments is V1-69 with 13 known alleles including duplication (Sasso et al., 1993). Polymorphism may be of functional significance. One obvious possibility is expansion of repertoire. Polymorphic V_H may affect the affinity of the antibody for its ligand as even mutations in FR residues of the Ig have been shown to influence the binding affinity (Foote and Winter, 1992). Furthermore, expression of particular allelic variants could influence the efficiency of H–L chain pairing or interaction with B cell super-antigens. It is important to test whether V_H polymorphisms are associated with disease susceptibility. Some reports have suggested the association of V_H polymorphisms with autoimmune diseases such as rheumatoid arthritis, systemic lupus erythematosus and multiple sclerosis (Yang et al., 1990; Walter et al., 1991b), while others have reported the absence of a clear association (Hashimoto et al., 1993; Shin et al., 1993a).

Physical mapping of D_H and J_H segments

The human V_H locus ends with a cluster of J_H gene segments lying just upstream of the C_μ gene (Fig. 3). The human J_H cluster contains three pseudo J_H segments interspersed among six functional J_H segments (Ravetch et al, 1981). A human counterpart to the murine DQ52 D_H segment exists about 100 bp 5′ of the first functional J_H (Ravetch et al., 1981). A number of additional human D_H segments have been identified, including ones homologous to the murine DFL16 segments as well as a number of those that are markedly dissimilar in size and sequence (Siebenlist et al., 1981; Schroeder et al., 1987; Buluwela et al., 1988; Ichihara et al., 1988a,b; Zong et al., 1988; Sonntag et al., 1989; Shin et al., 1993b). Initially, a family of D_H segments

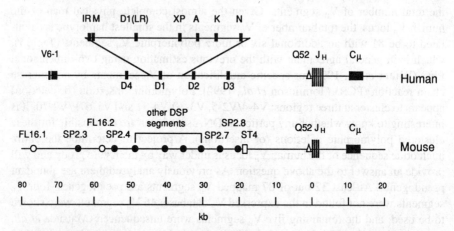

Fig. 3. Comparative map of human and mouse D–J_H loci. Mapping information of human V and D segments is from Siebenlist et al. (1981), Matsuda et al. (1988), Sato et al. (1988) and Ichihara et al. (1988a,b). Mouse part is modified from Feeney and Riblet (1993).

(D1–D4) was shown to be encoded at 9-kb regular intervals between V_H and J_H clusters (Siebenlist et al., 1981). Subsequently, six novel D_H segments (IR, M, XP, A, K and N) were identified by nucleotide sequencing analysis of a 15-kb DNA fragment containing the D1 segment and its surrounding region (Ichihara et al., 1988a,b) (Fig. 3). Each D_H cluster (D1–D4) appears to contain seven D_H segments (Buluwela et al., 1988; Ichihara et al., 1988a,b; Shin et al., 1993b). Taken together, the estimated total number of D_H segments on chromosome 14 would be about 28. Frequent polymorphic deletion of the D1 segment was found in the Japanese population (Zong et al., 1988). The most proximal known D_H segment of a human D_H family (D21/9) is 20 kb upstream of the DQ52/J_H cluster (Buluwela et al., 1988) in a position analogous to that of the proximal DSP2 segment of the mouse. Comparison of the nucleotide sequence of *Suncus murinus*, human and mouse from a position upstream of DQ52 to the S_μ region indicated that D_H and J_H segments, consisting of coding and signal regions, are highly conserved (Okamura et al., 1993). Moreover, although extensive sequence homology in the region between J_H and S_μ was observed between mouse and human, only core portions of the enhancer region of *Suncus murinus* exhibited homology to those of mouse and human. Sequence conservation of J_H segments in *Suncus murinus*, mouse and human was observed not only at the amino acid level but also at the nucleotide level including the third letters of the codons, which is difficult to explain by selection of the protein structure.

V_H and D_H segments on chromosome 15 and 16

Although the V_H locus is located at the telomere end of chromosome 14q, several V_H and two D_H clusters remained unmapped for some time. The first evidence that a D_H segment is located on chromosome 15 was obtained by *in situ* hybridization (Chung et al., 1984). Subsequently, studies using *in situ* hybridization as well as human/rodent somatic hybrid cells (Cherif and Berger, 1990; Matsuda et al., 1990; Nagaoka et al., 1994; Tomlinson et al., 1994) identified two V_H orphon loci on chromosome 15q11 and chromosome 16p11.

Studies on cosmid and YAC clones derived from these orphon loci revealed several striking findings (Matsuda et al., 1990; Nagaoka et al., 1994). First, about 40% of V_H segments in both loci (three out of seven V_H on chromosome 16 and one out of three V_H on chromosome 15) are apparently functional. A totally different approach based on PCR, using somatic cell hybrid DNAs as templates, specifically amplified 24 V_H segments including 10 apparently functional ones on chromosomes 15 and 16 (Tomlinson et al., 1994). Second, putative origins for the orphon V_H segments on chromosomes 15 and 16 were found in the 0.43–0.25 Mb J_H-proximal V_H region on chromosome 14 (Fig. 4). Comparison of the corresponding V_H segments suggests that a DNA fragment of more than 100 kb might have been translocated simultaneously to chromosomes 15 and 16 approximately 20 million years ago. Four overlapping YAC clones covering the V_H orphon locus on chromosome 16 were isolated, and seven V_H segments were identified and sequenced. All of

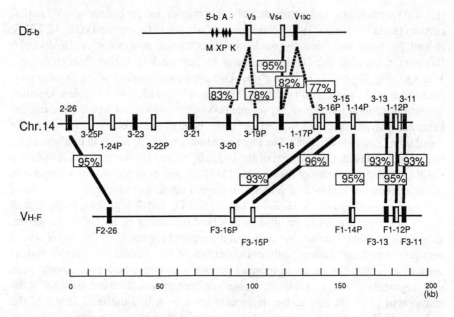

Fig. 4. Comparison of V_H segments on chromosomes 15 and 16 with their counterparts on chromosome 14. (Modified from Nagaoka et al., 1994.)

seven orphon V_H segments have more than 93% identity with the corresponding V_H segments on chromosome 14. The most remarkable homology was found between two truncated pseudogenes, VF1-12P and V1-12P, in which the homology extends into the region 3′ to the truncation site. The homology between the orphon V_H segments on chromosome 15 and the corresponding V_H segments on chromosome 14 is less remarkable except for one pair (V54/V1-18). *In situ* hybridization studies using cosmid clones confirmed that two orphon loci are located on chromosome 15q11–q12 and chromosome 16p11 (Nagaoka et al., 1994; Tomlinson et al., 1994). The orphon locus on chromosome 15 appears to contain at least four clusters of D_H segments, each of which consists of five D_H segments (Matsuda et al., 1990; Nagaoka et al., 1994). One of the D_H clusters (D5-b) is flanked by three V_H segments. Interestingly, these three V_H segments are located 3′ to the D5-b cluster. The polarity of one of them (V3) (Matsuda et al., 1988, 1990; Nagaoka et al., 1994) was determined and shown to have the same transcriptional orientation relative to D_H.

V_H segment usage and repertoire formation

Compelling evidence indicates that V_H, D_H and J_H segments are not used equally. Biased usage of particular segments during early phases of ontogeny was first reported in mouse (Yancopoulos et al., 1984; Reth et al., 1986). Similarly, dominant expression of V1-6, $D_H Q52$ and $J_H 4$ (Berman *et al.*, 1991; Pascual *et al.*, 1993) in early ontogeny was demonstrated in human. One study examined the V_H segments expressed in 14 and 10 independent H-chain cDNA sequences isolated from 130-day and 104-day human fetal liver cDNA libraries, respectively (Schroeder *et al.*, 1987; Schroeder and Wang, 1990). Notably, six of these sequences employed an identical V_H segment of the $V_H 3$ family (56P1 or V3-30), indicating that the early human repertoire is biased.

Utilization of V_H segments may be influenced by a number of factors that can be grouped into (a) those affecting the recombination frequency and (b) those affecting selection of B cells expressing that particular V_H segment. Group (a) includes distance between D_H (or J_H) and V_H segments, variation in the recombination signal sequence, and locations that favour the recombinase accessibility. Group (b) includes self-antigens and bacterial super-antigens.

The initial observation of the preferential usage of J_H-proximal V_H segments in mouse led to the hypothesis that the proximity of V_H segments to J_H favours biased expression of V_H segments in early stages of ontogeny. The complete physical mapping has clearly shown that the location of V_H segments within the locus has little association with the frequency of V_H usage in human. V_H segments often used preferentially in the early stages of ontogeny are V6-1, V1-2, V2-5, V3-13, V3-15, V3-23, V3-30, V5-51, V3-53 and V4-59 segments (Matsuda *et al.*, 1993). The V1-69 segment is also used frequently in peripheral B cells (Schwartz and Stoller, 1994), B-cell leukaemia, and autoantibodies (Zouali, 1992). This V_H segment is highly related to 51P1 cDNA, which is also found in the fetal repertoire and is localized approximately 900 kb upstream of the J_H cluster (Cook *et al.*, 1994). The results indicate that V_H segments preferentially used in early stages of ontogeny do not necessarily cluster in the J_H-proximal region.

Matsuda *et al.* (1993) have identified several germline V_H segments that were often used for autoantibodies, although it is premature to conclude that only limited V_H segments could be used for autoantibodies. The V3-30 sequence was homologous to cDNA for rheumatoid factors, RF-TS2, RF-SJ1 and RF-SJ2 (Pascual *et al.*, 1990). Also, this germline V_H segment was 99.7% identical to cDNA for Kim 1.6 autoantibody (Cairns *et al.*, 1989), which has DNA-binding activity. The V3-30 segment is also the germline counterpart of 56P1 cDNA, which is most frequently expressed in the fetal repertoire (Schroeder *et al.*, 1987; Schroeder and Wang, 1990). Similarly, the V3-15 segment, which is the germline gene of 20P1 cDNA expressed in fetal liver, is 99.7% identical to cDNA for 4B4, an anti-Sm antibody (Sanz *et al.*, 1989). The V3-23 segment is identical to 18/2 (an anti-DNA autoantibody) (Dersimonian *et al.*, 1987) and 30P1 cDNA found in fetal liver (Schroeder *et al.*, 1987; Schroeder and Wang, 1990). Such correlation between autoantibody V_H and early repertoire V_H may indicate that the preferred usage of V_H segments in early stages of ontogeny is due to positive selection by self-antigens rather than J_H-proximal location of the V_H segments.

In any case, the complete physical map of the human V_H locus has contributed to the identification of germline origins of autoantibodies. Comparison of V_H usage among polymorphic individuals may also shed light on mechanisms for biased V_H usage.

Pseudogenes and gene conversion

Are abundant, conserved pseudo V_H and orphon V_H segments of any functional significance? Since some orphon V_H segments are apparently functional, they can be joined, in theory, to J_H segments on chromosome 14 through interchromosomal recombination. As a V_H–D_H fusion product was isolated from a human B-cell line (Shin et al., 1993b), a similar fusion of V_H–D_H can be formed on chromosome 15. In addition, germline transcripts of orphon V_H have been identified in human fetal liver (Cuisinier et al., 1993), suggesting that orphon V_H loci might be targets of recombinase. Unfortunately, however, no direct evidence for the expression or recombination of orphon V_H segments has been reported so far. Conserved pseudogenes have been already shown to serve as sequence donors for gene conversion in other species. Somatic gene conversion (or double unequal crossing-over) has been shown to take place to amplify the V-region repertoire in chicken (Raynaud et al., 1987; Tompson and Neiman, 1987) and rabbit (Becker and Knight, 1990) but not in mouse and human.

To explain the extensive conservation of V_H pseudogenes, evidence for germline gene conversion was looked for among V_H segments located in the 3' 0.8-Mb region (Haino et al., 1994). To screen the candidates of gene conversion events, the substitution rates in the intron and the synonymous position of the coding region were compared in pairs of V_H segments. This is because the introns and the synonymous positions have been shown to evolve at high and remarkably similar rates in different genes (Miyata et al., 1980; Hayashida and Miyata, 1983). In addition, both introns and synonymous positions behave like clocks as they accumulate base substitutions at approximately constant rates with respect to geological time. A clear difference in substitution rates between the intron and synonymous position of a given pair of V_H segments suggests some recombination events.

Haino et al. (1994) selected for comparison several pairs of V_H segments that appear to be recently duplicated. DNA sequence homology of each pair (V3-62P/V3-60P and V4-61P/V4-59) of the tandemly duplicated segments (Kodaira et al., 1986) is greater than 94%, which is much higher than the homology to other published V_H segments belonging to the same V_H family. V3-62P and V3-60P have the same mutation in the heptamer signal sequence and the same 3-bp deletion in the 23-bp spacer of the recombination signal, suggesting that the internal duplication event occurred after these deleterious mutations. When V3-62P and V3-60P are compared using the above evolutionary molecular clock, the nucleotide difference (0.1637 ± 0.0461) at the intron (K^c_i) was significantly greater than that (0.0821 ± 0.0339) at the synonymous position of the coding region (K^c_s), indicating that the segmental change in the intron occurred in either V3-60P or V3-62P. To look for the origin of the modified sequence, intron sequences of V3-60P, V3-62P and other V_H segments were compared. When V3-43 and V3-62P are compared, K^c_i (0.0820 ± 0.0280) was very much smaller than K^c_s (0.4165 ± 0.0761). Nucleotide sequences of the leader, intron and FR1 sequences of V3-43 and V3-62P are highly homologous whereas their 3' halves are diverged. Such unusual homology of the 5' half region, including the intron between V3-43 and V3-62P, is most likely explained by germline gene conversion because the 5' half sequence of V3-62P, which must have been similar to that of the duplicated partner V3-60P, appears to be unidirectionally corrected by V3-43. This example supports the hypothesis that gene conversion contributes to the maintenance of the pseudogene structure. Since many conserved pseudogenes can be either acceptors or donors of germline gene conversion, the high percentage of germline pseudogenes in the V_H segments should also contribute to germline V_H diversity. The authors cannot completely exclude the possibility that double unequal crossing-over took place between V3-43 and V3-60P and a modified allele of V3-43 had been lost.

Evolution of the V_H loci

Almost all animals that have Ig carry multiple V_H segments, indicating that duplication of V_H segments must have started quite a long time ago. It is also important to realize that reorganization of the V_H locus is still continuing as evidenced by dramatic differences in V_H locus organization between mouse and human. The recent translocation of V_H and D_H segments to chromosomes 15 and 16 is further evidence for dynamic reshuffling of the V_H locus.

Matsumura et al. (1994) looked for genetic traits that may allow steps of DNA rearrangement within the human V_H loci to be traced. They have used 14 non-repetitive intergenic probes that can detect two to seven cross-hybridizing bands within the 3' 0.8-Mb region of the V_H locus on chromosome 14. Most of these probes also detected a few bands on chromosome 15 or 16, further confirming the recent translocation of the orphon loci. Such studies identified several pairs of regions that are hybridized by an identical set of non-repetitive probes. Each of five different sets of probes hybridized as clusters to two or three regions in a dispersed manner. In most of the cases, V_H segments adjacent to homologous clusters are closely related. The dispersed appearance of these clusters of non-repetitive sequences indicated that translocation of DNA fragments frequently took place in the human V_H locus. Careful comparison of V_H sequences revealed obvious tandem duplication of sets of V_H segments: V3-33–V4-34/V3-30–V4-28 and V3-62P–V4-61/V3-60P–V4-59 (Kodaira et al., 1986; Matsuda et al., 1993). The longer duplication set is also associated with duplication of non-repetitive probes (Matsumura et al., 1994). Among all these pairs of translocation and duplication none were inverted. Such non-repetitive probes should be useful not only to trace the evolution of the V_H locus but also to investigate polymorphisms of this locus. Extensive RFLP analyses using the probes isolated in this study would give us many markers to test the genetic linkage between RFLPs and susceptibility to immune disorders.

The human genome contains a large fraction of interspersed repetitive sequences such as short interspersed elements (*Alu* repeats) and long interspersed elements (L1 repeats). The total numbers of *Alu* and L1 repeats in the human genome have been estimated to be $3–5 \times 10^5$ and $10^4–10^5$, respectively. Assuming random distribution of these repeats throughout the genome, it is expected that there are 70–110 *Alu* repeats and 2–20 L1 repeats within the 730-kb region analysed in this study. Because recent studies have demonstrated that some repetitive sequences could be hotspots for recombination in the genome (Hyrien *et al.*, 1987; Devlin *et al.*, 1990), the frequent reorganization of the human V_H locus may be associated with the content and distribution of these repetitive sequences.

However, comparison between the homologous (and possibly translocated) V_H segments and distribution of repetitive sequences has shown that flanking regions of closely related V_H segments do not have a similar distribution to *Alu* and L1 repeats and that these V_H surrounding regions are not necessarily abundant in repetitive sequences. These studies failed to provide evidence that the homologous recombination mediated by repetitive sequences might be the main driving force of frequent reorganization of the V_H locus. Most *Alu* repeats were reported to have been amplified within the last 60 million years (Shen *et al.*, 1991). In particular, members of the HS subfamily, a subfamily of human *Alu* repeats, have spread after the divergence of chimpanzee and human (within the last 5 million years) (Shen *et al.*, 1991). Because the closely related V_H segments were estimated to have been generated 55 to 22 million years ago (Matsumura *et al.*, 1994), the most likely explanation in this case is that many *Alu* repetitive elements transposed into random positions after V_H reorganization.

MOUSE V_H LOCUS

V_H subgroup and family

Murine V_H gene segments were originally divided into three major subgroups on the basis of protein sequences (reviewed by Kabat *et al.*, 1991). These three protein subgroups have been further subdivided to yield 14 distinct V_H families based on nucleotide sequence relatedness according to the definition described for the human V_H family (Brodeur and Riblet, 1984; Winter *et al.*, 1985; Kofler *et al.*, 1992). Human subgroups I, II and III are homologous to mouse subgroups II, I and III, respectively (Table II). Human V_H4, V_H2, V_H1 and V_H7 families correspond to mouse 3660, 3609, J558 and VGAM 3–8 families, respectively, based on 70% nucleotide sequence homology as described previously (Lee *et al.*, 1987; Berman *et al.*, 1988; Haino *et al.*, 1994). The human V_H3 family appears to correspond to most of mouse subgroup III families, including 7183, T15, J606, X24, DNA4, CP3 and 3609N. The V_H6 family is closest to the 3660 family. The human V_H5 family does not have any mouse counterparts that are more than 70% homologous. Conversely, the mouse Q52 family does not show more than 66% homology with any of the human V_H families. It is interesting to note that the 7183 family members are more than 80% homologous to human V_H3 family members except for V3-15 and V3-49. The V3-15 and V3-49 sequences are 79 and 82%, respectively, homologous to the V11 sequence of the

S107 family. In addition, almost all human V_H3 family members are more than 70% homologous to mouse subgroup III families. The results indicate that the V_H3 family is more conserved than other families between mouse and human, which could be due to some functional constraint.

Since few physical mapping data are available, the estimation of mouse V_H segments is far from accurate. Initial estimations were based on the count of restriction fragments hybridizing to a given V_H probe. Such estimation assumes that each restriction fragment of 10–20 kb may contain a single V_H and overlapping restriction fragments may be negligible. By this approach, the J558 V_H family consisted of 50–100 or more specific hybridizing fragments (Brodeur and Riblet, 1984; Schiff *et al.*, 1985) whereas other families such as S107 have only a few hybridizing fragments (Crews *et al.*, 1981; Brodeur and Riblet, 1984). These approaches have led to estimates of the size of the murine germline V_H repertoire of approximately 100 or so members (Brodeur and Riblet, 1984). A very rough estimate of members belonging to the 14 known V_H families is shown in Table II. However, this quantitative approach may give an underestimate, as suggested by observations that V_H sequences may sometimes be more closely spaced and by the existence of multiple unique V_H-containing restriction fragments of the same size (Bothwell *et al.*, 1981; Siekevitz *et al.*, 1983; Schiff *et al.*, 1985; Berman *et al.*, 1988; Rathbun *et al.*,

Table II V_H classifications and V_H gene repertoire

Mouse V_H families*	Protein subgroups†	Complexity†	Related human V_H families
Q52 (V_H2)	I	15	
3660 (V_H3)	I	5–8	V_H4, V_H6
3609 (V_H8)	I	7–10	V_H2
CH27 (V_H12)	I	1	V_H4
J558 (V_H1)	II	60–1000	V_H1
VGAM3-8 (V_H9)	II	5–7	V_H7
SM7 (V_H14)	II	3–4	V_H1
X-24 (V_H4)	III	2	
7183 (V_H5)	III	12	
J606 (V_H6)	III	10–12	
S107 (V_H7)	III	2–4	V_H3
MRL-DNA4 (V_H10)	III	2–5	
CP3 (V_H11)	III	1–6	
3609N (V_H13)	III	1	

* V_H gene families 1–7 (Brodeur and Riblet, 1984), 8 and 9 (Winter *et al.*, 1985), 10 (Kofler, 1988), 11 (Reininger *et al.*, 1988), 12 (Pennell *et al.*, 1989), 13 and 14 (Tutter *et al.*, 1991). A prototype member for each family is given with the family number in parentheses. In this study, V_H families 3609N and SM7 have been tentatively termed V_H13 and 14, respectively. The families have been organized into three phylogenetically related groups (Tutter and Riblet, 1988).
† According to Kabat *et al.* (1991).
‡ Estimated number of V_H genes per family in the germline. For references see footnote* and Livant *et al.* (1986), Dzierzak *et al.* (1986), Perlmutter *et al.* (1984), Siu *et al.* (1987) for V_H1, 3, 6 and 7, respectively.

1988). Other hybridization-based approaches have included solution hybridization experiments, which suffer from a lack of ideal kinetics when measuring hybridization to a spectrum of partially related sequences (Livant et al., 1986), and plaque hybridization experiments, which determine the number of clones from a given V_H family relative to single copy sequence clones in a single-genome equivalent of a genomic DNA library (Livant et al., 1986; Berman et al., 1988). The latter approaches have suggested that the murine V_H locus, at least in some strains, may contain 1000 or more members, most of these being contributed by the exceptionally large J558 V_H family, which in BALB/c mice may contain 500–1000 or more members by itself.

While the size of the V_H gene repertoire is relatively conserved between different inbred strains of mice, there are still some important inter-strain differences in certain families, particularly J558 (discussed in Meek et al., 1990). The exact number of members in a particular strain is known for some V_H gene families, and for most others estimates are within a relatively narrow range (Table II). Only the size of the largest family (J558) is still controversial, with estimates varying between 60 (Brodeur and Riblet, 1984) to >1000 (Livant et al., 1986) members. However, since only about 30–50% of adult mitogen-stimulated splenocytes express the J558 family, the actual J558 family size might be closer to 60 than to 1000. Alternatively (or in addition), there may be multiple non-functional or essentially identical J558 family segments, which would explain the relative under-representation of this family in the expressed repertoire.

V_H segment organization

Murine V_H families are generally organized into clusters of related V_H genes. The suggestion that mouse V_H segments belonging to the same family are clustered was first made by Kemp et al. (1981) after they isolated distinct genomic DNA clones, each bearing pairs of highly related V_H sequences (S107 and J606). Bothwell et al. (1981) and Givol et al. (1981) reported similar findings for members of the J558 family. So-called deletion-mapping analyses of B cell lines that had rearranged known V_H segments (Kemp et al., 1981; Rechavi et al., 1982; Reth et al., 1986; Blankenstein and Krawinkel, 1987; Rathbun et al., 1987) were consistent with a clustered organization of V_H segment family members. V_H segment analyses of IgH-recombinant mouse strains also indicated a generally clustered organization of V_H families (Brodeur et al., 1984). However, other studies revealed that several V_H family members are interspersed (Crews et al., 1981). In particular, pairs of families mapped at both ends of the V_H locus, namely J558 and 3609 at the 5' end and Q52 and 7183 at the 3' end, were shown to be intermingled extensively with each other (Reth et al., 1986; Blankenstein and Krawinkel, 1987; Rathbun et al., 1987). Furthermore, J558-related segments appear to be dispersed over a wide range in the V_H locus.

Deletion-mapping analyses together with strain-specific RFLP analyses of V_H families in recombinant inbred strains were used to order eight V_H families and

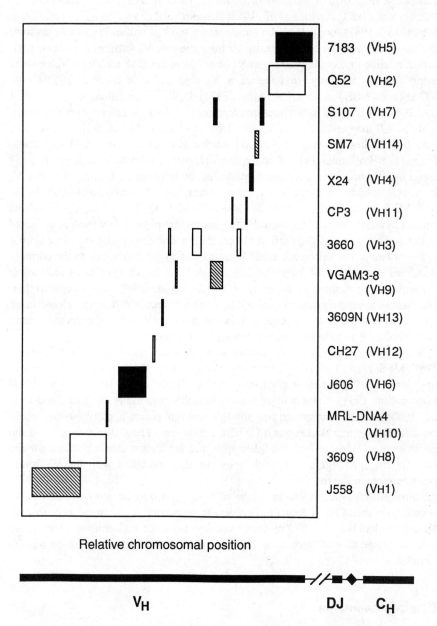

Fig 5. Relative chromosomal position of 14 murine V_H families. Mapping information of V_H10, 11, 12, 13 and 14 families is from Meek *et al.* (1990), Pennell *et al.* (1989) and Tutter *et al.* (1991). V_H families corresponding to V_H1, 2 and 3 protein subgroups are indicated by open, shaded and filled rectangles, respectively. (Modified from Brodeur *et al.*, 1988.)

generate a map of 5'–(3609, J606, 3660, X24)–J558–S107–Q52–7183–D_H–J_H–3' (Kemp et al., 1981; Rechavi et al., 1982; Brodeur et al., 1984; Mäkelä et al., 1984; Riblet et al., 1987) for the BALB/c mouse strain; the V_H families in parentheses were not mapped relative to each other. In A/J mice the nine V_H families have been positioned relative to specific rearranged V_H genes from the J558 and 3660 V_H clusters, using deletion mapping, to generate a V_H map order of 5'–3609–J558–(J606, VGAM3–8, S107)–3660–(X24, Q52, 7183)–D_H–J_H–3' (Rathbun et al., 1987); BALB/c and C57BL/6 B-cell lines that rearranged J558 V_H segments were consistent with the A/J map order. More recently, Brodeur et al. (1988) mapped 13 clusters of nine V_H families utilizing 32 Abelson murine leukaemia virus (A-MuLV)-transformed pre-B cell lines that had undergone VDJ recombination. In these cell lines 51 chromosomes had rearranged and provided useful information for mapping nine V_H families, which showed that the order is 5'–J558–3609–J606–3660–VGAM3.8–3660–S107–VGAM3.8–3660–X24–S107–Q52–7183–D_H–J_H–3'. These authors have also shown that several V_H segments are split into few clusters as shown in Fig. 5. This map is slightly different from their previous mapping and more similar to the A/J map. The difference could be explained by confusion due to the partially dispersed distribution of large families like J558. A larger number of rearranged chromosomes examined in the study by Brodeur et al. (1988) lend weight to their conclusion. Strain polymorphism could be another source of difference. However, as will be discussed below, the general organization of the V_H locus appears to be more conserved than previously expected among haplotypes.

Subsequently, five more new families were identified and mapped: V_H10 (Kofler, 1988; Meek et al., 1990), V_H11 (Reininger et al., 1988; Hardy et al., 1989; Meek et al., 1990), V_H12 (Pennell et al., 1989), V_HSM7 (Tutter et al., 1991) and V_H3609N (Tutter et al., 1991). A summary of these results is shown in Fig. 5. Since Brodeur et al. (1988) utilized F_1 mice carrying the Igh[a] and Igh[b] parent haplotypes, they could simultaneously map BALB/c and C57/BL haplotypes. These data indicate that the general organization of Igh[a] and Igh[b] haplotypes are almost identical. More limited data for the IgH[n], IgH[e], IgH[i] haplotypes are also consistent with the V_H family positions shown above.

Unfortunately, the resolution of the technique utilized for studying mouse V_H segment organization is limited. The details of murine V_H segment organization should be elucidated by PFG mapping and direct-linkage studies using overlapping cosmid, phage or YAC clones, as has been done more extensively for human V_H families.

D_H and J_H segments

The heavy-chain V-gene segments (V_H, D_H and J_H) are arranged in contiguous but separate clusters on chromosome 12 in the mouse genome, although there is no direct physical linkage between V_H and D_H segments (Fig. 3). Four J_H segments lie at several hundred base pair intervals approximately 7 kb upstream of the C_μ gene. The

D_H segments have been subdivided into three families on the basis of coding and flanking region relatedness (Kurosawa and Tonegawa, 1982). A single DQ52 segment resides about 750 bp 5' of J_H1; nine DSP2 segments are positioned 10–80 kb upstream of DQ52 (Wood and Tonegawa, 1983). The most 5' DSP2 (DSP2.3) segment is positioned between the two DFL16 segments; DFL16.1 (in BALB/c) has been identified as the most upstream D_H segment characterized to date. Recently, a new functional D_H segment (DST4), which is not related to any of the known D_H families, was identified and mapped between the 3'-end DSP2 (DSP2.8) and DQ52 segments (Feeney and Riblet, 1993). All known murine V_H segments are located upstream of DFL16.1 and to date there is no evidence for additional D_H segments further upstream (within the V_H locus) or elsewhere in the genome (Ichihara et al., 1989; Feeney and Riblet, 1993).

V_H segment usage and repertoire formation

The first evidence that the J_H-proximal V_H (7183) is preferentially used came from studies on nine A-MuLV-transformed pre-B lines that were supposed to be generated without selection of surface Ig as pre-B cells do not produce L chains and no surface Ig. Yancopoulos et al. (1984) examined nine A-MuLV-transformed BALB/c fetal liver-derived pre-B cell lines and documented 12 $V_H D_H J_H$ rearrangements in these cell lines, of which 11 involved the 7183 family. Similar results were obtained by Perlmutter et al. (1985), who found that seven of nine fetal liver hybridomas express the 7183 family. In contrast to initial reports, several groups (Reth et al., 1986; Lawler et al., 1987; Sugiyama et al., 1987; Osman et al., 1988) have observed the preferential utilization of both 7183 and Q52 family members in pre-B cells. In particular analyses of A-MuLV-transformed pre-B lines that actively continued V_H to $D_H J_H$ joining in culture provided a useful system for the study of V_H utilization in the absence of selection by the environment. Such lines from several different mouse strains all showed frequent utilization of the J_H-proximal V_H (Reth et al., 1986). Taken together, the findings (Reth et al., 1986; Rathbun et al., 1987) that 7183 and Q52 family members are interspersed in a variety of haplotypes, and that both families are preferentially utilized in pre-B cells, support the suggestion of Yancopoulos et al. (1984) of a position-dependent rearrangement of V_H segments in early ontogeny. This view, however, is not generally accepted and a locus outside of the V_H region has been suggested as being responsible for V_H gene family utilization (Wu and Paige, 1988; Atkinson et al., 1991). Unfortunately, no parallel results were obtained in human V_H usage except that the J_H-proximal V_H (i.e. V6-1) is preferentially used in early ontogeny (Berman et al., 1991). The difference will be resolved only when the precise physical map of the murine V_H locus is available.

In contrast to V_H segments, V_κ segment rearrangement in the newly generated repertoire appears position independent, although not entirely stochastic (Lawler et al., 1989). In adult bone marrow-derived pre-B cell lines, a strong bias for rearrangement of genes from the $V_\kappa 4/5$ family has been reported (Kalled and Brodeur, 1990).

Thus, primary V-gene selection might be biased in differentiating pre-B cells at all stages of ontogeny, although the as yet unidentified molecular mechanisms acting on the V_H and V_κ loci might be distinct.

In contrast to the results from pre-B cell lines, hybridomas or B-cell colonies derived from bacterial lipopolysaccharide (LPS)-activated spleen cells of adult mice had 'random' utilization of V_H families; thus, family representation depended on the family complexity (Dildrop et al., 1985; Schulze and Kelsoe, 1987). Assays of V_H utilization patterns in RNA prepared from organs of normal B-cell differentiation (neonatal liver versus adult spleen) indicate contrasting V_H usage profiles (non-random versus random) in agreement with studies using cell lines. Several other groups have also reported that V_H family utilization is normalized in the adult peripheral repertoire and roughly corresponds to the germline complexity of these families (Dildrop et al., 1985; Schulze and Kelsoe, 1987; Jeong and Teale, 1988; Yancopoulos et al., 1988). The issue is, however, still controversial: non-random V_H gene expression in the peripheral repertoire was shown in at least one report (Lawler et al., 1987), and suggested on the basis of inter-strain repertoire comparisons in others (Jeong et al., 1988; Sheehan and Brodeur, 1989). Moreover, Wu and Paige (1986) did not find significant differences between newly generated and functional peripheral repertoires.

C_H LOCUS

Structure of C_H genes

All the human and murine C_H genes have been isolated and sequenced completely; the references for the complete C_H gene sequences are summarized in Table III. The mouse C_H genes for secretory forms are composed of three (α) or four (μ, δ, γ and ϵ) exons, each encoding a functional and structural unit of the H chain, namely a domain (Edelman et al., 1969) or hinge region. In addition, one (α) or two (other) separate exons encode the hydrophobic transmembrane and short intracytoplasmic segments that are used for a membrane-form Ig. The C_α genes are exceptional because the hinge region is encoded by the C_H2 exon. The other exception is the C_δ gene, which has two additional exons, the most 3' of which encodes a C-terminal tail for the secretory form 1 kb 5' to the membrane exons. The size of each C_H exon is similar to that of the C_L exon, suggesting that the C_H gene evolved by duplication of a primordial single exon gene like the C_L gene. Such exon–intron organization of the C_H gene is consistent with the domain hypothesis that the H-chain protein consists of a tandem array of three or four functional units (Edelman et al., 1969). The total length of each C_H gene is therefore variable, ranging from 3 to 7 kb.

Expression of the membrane exons is controlled by differential splicing. Transcripts of the membrane exons are spliced to the domain exons nearest the 3' end

Table III References for complete nucleotide sequences of C_H genes*

Genes	Reference
Mouse	
μ	Kawakami et al. (1980); Rogers et al. (1980)
δ	Tucker et al. (1980); Cheng et al. (1982)
γ1	Honjo et al. (1979); Tyler et al. (1982)
γ2a	Yamawaki-Kataoka et al. (1981, 1982)
γ2b	Yamawaki-Kataoka et al. (1980, 1982)
γ3	Wels et al. (1984); Komaromy et al. (1983)
ε	Ishida et al. (1982); Liu et al. (1982)
α	Tucker et al. (1981); Word et al. (1983)
Human	
δ	Milstein et al. (1984); White et al. (1985)
γ1	Ellison et al. (1982)
γ2	Ellison and Hood (1982)
γ3	Huck et al. (1986)
γ4	Ellison et al. (1981)
ψγ	Bensmana et al. (1988)
ε3, ε2, ε3	Max et al. (1982); Ueda et al. (1982)
α1, α2	Flanagan et al. (1984)

* Corrected sequences are found in Kabat et al. (1991).

by removing the last few residues of the secreted Ig tail. All the membrane segments except C_α are encoded by two exons. The hydrophobic transmembrane segment of 26 residues is relatively conserved among all the H chains, suggesting the possibility that membrane-form Ig is anchored by a common membrane protein (Yamawaki-Kataoka et al., 1982). Since the intracytoplasmic segments of the membrane-form Ig are too short (27 residues for C_γ and C_ε chains, 14 residues for C_α and two residues for C_μ and C_δ) to catalyse any enzyme activity such as phosphorylation, transduction of the triggering signal of the antigen–antibody interaction may require involvement of at least one other protein. This hypothesis has been verified by subsequent identification of Igα and Igβ proteins (see Chapter 6).

Allotypes of Ig are mostly explained by polymorphism in C_H regions. Allotypes were originally defined by antigenic differences in Ig between different strains of mice and it is not always clear which difference in the amino acid sequence is responsible for a particular allotype defined by an antibody. The typical cases are the comparison of a and b haplotypes of the $C_\gamma 2a$ and $C_\gamma 2b$ genes, which revealed differences in 54 and 4 residues, respectively (Ollo and Rougeon, 1982, 1983). Further studies using *in vitro* mutagenesis and expression of mutants in culture cells will define polymorphic differences responsible for antigenicity for various antibodies. The Am determinants in human and allotypic determinants in rabbit IgG were assigned to one or a few residue differences (Flanagan et al., 1984; Martens et al., 1984).

Organization of C_H genes

The mouse C_H gene locus, which is mapped to chromosome 12 (D'eustachio et al., 1980), consists of eight genes that cluster in a 200 kb region (Shimizu et al., 1982a). The order of the mouse C_H genes, $5'-J_H-(6.5\text{ kb})-C_\mu-(4.5\text{ kb})-C_\delta-(55\text{ kb})-C_\gamma 3-(34\text{ kb})-C_\gamma 1-(21\text{ kb})-C_\gamma 2b-(15\text{ kb})-C_\gamma 2a-(14\text{ kb})-C_\varepsilon-(12\text{ kb})-C_\alpha-3'$ (Shimizu et al., 1982a), is consistent with the order proposed by the deletion profile of C_H genes in myelomas producing different Ig classes (Honjo and Kataoka, 1978). Unlike the C_λ gene (Miller et al., 1982), the C_H genes share one set of J_H segments (Shimizu et al., 1982a), which allows them to retain the same V_H gene during class switching. The mouse genome does not contain any well-conserved pseudogene of the C_H genes.

The general organization of the C_H gene locus is similar among laboratory strain mice, though there are many polymorphic differences. Some wild mice, however, have duplicated $C_\gamma 2a$ genes (Shimizu et al., 1982b; Fukui et al., 1984). Of 31 Japanese and Chinese wild mice screened 17 had similar duplication, suggesting that this duplication took place relatively recently.

The human C_H gene family is mapped to the q32 band of chromosome 14 (Kirsch et al., 1982) and consists of nine functional genes and two pseudogenes. The organization of the human C_H gene cluster is different from that of mouse in that the $C_\gamma-C_\gamma-C_\varepsilon-C_\alpha$ unit is duplicated downstream of the $C_\mu-C_\delta$ genes. In addition, a pseudo C_γ gene has been genetically mapped between the duplication unit (Bech-Hansen et al., 1983). The 5' C_ε or $C_\varepsilon 2$ gene is a truncated pseudogene. The other pseudogene $C_\varepsilon 3$ is processed and translocated to chromosome 9 (Battey et al., 1982). The organization of the human C_H locus is as follows: $5'-J_H-(8\text{ kb})-C_\mu-(5\text{ kb})-C_\delta-(\sim 60\text{ kb})-C_\gamma 3-(26\text{ kb})-C_\gamma 1-(19\text{ kb})-C\varepsilon 2-(13\text{ kb})-C_\alpha 1-(\sim 35\text{ kb})-\psi C_\gamma-(\sim 45\text{ kb})-C_\gamma 2-(18\text{ kb})-C_\gamma 4-(23\text{ kb})-C_\varepsilon 1-(10\text{ kb})-C_\alpha 2-3'$ (Ravetch et al., 1981; Flanagan and Rabbitts, 1982; Bottaro et al., 1989; Hofker et al., 1989).

Several deletion mutations in the C_H gene locus have been reported; deleted regions are $C_\gamma 1-C_\gamma 4$ (Lefranc et al., 1982), $C_\gamma 2-C_\gamma 4$ (Migone et al., 1984) or $C_\alpha 1-C_\varepsilon 1$ (Migone et al., 1984). It is rather surprising that individuals with deletions of several C_H genes have not shown any severe clinical symptoms, suggesting that C_γ and C_α subclass genes are capable of substituting for each other and that the C_γ genes might not be obligatory but might facilitate efficient protection from parasite infection.

Other mammals

Rat C_H gene organization was strikingly homologous to that of mouse. The order of the C_H genes is $5'-J_H-C_\mu-C_\delta-(C_\gamma 2c/C_\gamma 2a)-C_\gamma 1-C_\gamma 2b-C_\varepsilon-C_\alpha-3'$. However, rat $C_\gamma 2a$ and $C_\gamma 1$ genes, which are very similar to each other, are most homologous to the mouse $C_\gamma 1$ gene. Rat $C_\gamma 2b$ gene is most homologous to mouse $C_\gamma 2b$ and $C_\gamma 2a$ genes (Brüggemann et al., 1986). These results suggest that although C_H loci of these species have the same number of C_γ genes, these C_γ genes have evolved very dynamically by duplication (and deletion) as shown in wild mouse (Shimizu et al.,

1982b; Fukui et al., 1984). The organization of rabbit Ig genes is described in detail in Chapter 13.

Acknowledgements

We are grateful to Ms Y. Nakamura for her assistance in preparation of the manuscript. This work was supported in part by grants from Creative Basic Research (Human Genome Program) of the Ministry of Education, Science and Culture of Japan and from the Science and Technology Agency of Japan. This work was done while TH was a Fogarty scholar-in-residence at National Institute of Health, USA.

REFERENCES

Atkinson, M.J., Michnick, D.A., Paige, C.J. and Wu, G.E. (1991). *J. Immunol.* **146**, 2805–2812.
Ballard, D.W. and Bothwell, A. (1986). *Proc. Natl Acad. Sci. USA* **83**, 9626–9630.
Battey, J., Max, E.E., McBride, W.O., Swan, D. and Leder, P. (1982). *Proc. Natl Acad. Sci. USA* **79**, 5956–5960.
Bech-Hansen, N.T., Linsley, P.S. and Cox, D.W. (1983). *Proc. Natl Acad. Sci. USA* **80**, 6952–6956.
Becker, R.S. and Knight, K.L. (1990). *Cell* **63**, 987–997.
Bensmana, M., Huck, S., Lefranc, G. and Lefranc, M.-P. (1988). *Nucleic Acids Res.* **16**, 3108.
Berman, J.E., Mellis, S.J., Pollock, R., Smith, C.L., Suh, H., Heinke, B., Kowal, C., Surti, U., Chess, L., Cantor, C.R. and Alt, F.W. (1988). *EMBO J.* **7**, 727–738.
Berman, J.E., Nickerson, K.G., Pollock, R.R., Barth, J.E., Schuurman, R.K.B., Knowles, D.M., Chess, L. and Alt, F.W. (1991). *Eur. J. Immunol.* **21**, 1311–1314.
Blankenstein, T. and Krawinkel, U. (1987). *Eur. J. Immunol.* **17**, 1351–1357.
Bothwell, A.L.M., Paskind, M., Reth, M., Imanishi-Kari, T., Rajewsky, K. and Baltimore, D. (1981). *Cell* **24**, 625–637.
Bottaro, A., de Marchi, M., Migone, N. and Carbonara, A.O. (1989). *Genomics* **4**, 505–508.
Brodeur, P.H. and Riblet, R. (1984). *Eur. J. Immunol.* **14**, 922–930.
Brodeur, P.H., Tompson, M.A. and Riblet, R. (1984). In 'Regulation of the Immune System' (H. Cantor, L. Chess and E. Sercarz, eds), pp. 445–453. Alan R. Liss, New York.
Brodeur, P.H., Osman, G.E., Mackle, J.J. and Lalor, T.M. (1988). *J. Exp. Med.* **168**, 2261–2278.
Brüggemann, M., Free, J., Diamond, A., Howard, J., Cobbold, S. and Waldmann, H. (1986). *Proc. Natl Acad. Sci. USA* **83**, 6075–6079.
Buluwela, L. and Rabbitts, T.H. (1988). *Eur. J. Immunol.* **18**, 1843–1845.
Buluwela, L., Albertson, D.G., Sherrington, P., Rabbitts, P.H., Spurr, N. and Rabbitts, T.H. (1988). *EMBO J.* **7**, 2003–2010.
Cairns, E., Kwong, P.C., Misener, V., Ip, P., Bell, D.A. and Siminovitch, K.A. (1989). *J. Immunol.* **143**, 685–691.
Cheng, H.-L., Blattner, F.R., Flitzmaurice, L., Mushinski, J.F. and Tucker, P.W. (1982). *Nature* **296**, 410–415.
Cherif, D. and Berger, R. (1990). *Genes, Chromosomes and Cancer* **2**, 103–108.
Chung, J.H., Siebenlist, U., Morton, C.C. and Leder, P. (1984). *Fed. Proc.* **43**, 1486.
Cook, G., Tomlinson, I.M., Walter, G., Riethman, H., Carter, N.P., Buluwela, L., Winter, G. and Rabbitts, T.H. (1994). *Nature Genet.* **7**, 162–168.
Crews, S., Griffin, J., Huang, H., Calame, K. and Hood, L. (1981). *Cell* **25**, 59–66.
Croce, C.M., Sander, M., Martins, J., Cicurel, L., D'Ancona, G.G., Dolby, T.W. and Koprowski, H. (1979). *Proc. Natl Acad. Sci. USA* **76**, 3416–3419.
Cuisinier, A.M., Gauthier, L., Boubli, L., Fougereau, M. and Tonnelle, C. (1993). *Eur. J. Immunol.* **23**, 110–118.

Dersimonian, H., Schwartz, R.S., Barrett, K.J. and Stollar, B.D. (1987). *J. Immunol.* **139**, 2496–2501.
D'eustachio, P., Pravtceva, D., Marcu, K. and Raddle, F.H. (1980). *J. Exp. Med.* **151**, 1545–1550.
Devlin, R.H., Deeb, S., Brunzell, J. and Hayden, M.R. (1990). *Am. J. Hum. Genet.* **46**, 112–119.
Dildrop, R., Krawinkel, U., Winter, E. and Rajewsky, K. (1985). *Eur. J. Immunol.* **15**, 1154–1156.
Dzierzak, E.A., Janeway, C.A. Jr, Richard, N. and Bothwell, A. (1986). *J. Immunol.* **136**, 1864–1870.
Eaton, S. and Calame, K. (1987). *Proc. Natl Acad. Sci. USA* **84**, 7634–7638.
Edelman, G.M., Cunningham, B.A., Gall, W.E., Gottliep, P.D., Rutishauser, U. and Waxdal, M.J. (1969). *Proc. Natl Acad. Sci. USA* **63**, 78–85.
Ellison, J. and Hood, L. (1982). *Proc. Natl Acad. Sci. USA* **79**, 1984–1988.
Ellison, J., Buxbaum, J. and Hood, L. (1981). *DNA* **1**, 11–18.
Ellison, J., Berson, B.J. and Hood, L.E. (1982). *Nucleic Acids Res.* **10**, 4071–4079.
Feeney, A.J. and Riblet, R. (1993). *Immunogenetics* **37**, 217–221.
Flanagan, J.G. and Rabbitts, T.H. (1982). *Nature* **300**, 709–713.
Flanagan, J.G., Lefranc, M.-P. and Rabbitts, T.H. (1984). *Cell* **36**, 681–688.
Foote, J. and Winter, G. (1992). *J. Mol. Biol.* **224**, 487–499.
Fukui, K., Hamaguchi, Y., Shimizu, A., Nakai, S., Moriwaki, K., Wang, C.H. and Honjo, T. (1984). *J. Mol. Cell. Immunol.* **1**, 321–330.
Givol, D., Zakut, R., Effron, K., Rechavi, G., Ram, D. and Cohen, J.B. (1981). *Nature* **292**, 426–430.
Green, L.L., Hardy, M.C., Maynard-Currie, C.E., Tsuda, H., Louie, D.M., Mendez, M.J., Abderrahim, H., Noguchi, M., Smith, D.H., Zeng, Y., David, N.E., Sasai, H., Garza, D., Brenner, D.G., Hales, J.F., McGuiness, R.P., Capon, D.J., Klapholz, S. and Jakobovits, A. (1994) *Nature Genet.* **7**, 13–21.
Haino, M., Hayashida, H., Miyata, T., Shin, E.K., Matsuda, F., Nagaoka, H., Matsumura, R., Taka-ishi, S., Fukita, Y., Fujikura, J. and Honjo, T. (1994). *J. Biol. Chem.* **269**, 2619–2626.
Hardy, R.R., Carmack, C.E., Shinton, S.A., Riblet, R.J. and Hayakawa, K. (1989). *J. Immunol.* **142**, 3643–3651.
Hashimoto, L.L., Walter, M.A., Cox, D.W. and Ebers, G.C. (1993). *J. Neuroimmunol.* **44**, 77–83.
Hayashida, H. and Miyata, T. (1983). *Proc. Natl Acad. Sci. USA* **80**, 2671–2675.
Hofker, M.H., Walter, M.A. and Cox, D.W. (1989). *Proc. Natl Acad. Sci. USA* **86**, 5567–5571.
Honjo, T. and Kataoka, T. (1978). *Proc. Natl Acad. Sci. USA* **75**, 2140–2144.
Honjo, T., Obata, M., Yamawaki-Kataoka, Y., Kataoka, T., Takahashi, N. and Mano, Y. (1979). *Cell* **18**, 559–568.
Huck, S., Crawford, D.H., Lefranc, M.-P. and Lefranc, G. (1986). *Nucleic Acids Res.* **14**, 1779–1789.
Hyrien, O., Debatisse, M., Buttin, G. and Robert de Saint Vincent, B. (1987). *EMBO J.* **6**, 2401–2408.
Ichihara, Y., Abe, M., Yasui, H., Matsuoka, H. and Kurosawa, Y. (1988a). *Eur. J. Immunol.* **18**, 649–652.
Ichihara, Y., Matsuoka, H. and Kurosawa, Y. (1988b). *EMBO J.* **7**, 4141–4150.
Ichihara, Y. Hayashida, H., Miyazawa, S. and Kurosawa, Y. (1989). *Eur. J. Immunol.* **19**, 1849–1854.
Ishida, N., Ueda, S., Hayashida, H., Miyata, T. and Honjo, T. (1982). *EMBO J.* **1**, 1117–1123.
Jeong, H.D. and Teale, J.M. (1988). *J. Exp. Med.* **168**, 589–603.
Jeong, H.D., Komisar, J.L., Kraig, E. and Teale, J.M. (1988). *J. Immunol.* **140**, 2436–2441.
Kabat, E.A., Wu, T.T., Perry, H.M., Gottesman, K.S. and Foeller, C. (1991). 'Sequences of Proteins of Immunological Interest', 5th edn. NIH publications, Washington DC.
Kalled, S.L. and Brodeur, P.H. (1990). *J. Exp. Med.* **172**, 559–566.
Kawakami, T., Takahashi, N. and Honjo, T. (1980). *Nucleic Acids Res.* **8**, 3933–3945.
Kemler, I., Schreiber, E., Müller, M.M., Matthias, P. and Schaffner, W. (1989). *EMBO J.* **8**, 2001–2008.
Kemp, D.J., Tyler, B., Bernard, O., Gough, N., Gerondakis, S., Adams, J.M. and Cory, S. (1981). *J. Mol. Appl. Genet.* **1**, 245–261.
Kirkham, P.M. and Schroeder, H.W. Jr (1994). *Semin. Immunol.* **6**, 347–360.
Kirsch, I.R., Morton, C.C., Nakahara, K. and Leder, P. (1982). *Science* **216**, 301–303.
Kodaira, M., Kinashi, T., Umemura, I., Matsuda, F., Noma, T., Ono, Y. and Honjo, T. (1986). *J. Mol. Biol.* **190**, 529–541.
Kofler, R. (1988). *J. Immunol.* **140**, 4031–4034.

Kofler, R., Geley, S., Kofler, H. and Helmberg, A. (1992). *Immunol. Rev.* **128**, 5–21.
Komaromy, M., Clayton, L., Rogers, S., Robertson, S., Kettman, J. and Wall, R. (1983). *Nucleic Acids Res.* **11**, 6775–6785.
Kurosawa, Y. and Tonegawa, S. (1982). *J. Exp. Med.* **155**, 201–218.
Lawler, A.M., Lin, P.S. and Gearhart, P.J. (1987). *Proc. Natl Acad. Sci. USA*, **84**, 2454–2458.
Lawler, A.M., Kearney, J.F., Kuehl, M. and Gearhart, P.J. (1989). *Proc. Natl Acad. Sci. USA* **86**, 6744–6748.
Lee, K.H., Matsuda, F., Kinashi, T., Kodaira, M. and Honjo, T. (1987). *J. Mol. Biol.* **195**, 761–768.
Lefranc, M.-P., Lefranc, G. and Rabbitts, T.H. (1982). *Nature* **300**, 760–762.
Liu, C.-P., Tucker, P.W., Mushinski, G.F. and Blattner, F.R. (1980). *Science* **209**, 1348–1353.
Liu, F.-T., Albrand, K., Sutchliffe, J.G. and Katz, D.H. (1982). *Proc. Natl Acad. Sci. USA* **79**, 7852–7856.
Livant, D., Blatt, C. and Hood, L. (1986). *Cell* **47**, 461–470.
Mäkelä, O., Seppala, I.J.T., Pekonen, J., Kaartinen, M., Cazenave, P.A. and Gefter, M.L. (1984). *Ann. Immunol. (Paris)* **135C**, 169–173.
Martens, C.L., Currier, S.J. and Knight, K.L. (1984). *J. Immunol.* **133**, 1022–1027.
Matsuda, F., Lee, K.H., Nakai, S., Sato, T., Kodaira, M., Zong, S.Q., Ohno, H., Fukuhara, S. and Honjo, T. (1988). *EMBO J.* **7**, 1047–1051.
Matsuda, F., Shin, E.K., Hirabayashi, Y., Nagaoka, H., Yoshida, M.C., Zong, S.Q. and Honjo, T. (1990). *EMBO J.* **9**, 2501–2506.
Matsuda, F., Shin, E.K., Nagaoka, H., Matsumura, R., Haino, M., Fukita, Y., Taka-ishi, S., Imai, T., Riley, J.H., Anand, R., Soeda, E. and Honjo, T. (1993). *Nature Genet.* **3**, 88–94.
Matsumura, R., Matsuda, F., Nagaoka, H., Shin, E.K., Fukita, Y., Haino, M., Fujikura, J. and Honjo, T. (1994). *J. Immunol.* **152**, 660–666.
Max, E.E., Battey, J., Mey, R., Kirsch, I.R. and Leder, P. (1982). *Cell*, **29**, 691–699.
Meek, K., Rathbun, G., Reininger, L., Jaton, J.C., Kofler, R., Tucker, P.W. and Capra, J.D. (1990). *Mol. Immunol.* **27**, 1073–1081.
Migone, N., Oliviero, S., de Lange, G., Delacroix, D.L., Boschis, D., Altruda, F., Silengo, L., DeMarchi, M. and Carbonara, A.O. (1984). *Proc. Natl Acad. Sci. USA* **81**, 5811–5815.
Miller, J., Selsing, E. and Storb, U. (1982). *Nature* **295**, 428–430.
Milstein, C.P., Deverson, E.V. and Rabbitts, T.H. (1984). *Nucleic Acids Res.* **12**, 6523–6535.
Miyata, T., Yasunaga, T. and Nishida, T. (1980). *Proc. Natl Acad. Sci. USA* **77**, 7328–7332.
Nagaoka, H., Ozawa, K., Matsuda, F., Hayashida, H., Matsumura, R., Haino, M., Shin, E.K., Fukita, Y., Imai, T., Anand, R., Yokoyama, K., Eki, T., Soeda, E. and Honjo, T. (1994). *Genomics* **22**, 189–197.
Okamura, K, Ishiguro, H., Ichihara, Y. and Kurosawa, Y. (1993). *Mol. Immunol.* **30**, 461–467.
Ollo, R. and Rougeon, F. (1982). *Nature* **296**, 761–763.
Ollo, R. and Rougeon, F. (1983). *Cell* **32**, 515–523.
Osman, G.E., Wortis, H.H. and Brodeur, P.H. (1988). *J. Exp. Med.* **168**, 2023–2030.
Pascual, V., Randen, I., Tompson, K., Sioud, M., Forre, O., Natvig, J. and Capra, J.D. (1990). *J. Clin. Invest.* **86**, 1320–1328.
Pascual, V., Verkruyse, L., Casey, M.L. and Capra, J.D. (1993). *J. Immunol.* **151**, 4164–4172.
Pennell, C.A., Sheehan, K.M., Brodeur, P.H. and Clarke, S.H. (1989). *Eur. J. Immunol.* **19**, 2115–2121.
Perlmutter, R.M., Klotz, J.L., Bond, M.W., Nahm, M., Davie, J.M. and Hood, L. (1984). *J. Exp. Med.* **159**, 179–192.
Perlmutter, R.M., Kearney, J.F., Chang, S.P. and Hood, L. (1985). *Science* **227**, 1597–1601.
Rathbun, G.A., Capra, J.D. and Tucker, P.W. (1987). *EMBO J.* **6**, 2931–2937.
Rathbun, G.A., Otani, F., Milner, E.C.B., Capra, J.D. and Tucker, P.W. (1988). *J. Mol. Biol.* **202**, 383–395.
Ravetch, J.V., Siebenlist, U., Korsmeyer, S., Waldmann, T. and Leder, P. (1981). *Cell* **27**, 583–591.
Raynaud, C.-A., Anques, V., Grimal, H. and Weill, J.C. (1987). *Cell* **48**, 379–388.
Rechavi, G., Bientz, B., Ram, D., Ben-Neriah, Y., Cohen, J.B., Zakut, R. and Givol, D. (1982). *Proc. Natl Acad. Sci. USA* **79**, 4405–4409.
Reininger, L., Kaushik, A., Izui, S. and Jaton, J.C. (1988). *Eur. J. Immunol.* **18**, 1521–1526.
Reth, M., Jackson, N. and Alt, F.W. (1986). *EMBO J.* **5**, 2131–2138.

Riblet, R., Brodeur, P., Tutter, A. and Tompson, M.A. (1987). In 'Evolution and Vertebrate Immunity: The Antigen Receptor and MHC Gene Families' (G. Kelsoe and D.H. Schultze, eds), p. 53. University of Texas, Austin.
Rogers, J., Early, P., Carter, C., Calame, K., Bond, M., Hood, L. and Wall, R. (1980). *Cell* **20**, 303–312.
Sanz, I., Dang, H., Takei, M., Talal, N. and Capra, J.D. (1989). *J. Immunol.* **142**, 883–887.
Sasso, E.H., Willems van Dijk, K., Bull, A.P. and Milner, E.C.B. (1993). *Proc. Natl Acad. Sci. USA* **89**, 10430–10434.
Sato, T., Matsuda, F., Lee, K.H., Shin, E.K. and Honjo, T. (1988). *Biochem. Biophys. Res. Commun.* **154**, 265–271.
Schiff, C., Milili, M. and Fougereau, M. (1985). *EMBO J.* **2**, 1225–1330.
Schroeder, H.W., Jr and Wang, J.Y. (1990). *Proc. Natl Acad. Sci. USA* **87**, 6146–6150.
Schroeder, H.W., Jr, Hillson, J.L. and Perlmutter, R.M. (1987). *Science* **238**, 791–793.
Schroeder, H.W., Jr, Walter, M.A., Hofker, M.H., Ebens, A., Willem van Dijk, K., Liao, L.C., Cox, D.W., Milner, E.C.B. and Perlmutter, R.M. (1988). *Proc. Natl Acad. Sci. USA* **85**, 8196–8200.
Schroeder, H.W., Jr, Hillson, J.L. and Perlmutter, R.M. (1990). *Int. Immunol.* **20**, 41–50.
Schulze, D.H. and Kelsoe, G. (1987). *J. Exp. Med.* **166**, 163–172.
Schwartz, R.S. and Stoller, B.D. (1994). *Immunol. Today* **15**, 27–32.
Sheehan, K.M. and Brodeur, P.H. (1989). *EMBO J.* **8**, 2313–2320.
Shen, A., Humphries, C., Tucker, P. and Blattner, F. (1987). *Proc. Natl Acad. Sci. USA* **84**, 8563–8567.
Shen, M.R., Batzer, M.A. and Deininger, P.L. (1991). *J. Mol. Evol.* **33**, 311–320.
Shimizu, A., Takahashi, N., Yaoita, Y. and Honjo, T. (1982a). *Cell* **28**, 499–506.
Shimizu, A., Hamaguchi, Y., Yaoita, Y., Moriwaki, K., Kondo, S. and Honjo, T. (1982b). *Nature* **298**, 82–84.
Shin, E.K., Matsuda, F., Nagaoka, H., Fukita, Y., Imai, T., Yokoyama, K., Soeda, E. and Honjo, T. (1991). *EMBO J.* **10**, 3641–3645.
Shin, E.K., Matsuda, F., Ozaki, S., Kumagai, S., Olerup, O., Strom, H., Melchars, I. and Honjo, T. (1993a). *Immunogenetics* **38**, 304–306.
Shin, E.K., Matsuda, F., Fujikura, J., Akamizu, T., Sugawa, H., Mori, T. and Honjo, T. (1993b). *Eur. J. Immunol.* **23**, 2365–2367.
Siebenlist, U., Ravetch, J.V., Korsmeyer, S., Waldmann, T. and Leder, P. (1981). *Nature* **294**, 631–635.
Siekevitz, M., Huang, S.Y. and Gefter, M.L. (1983). *Eur. J. Immunol.* **13**, 123–132.
Siu, G., Springer, E.A., Huang, H.V., Hood, L.E. and Crews, S.T. (1987). *J. Immunol.* **138**, 4466–4471.
Sonntag, D., Weingartner, B. and Grutzmann, R. (1989). *Nucleic Acids Res.* **17**, 1267.
Sugiyama, H., Maeda, T., Tani, Y., Miyake, S., Oka, Y., Komori, T., Ogawa, H., Soma, T., Minami, Y., Sakato, N. and Kishimoto, S. (1987). *J. Exp. Med.* **166**, 607–612.
Taylor, L.D., Carmack, C.E., Huszar, D., Higgins, K.M., Mashayekh, R., Sequar, G., Schramm, S.R., Kuo, C.C., O'Donnell, S.L., Kay, R.M., Woodhouse, C.S. and Lonberg, N. (1994). *Int. Immunol.* **6**, 579–591.
Tomlinson, I.M., Walter, G., Marks, J.D., Llewelyn, M.B. and Winter, G. (1992). *J. Mol. Biol.* **227**, 776–798.
Tomlinson, I.M., Cook, G.P., Carter, N.P., Elaswarapu, R., Smith, S., Walter, G., Buluwela., L., Rabbitts, T.H. and Winter, G. (1994). *Hum. Mol. Genet.* **3**, 856–860.
Tompson, C.B. and Neiman, P.E. (1987). *Cell* **48**, 369–378.
Tucker, P.W., Liu, C.P., Mushinski, J.F. and Blattner, F.R. (1980). *Science* **209**, 1353–1360.
Tucker, P.W., Slighton, J.L. and Blattner, F.R. (1981). *Proc. Natl Acad. Sci. USA* **78**, 7684–7688.
Tutter, A. and Riblet, R. (1988). *Curr. Top. Microbiol. Immunol.* **137**, 107–115.
Tutter, A., Brodeur, P., Shlomchik, M. and Riblet, R. (1991). *J. Immunol.* **147**, 3215–3223.
Tyler, B.M., Cowman, A.F., Genondakis, S.D., Adams, J.M. and Bernerd, O. (1982). *Proc. Natl Acad. Sci. USA* **79**, 2008–2012.
Ueda, S., Nakai, S., Nishida, Y., Hisajima, H. and Honjo, T. (1982). *EMBO J.* **1**, 1539–1544.
van Dijk, K.W., Mortari, F., Kirkham, P.M., Schroeder, H.W. and Milner, E.C.B. (1993). *Eur. J. Immunol.* **23**, 832–839.
Walter, G., Tomlinson, I.M., Cook, G.P., Winter, G., Rabbitts, T.H. and Dear, P.H. (1993). *Nucleic Acids Res.* **21**, 4524–4529.

Walter, M.A., Surti, U., Hofker, M.H. and Cox, D.W. (1990). *EMBO J.* **9**, 3303–3313.
Walter, M.A., Dosch, H.M. and Cox, D.W. (1991a). *J. Exp. Med.* **174**, 335–349.
Walter, M.A., Gibson, W.T., Ebers, G.C. and Cox, D.W. (1991b). *J. Clin. Invest.* **87**, 1266–1273.
Wels, J.A., Word, C.J., Rimm, D., Der-Balan, G.P., Martinez, H.M., Tucker, P.W. and Blattner, F.R. (1984). *EMBO J.* **3**, 2041–2046.
White, M.B., Shen, A.W., Word, J.C., Tucker, P.W. and Blattner, F.R. (1985). *Science* **228**, 733–737.
Winter, E., Radbruch, A. and Krawinkel, U. (1985). *EMBO J.*, **4**, 2861–2867.
Wood, C. and Tonegawa, S. (1983). *Proc. Natl Acad. Sci. USA* **80**, 3030–3034.
Word, J.C., Munshinski, J.F. and Tucker, P.W. (1983). *EMBO J.* **2**, 887–898.
Wu, G.E. and Paige, C.J. (1986). *EMBO J.* **5**, 3475–3481.
Wu, G.E. and Paige, C.J. (1988). *J. Exp. Med.* **167**, 1499–1504.
Yamawaki-Kataoka, Y., Kataoka, T., Takahashi, N., Obata, M. and Honjo, T. (1980). *Nature* **283**, 786–789.
Yamawaki-Kataoka, Y., Miyata, T. and Honjo, T. (1981). *Nucleic Acids Res.* **9**, 1365–1381.
Yamawaki-Kataoka, Y., Nakai, S., Miyata, T. and Honjo, T. (1982). *Proc. Natl Acad. Sci. USA* **79**, 2623–2627.
Yancopoulos, G.D., Desiderio, S.V., Pasking, M., Kearney, J.F., Baltimore, D. and Alt, F.W. (1984). *Nature* **311**, 727–733.
Yancopoulos, G.D., Maylnn, B.A. and Alt, F.W. (1988). *J. Exp. Med.* **168**, 417–435.
Yang, P.-H., Olsen, N.J., Siminovitch, K.A., Olee, T., Kozin, F., Carson, D.A. and Chen, P.P. (1990). *Proc. Natl Acad. Sci. USA* **88**, 7907–7911.
Zong, S.Q., Nakai, S., Matsuda, F., Lee, K.H. and Honjo, T. (1988). *Immunol. Lett.* **17**, 329–334.
Zouali, M. (1992). *Immunol. Rev.* **128**, 73–99.

8

The human immunoglobulin κ genes

Hans G. Zachau

Institut für Physiologische Chemie der Universität München, München, Germany

The elucidation of the human κ locus 173
The V_κ genes of the locus 175
The B3–J_κ–C_κ–κde region 178
The structural organization of the κ locus in germline DNA 179
Polymorphisms in the κ locus 180
Rearranged V_κ genes 181
Dispersed V_κ genes 184
Repetitive and unique sequences in the κ locus 185
Evolution of the V_κ genes 186
Biomedical implications 187
Miscellaneous and concluding remarks 187

While the chapter on the immunoglobulin κ genes in the first edition of this book covered the information available in 1987 on the genes of human and mouse (Zachau, 1989a), the present review deals with the human κ genes only and concentrates on the more recent results. In fact, quotations to the older literature will not be repeated here. For the mouse V_κ gene families, their complexity, polymorphism and use in non-autoimmune responses the reader is referred to a recent review by Kofler *et al.* (1992). Other aspects are dealt with in the respective chapters of this book.

THE ELUCIDATION OF THE HUMAN κ LOCUS

The single human C_κ gene and the five J_κ genes were cloned and characterized by P. Leder's group (Hieter *et al.*, 1980, 1982) and the first human V_κ genes were isolated and sequenced by Bentley and Rabbitts (1980, 1981, 1983). In our group, the κ genes have been studied since the early 1980s, and some aspects of the work were reviewed in lecture reports (Zachau, 1990, 1993) and in a recent survey (Zachau, 1995).

The structural work in our laboratory was based on 440 cosmid and 30 phage λ clones that were isolated from various libraries of germline DNA and mapped with 5–12 different restriction nucleases each. Initially, the search clones were $V_κ$ gene probes and specific chromosomal walking probes at later stages of the work. The first indications that there may be two copies of the κ locus came from an apparently duplicated $V_κ$ gene (Bentley and Rabbitts, 1983) and then from the systematic cloning studies of Pech et al. (1985). The extent of the duplication became known with some certainty only at a late state of the work. It is now clear that the so-called $C_κ$ proximal (p) copy of the locus contains in a 600-kb contig, in addition to the $J_κ$–$C_κ$ region, 40 $V_κ$ genes and pseudogenes, and the distal (d) copy contains in a 440-kb contig 36 $V_κ$ genes and pseudogenes. A scheme of the κ locus and its surroundings is presented for general orientation (Fig. 1). The two large contigs were assembled from smaller ones that had been studied separately before: Op/Od (Pargent et al., 1991a), Ap/Ad (Lautner-Rieske et al., 1992), Lp/Ld (Huber et al., 1993a,b), B (Lorenz et al., 1988) and $J_κ$–$C_κ$–κde (Klobeck and Zachau, 1986; Klobeck et al., 1987a). An SnaBI map of the contigs was reported by Ermert (1994). The cloning of the gaps between these original

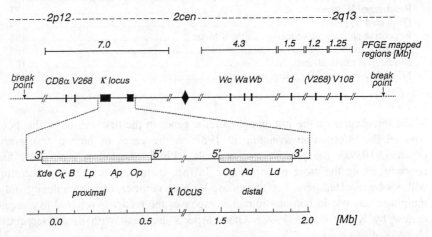

Fig. 1. Schematic representation of the central part of chromosome 2. The cloned regions of the κ locus are shown as black bars or stippled boxes. Vertical lines indicate cloned orphon regions (V268, Wa–Wc, V108; see p. 185) and fragments hybridizing to the following probes: CD8α and 273–2, derived from the V268 region and therefore designated (V268) (Weichhold et al., 1993b; Huber et al., 1994); d, homologous to the κde region (Graninger et al., 1988). The organization of the κ locus is depicted according to Weichhold et al. (1993a). In the scale $C_κ$ is taken as zero, while for the contigs counting starts at the 5′ ends (Pargent et al., 1991a). The orphons on the long arm and the breakpoints of pericentric inversion(s) are described by Lautner-Rieske et al. (1993).

small contigs by chromosomal walking was a lengthy and cumbersome process, since the clones required for linking were highly underrepresented in the libraries. Structural reasons for this did not become apparent when the linking was achieved.

THE V_κ GENES OF THE LOCUS

Genes and pseudogenes, subgroups, polymerase chain reaction (PCR) primers

The 76 V_κ genes and pseudogenes of the locus were sequenced and their transcriptional polarities were determined within the maps. An outline of the V_κ genes of the locus is given in Fig. 2. There are 10 solitary genes and 33 gene pairs whose sequences are 95–100% identical in the coding regions. Of the V_κ genes 32 are potentially functional, 16 have minor defects and 25 are pseudogenes; three genes were found to occur both as potentially functional and slightly defective alleles. The minor defects are defined as one or two 1-bp alterations in a gene, for instance the occurrence of a stop codon and/or a deviation from the canonical sequences of regulatory elements, splice sites or hepta- and nona-nucleotide recognition sequences. The 16 genes with minor defects are defined as a separate class of genes, since, as for the three genes mentioned above, potentially functional alleles may exist in the human population. This is not to be expected for the pseudogenes, which usually carry several defects each. All human V_κ gene, pseudogene and orphon sequences including all alleles known to us were compiled (Schäble and Zachau, 1993) and, in addition, some of the pseudogenes were dealt with specifically by Schäble et al. (1994). Recently, Cox et al. (1994) amplified from genomic DNA by PCR FR1–CDR3 sequences and called four of them 'new gene segments'. However, the sequences do not correspond to new gene loci but have to be considered alleles of published V_κ genes and orphons, as far as this can be concluded in the absence of intron sequences and data on the genomic context (Klein and Zachau, 1995). The systematic nomenclature of V_κ genes (Fig. 2) is used throughout this chapter and in all recent reports from our laboratory, but it is expected that some alternative designations will be further employed in the literature. The different designations of the various genes were compiled by Schäble and Zachau (1993).

There is good circumstantial evidence that we have now cloned all or most V_κ genes of the locus. The reservations inevitably connected with such a statement have been discussed by Meindl et al. (1990a), Huber et al. (1993a,b) and Klein et al. (1993). One prerequisite was, of course, that we were able to close all gaps within the p and the d copies of the locus and to extend the two contigs in both directions by 50–80 kb without finding additional genes, an effort that is at present being continued with YAC clones (I. Zocher and J. Brensing-Küppers, unpublished data). Recently, one still Unidentified Hybridizing Object (UHO) was detected in one of the YAC clones (J. Brensing-Küppers, unpublished). However, previously identified UHOs were either orphons (p. 184) or turned out on sequencing not to contain V_κ like structures but a LINE1 sequence and, in some cases, M13 vector sequences, the cross-hybridizations of which with human DNA are known (Vassart *et al.*, 1987); this can be taken as an indication that the search for additional V_κ genes by hybridization was carried to the limit (Röschenthaler *et al.*, 1992; Schäble *et al.*, 1994).

The classification of κ proteins into four subgroups (Kabat *et al.*, 1991 and earlier editions) was fully confirmed when the V_κ gene sequences were aligned (Schäble and Zachau, 1993). The similarity between potentially functional V_κ gene segments is

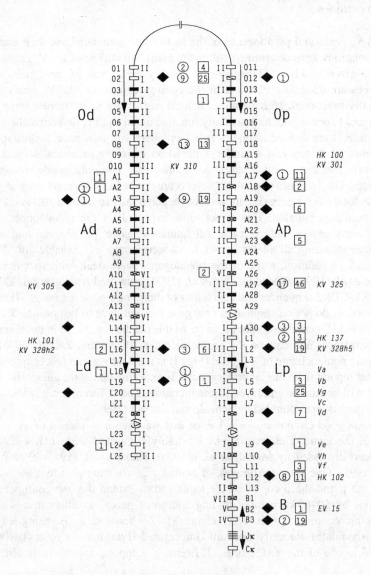

higher than 84% among the members of subgroups I–III and between 57 and 78% when members of different subgroups including pseudogenes are compared. While the one gene of subgroup IV is transcribed and translated, no proteins are known for the genes of subgroups V–VII. However, transcripts of $V_\kappa V$ and $V_\kappa VI$ genes have been found recently (Marks *et al.*, 1991). There are only five V_κ genes altogether in subgroups V–VII (Straubinger *et al.*, 1988a), but the definition of separate subgroups for them seems unavoidable if the members of a subgroup should be at least 80% similar to each other.

The alignment of the sequences of the V_κ gene regions (Schäble and Zachau, 1993) also served two other purposes. PCR primer combinations were derived, which should allow the reliable amplification of certain groups of germline genes and also some single genes. The other aim was to define and evaluate conserved sequence elements.

Conserved sequence elements

In the upstream region there is, in addition to the rather variable TATAA-like sequence, the decanucleotide (dc) sequence TNATTTGCAT, which was early recognized as a functional promoter (Falkner and Zachau, 1984). Independently, the octanucleotide sequence ATTTGCAT was defined as a conserved sequence (Parslow *et al.*, 1984). It is now seen in the alignments of the human V_κ gene sequences that the dc sequence is very largely conserved among potentially functional V_κ genes and that the heptanucleotide TTTGCAT is fully conserved. This is in line with the observation that these seven nucleotides are essential for promoter activity, while alterations in the first and third position of dc allow reduced transcription (Wirth *et al.*, 1987). A 15-mer or pd element (Falkner and Zachau, 1984) is found 17 bp 5′ of dc in all $V_\kappa I$ genes and about 150 bp 5′ of the $V_\kappa VI$ genes A10 and A26, while it is not seen in V_κ genes of the other subgroups. pd is not a promoter element essential for transcription (Bergman *et al.*, 1984) but it seems to have a supportive activity (Sigvardsson *et al.*, 1995). Another possibly

Fig. 2. Outline of the human κ locus. Open boxes represent potentially functional V_κ genes and the single C_κ gene, filled boxes the V_κ pseudogenes. Boxes with crossed lines designate genes with minor defects (as defined in the text) and boxes with one diagonal line the three genes for which potentially functional and slightly defective alleles are known. Roman numerals refer to the subgroups of the respective V_κ genes. The deletions in the A and L regions -(Δ)- are described on pp. 179 and 186. The drawing is not to scale. Arrows show the direction of transcription. The boxes, circles and rhomboids beside the V_κ genes or between undistinguishable gene pairs refer to transcription products, κ proteins and genomic joints, respectively. The figures in the boxes and circles are the numbers of different gene products found. Alternative designations of some V_κ genes used in the literature are shown in italics; a complete listing of such designations is given in Schäble and Zachau (1993). The figure is similar to previous published versions (Klein *et al.*, 1993; Klein and Zachau, 1995) and further details are described therein.

supportive element in the dc region is CCCT (Högbom et al., 1991). An ACCC element nearby was found to bind nuclear proteins (Mocikat et al., 1988). It is a matter of definition whether some of the elements, notably the dc element itself which is found in the opposite 5', 3' polarity upstream of V_H genes (Falkner and Zachau, 1984), serve a promoter or an enhancer function. The work on the upstream elements of κ genes was compiled by Mocikat et al. (1989) and a comprehensive review of the regulatory elements in immunoglobulin genes was given by Staudt and Lenardo (1991).

Another outcome of the V_κ alignments (Schäble and Zachau, 1993) is that within the major subgroups the leader segments are more similar to each other than the V gene segments. Some conserved regions in the introns are related to functions in the splice process. The intron sizes are remarkably similar within subgroups and show pronounced differences between subgroups. Deviations from the recombination signal sequences are compiled in Schäble et al. (1994).

THE B3–J_κ–C_κ-κde REGION

The 23-kb region between the single V_κIV gene of the locus, which is called B3, and $J_\kappa 1$ was found to be free of V_κ gene-like structures in the DNAs of several individuals. However, a sequence of about 0.5 kb was found in the middle of the region that has a counterpart called homox on another chromosome but otherwise does not hybridize to genomic DNA (Klobeck et al., 1989). No function is known for the two sequences, which are 96% identical. The finding of spliced J_κ–C_κ transcripts without a V_κ gene (Martin et al., 1991) points to the existence of a promoter and a transcription start site about 4 kb upstream of $J_\kappa 1$, which may function in a prelude to V_κ–J_κ rearrangements.

Although the sequences of the J_κ and C_κ genes have been known for a long time (Hieter et al., 1980, 1982), the sequence of the whole J_κ–C_κ region of more than 5 kb has been made available only recently (Whitehurst et al., 1992). This was done in the context of further defining the location of the matrix association region (MAR) in the J_κ–C_κ intron. Slightly upstream of MAR lies the κde target sequence (see below) and slightly downstream the intron enhancer, which was characterized by Gimble and Max (1987). The various enhancing and silencing sequence motifs were reviewed by Staudt and Lenardo (1991). An additional silencing element immediately upstream of the NF-$_\kappa$B binding site of the intron was recently described for the mouse and human systems (Saksela and Baltimore, 1993).

The C_κ allotypes, which were originally defined serologically and by protein sequencing, were now studied by PCR permitting easy detection of the association to other polymorphic markers of the region (Moxley and Gibbs, 1992). At a position 12 kb to the 3' side of C_κ lies the so-called downstream enhancer, which was identified by sequence homology to the corresponding mouse enhancer (Müller et al., 1990). Its functional characteristics were then defined by Judde and Max (1992).

Another 12 kb downstream lies the κde (C_κ deleting) element, which was recognized in the human system by Siminovitch *et al.* (1985) and localized by Klobeck and Zachau (1986). Its hepta- and nona-nucleotide recognition signals (located at 23 bp distance) recombine with the complementary signal sequences in the intron (located at 29 bp distance) leading to the excision of C_κ and the enhancers in some λ chain-producing B cells. Nothing is known yet about the enzymology and regulation of this process.

THE STRUCTURAL ORGANIZATION OF THE κ LOCUS IN GERMLINE DNA

The κ locus is located on the short arm of chromosome 2 at 2cen–p12 (Malcolm *et al.*, 1982; McBride *et al.*, 1982) or, more specifically, at 2cen–p11.2 (Lautner-Rieske *et al.*, 1993). Mapping by pulsed field gel electrophoresis (PFGE) with the help of 13 rare-cutter restriction nucleases and 15 unique hybridization probes led to a detailed picture of the locus and its surroundings (Weichhold *et al.*, 1993a). The p and d contigs are arranged in opposite 5′, 3′ polarity (Fig. 1). The still uncloned region of 800 kb between the contigs appears not to contain further V_κ genes (see p. 175). The structure is largely symmetrical starting from a centre in the uncloned region and extending for about 850 kb to each side, i.e. to the duplicate gene pair L10/L25 as depicted in Fig. 2. The map of the κ locus, comprising about 2 Mb, extends for another 1.5 Mb towards the centromere, but no marker is known between the locus and the centromere. The map of 3.5 Mb towards the telomere includes the orphon V_κ gene V268 (Huber *et al.*, 1994) and, at a distance of 2–2.2 Mb from C_κ, the CD8α locus (Weichhold *et al.*, 1993b). Some detours and artefacts in establishing the PFGE map were also described (Weichhold *et al.*, 1993c).

The 5′ termini of the p and d contigs, i.e. the regions towards the uncloned central part beyond the genes O1 and O10 (Fig. 2), are rich in repetitive DNA sequences (Pargent *et al.*, 1991a). The same seems to be the case at the 3′ end of the d contig beyond L25 (Huber *et al.*, 1993a). These sequences have possibly played a role in confining the κ locus to its present limits. The inverted duplicated structure of the κ locus may have been formed in reactions similar to those described in models of gene amplification (for a discussion see Weichhold *et al.*, 1993a).

Three regions in the cloned parts of the locus, i.e. between A29 and A30, L8 and L9, and L22 and L23 (Fig. 2), are not represented on the opposite copy. Since an artefactual loss of these regions on cloning can be excluded, they must have been deleted during evolution, albeit after the duplication of the locus. All deletion breakpoints were sequenced. The sequence motif CCAG/CTGG found by Chou and Morrison (1993) to occur commonly near (somatic) non-homologous recombination breakpoints involving immunoglobulin gene sequences was observed rather frequently in our sequences, but no accumulation near the breakpoints was seen.

POLYMORPHISMS IN THE κ LOCUS

Allelic differences and haplotypes

Although the V_κ gene probes of the major subgroups hybridize to many related genes and complicated patterns ensue, it has been possible to define some V_κ gene-related allelic polymorphisms (Turnbull *et al.*, 1987). One such polymorphism was linked to rheumatoid arthritis with a relative risk of 5 (Meindl *et al.*, 1990b). However, for systematic studies of polymorphisms single-copy probes are much preferred. The detection of allelic differences by restriction fragment length polymorphism (RFLP) studies is relatively straightforward in the non-duplicated part of the locus, as in the B3 region (Klobeck *et al.*, 1987a), the B3–J_κ intergenic region (Klobeck *et al.*, 1989) and in the C_κ region (Field *et al.*, 1987; Klobeck *et al.*, 1987a; Moxley and Gibbs, 1992; and earlier literature on C_κ allotypes). In the duplicated part of the locus most specific probes recognize the homologous parts of both copies. Although they are, therefore, not truely unique they are included for the present discussion in the group of single-copy probes. Five RFLPs were established in the duplicated O regions, which together with the three RFLPs of the B3–C_κ region served to define three basic and several derived haplotypes of the κ locus (Pargent *et al.*, 1991b).

There is little allelic variation in the gene regions: no variants at all were found in the C_κ genes of 50 unrelated individuals. 12 variants were identified in the B3 genes of 26 individuals, but all of them were located in the intron (Kurth and Cavalli-Sforza, 1994); the 1-bp difference between all their germline B3 gene sequences and the published sequence, which is pointed out by Kurth and Cavalli-Sforza (1994), results from aligning to the sequence of a rearranged and mutated B3 gene but not to the germline B3 gene sequence described in the same paper (Klobeck *et al.*, 1985). Schäble and Zachau (1993) compiled 22 alleles of 19 other V_κ genes. Of 27 different V_κ gene sequences with open reading frames reported by Cox *et al.* (1994) 26 were found to be identical to previously published sequences (that had been determined in the DNA of various individuals) and one had a 1-bp difference; an allele of a pseudogene pair and sequences related to two orphons were also reported. This is our interpretation of the respective data as described in Klein and Zachau (1995). The implications of allelic variation for mutation studies in V_κ genes are discussed on p. 183.

Duplication differentiating polymorphisms

If a hybridization probe recognizes homologous p- and d-copy derived fragments and the fragments are of different sizes, it defines a duplication differentiating polymorphism (DDP). With some DDP probes RFLPs were also detected but for the majority no allelic differences have been found as yet. The extent of duplication of the κ locus was determined with the help of 16 DDPs distributed over the entire locus (Pargent *et al.*, 1991b) and, of course, by the PFGE work (Lorenz *et al.*, 1987;

Weichhold et al., 1993a). The DDPs were essential in the structural work on the κ locus, since every newly isolated phage or cosmid clone had to be assigned to the p or d copy.

Haplotype 11

For most RFLPs and DDPs it is not known whether the underlying appearance or disappearance of a restriction site is caused by a base change, a deletion, insertion or by another structural change. One haplotype, however, is known to differ from the 'normal' haplotype N by the absence of the whole d copy of the locus. This so-called haplotype 11 was found in an individual homozygous for it (Straubinger et al., 1988b). The haplotype was characterized by hybridization to DDP probes across the locus (Pargent et al., 1991b). In a group of 23 caucasoid individuals there was, in addition to the homozygous one, one heterozygous individual. In a group including individuals of African and Asian origin 2 of 41 individuals were found to be heterozygous for haplotype 11 (Schaible et al., 1993). In PFGE experiments it was shown that about 1.0 Mb, including the whole d contig, is absent from the DNA of the homozygous individual, and indirect evidence indicates that this is due to a deletion rather than to the persistence of an evolutionarily early non-duplicated state (Weichhold et al., 1993a).

REARRANGED V_κ GENES

Since this topic is also dealt with in other chapters of this book, only some aspects related to the structure of the human κ locus are covered here. The available data on the mechanism of V(D)J joining and on V–J, V–D and DJ junctions were comprehensively reviewed by Lewis (1994). Among the reviews on hypermutation the recent ones by Berek (1993) MacLennan (1994) and Hengstchläger et al. (1995) should be mentioned.

V_κ–J_κ rearrangements

The 5′, 3′–polarity of the V_κ genes within the locus (arrows in Fig. 2) determines the type of rearrangement: the two J_κ-proximal genes B2 and B3, whose polarity is opposite to that of the J_κ–C_κ segment, are rearranged by an inversion mechanism (Klobeck et al., 1987a; Lorenz et al., 1988), while the other V_κ genes of the p copy rearrange by deletion of the stretch of DNA between the V_κ and J_κ genes (Weichhold et al., 1990). The genes of the d copy are located 1.35–1.8 Mb from J_κ–C_κ and the polarities of all of them are opposite to the polarity of J_κ–C_κ. For one of the d-copy genes the rearrangment by inversion was proven by PFGE experiments (Weichhold et al., 1990). Since for all p- and d-copy genes the polarities were determined by sequencing and detailed restriction mapping, their mode of rearrangement can be inferred from the cases that had been studied in detail.

The reciprocal products to the V_κ–J_κ joints are the signal joints, in which the hepta- and nona-nucleotide recombination sequences are linked back to back. The first signal joint was found in genomic DNA by Steinmetz et al. (1980) and was interpreted by Lewis et al. (1982) as the product of an inversion. Several signal joints have been found in the human κ locus (reviewed in Zachau, 1989a and Klein et al., 1993). When V and J gene segments are joined by the deletion mechanism, the excised material is lost from the cells or found as circular DNA. Such circles also carrying signal joints have been found in several systems including the mouse κ system (e.g. Hirama et al., 1991) but, possibly for technical reasons, not yet for the human κ genes.

B cells can undergo consecutive V_κ–J_κ rearrangements until all J_κ elements have been used. Examples of an inversion followed by a deletion and of two subsequent inversions have been reported (e.g. Klobeck et al., 1987a; Lorenz et al., 1988). In one cell line all products of a deletional V_κ–J_κ joining (in combination with a t(2;8) translocation; Klobeck et al., 1987b) and two consecutive inversions were cloned and sequenced. In the second rearrangement a productive V_κ–J_κ–C_κ joint was produced and, contrary to the common assumptions, this did not prevent a further recombination, which in this case was an aberrant one (Huber et al., 1992).

Apparently, the recombination machinery can handle inversions of 25–42-kb fragments for the genes B3 and B2, and of megabase-sized, that is millimetre-long, fragments for the d-copy genes. An intermediate formation of looped chromatin threads would have to be assumed. Clearly any deletional V_κ–J_κ joining leads to a loss of V_κ genes from the genome, while in an inversional joining all V_κ genes stay in the genome and can, in principle, be used in a second round of recombination. Also, genes other than V_κ, whose existence in the 800 kb between the p and d copies cannot be excluded (see below), would be kept in the genome on the (inversional) rearrangement of d-copy genes.

Which V_κ genes of the locus are rearranged, transcribed and translated?

This question was addressed by Klein et al. (1993) and Klein and Zachau (1995) on the basis of 70 of our own cDNA sequences and numerous nucleic acid and protein sequences from the literature. The results are shown in Fig. 2. It can be seen that some germline genes give rise to many transcripts and proteins and others to much fewer ones. For eight genes that, by definition (see p. 175), are potentially functional no transcripts or proteins were found. In general, fewer products were found to be derived from the d-copy than from the p-copy genes, although some cDNAs or proteins, whose sequences fit both duplicated germline genes (middle section in Fig. 2), may well be derived from d-copy genes. Figure 2 represents our current state of knowledge. Other genomic V_κ–J_κ joints and κ proteins may still be found, since the currently known ones are not the outcome of systematic searches. Transcription products (cDNAs), on the other hand, have been screened for in several laboratories. However, here also the absence of products for a potentially functional gene does not necessarily reflect its inability to be rearranged and transcribed, since the cDNA libraries are, of course, the outcome of immunological selection in the particular B cell repertoires under study; also in some cases an experimental bias in screening of the libraries cannot be excluded. At present, of the 76 V_κ genes of the locus 22 genes and five pairs of duplicated identical genes are known to be transcribed. The corresponding numbers for rearranged genomic V_κ genes and for full-length κ proteins are 17 plus 4 and 7 plus 7.

Somatic mutation

In considering the extent and type of somatic mutations, the definite assignment of the rearranged V_κ genes to certain germline genes is essential. The first mutated V_κ genes found (in the mouse system) could be defined, because the genomic surroundings of the rearranged and the unrearranged genes helped with the assignments (Pech et al., 1981). The assignment of cDNAs to germline genes is more difficult. In the human system this became possible only after it could be reasonably assumed that all functional germline genes of the locus were known. Because of allelic variation, the sequence of the unrearranged germline gene of the same individual should be known from whom the mutated V_κ–J_κ gene or cDNA is derived (see p. 180). This demand has to be taken seriously if one is interested in the mutation behaviour of specific single V_κ genes. It is less important if one considers the average extent and type of mutations of large numbers of rearranged genes or cDNAs.

There is a wide range of numbers of mutations per gene: for instance a cDNA with no mutation was isolated from the same library as other cDNAs with 25 mutations, which were derived from the same germline gene pair (Klein et al., 1993). The survey of the cDNA sequences from our laboratory and the data from the literature gave information on the type of mutations and on various other features of the mutation process, but only one feature should be mentioned here — the fact that mutations in adjacent nucleotides are found about twice as often as expected statistically. On a nucleotide basis, 20–25% of the mutations are in such blocks. About half of our cDNA clones and about 40% of the human V_κ sequences from the literature carry block mutations. They may have arisen from independent mutations in adjacent nucleotides and/or from combined exchange processes. The occurrence of blocks of altered nucleotides may be a feature of the maturation of the immune response in human and mouse. Certainly, block mutations that lead to amino acid replacements would be subject to selection. Somatic gene conversion appears not to play a major role in introducing single or block mutations in the human κ system. The mechanistic features possibly involved in the hypermutation and block mutation processes of V_κ genes were discussed by Klein et al. (1993), Klein and Zachau (1995) and Zachau (1995).

Sequences within or downstream of the J_κ–C_κ region are probably important for the hypermutation process since in an aberrantly rearranged κ gene, which is broken in V_κ by a t(2;8) translocation, the mutations extend to the adjacent chromosome 8 sequences but are not found in the 5' part of the V_κ gene (Klobeck et al., 1987b). A more detailed study in the mouse system was reported by Betz et al. (1994). Somatic mutations in an aberrantly rearranged V_κ gene had been found previously (Pech et al., 1981).

V_κ–J_κ junctions

There is an accumulation of base changes in the V_κ gene sequences close to the junction that is caused, at least in part, by the truncation and repair processes found in many V(D)J joining systems (review by Lewis, 1994). Therefore, the assignment of the V-gene moiety of the junction to a certain germline gene has to be based on a comparison of full-length sequences. In the survey of cDNA sequences from our laboratory and of data from the literature (Klein et al., 1993; Klein and Zachau, 1995) about one-fifth of the V_κ–J_κ junctions contained additional nucleotides between V_κ and J_κ. These nucleotides code for amino acids 95A and 95B or, if their number is different from 3 or 6, destroy the reading frame. The additional nucleotides do not have the characteristics of P or palindromic elements (Lafaille et al., 1989; Roth et al., 1992). They may be derived from the germline nucleotides between the last canonical codon, i.e. codon 95, and the heptanucleotide on the one side and between the complementary heptanucleotide and the first J_κ codon on the other side. The sequences of the additional nucleotides fit those germline sequences in several cases either fully or with one base change, which would have to be attributed to somatic mutations. Although terminal deoxynucleotidyl transferase (TdT) is generally not detected in B cells at the time the κ genes are rearranged, the presence of N segments (Alt and Baltimore, 1982) in these genes has been reported repeatedly (Klobeck et al., 1987b; Martin et al., 1991; H. Schroeder, personal communication). About 80% of the additional nucleotides in V_κ–J_κ junctions are C and G residues. This fits what would be expected for TdT-catalysed insertions, but it also fits the composition of germline nucleotides adjacent to those V_κ and J_κ genes that are frequently found in V_κ–J_κ joints (Klein et al., 1993; Victor and Capra, 1994). Therefore, it has to be checked in every case whether it is more likely that the additional nucleotides are germline or TdT derived or inserted by still another mechanism.

DISPERSED V_κ GENES

V_κ genes that are located outside the κ locus were called orphons in analogy to the histone and ribosomal RNA genes found outside the respective loci (Childs et al., 1981). The V_κ orphons were discovered when it proved impossible to link by chromosomal walking certain V_κ gene-containing cosmid clones to the existing contigs of the κ locus (Lötscher et al., 1986). The true locations of the orphon V_κ clones were first shown with the help of panels of human–rodent cell hybrid DNAs and later by in situ hybridization. Twenty-four orphon V_κ genes have been cloned and sequenced. One of them is localized on chromosome 1 and a cluster of five V_κ genes on chromosome 22 (Lötscher et al., 1986, 1988a). For five $V_\kappa I$ orphons it is only known that they are very similar to each other in sequence but not identical and that they are located on chromosomes other than chromosome 2 (Straubinger et al., 1988c; Röschenthaler et al., 1992). This so-called Z family of $V_\kappa I$ orphons may, in fact, have several more members (Meindl et al., 1990a). Two yeast artificial chromosome (YAC) clones and three cosmid clones with restriction maps and hybridization properties similar to those of Z-orphon clones were isolated but not studied in detail (Pargent, 1991; Huber, 1993). Because of the high sequence similarity between the Z orphons it would require much effort to specify whether newly isolated clones are derived from independent loci or whether they are alleles of already known orphons (Röschenthaler et al., 1992).

Of the 24 sequenced orphons 13 are localized on chromosome 2. One of them is located 1.5 Mb 3' of C_κ (V268 in Fig. 1). According to its sequence this V_κ gene is potentially functional (Huber et al., 1994) and a V_κ–J_κ rearrangement by inversion would not involve larger fragments than the rearrangements of V_κ genes of the d copy of the locus. However, no rearrangement products have been found yet. Because of its location outside the locus the gene is classified as an orphon. Another V_κ orphon without sequence defects is located on the long arm of chromosome 2 (V108 in Fig. 1; Huber et al., 1990). The 11 V_κ orphons of the W regions (Fig. 1), on the other hand, are pseudogenes also according to their sequences (Zimmer et al., 1990a). The three groups of W orphons were mapped to a 4.3-Mb region (Fig. 1; Weichhold et al., 1992). They were probably derived from gene regions of the κ locus by a pericentric inversion and subsequent amplification events (Zimmer et al., 1990b).

All V_κ orphons contain introns and, therefore, should have been dispersed on the DNA and not on the RNA and retrotranscript level; also, germline V_κ genes are believed not to be transcribed. Since, according to their sequences, all orphons, with one exception, have their closest relatives in the O regions (Schäble et al., 1994), they may have been derived in evolution from a common precursor. However, the structural similarities between orphons and κ locus regions are not high enough to allow a duplicative mechanism of orphon formation to be postulated. The attempts to specify other features of the dispersion mechanism(s) have not been very successful. One such feature is the presence of sequences in the neighbourhood of some orphons that were supposed to bind replication and/or transcription factors (Lötscher et al., 1988b), but no convincing arguments could be derived from that. At break-off points of homology between different orphon regions, which are probably junctions between translocated and receiving structures, direct and inverted repeats and an Alu element were found (Borden et al., 1990). However, since the sequence features differed from one insertion break-point to the other, no unique mechanism of translocation could be proposed.

REPETITIVE AND UNIQUE SEQUENCES IN THE κ LOCUS

The κ locus was not specifically investigated for repetitive elements but 15 LINE1 and 25 *Alu* sequences were detected in hybridization experiments and/or sequence comparisons. The properties of the elements and some evolutionary considerations were compiled by Schäble *et al.* (1994). In the same report the unique sequences, which qualify as sequence-tagged sites (STS) as defined by Olson *et al.* (1989), are described. Such sequences are an important feature of the Human Genome Project, since they should allow the reproducible detection via PCR of certain chromosomal sites or the isolation of the respective clones from libraries. The STS sequences are distributed fairly well across the κ locus.

EVOLUTION OF THE VK GENES

A crucial event in the recent history of the κ locus was its duplication. The sequenced regions of the p- and d-copy genes and pseudogenes differ on average by about 1% (404 of 38 136 bp; Schäble and Zachau, 1993). If one assumes 1% of divergence to correspond to 1 million years of evolution (Wilson *et al.*, 1987), this should be the age of the duplicated locus. However, there are various caveats. First, the extent of divergence is highly uneven across the locus ranging from 0 to 3.7% for different gene regions, which may be interpreted in terms of a surveillance mechanism counteracting in certain regions the mutational divergence. In addition, the basic relation of 1% mutation per 10^6 years is certainly not undisputed. However, the postulated date of duplication may be roughly right since the κ locus of the chimpanzee seems not to be duplicated (Ermert *et al.*, 1995) and the time point in evolution when human and chimpanzee clades diverged may have been 4–5 million years ago. $V_κ$ and $C_κ$ sequences of non-human primates are very similar to human ones, e.g. the $C_κ$ sequences of the chimpanzee and human are 99.6% identical.

Many events in the evolution of the κ locus occurred long before its duplication, e.g. the interdigitation of $V_κ$ genes of different subgroups (Pech and Zachau, 1984; review Zachau, 1989b) and the duplication of a group of three $V_κ$ genes (Huber *et al.*, 1993a). Also most changes, which converted $V_κ$ genes to pseudogenes, happened before the duplication of the locus. Another old feature of the κ locus is the dispersion of $V_κ$ genes to other parts of the genome. Since cosmid clones from the orphon regions of chromosomes 1 and 22 hybridized *in situ* to the assumed homologous chromosome bands of all great apes (Arnold *et al.*, 1995), the translocations may have happened very early in primate evolution. In the same study, a cosmid clone from one of the W regions was found to hybridize to a site that was pericentrically inverted in the corresponding chromosome of the chimpanzee, but not in that of the gorilla. Accordingly, the transposition occurred after the gorilla and before the chimpanzee clades diverged from the human evolutionary tree. This is also the time when the V108 region (see above; Fig. 1) became dissociated from the κ locus (Ermert *et al.*, 1995). The amplification of the transposed W region then occurred in at least two steps, which are postulated to have taken place 2×10^6 and 10^5 years ago, respectively (Zimmer *et al.*, 1990a).

Events that have to be dated after the duplication of the κ locus are the deletions of parts of Ap, Ld and Lp (Lautner-Rieske *et al.*, 1992; Huber *et al.*, 1993a), the insertion of an *Alu* element into one but not the other copy of the locus (Lautner-Rieske *et al.*, 1992; Schäble *et al.*, 1994) and at least some of the gene conversion-like events in the L regions (Huber *et al.*, 1993b). Most events that led to the divergence of the copies of the κ locus were point mutations. Not surprisingly, twice as many transitions as transversions are observed (Schäble and Zachau, 1993).

BIOMEDICAL IMPLICATIONS

Since there is reasonable certainty by now that all functional germline V_κ genes of the locus are known, conclusions are possible as to which part of the repertoire is expressed at which time in the development of the immune response. The κ chains found in pathological conditions, such as as autoimmune diseases or lymphomas, can be assigned to certain germline V_κ genes (compiled in Klein *et al.*, 1993; Klein *et al.*, 1995).

The individual who lacks, in a homozygous fashion, the d copy of the κ locus with its 36 V_κ genes (individual and haplotype 11; see p. 181) is apparently healthy and his κ chain/λ chain ratio is not altered (Schäble and Zachau, 1993). In general, the d-copy genes are not expressed to a great extent (Fig. 2) but it is known that the d-copy gene A2 codes for the most common light chain in the *Haemophilus influenzae* response (Scott *et al.*, 1989). Vaccination of individual 11 with the appropriate carbohydrate vaccine gave rise to antibodies whose light chains were derived, of course, from p-copy genes, but these light chains contained more somatic mutations than the usual A2-derived light chains (Scott *et al.*, 1991; review Scott *et al.*, 1992).

There is a 0.1% incidence of pericentric inversions of chromosome 2 in the present-day population. They have been formed in a process that is supposedly still going on. Heterozygous and homozygous (Gelman-Kohan *et al.*, 1993) individuals are apparently healthy. The inverted chromosomes carry the κ locus on the long arm and the W orphons on the short arm. The breakpoint on the long arm is, at the present level of analysis, indistinguishable from the one of the pericentric inversion that occurred in evolution (Lautner-Rieske *et al.*, 1993; previous literature quoted therein).

MISCELLANEOUS AND CONCLUDING REMARKS

The κ locus and its immediate surroundings comprise 2–3 Mb (Fig. 1), i.e. somewhat less than 0.1% of the human genome; 1040 kb of the locus and 800 kb of κ-related sequences from outside the locus have been cloned and mapped at high resolution; 160 kb and 90 kb respectively of the clones have been sequenced, mostly gene regions, deletion breakpoints and the like. An additional 11 Mb in the neighbourhood of the κ locus and the orphons have been mapped at medium or low resolution. Further work on the κ locus, particularly large-scale sequencing studies, will contribute to the general understanding of genome structure and evolution, and there will always be surprising results in this type of work. One of the more interesting questions is whether there are non-V_κ open reading frames between the known V_κ genes and particularly in the still uncloned part between the p and d copies of the locus. The related question of detecting non-V_κ transcripts from the germline κ locus was studied recently and the results were clearly negative for a number of cell lines (Lautner-Rieske et al., 1995). In the gap between the p and d copies clusters of rare-cutter restriction sites were observed (Weichhold et al., 1993a) as they are found in CpG islands adjacent to housekeeping genes. However, only a combination of further cloning, sequencing and expression studies can reveal whether there are non-V_κ genes located within or near the κ locus.

The study of the κ genes has advanced to a state where the germline and expressed repertoires are largely known. Some open questions are general problems of immunogenetics, such as the enzymology and mechanisms of V(D)J joining, V-gene maturation by hypermutation and selection, and the switch from κ to λ gene expression. The known structures provide a basis for further mechanistic studies aimed at answering these questions.

NOTE ADDED IN PROOF

This review was concluded at the time of submission in March 1994 and only partially updated in the proofs in June 1995.

Acknowledgements

I thank the members of our group for their contributions and for the discussion of the present manuscript. The work of our laboratory was supported by the Bundesministerium für Forschung und Technologie, Center Grant 0316200A, and the Fonds der Chemischen Industrie.

REFERENCES

Alt, F.W. and Baltimore, D. (1982). *Proc. Natl Acad. Sci. USA.* **79**, 4118–4122.
Arnold, N., Wienberg, J., Ermert, K. and Zachau, H.G. (1995). *Genomics* **26**, 147–150.
Bentley, D.L. and Rabbitts, T.H. (1980). *Nature* **288**, 730–733.
Bentley, D.L. and Rabbitts, T.H. (1981). *Cell* **24**, 613–623.
Bentley, D.L. and Rabbitts, T.H. (1983). *Cell* **32**, 181–189.

Berek, C. (1993). *Curr. Opin. Immunol.* **5**, 218–222.
Bergman, Y., Rice, D., Grosschedl, R. and Baltimore, D. (1984). *Proc. Natl Acad. Sci. USA.* **81**, 7041–7045.
Betz, A.G., Milstein, C., González-Fernandez, A., Pannell, R., Larson, T. and Neuberger, M.S. (1994). *Cell* **77**, 239–248.
Borden, P., Jaenichen, R. and Zachau, H.G. (1990). *Nucleic Acids Res.* **18**, 2101–2107.
Childs, G., Maxson, R., Cohn, R.H. and Kedes, L. (1981). *Cell* **23**, 651–663.
Chou, C.L. and Morrison, S.L. (1993). *J. Immunol.* **150**, 5350–5360.
Cox, J.P.L., Tomlinson, I.M. and Winter, G. (1994). *Eur. J. Immunol.* **24**, 827–836.
Ermert, K. (1994). Doctoral Thesis, Fakultät für Chemie und Pharmazie, Universität München.
Ermert, K., Mitlöhner, H., Schempp, W. and Zachau, H.G. (1995). *Genomics* **25**, 623–629.
Falkner, F.G. and Zachau, H.G. (1984). *Nature* **310**, 71–74.
Field, L.L., Tobias, R. and Bech-Hansen, T. (1987). *Nucleic Acids Res.* **15**, 3942.
Gelman-Kohan, Z., Rosensaft, J., Nisani Ben-Cohen, R. and Chemke, J. (1993). *Hum. Genet.* **92**, 427.
Gimble, J.M. and Max, E.E. (1987). *Mol. Cell. Biol.* **7**, 15–25.
Graninger, W.B., Goldman, P.L., Morton, C.C., O'Brien, S.J. and Korsmeyer, S.J. (1988). *J. Exp. Med.* **167**, 488–501.
Hengstschläger, M., Maizels, N. and Leung, H. (1995) In 'Progress in Nucleic Acid Research and Molecular Biology' (W.E. Cohn and K. Moldave, eds.) Vol. 50, pp. 67–99. Academic Press, New York.
Hieter, P.A., Max, E.E., Seidman, J.G., Maizel, J.V. and Leder, P. (1980). *Cell* **22**, 197–207.
Hieter, P.A., Maizel, J.V. and Leder, P. (1982). *J. Biol. Chem.* **257**, 1516–1522.
Hirama, T., Takeshita, S., Yoshida, Y. and Yamagishi, H. (1991). *Immunol. Lett.* **27**, 19–24.
Högbom, E., Magnusson, A.-C. and Leanderson, T. (1991). *Nucleic Acids Res.* **19**, 4347–4354.
Huber, C. (1993). Doctoral Thesis, Fakultät für Chemie und Pharmazie, Universität München.
Huber, C., Thiebe, R., Hameister, H., Smola, H., Lötscher, E. and Zachau, H.G. (1990). *Nucleic Acids Res.* **18**, 3475–3478.
Huber, C., Klobeck, H.-G. and Zachau, H.G. (1992). *Eur. J. Immunol.* **22**, 1561–1565.
Huber, C., Huber, E., Lautner-Rieske, A., Schäble, K.F. and Zachau, H.G. (1993a). *Eur. J. Immunol.* **23**, 2860–2867.
Huber, C., Schäble, K.F., Huber, E., Klein, R., Meindl, A., Thiebe, R., Lamm, R. and Zachau, H.G. (1993b). *Eur. J. Immunol.* **23**, 2868–2875.
Huber, C., Thiebe, R. and Zachau H.G. (1994). *Genomics* **22**, 213–215.
Judde, J.-G. and Max, E.E. (1992). *Mol. Cell. Biol.* **12**, 5206–5216.
Kabat, E.A., Wu, T.T., Perry, H.M., Gottesman, K.S. and Foeller, C. (1991). 'Sequences of Proteins of Immunological Interest.' NIH Publication No. 91-3242.
Klein, R. and Zachau, H.G. (1995). *Ann. N.Y. Acad. Sci.*, in press.
Klein, R., Jaenichen, R. and Zachau, H.G. (1993). *Eur. J. Immunol.* **23**, 3248–3271.
Klein, U., Klein, G., Ehlin-Henriksson, B., Rajewsky, K. and Küppers, R. (1995). *Molecular Medicine*, in press.
Klobeck, H.-G. and Zachau, H.G. (1986). *Nucleic Acids Res.* **14**, 4591–4603.
Klobeck, H.-G., Bornkamm, G.W., Combriato, G., Mocikat, R., Pohlenz, H.D. and Zachau, H.G. (1985). *Nucleic Acids Res.* **13**, 6515–6529.
Klobeck, H.-G., Zimmer, F.-J., Combriato, G. and Zachau, H.G. (1987a). *Nucleic Acids Res.* **15**, 9655–9665.
Klobeck, H.-G., Combriato, G. and Zachau, H.G. (1987b). *Nucleic Acids Res.* **15**, 4877–4888.
Klobeck, H.-G., Combriato, G. and Zachau, H.G. (1989). *Biol. Chem. Hoppe-Seyler* **370**, 1007–1012.
Kofler, R., Geley, S., Kofler, H. and Helmberg, A. (1992). *Immunol. Rev.* **128**, 5–21.
Kurth, J.H. and Cavalli-Sforza, L.L. (1994). *Am. J. Hum. Genet.* **54**, 1037–1041.
Lafaille, J.J., DeCloux, A., Bonneville, M., Takagaki, Y. and Tonegawa, S., (1989). *Cell* **59**, 859–870.
Lautner-Rieske, A., Thiebe, R. and Zachau, H.G. (1995). *Gene*, **159**, 199–202.
Lautner-Rieske, A., Huber, C., Meindl, A., Pargent, W., Schäble, K.F., Thiebe, R., Zocher, I. and Zachau, H.G. (1992). *Eur. J. Immunol.* **22**, 1023–1029.
Lautner-Rieske, A., Hameister, H., Barbi G. and Zachau, H.G. (1993). *Genomics* **16**, 497–502.

Lewis, S., Rosenberg, N., Alt, F. and Baltimore, D. (1982). *Cell* **30**, 807–816.
Lewis, S.M. (1994). *Adv. Immunol.* **56**, 27–150.
Lorenz, W., Straubinger, B. and Zachau, H.G. (1987). *Nucleic Acids Res.* **15**, 9667–9676.
Lorenz, W., Schäble, K.F., Thiebe, R., Stavnezer, J. and Zachau, H.G. (1988). *Mol. Immunol.* **25**, 479–484.
Lötscher, E., Grzeschik, K.-H., Bauer, H.-G., Pohlenz, H.-D., Straubinger, B. and Zachau, H.G. (1986). *Nature* **320**, 456–458.
Lötscher, E., Zimmer, F.-J., Klopstock, T., Grzeschik, K.-H., Jaenichen, R., Straubinger, B. and Zachau, H.G. (1988a). *Gene* **69**, 215–223.
Lötscher, E., Siwka, W., Zimmer, F.-J., Grummt, F. and Zachau, H.G. (1988b). *Gene* **69**, 225–236.
McBride, O.W., Hieter, P.A., Hollis, G.F., Swan, D., Otey, M.G. and Leder, P. (1982). *J. Exp. Med.* **155**, 1480–1490.
MacLennan, I.C.M. (1994). *Cur. Biol.* **4**, 70–72.
Malcolm, S., Barton, P., Murphy, C., Ferguson-Smith, M.A., Bentley, D.L. and Rabbitts, T.H. (1982). *Proc. Natl Acad. Sci. USA* **79**, 4957–4961.
Marks, J.D., Tristem, M., Karpas, A. and Winter G. (1991). *J. Immunol.* **21**, 985–991.
Martin, D., Huang, R., LeBien, T. and Van Ness, B. (1991). *J. Exp. Med.* **173**, 639–645.
Meindl, A., Klobeck, H.-G., Ohnheiser, R. and Zachau, H.G. (1990a). *Eur. J. Immunol.* **20**, 1855–1863.
Meindl, A., Kellner, B., Schattenkirchner, M. and Zachau, H.G. (1990b). *Exp. Clin. Immunogenet.* **7**, 20–25.
Mocikat, R., Pruijn, G.J.M., van der Vliet, P.E. and Zachau, H.G. (1988). *Nucleic Acids Res.* **16**, 3693–3704.
Mocikat, R., Falkner, F.-G. and Zachau, H.G. (1989). In 'Tissue Specific Gene Expression' (R. Renkawitz ed.), pp. 73–85. VCH Verlag, Weinheim.
Moxley, G. and Gibbs, R.S. (1992). *Genomics* **13**, 104–108.
Müller, B., Stappert, H. and Reth, M. (1990). *Eur. J. Immunol.* **20**, 1409–1411.
Olson, M., Hood, L., Cantor, C. and Botstein, D. (1989). *Science* **245**, 1434–1435.
Pargent, W. (1991). Doctoral Thesis, Fakultät für Biologie, Universität München.
Pargent, W., Meindl, A., Thiebe, R., Mitzel, S. and Zachau, H.G. (1991a). *Eur. J. Immunol.* **21**, 1821–1827
Pargent, W., Schäble, K.F. and Zachau, H.G. (1991b). *Eur. J. Immunol.* **21**, 1829–1835.
Parslow, T.G., Blair, D.L., Murphy, W.J. and Granner, D.K. (1984). *Proc. Natl Acad. Sci. USA* **81**, 2650–2654.
Pech, M. and Zachau, H.G. (1984). *Nucleic Acids Res.* **12**, 9229–9236.
Pech, M., Höchtl, J., Schnell, H. and Zachau, H.G. (1981). *Nature* **291**, 668–670.
Pech, M., Smola, H., Pohlenz, H.-D., Straubinger, B., Gerl, R. and Zachau, H.G. (1985). *J. Mol. Biol.* **183**, 291–299.
Röschenthaler, F., Schäble, K.F., Thiebe, R. and Zachau, H.G. (1992). *Biol. Chem. Hoppe-Seyler* **373**, 177–186.
Roth, D.B., Menetski, J.P., Nakajima, P.B., Bosma, M.J. and Gellert, M. (1992). *Cell* **70**, 983–991.
Saksela, K. and Baltimore, D. (1993). *Mol. Cell. Biol.* **13**, 3698–3705.
Schäble, K.F. and Zachau, H.G. (1993). *Biol. Chem. Hoppe-Seyler* **374**, 1001–1022.
Schäble, K.F., Thiebe, R., Meindl, A., Flügel, A. and Zachau, H.G. (1994). *Biol. Chem. Hoppe-Seyler* **375**, 189–199.
Schaible, G., Rappold, G.A., Pargent, W. and Zachau, H.G. (1993). *Hum. Genet.* **91**, 261–267; Erratum **92**, 195.
Scott, M.G., Crimmins, D.L., McCourt, D.W., Zocher, I., Thiebe, R., Zachau, H.G. and Nahm, M.H. (1989). *J. Immunol.* **143**, 4110–4116.
Scott, M.G, Crimmins, D.L., McCourt, D.W., Chung, G., Schäble, K.F., Thiebe, R., Quenzel, E.-M., Zachau, H.G. and Nahm, M.H. (1991). *J. Immunol.* **147**, 4007–4013.
Scott, M.G., Zachau, H.G. and Nahm, M.H. (1992). *Int. Rev. Immunol.* **9**, 45–55.
Sigvardsson, M., Bemark, M., and Leanderson, T. (1995). *Mol. Cell Biol.* **15**, 1343–1352.
Siminovitch, K.A., Bakhshi, A., Goldman, P. and Korsmeyer, S.J. (1985). *Nature* **316**, 260–262.
Staudt, L.M. and Lenardo, M.J. (1991). *Annu. Rev. Immunol.* **9**, 373–398.
Steinmetz, M., Altenburger, W. and Zachau, H.G. (1980). *Nucleic Acids Res.* **8**, 1709–1720.

Straubinger, B., Thiebe, R., Huber, C., Osterholzer, E. and Zachau, H.G. (1988a). *Biol. Chem. Hoppe-Seyler* **369**, 601–607.
Straubinger, B., Huber, E., Lorenz, W., Osterholzer, E., Pargent, W., Pech, M., Pohlenz, H.-D., Zimmer, F.-J. and Zachau, H.G. (1988b). *J. Mol. Biol.* **199**, 23–34.
Straubinger, B., Thiebe, R., Pech, M. and Zachau, H.G. (1988c). *Gene* **69**, 209–214.
Turnbull, I.F., Sriprakash, K.S. and Mathews, J.D. (1987). *Immunogenetics* **25**, 193–199.
Vassart, G., Georges, M., Monsieur, R., Brocas, H., Lequarre, A.S. and Christophe, D. (1987). *Science* **235**, 683–684.
Victor, K.D. and Capra, J.D. (1994). *Mol. Immunol.* **31**, 39–46.
Weichhold, G.M., Klobeck, H.-G., Ohnheiser, R., Combriato, G. and Zachau, H.G. (1990). *Nature* **347**, 90–92.
Weichhold, G.M., Lautner-Rieske, A. and Zachau, H.G. (1992). *Biol. Chem. Hoppe-Seyler* **373**, 1159–1164.
Weichhold, G.M., Ohnheiser, R. and Zachau, H.G. (1993a). *Genomics* **16**, 503–511.
Weichhold, G.M., Huber, C., Parnes, J.R. and Zachau H.G. (1993b). *Genomics* **16**, 512–514.
Weichhold, G.M., Ohnheiser, R. and Zachau, H.G. (1993c). In 'Progress in Immunology', vol. 8 (J. Gergely *et al.*, eds), pp. 115–119. Springer Verlag, Berlin.
Whitehurst, C., Henney, H.R., Max, E.E., Schroeder, H.W. Jr, Stüber, F., Siminovitch, K.A. and Garrard, W.T. (1992). *Nucleic Acids Res.* **20**, 4929–4930.
Wilson, A.C., Ochman, H. and Prager, E.M. (1987). *Trends Genet.* **3**, 241–247.
Wirth, T., Staudt, L. and Baltimore, D. (1987). *Nature* **329**, 174–178.
Zachau, H.G. (1989a). In 'Immunoglobulin Genes' (T. Honjo, F.W. Alt and T.H. Rabbitts, eds), pp. 91–109. Academic Press, London.
Zachau, H.G. (1989b). In 'Evolutionary Tinkering in Gene Expression' (M. Grunberg-Manago, B.F.C., Clark and H.G. Zachau, eds), pp. 111–119. Plenum Publishing, London.
Zachau, H.G. (1990). *Biol. Chem. Hoppe-Seyler* **371**, 1–6.
Zachau, H.G. (1993). *Gene* **135**, 167–173.
Zachau, H.G. (1995). *The Immunologist*, in press.
Zimmer, F.-J., Huber, C., Quenzel, E.-M., Schek, H., Stiller, C., Thiebe, R. and Zachau, H.G. (1990a). *Biol. Chem. Hoppe-Seyler* **371**, 283–290.
Zimmer, F.-J., Hameister, H., Schek, H. and Zachau, H.G. (1990b). *EMBO J.* **9**, 1535–1542.

9

Immunoglobulin λ genes

Erik Selsing and Loren E. Daitch

Department of Pathology, Tufts University School of Medicine, Boston, MA, USA

λ gene organization 193
Mice that express low levels of λ chains 195
λ gene recombinations 196
The order of light-chain gene recombination 197
C_κ deletion and recombination of RS DNA 200

Two classes of immunoglobulin light-chain polypeptides, designated as kappa (κ) and lambda (λ), are found among serum antibodies, although individual antibody molecules contain only one of these classes, either κ or λ. Both κ and λ light chains are found in most vertebrates; however, the κ/λ ratio varies widely among different species. In mice and humans, κ and λ chains are encoded by gene families located on separate chromosomes (chromosomes 6 (κ) and 16 (λ) in the mouse; chromosomes 2 (κ) and 22 (λ) in the human). No immunological effector functions that are specific for either κ or λ light-chains have been yet found. Instead, the two classes of light chains appear to reflect an evolutionary duplication and divergence process that has served to provide additional V-region diversity for the antibody repertoire.

λ GENE ORGANIZATION

The λ gene family found in most laboratory mouse strains is one of the smallest immunoglobulin gene systems; all of the gene segments that appear to be involved in the synthesis of λ light chains in the BALB/c mouse have been isolated, characterized and sequenced (Bernard *et al.*, 1978; Tonegawa *et al.*, 1978; Arp *et al.*, 1982; Blomberg and Tonegawa, 1982; Miller *et al.*, 1982; Dildrop *et al.*, 1987;

Sanchez and Cazenave, 1987). BALB/c mice have three V_λ genes and three C_λ genes that are functional. The small number of murine V_λ genes correlates with the low levels of λ light chains found in mice; only 5% of mouse serum light chains are of the λ class. Each C_λ segment has an associated upstream J_λ segment, and the $J_\lambda C_\lambda$ segments are arranged in two clusters, each of which contains two $J_\lambda C_\lambda$ units. The physical linkage between the λ gene segments has been described (Storb *et al.*, 1989) and the organization of the locus is shown in Fig. 1. The two clusters of $J_\lambda C_\lambda$ genes in the mouse appear to have arisen by two sequential gene duplications, the first duplication occurring at about the time the first mammals appeared and the second at about the time mice and humans diverged as separate lineages (Selsing *et al.*, 1982). Although most laboratory mouse strains display this λ gene organization, some wild mice have larger numbers of λ genes that appear to represent further gene duplications (Scott *et al.*, 1982; Scott and Potter, 1984a,b). Some inbred strains recently derived from wild mice also show larger complements of λ genes, suggesting that the inbreeding process does not result in the loss of λ genes (Kindt *et al.*, 1985). Within the λ locus, the $J_\lambda 4 C_\lambda 4$ gene segment appears to be a pseudogene and no λ4 protein products have yet been detected. Defects have been found in the $J_\lambda 4$ recombination–recognition site, in the $J_\lambda 4$ RNA splice site and in the $C_\lambda 4$ exon (Blomberg and Tonegawa, 1982; Miller *et al.*, 1982; Selsing *et al.*, 1982). Because some wild mice appear to have an intact $J_\lambda 4$ splice site (Mami and Kindt, 1987), it appears that the defects in the $J_\lambda 4$ recognition site or in the $C_\lambda 4$ exon are more likely to have initially inactivated the $J_\lambda 4 C_\lambda 4$ gene with other defects subsequently accumulating during evolution.

A murine gene (designated as λ5) that has extensive homology to mouse λ genes and that is selectively expressed in pre-B cell lines has also been described (Kudo *et al.*, 1987). The λ5 gene does not recombine during B-cell development, but is impor-

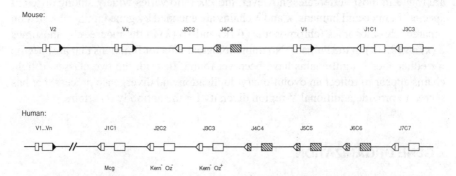

Fig. 1. The organization of V, J and C region gene segments for the λ genes in mice and humans. Recombination signal sequences are depicted by triangles. Open boxes represent gene segments that have open reading frames and are, or may be, functional, whereas hatched boxes represent known pseudogenes. Gene segments that have not been physically linked are indicated by broken lines. Distances are not to scale.

tant in pre-B cell differentiation (Kitamura *et al.*, 1992), perhaps due to interactions with V_{pre-B} and H-chain proteins produced in pre-B cells after H-chain VDJ recombination.

In humans, about 40% of serum antibodies contain λ light chains. Although the number of germline V_λ genes in the human is not known, human λ proteins show more V-region diversity than mouse λ chains and humans clearly have a larger number of V_λ genes than mice (Anderson *et al.*, 1984). Seven clustered human C_λ genes have been isolated and characterized (Heiterr *et al.*, 1981a; Vasicek and Leder, 1990). Each human C_λ gene has a single upstream J_λ segment (Udey and Blomberg, 1987). Of the seven C_λ genes in the human cluster (Fig. 1), three are known to produce functional λ chains having either the Mcg, Kern⁻Oz⁻ or Kern⁻Oz⁺ serological markers (Heiterr *et al.*, 1981a), three are non-functional pseudogenes (Vasicek and Leder, 1990), and the remaining gene appears functional but has not yet been found among myeloma proteins (Vasicek and Leder, 1990; Bauer and Blomberg, 1991). Several additional human C_λ-like genes have been isolated and characterized. Two of these genes have open reading frames that could encode λ-like proteins (Chang *et al.*, 1986). None of these C_λ-like genes undergo recombination but evidence suggests one or more of these C_λ-like genes might represent the human equivalents to the murine λ5 gene expressed selectively in pre-B cells (Hollis *et al.*, 1989; Schiff *et al.*, 1989).

MICE THAT EXPRESS LOW LEVELS OF λ CHAINS

Certain strains of mice express lower serum levels of λ chains than are found in most inbred laboratory strains. In SJL mice, for instance, λ1 serum levels are about 50 times below the values in BALB/c or C57BL/6 mice (Geckler *et al.*, 1978). Analyses of relevant λ1 DNA sequences in SJL and BALB/c mice have shown only a single base-pair difference that replaces a glycine at amino acid position 155 in BALB/c with a valine in SJL (Arp *et al.*, 1982; Kim *et al.*, 1994). This difference also causes a *Kpn*I site in the BALB/c $C_\lambda 1$ gene to be lost in SJL. A number of inbred, recombinant-inbred and random-bred strains show a complete correlation of the $C_\lambda 1$ *Kpn*I site polymorphism with low λ1 serum levels (Arp *et al.*, 1982; Epstein *et al.*, 1983; Ju *et al.*, 1986), suggesting that the Gly→Val polymorphism is directly involved in the low λ1 phenotype. Studies of κ-knockout mice carrying the SJL locus indicate that the SJL defect is λ1 specific because expression of the linked λ3 gene is normal (Kim *et al.*, 1994). The mechanisms responsible for the low λ1 levels in SJL mice have not been straightforward to elucidate. Although adult SJL mice have low numbers of λ1-bearing B cells, correlating with their low serum λ1 levels, newborn SJL mice have approximately the same number of λ1-bearing B cells as newborn BALB/c or C57BL/6 mice (Takemori and Rajewsky, 1981). In addition, a single γ2b, λ1-producing hybridoma, isolated from an SJA/20 mouse, exhibits normal gene transcription, mRNA translation and protein secretion of the SJL λ1 allele (Weiss *et al.*, 1985). This hybridoma λ1 chain also combines with the γ2b heavy chain, suggesting that the low levels of λ1-bearing IgG antibodies in SJL mice are not due to a defect in the association of SJL λ1 chains with H-chains other than μ. Models proposed for the SJL λ1 defect include an effect of the Gly→Val exchange on the ability of λ-producing B cells to be stimulated by antigen binding or an effect of the Gly→Val exchange on the regulation (perhaps by T cells) of λ-producing B cells in SJL and SJL-like strains.

Some wild mice also exhibit low levels of λ chains in the serum. Unlike the laboratory strains, however, these wild mice do not show the *Kpn*I site polymorphism indicative of an SJL-like allele, but instead exhibit λ1 genes that appear similar to those in BALB/c mice (Ju *et al.*, 1986). Thus, mechanisms that are not related to the Gly→Val exchange may regulate λ expression in some mice.

λ GENE RECOMBINATIONS

As with other antibody genes, λ genes require V–J recombination to produce functional proteins. Following the empirical '1-turn spacer/2-turn spacer' rule for immunoglobulin gene-recognition-site recombinations, almost all of the allowable gene recombinations in the murine λ system have been observed. The $V_\lambda 2$ gene can recombine with either $J_\lambda 2$, or $J_\lambda 3$ or $J_\lambda 1$ whereas the $V_\kappa 1$ gene can recombine with either $J_\lambda 3$ or $J_\lambda 1$. The $V_\lambda x$ gene has to date only been found to recombine with $J_\lambda 2$ (Dildrop *et al.*, 1987; Sanchez and Cazenave, 1987). Consistent with the observation that all the λ gene segments are organized in the same transcriptional orientation within the locus, no evidence for the retention of 'signal joint' reciprocal DNA segments, such as are found in the κ light-chain gene system (e.g. Lewis *et al.*, 1982; Selsing *et al.*, 1984), has been seen in λ-producing cells (Persiani *et al.*, 1987). This indicates that gene inversions do not occur within the small murine λ gene family.

B cells producing λ1, λ2 or λ3 proteins are present in ratios of approximately 3:2:1 among mouse fetal liver or adult spleen cell populations (Takemori and Rajewsky, 1981; Reilly *et al.*, 1982). However, the first extensive analysis of an Abelson murine leukemia virus (A-MuLV)-transformed cell line capable of active λ gene recombination (ABC-1) indicated that $V_\lambda 1 \rightarrow J_\lambda 1$ and $V_\lambda 1 \rightarrow J_\lambda 3$ recombinations occur at roughly equal frequency, whereas $V_\lambda 2$ recombinations occur much less frequently (Persiani *et al.*, 1987). The recombination frequencies of the λ isotypes within the ABC-1 cell line correlate with the relative distances between the respective V and J segments within the λ locus (Storb *et al.*, 1989). However, more limited analyses of other cell lines capable of active λ rearrangement indicate that $V_\lambda 1$ recombinations do not predominate (Muller and Reth, 1988; Chen *et al.*, 1994). This raises the possibility that pre-B cells in different subpopulations or at different stages of development exhibit different recombinational activities for the various λ isotypes. Alternately, recombination frequencies in some transformed cell lines may not accurately reflect the process in normal pre-B cells.

THE ORDER OF LIGHT-CHAIN GENE RECOMBINATION

In general, B cells express only one L-chain allele despite the presence of two alleles at both the κ and λ loci. In addition, in λ-producing cells, only one of the three λ isotypes is expressed. Although some studies have suggested that L-chain allotype and isotype exclusion might not be absolute (Hardy *et al.*, 1986; Gollahon *et al.*, 1988; Harada and Yamagishi, 1991), most B cells appear to be committed to expression of either κ or λ chains during the pre-B stages of development.

Early studies of L-chain gene recombination led to two contrasting models to explain how differentiating B cells are committed to either κ or λ gene expression (Alt *et al.*, 1981; Coleclough *et al.*, 1981; Korsmeyer *et al.*, 1981). One model proposed that both κ and λ genes are activated for recombination concurrently during pre-B cell development and that L-chain isotype expression is determined, on a stochastic basis, by the relative frequencies of forming either functional κ genes or λ genes. The second model proposed an ordered hierarchy of L-chain gene recombination in which κ genes are recombined first in maturing B cells and, in those cells that fail to produce a functional κ gene, λ gene recombinations are subsequently activated.

One source of support for a stochastic L-chain recombination model is the observed correlation, in mice and humans, between the proportions of κ and λ chains in the serum and the relative number of V_κ and V_λ genes that are generally thought to be present in the germline. This correlation seems compatible with the notion that the number of B cells committed to either κ or λ production reflects the relative number of genes that can be potentially recombined.

One difficulty with the stochastic model is that κ-producing cell lines generally have λ genes in the germline (unrecombined) configuration whereas λ-producing cell lines often have κ alleles that are recombined (Coleclough *et al.*, 1981; Hieter *et al.*, 1981b). Because a significant number of κ-producing cells frequently exhibit one recombined allele and one germline allele (Coleclough *et al.*, 1981) and, therefore, appear to have undergone only a single L-chain gene recombination event, a simple stochastic model would predict similar results for λ-producing cells. Thus, although mouse λ-producing B cells represent a small proportion of the total B-cell population, those λ-producing cells should also frequently have undergone a single recombination event and, therefore, exhibit two germline κ alleles. Yet, almost all murine λ-producing B cells have at least one, if not both, κ alleles recombined (Alt *et al.*, 1980; Coleclough *et al.*, 1981; Nadel *et al.*, 1990). In humans, where the number of κ and λ genes are more nearly equal, the situation is even more dramatic; almost all λ-producing cells have both κ alleles recombined (Hieter *et al.*, 1981b). These results are consistent with, and indeed provided the initial impetus for, the ordered model for L-chain gene recombination. Still, the stochastic model can be reconciled with these results by assuming that the rates of gene recombination differ for the κ and λ gene families. Supporting this notion, studies of recombination substrates introduced into an Abelson pre-B cell line have indicated that the V(D)J recombination signal sequences for the murine λ1 genes are inefficient for promoting rearrangement (Ramsden and Wu, 1991).

Studies of A-MuLV-transformed pre-B cell lines (Abelson lines) have provided many insights into the recombinational capacity of immature B cells (e.g. Alt *et al.*, 1981; Lewis *et al.*, 1982). Most Abelson lines have recombined heavy-chain genes but maintain L-chain genes in a germline configuration. However, a small number of Abelson lines exhibit active L-chain gene recombination during passage in culture (Lewis *et al.*, 1982; Reth *et al.*, 1985; Persiani *et al.*, 1987). Among these Abelson lines, almost all exhibit only κ gene recombinations in culture; λ gene recombinations are not detected or detected only at very low levels. In keeping with the notion that 'open' (DNase I-sensitive) chromatin structures and transcriptional activity may be involved in regulating antibody gene recombination (Blackwell *et al.*, 1986; Yancopoulos *et al.*, 1986), the C_κ genes are in DNAase I-sensitive, 'open' chromatin in the Abelson lines capable of κ gene recombination, whereas the non-recombining C_λ genes are in DNAase I-insensitive, 'closed' chromatin (Persiani and Selsing, 1989).

The finding that some Abelson lines recombine κ genes but not λ genes lends support for the ordered model of L-chain recombinations in pre-B cell differentiation. However, in one Abelson line, ABC-1, active κ and λ gene recombination occur concurrently (Persiani *et al.*, 1987) and both C_κ and C_λ genes are in 'open' chromatin structures (Persiani and Selsing, 1989). Subclone analyses of ABC-1 suggest that the relative frequencies of recombination for the κ and λ families are not greatly different in this cell line and indicate that, because both κ and λ genes undergo active recombination, a strictly ordered model for L-chain gene recombination in maturing pre-B cells is not correct (Persiani *et al.*, 1987). Nevertheless, even in the ABC-1 line, the pattern of L-chain gene recombination is not random because all cells that have recombined λ genes also have recombined κ alleles, whereas some cells have recombined κ genes with no detectable recombined λ alleles (Persiani *et al.*, 1987).

Abelson cell lines have also been produced using A-MuLVs that express a temperature-sensitive v-*abl* transforming protein (Engelman and Rosenberg, 1987). At permissive temperatures these *ts* Abelson lines, like most Abelson lines produced using the wild-type virus, exhibit H-chain gene rearrangement but little or no L-chain recombination. However, when the *ts* Abelson lines are shifted to the restrictive temperature, L-chain gene recombination is induced (Chen *et al.*, 1994). The temperature shift induces rearrangement of both κ and λ genes, suggesting that the two gene loci are concurrently activated for recombination. These results clearly suggest that κ and λ rearrangements in maturing pre-B cells are not initiated at separate developmental stages.

Studies of 'κ-knockout' mice, in which the κ locus has been inactivated to prevent either recombination or expression, have also provided insights into the regulation of L-chain expression (Chen *et al.*, 1993; Takeda *et al.*, 1993; Zou *et al.*, 1993). These mutant mice, which are not capable of producing κ-chains, have about half the number of B cells found in normal mice and all of these B cells express λ chains. In the κ-knockout mutants that cannot undergo κ recombination, large numbers of λ-producing B cells are generated, showing that recombinations of the κ locus (including recombinations of the RS element; see below) are clearly not required for the activation of λ recombination. In addition, the pre-B and B-cell numbers in the mutants are relatively similar to normal levels suggesting that those cells that would give rise to κ-producers in normal mice instead give rise to λ-producers in the κ-knockout mutants.

It appears that the results from a variety of experimental systems are leading to a picture of L-chain gene recombination that represents a blend of the original stochastic and ordered models. Although certainly speculative at this point, we would like to advance the following model for L-chain rearrangement that might account for the evidence which has been obtained from studies of both human and mouse L-chain gene rearrangement. In this model, we presume that, within most developing pre-B cells, the production of a functional H-chain allele signals the activation of L-chain recombination, as has been indicated by a number of studies (e.g. Nussenzweig et al., 1987; Reth et al., 1987). We suggest that the κ and λ L-chain loci are both activated for rearrangement concurrently but that, at the initial stages of L-chain gene recombination, the frequency of κ gene recombination is greater than λ gene recombination. As L-chain rearrangement progresses within an individual pre-B cell, we propose that the relative frequencies of κ and λ recombination become more equal. During L-chain recombination, we assume that production of a functional κ or λ L-chain protein in general signals the cessation of V(D)J recombination by the formation of an intact antibody molecule having the membrane-bound version of the μ chain.

The notion that the relative frequencies of κ and λ gene recombinations change during pre-B cell development deserves further comment. This feature of the model is included predominantly to account for the κ locus recombinations that are found in all human λ-producing cells, and to account for the murine Abelson cell lines that predominantly recombine κ genes with little or no concomitant λ gene rearrangement. The high frequency of recombined κ alleles in human λ-producing B cells has always been particularly difficult to reconcile with stochastic models. Although the presence of recombined κ alleles in murine λ producers could be explained by a low efficiency for λ recombination signal sequences (Ramsden and Wu, 1991), this explanation seems less likely for human B cells considering the larger percentages of human λ-producers, the larger numbers of human V_λ genes, and the apparent similarity of human κ and λ recombination signal sequences.

Several mechanisms, separately or together, could account for the proposed relative changes in κ and λ recombination frequencies during pre-B cell development. For instance, if κ and λ genes compete for some component of the recombinational machinery, the loss of κ V and J genes during earlier stages of rearrangement might allow λ genes to subsequently compete more effectively. The levels of an initially limiting component of the recombination machinery could also increase during L-chain rearrangement, thus decreasing competition between κ and λ genes. Furthermore, the chromatin organization of the λ locus might change, increasing the accessibility of the λ genes to the recombinational machinery.

In the proposed model, we have assumed that similar mechanisms regulate L-chain expression in mice and humans. It might be that the control of L-chain expression differs between species but, at present, the notion of a common mechanism is attractive. More detailed studies of species that express predominately λ chains (such as horses and cows) might provide new insights into the control of L-chain isotype exclusion.

C_κ DELETION AND RECOMBINATION OF RS DNA

As mentioned above, many λ-producing B cells display recombinations of κ chromosomes, some of which represent V–J joined genes that are non-functional owing to frameshifts introduced by the imprecise antibody gene recombination mechanism. However, many λ-producing B cells exhibit deletions of either the C_κ exon or the entire J_κ–C_κ region. These deletions are due to the recombination of a DNA sequence (designated as recombining sequence or RS DNA) that is located 25 kb downstream of the C_κ exon on chromosome 6 of the mouse (Durdik et al., 1984; Moore et al., 1985; Muller et al., 1990). Recombinations of RS DNA involve either a joining to a site located in the intron sequence separating J_κ and C_κ or joining to a V_κ gene (Fig. 2). The former RS recombinations result in a deletion of the C_κ region, whereas the latter recombinations delete the entire J_κ–C_κ region. Because a 2-turn spacer recognition site is found contiguous to RS DNA sequences, and because a partial recognition site is found in the J_κ–C_κ intron at the RS recombination point, all RS recombinations involve recombinase–recognition sites and appear to be mediated by the same enzymes that are involved in V(D)J joining (Durdik et al., 1984; Moore et al., 1985).

Analyses of hybridomas have shown that RS DNA recombinations are frequently found in λ-producing B cells but are rare or absent in κ-producing B cells (Durdik et al., 1984; Nadel et al., 1990). RS recombinations are also found to occur in Abelson cell lines that actively recombine L-chain genes; in those Abelson lines that exhibit predominantly κ gene recombination, RS rearrangements are observed but occur at a

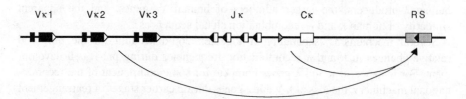

Fig. 2. RS DNA recombinations. Open and filled boxes represent J_κ, C_κ and V_κ gene segments, respectively. The stippled box represents the homology region found between the mouse RS segment and the human kde segment (see text). Open and filled triangles depict 1-turn and 2-turn recombination signal sequences, respectively. The hatched triangle represents a partial signal sequence found in the J_κ–C_κ intron. Arrows show the two types of RS DNA recombinations that have been observed. Distances are not to scale.

lower frequency than $V_\kappa J_\kappa$ joining events (Persiani *et al.*, 1987). The basis of the lower RS recombination frequency is not known, but could be related to the relative positioning of $V_\kappa J_\kappa$ and RS sequences on chromosome 6, similar to the chromosomal position effects on V(D)J recombination observed for other immunoglobulin gene segments (Alt *et al.*, 1984; Perlmutter *et al.*, 1985).

In the ABC-1 line, where active λ gene recombination is observed, almost all cells that have recombined λ genes also have recombined RS sequences (Persiani *et al.*, 1987). In fact, almost all ABC-1 cells have at least one recombined RS allele, regardless of the status of their λ genes. The presence of one recombined RS allele in an ABC-1 cell does not prevent either VJ joining or RS recombination on the remaining κ allele (Persiani *et al.*, 1987).

In humans, a DNA element located 24 kb downstream of the C_κ exon has also been found to mediate the deletion of C_κ or J_κ-C_κ regions in λ-producing B cells (Siminovitch *et al.*, 1985; Klobeck and Zachau, 1986). This element (designated as kde, for kappa-deleting-element) undergoes recombinations analogous to the mouse RS recombinations described above. The kde and RS elements share DNA homology in the regions directly contiguous to the identical recombinase-recognition sites present in each element (Siminovitch *et al.*, 1987). A portion of the RS/kde homology region (300 bp) is located downstream of a V_κ promoter after RS recombination. RNA transcripts containing sequences from the region of RS/kde homology have been detected in some mouse cell lines that have RS recombinations, but no function has yet been found for these transcripts (Daitch *et al.*, 1992). In the human and mouse DNA segments that separate the C_κ exons and the RS/kde elements, the only region of homology detectable by heteroduplex analyses corresponds to the enhancer located 3' of C_κ (Muller *et al.*, 1990).

The role of RS recombination in B-cell development has not yet been determined. The linkage between RS recombination and λ gene rearrangement noted in early studies led to the hypothesis that RS recombinations might cause the production of a factor that could activate or enhance gene rearrangement (Persiani *et al.*, 1987). However, gene transfer studies in Abelson cell lines have not supported this model (Daitch *et al.*, 1992) and analysis of κ-knockout mice have clearly shown that RS recombinations are not required for λ gene assembly (Chen *et al.*, 1993; Takeda *et al.*, 1993). It is possible that *trans*-acting negative factors which repress λ rearrangement might be deleted by RS recombination. However, this possibility could only be correct if the production of the negative factor is also disrupted in all the reported κ-knockout mutant mice.

On the other hand, an obvious consequence of RS recombination is the deletion of the C_κ region of the κ locus. Such C_κ region deletions could be important during B-cell development. For instance, C_κ deletions could prevent the useless transcription of non-functional κ genes and could serve to prevent the synthesis of C_κ fragment proteins in λ-producing cells. In addition, C_κ deletions during pre-B cell maturation would decrease the number of κ gene substrates competing for the recombinational machinery and, therefore, might increase the relative frequency of λ gene rearrangement in maturing pre-B cells as well as reduce the likelihood of functional κ genes arising in cells that are recombining λ genes. Furthermore, studies have suggested that self-reactive B cells which arise during pre-B cell differentiation can turn off L-chain expression and reactivate L-chain gene rearrangement; this process has been called 'receptor editing' (Radic *et al.*, 1993; Tiegs *et al.*, 1993). The deletion of the C_κ region by RS recombination would be one obvious way of inactivating functional κ genes to allow receptor editing. The production of mutant mice that cannot undergo RS recombination would appear the most direct approach towards illuminating the physiological role of C_κ deletion during B cell maturation.

REFERENCES

Alt, F.W., Enea, V., Bothwell, A.L.M. and Baltimore, D. (1980). *Cell* **21**, 1–12.
Alt, F., Rosenberg, N., Lewis, S., Thomas, E. and Baltimore, D. (1981). *Cell* **27**, 381–390.
Alt, F.W., Yancopoulos, G.D., Blackwell, K.T., Wood, C., Thomas, E., Boss, M., Coffman, R., Rosenberg, N., Tonegawa, S. and Baltimore, D. (1984). *EMBO J.* **3**, 1209–1219.
Anderson, M.L.M., Szajnert, M.F., Kaplan, J.C., McColl, L. and Young, B.D. (1984). *Nucleic Acids Res.* **12**, 6647–6661.
Arp, B., McMullen, M.D. and Storb, U. (1982). *Nature* **298**, 184–187.
Bauer, T.R., Jr and Blomberg, B. (1991). *J. Immunol.* **146**, 2813–2820.
Bernard, O., Hozumi, N. and Tonegawa, S. (1978). *Cell* **15**, 1133–1144.
Blackwell, T.K., Moore, M.W., Yancopoulos, G.D., Suh, H., Lutzker, S., Selsing, E. and Alt, F.W. (1986). *Nature* **324**, 585–589.
Blomberg, B. and Tonegawa, S. (1982). *Proc. Natl Acad. Sci. USA* **79**, 530–533.
Chang, H., Dmitrovsky, E., Hieter, P.A., Mitchell, K., Leder, P., Turoczi, L., Kirsch, I.R. and Hollis, G.F. (1986). *J. Exp. Med.* **163**, 425–435.
Chen, J., Trounstine, M., Kurahara, C., Young, F., Kuo, C.C., Xu, Y., Loring, J., Alt, F.W. and Huszar, D. (1993). *EMBO J.* **3**, 821–830.
Chen, Y.Y., Wang, L.C., Huang, M.S. and Rosenberg, N. (1994). *Genes Dev.* **8**, 688–697.
Coleclough, C., Perry, R.P., Karjalainen, K. and Weigert, M. (1981). *Nature* **290**, 372–378.
Daitch, L.E., Moore, M.W., Persiani, D.M., Durdik, J.M. and Selsing, E. (1992). *J. Immunol.* **149**, 832–840.
Dildrop, R., Grause, A., Muller, W. and Rajewsky, K. (1987). *Eur. J. Immunol.* **17**, 731–734.
Durdik, J., Moore, M.W. and Selsing, E. (1984). *Nature* **307**, 749–752.
Engelman, A. and Rosenberg, N. (1987). *Proc. Natl Acad. Sci. USA* **84**, 8021–8025.
Epstein, R., Lehmann, K. and Cohn, M. (1983). *J. Exp. Med.* **157**, 1681–1686.
Geckler, W., Faversham, J. and Cohn, M. (1978). *J. Exp. Med.* **148**, 1122–1136.
Gollahon, K., Hagman, J., Brinster, R. and Storb, U. (1988). *J. Immunol.* **141**, 2771–2780.
Harada, K. and Yamagishi, H. (1991). *J. Exp. Med.* **173**, 409–415.
Hardy, R.R., Dangl, J.L., Hayakawa, K., Jager, G., Herzenberg, L.A. and Herzenberg, L.A. (1986). *Proc. Natl Acad. Sci. USA* **83**, 1438–1442.
Hieterr, P.A., Hollis, G.F., Korsmeyer, S.J., Waldmann, T.A. and Leder, P. (1981a). *Nature* **294**, 536–540.
Hieterr, P.A., Korsmeyer, S.J., Waldmann, T.A. and Leder, P. (1981b). *Nature* **290**, 368–372.
Hollis, G.F., Evans, R.J., Stafford-Hollis, J.M., Korsmeyer, S.J. and McKearn, J.P. (1989). *Proc. Natl Acad. Sci. USA* **86**, 5552–5556.
Ju, S.-T., Selsing, E., Huang, M.-C., Kelly, K. and Dorf, M.E. (1986). *J. Immunol.* **136**, 2684–2688.
Kim, J.Y., Kurtz, B., Huszar, D. and Storb, U. (1994). *EMBO J.* **13**, 827–834.
Kindt, T.J., Gris, C., Guenet, J.L., Bonhomme, F. and Cazenave, P.-A. (1985). *Eur. J. Immunol.* **15**, 535–540.

Kitamura, D., Roes, J., Kuhn, R. and Rajewsky, K. (1991). *Nature* **350**, 423–426.
Kitamura, D., Kudo, A., Schaal, S., Muller, W., Melchers, F. and Rajewsky, K. (1992). *Cell* **69**, 823–831.
Klobeck, H.G. and Zachau, H.G. (1986). *Nucleic Acids Res.* **14**, 4591–4603.
Korsmeyer, S.J., Hieter, P.A., Ravetch, J.V., Poplack, D.G., Waldmann, T.A. and Leder, P. (1981). *Proc. Natl Acad. Sci. USA* **78**, 7096–7100.
Kudo, A., Sakaguchi, N. and Melchers, F. (1987). *EMBO J.* **6**, 103–107.
Lewis, S., Rosenberg, N., Alt, F. and Baltimore, D. (1982). *Cell* **30**, 807–816.
Mami, F. and Kindt, T.J. (1987). *J. Immunol.* **138**, 3980–3985.
Miller, J., Selsing, E. and Storb, U. (1982). *Nature* **295**, 428–430.
Moore, M.W., Durdik, J., Persiani, D.M. and Selsing, E. (1985). *Proc. Natl Acad. Sci. USA* **82**, 6211–6215.
Muller, B. and Reth, M. (1988). *J. Exp. Med.* **168**, 2131–2137.
Muller, B., Stappert, H. and Reth, M. (1990). *Eur. J. Immunol.* **20**, 1409–1411.
Nadel, B., Cazenave, P.A. and Sanchez, P. (1990). *EMBO J.* **9**, 435–440.
Nussenzweig, M.C., Shaw, A.C., Sinn, E., Danner, D.B., Holmes, K.L., Morse, H.C., III and Leder, P. (1987). *Science* **236**, 816–819.
Perlmutter, R., Kearney, J., Chang, S. and Hood, L. (1985). *Science* **227**, 1597–1601.
Persiani, D.M. and Selsing, E. (1989). *Nucleic Acids Res.* **17**, 5339–5348.
Persiani, D.M., Durdik, J. and Selsing, E. (1987). *J. Exp. Med.* **165**, 1655–1674.
Radic, M.Z., Erikson, J., Litwin, S. and Weigert, M. (1993). *J. Exp. Med.* **177**, 1165–1173.
Ramsden, D.A. and Wu, G.E. (1991). *Proc. Natl Acad. Sci. USA* **88**, 10721–10725.
Reilly, E.B., Frackelton, A.R., Jr and Eisen, H. (1982). *Eur. J. Immunol.* **12**, 552–557.
Reth, M.G., Ammirati, P., Jackson, S. and Alt, F.W. (1985). *Nature* **317**, 353–355.
Reth, M., Petrac, E., Wiese, P., Lobel, L. and Alt, F.W. (1987). *EMBO J.* **6**, 3299–3305.
Sanchez, P. and Cazenave, P.-A. (1987). *J. Exp. Med.* **166**, 265-270.
Schiff, C., Milili, M. and Fougereau, M. (1989). *Eur. J. Immunol.* **19**, 1873–1878.
Scott, C. and Potter, M. (1984a). *J. Immunol.* **132**, 2630–2637.
Scott, C. and Potter, M. (1984b). *J. Immunol.* **132**, 2638–2643.
Scott, C.L., Mushinski, J.F., Huppi, K., Weigert, M. and Potter, M. (1982). *Nature* **300**, 757–760.
Selsing, E., Miller, J., Wilson, R. and Storb, U. (1982). *Proc. Natl Acad. Sci. USA* **79**, 4681–4685.
Selsing, E., Voss, J. and Storb, U. (1984). *Nucleic Acids Res.* **12**, 4229–4245.
Siminovitch, K.A., Bakhshi, A., Goldman, P. and Korsmeyer, S.J. (1985). *Nature* **316**, 260–261.
Siminovitch, K.A., Moore, M.W., Durdik, J. and Selsing, E. (1987). *Nucleic Acids Res.* **15**, 2699–2705.
Storb, U., Haasch, D., Arp, B., Sanchez, P., Cazenave, P.-A. and Miller, J. (1989). *Mol. Cell. Biol.* **9**, 711–718.
Takeda, S., Zou, Y.-R., Bluethmann, H., Kitamura, D., Muller, U. and Rajewsy, K. (1993). *EMBO J.* **12**, 2329–2336.
Takemori, T. and Rajewsky, K. (1981). *Eur. J. Immunol.* **11**, 618–625.
Tiegs, S.L., Russell, D.M. and Namazee, D. (1993). *J. Exp. Med.* **177**, 1009–1020.
Tonegawa, S., Maxam, A.M., Tizard, R., Bernard, O. and Gilbert, W. (1978). *Proc. Natl Acad. Sci. USA* **75**, 1485–1489.
Udey, J.A. and Blomberg, B. (1987). *Immunogenetics* **25**, 63–70.
Vasicek, T.J. and Leder, P. (1990). *J. Exp. Med.* **172**, 609–620.
Weiss, S., Lehmann, K. and Cohn, M. (1985). *Eur. J. Immunol.* **15**, 768–772.
Yancopoulos, G.D., Blackwell, T.K., Suh, H., Hood, L. and Alt, F.W. (1986). *Cell* **44**, 251–259.
Zou, Y.R., Takeda, S. and Rajewksy, K. (1993). *EMBO J.* **12**, 811–820.

10

The variable region gene assembly mechanism

Ami Okada and Frederick W. Alt

The Howard Hughes Medical Institute Research Laboratories, The Children's Hospital, Boston, MA, USA

Introduction 205
Model systems for studying V(D)J recombination 207
Targeting the site-specific recombination: the recombination signal sequences........209
Double-strand cleavage and the joining of the coding sequence and RSS 213
Interversional vs. deletional joining 216
Modification of the coding end during rearrangement 218
Lymphocyte-specific genes and activities involved in V(D)J recombination...........222
Generally expressed activities involved in V(D)J recombination..............225
Control of V(D)J recombination 228

INTRODUCTION

B- and T-cells must be able to recognize a virtually unlimited set of antigen molecules with a high degree of specificity (Tonegawa, 1983). Almost every B- and T-cell expresses antigen receptors on the surface that are unique to that cell. The primary mechanism responsible for generating diversity of antigen receptors is the assembly of the variable (V) region of each antigen receptor chain by the somatic recombination of variable (V), joining (J) and, in some cases, diversity (D) gene segments (Fig. 1) (Tonegawa, 1983). This somatic recombination event, or V(D)J recombination, occurs during early stages of lymphocyte differentiation (Tonegawa, 1983), and is tightly associated with the program of lymphocyte development (reviewed in Chen *et al.*, 1993). Since the V, D, and J segments that encode each antigen receptor V region can have several to several hundred members, a significant level of diversity can be generated from the combinatorial assembly of these segments (Davis *et al.*, 1988).

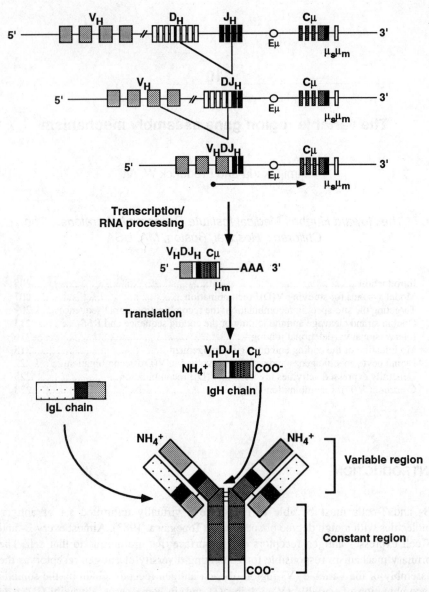

Fig. 1. Immunoglobulin heavy (IgH) chain gene assembly and expression. The order of events that lead to the assembly of the IgH chain into a complete immunoglobulin molecule is depicted. H chain gene assembly begins with the rearrangement of D segment (white box) to J segment (black box), followed by the rearrangement of the V segment (gray box) to the pre-assembled D–J segment. Transcripts originating from the V_H promoter that encode the V_H, D, J_H, and C_μ (stippled box) gene segments are differentially spliced and give rise to both the membrane and secreted form of IgM. An IgH chain protein combines with an IgL chain protein to form a typical monomeric subunit of an Ig molecule.

Additional diversity of the antigen receptor can also be generated during the somatic recombination event itself. The junction of the assembled V-region gene segments encodes the third complementarity determining region (CDR3) of the immunoglobulin (Ig) V region that contacts the antigen, and an analogous antigen contact region of the T-cell receptor (TCR) (Davis et al., 1988). Thus, features of the recombination machinery that modify the joining segment ends during assembly can significantly contribute to increasing the diversity of the antigen receptor repertoires (Alt et al., 1992). In this chapter, the current understanding of the mechanism of V(D)J recombination is summarized.

MODEL SYSTEMS FOR STUDYING V(D)J RECOMBINATION

The sequence of events that are believed to occur during the assembly of antigen receptor V-region genes has primarily been deduced from the analysis of the products of such recombination events, as compared to the germline unrearranged V-region gene segment counterparts (Tonegawa, 1983). The first recombined V(D)J genes were isolated from myelomas and plasmacytomas that secrete Ig (Hozumi and Tonegawa, 1976; Bernard et al., 1978). Subsequently, Abelson murine leukemia virus (A-MuLV)-transformed pre-B cell lines that continue to actively rearrange Ig loci in culture were useful for isolating V(D)J rearrangements (Alt et al., 1981; Baltimore, 1981). In particular, these A-MuLV-transformed cell lines have been useful for analyzing the progression of rearrangement events during pre-B cell differentiation and as hosts for V(D)J recombination substrates (reviewed by Alt et al., 1987).

Cell lines and mice

The recent isolation of recombination activating genes (*RAG*) 1 and 2, which are genes that can activate V(D)J rearrangement when expressed in vertebrate cell lines, have greatly aided elucidation of the V(D)J recombination mechanism (Schatz et al., 1989; Oettinger et al., 1990). Isolation of the *RAG* genes have enabled the study of the efficiency and accuracy of V(D)J recombination in any cell line; thus mutations found in non-lymphoid cell lines that affect generally expressed components of the recombination machinery have been able to be characterized (Pergola et al., 1993; Taccioli et al., 1993, 1994a,b). The analysis of antigen receptor rearrangements in mice with naturally occurring or gene-targeted mutations that affect V(D)J recombination (i.e. *RAG*-1, *RAG*-2, terminal deoxynucleotidyl transferase (TdT), severe combined immunodeficiency (SCID)) also have contributed significantly to the current understanding of the mechanism of V(D)J recombination (Bosma and Carroll, 1991; Mombaerts et al., 1992; Shinkai et al., 1992; Gilfillan et al., 1993; Komori et *al.*, 1993). Finally, the ability to isolate increasingly discrete populations of cells and to analyze the rearrangement events from even a single cell in normal and mutant mice by the polymerase chain reaction (PCR) has resulted in the rapid analysis of lymphoid compartments that have distinct antigen receptor rearrangements (e.g. Asarnow *et al.* 1989; Jacob *et al.*, 1991; Kuppers *et al.*, 1993; Li *et al.*, 1993).

Recombination substrates

The development of 'recombination substrates'—exogenous DNA segments containing the target sequences of the V(D)J recombination machinery that undergo rearrangement when introduced into cell lines with V(D)J recombinase activity—has also contributed to the understanding of mechanisms involved in V(D)J rearrangement (Fig, 2; Blackwell and Alt, 1984; Lewis *et al.*, 1984; Yancopoulos *et al.*, 1986; Hendrickson *et al.*, 1988, 1991; Engler *et al.*, 1991b; Asarnow *et al.*, 1993). In particular, the development of extrachromosomal recombination substrates has

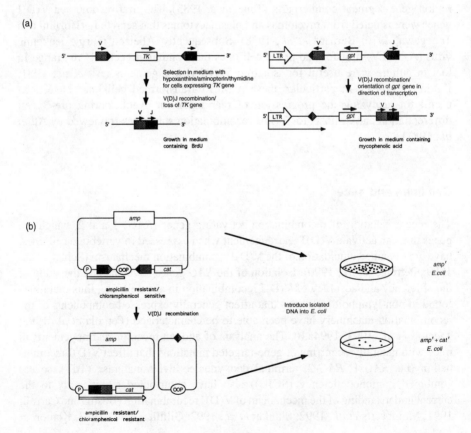

enabled the rapid analysis of recombination efficiency, and has enabled a detailed characterization of *cis*-acting sequences that effect rearrangement (Fig. 2b) (Hesse *et al.*, 1987, 1989; Lieber *et al.*, 1988; Gauss and Leiber, 1992; Gerstein and Lieber, 1993a,b Wei and Leiber, 1993). These extrachromosomal substrates are designed so that upon recombination of V-region segments within the substrate, a bacterial transcriptional stop signal is removed. Thus, rearrangement can be examined by introduction of plasmid harvested from the cells into bacteria, and assaying for the expression of a drug-resistance (chloramphenicol) gene.

TARGETING THE SITE-SPECIFIC RECOMBINATION: THE RECOMBINATION SIGNAL SEQUENCES

Nucleotide composition of the RSS

The putative events resulting in the assembly of an antigen receptor gene are outlined in Fig. 3, and are described in more detail in the text (Gellert, 1992; Tonegawa, 1983). The V(D)J recombination process is apparently initiated by the recognition of recombination signal sequences (RSSs) by the V(D)J recombination machinery, followed by the introduction of a double-strand cleavage between the nucleotides encoding the RSS and the adjacent 'coding' sequences. Two RSSs (and two coding segments) are usually involved, and the subsequent joining of the excised DNA

Fig. 2. Recombination substrates. (A) Chromosomally integrated V(D)J recombination substrates. A combination of V-region gene segments (black or gray squares) with recombination signal sequences (RSS; black or gray triangles) can be assembled and introduced into the genome to assay for their ability to undergo V(D)J rearrangement. Genes that confer drug resistance (i.e. the herpes simplex thymidine kinase gene (*TK*) or the *E. coli* gene encoding xanthine-guanine phosphoribosyl transferase gene (*gpt*; denoted as open squares) can be designed into such recombination substrates so that cells that have recombined the substrate can be selected by resistance to drugs. The long terminal repeat sequences (LTR) from Moloney murine leukemia virus can provide transcriptional activity in most cell types. (B) Extrachromosomal recombination substrates. Plasmid vectors harboring RSS can undergo site-specific rearrangement without integration into the genome (Hesse *et al.*, 1987). Such substrates harbor a prokaryotic promoter (P) and drug resistance or color marker gene (i.e. chloramphenicol resistance (*cat*) gene). These elements are separated by a prokaryotic transcription terminator (OOP) flanked by RSS. The site-specific recombination of RSS removes the transcription terminator sequence, and the plasmid formed can transcribe the drug resistance or marker gene in bacteria. To isolate rearranged products and to assay for recombination efficiency, the extrachromosomal substrates are isolated from the cells and introduced into bacteria. The relative number of chloramphenicol-resistant bacterial colonies (rearranged colonies) to ampicillin-resistant colonies (total input plasmid) reflects the efficiency of rearrangement of the cell line. The RSS can be placed in different orientations to assay for the formation of 'RSS' or 'coding' joins. (Adapted from Hesse *et al.* 1987.)

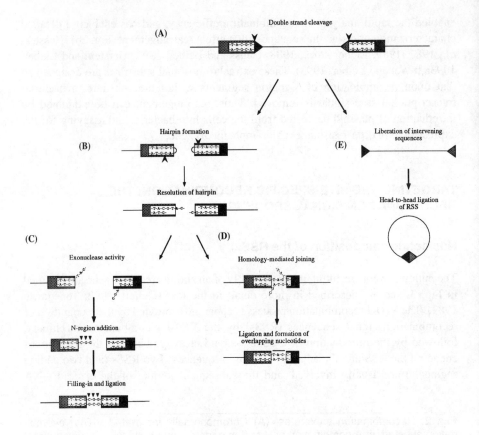

Fig. 3. Overview of the mechanism of V(D)J recombination. (A) V(D)J recombination is initiated with the cleavage of the phosphodiester bond adjacent to the RSS on both strands; the RSS are denoted as black or gray triangles, example coding sequences are surrounded by gray or black boxes, and the break is denoted with arrows. (B) The coding sequences could be modified in a number of different ways prior to their ligation (detailed in Fig. 5.). The coding and complementary strands of the liberated coding sequences form a covalent bond, resulting in a 'hairpin' structure at the end; the hairpin is resolved with a single strand nick near the apex (denoted with small arrows), sometimes resulting in P-nucleotide addition (nucleotides in bold). (C) Nucleotides from the ends are removed by an exonuclease activity. N-regions (denoted with black triangles) can be added, if TdT is expressed. Gaps in the two sequences are filled in, and ligated. (D) If short homologies exist between the terminal nucleotides of the joining segments, these could be used to align the segments, and initiate polymerization; 'overlapping' nucleotides are denoted with a line. (E) The two recombining RSSs are usually ligated head-to-head precisely, without modification of the ends before ligation. Depending on the orientation of the RSS, the intervening sequences flanked by the RSS can be deleted as a circle, or are inverted (see Fig. 7 for details).

ends—the two RSSs to each other and the two coding segments to each other—results in the site-specific recombination of the V-region gene segments.

All functional germline Ig and TCR V-region gene segments are flanked by the conserved RSS (Early et al., 1980; Tonegawa, 1983). The consensus RSS consists of a conserved heptamer and an A–T rich nonamer separated by either 12 ± 1 or 23 ± 1 bp called 'spacer' sequences (Fig. 4; Early et al., 1980; Sakano et al., 1981; Akira et al., 1987). V-gene segments flanked by RSSs with 12-bp spacers efficiently recombine with sequences flanked by RSSs with 23-bp spacers (the 12–23 rule; Fig. 4) (Early et al., 1980). The consensus sequences for the heptamer and nonamer of the RSS are CACAGTG and ACAAAAACC respectively (Early et al., 1980; Akira et al., 1987; Hesse et al., 1989; Hendrickson et al., 1991a). Mutational studies of the RSS revealed that the three residues of the heptamer closest to the cleavage site are particularly critical for efficient V(D)J recombination (Hunkapillar and Hood, 1986; Akira et al., 1987). When these nucleotides are altered from the consensus, the efficiency of rearrangement is reduced up to 100 fold in assays of recombination substrates (Hesse et al., 1989). While the nonamer sequence can vary more than the heptamer sequence, the placement of a G residue at position 6 or 7 in the nonamer has been shown to reduce the efficiency of rearrangement significantly (Hesse et al., 1989).

The nucleotide composition of the spacer sequences is not conserved between different RSSs, and mutation of individual nucleotides within the spacer does not appear to affect the efficiency of recombination (Hesse et al., 1989; Wei and Lieber, 1993). However, altering the 12 or 23 bp spacing between the heptamer and nonamer diminishes the rearrangement efficiency, and complementary changes in the spacing of the corresponding RSS do not restore this loss of function (Hesse et al., 1989; Wei and Lieber, 1993). One suggested explanation for this exact spacing requirement is that the 12 and 23 intervening bases permit one or two complete turns of the helix between the heptamer and nonamer, enabling putative proteins bound to the heptamer and nonamer to potentially interact within each RSS (Blackwell et al., 1989). Heptamer and nonamer binding activities in nuclear extracts have been described (Aguilera et al., 1987; Matsunami et al., 1989; Thompson, 1992) and cDNAs encoding these factors have been cloned (Aguilera et al., 1987; Matsunami et al., 1989; Shirakata et al., 1991); however, the physiological relevance of these activities has not been demonstrated.

V(D)J recombination to isolated heptamer sequences

A solitary heptamer can also target rearrangement events when paired with a normal RSS, indicating that nonamer sequences are not absolutely required for efficient rearrangement to take place. The deletion of the Ig C_κ region and replacement of V_H sequences are heptamer-mediated rearrangements that occur during normal lymphoid development (Durdick et al., 1984; Reth et al., 1986; Covey et al., 1990; Usuda et al. 1992). In addition, chromosomal translocations isolated from many lymphomas appear to have been caused by aberrant V(D)J recombinase-mediated

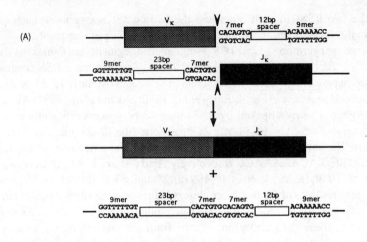

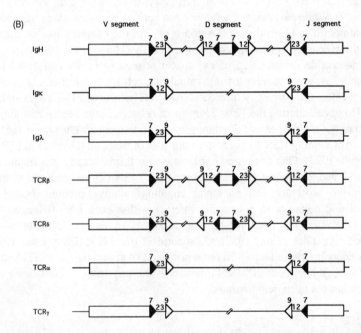

Fig. 4. The '12–23' rule. (A) Two V-region gene segments (denoted by grey or black squares) are depicted: one that is flanked by RSS with 23-bp spacer sequences, and the other flanked by RSS with 12-bp spacer sequence. V(D)J rearrangement results in the joining of 'coding' sequences that flank the RSS with the 23-bp spacer with sequences adjacent to the RSS with the 12-bp spacer, and the ligation of the two RSSs to each other. (B) The configuration of the spacing between the heptamer and the nonamer of RSSs flanking the Ig and TCR V-region gene segments. The heptamer is denoted as a closed triangle and the nonamer is denoted as an open triangle; the number in between the heptamer and nonamer denotes the number of nucleotides in the 'spacer' of the RSS.

recombination between a RSS and cryptic heptamers in the genome (Haluska et al., 1987; Finger et al., 1989). To this effect, the pairing of one normal RSS together with a normal or mutated RSS seems to be necessary to induce double-strand cleavage of the DNA to initiate V(D)J rearrangement. It has been shown in recombination substrates that sequences adjacent to a singular consensus RSS will be cleaved and rejoined to form an 'open and shut' join (described below), if a cryptic or mutated heptamer is present in *cis* (Hendrickson et al., 1991a; Lewis and Hesse, 1991); without the heptamer-like sequences, the wild-type RSS does not seem to be recognized or cleaved (Hendrickson et al., 1991; Lewis and Hesse, 1991).

Nucleotide composition of the coding segment can affect V(D)J recombination efficiency

The sequences adjacent to the RSS or in the spacer are apparently not conserved (Wei and Lieber, 1993). However, the nucleotide composition of the coding segment that directly flanks the RSS can affect the efficiency of V(D)J recombination (Boubnov et al., 1993; Gerstein and Lieber, 1993b). Extrachromosomal substrates with an A or T next to the heptamer in the coding sequence rearrange less efficiently than similar substrates with G or C residues in this position (Gerstein and Lieber, 1993b). The nucleotide composition of sequences adjacent to the recognition sequence of site-specific recombination systems in bacteria and yeast also have been found to affect the rate of recombination (Cox, 1988; Wu and Chaconas, 1992). In the FLP recombination system in yeast, the product of the FLP gene encoded by the yeast plasmid 2-microns recognizes inverted repeats within the plasmid and catalyzes their efficient recombination. The composition of nucleotide sequences bordering the FLP binding site apparently affects the efficiency of recombination, without affecting the ability of FLP to bind to the recognition sequence. Rather, the nucleotide composition of the sequences flanking the recombinase-binding sites is believed to influence the recombination efficiency by hindering the initiation of the cleavage step, perhaps by altering the conformation of the nucleotide residues in the binding site or by affecting the ability of the region to unwind (Cox, 1988). The nucleotides in the V-region coding segment potentially could affect the efficiency of V(D)J recombination in a similar manner (Gerstein and Lieber, 1993b).

DOUBLE-STRAND CLEAVAGE AND THE JOINING OF THE CODING SEQUENCE AND RSS

Coding and RSS joins

The coding segments and RSS are distinctly manipulated by the V(D)J recombination machinery prior to their ligation. While the ends of the coding segments are often altered by a combination of several different activities prior to their joining (e.g. N-region addition, P-nucleotide addition, and exonuclease activity), the two RSS ends are most often ligated head-to-head without modification (Fig. 5; Alt and Baltimore, 1982; Hochtl and Zachau, 1983; Lewis et al., 1985). These qualitative differences in the recombination products of RSS compared to the recombination products of the coding segments, together with the absence of significant homology in the joining sequences, suggested that V(D)J recombination occurs by a site-specific mechanism that requires the double-strand cleavage of the DNA, rather than

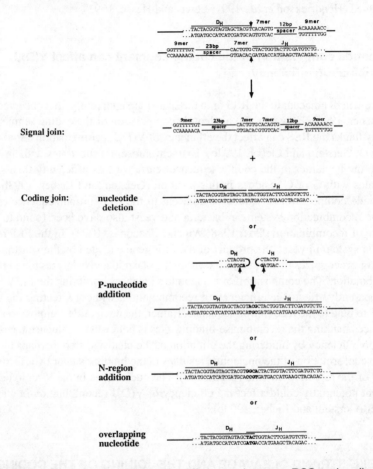

Fig. 5. RSS vs. coding joins. Qualitative differences between RSS and coding joins are depicted using D_H and J_H sequences. The D and J segments are cleaved (small arrows) adjacent to the heptamer of the RSS. The RSS join is often the perfect ligation of the heptamers of the recombining RSSs to each other. Several different events can modify the ends of the coding sequences prior to their ligation: deletion, P-nucleotide addition, N-region addition, and alignment of the two joining segments with homologies. These events are depicted here as independent, but occur concurrently during V(D)J recombination.

by homology-directed recombination (Alt and Baltimore, 1982). Consistent with this prediction for double-strand cleaved DNA intermediates in V(D)J recombination, DNA fragments with site-specific double-strand breaks near the RSS of $TCR_\delta D$ and J segments in murine thymus (Roth *et al.*, 1992b), and near the RSS of Ig J_H and J_κ in bone-marrow (Schlissel *et al.*, 1993) have been detected. Analysis of J_H and J_κ recombination intermediates revealed that the ends of the double-strand breaks generated by the V(D)J recombination machinery appear to be blunt and phosphorylated at the 5' end (Schlissel *et al.*, 1993).

V(D)J recombination culminates in the joining of the cleaved ends of the two coding sequences and two RSSs (Figs 3 and 5). Unlike the precise joins formed by the ligated RSSs, the terminal nucleotides of the coding sequences often are removed and template-independent nucleotides, N-regions, are sometimes added before their joining (see section below on N-region addition; Alt and Baltimore, 1982). Where nucleotides have not been removed by exonucleolytic activity, the addition of nucleotides palindromic to the terminal nucleotides of the coding sequences, P-nucleotides, can also be detected (see section below on P-nucleotides) (Lafaille *et al.*, 1989). Notably, the efficiency with which the coding segment end is modified appears to be influenced by its nucleotide composition (Boubnov *et al.*, 1993). These differences in the processing of coding versus RSS have been reinforced by the isolation of mice and cell lines with mutations (i.e. *scid*) that affect coding but not RSS join formation (Alt and Baltimore, 1982; Hendrickson *et al.*, 1988; Lieber *et al.*, 1988; Malynn *et al.*, 1988; Blackwell *et al.*, 1989; Ferrier *et al.*, 1990).

'Hybrid' and 'open-and-shut' joins

Although less frequently observed than conventional V(D)J recombination events, liberated RSS and coding-sequences can be joined in other combinations. The site-specific cleavage adjacent to RSS, followed by the processing of the ends and their religation, without strand exchange results in 'open-and-shut' joins (Fig. 6) (Lewis *et al.*, 1988; Morzycka-Wroblewska *et al.*, 1988; Lewis and Hesse, 1991). The end of one RSS and the coding sequence next to its partner RSS can also religate, to form 'hybrid joins' (Fig. 6) (Seidman and Leder, 1980; Lewis *et al.*, 1988; Okazaki *et al.*, 1988). 'Hybrid joins' occur rather frequently in rearrangements of extrachromosomal recombination substrates, which has been interpreted to suggest that the four recombining RSS and sequences coding are closely positioned during intermediate stages of V(D)J recombination, such that the strands can be readily exchanged (Lewis *et al.*, 1988). The RSS and coding sequences are also differentially processed even in these unconventional 'open-and-shut' and 'hybrid' joins. In general, nucleotide removal and addition still occurs preferentially on the coding join ends, whereas the joined RSSs remain relatively intact (Lieber et al., 1988a and b).

'Hybrid' and 'open-and-shut' recombination events in germ cells have been suggested to account for some aspects of the evolution of the antigen receptor loci. For example, 'D' segments could potentially evolve from an open-and-shut

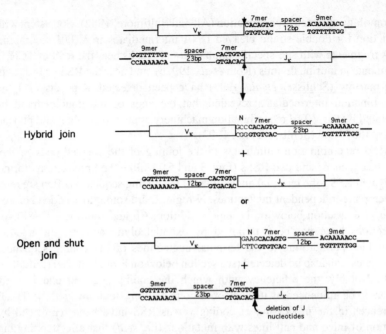

Fig. 6. Unusual joins. The unconventional 'hybrid' and 'open-and-shut' type joins are depicted, using V_κ and J_κ sequences. The starting segments are labeled as in Fig. 3. 'Open-and-shut' join is the formation of a double-strand break adjacent to the RSS, followed by its religation. Open-and-shut joins can only be detected when an end of the cleaved segment is modified. 'Hybrid' joins form when a coding segment (V) is religated to the RSS of its partner (J), and vice versa. Examples where N-region has been added (N) or nucleotide deleted (black zig-zag) from the joined coding segment are shown.

rearrangement with N-region addition taking place at a join (Lewis et al., 1988). In addition, 'hybrid joining' during the evolution of the κ locus has been suggested to account for the phenomena that while V_κ segments are flanked by RSSs with a 12-bp spacer sequence, the V segments of other antigen receptor loci are flanked by RSSs with a 23-bp spacer (Fig. 4; Lewis et al., 1988).

INVERSIONAL VS. DELETIONAL JOINING

Upon V(D)J recombination, sequences that lie between the rearranging V-region gene segments can be either deleted or inverted; the relative orientation of the interacting RSSs adjacent to the V-region gene segments determines whether the sequences that lie between the RSS will undergo these events (Fig. 7; Alt and Baltimore, 1982). When the two recombining RSSs face the same orientation, the intervening segment is inverted and maintained in the chromosome—the two coding sequences are fused together on one end of the inverted sequence, while the two RSSs are fused together on the other. When the RSSs initially face in the opposite orientation, the intervening segment is deleted. In deletional rearrangement, either the coding (deletional join) or RSS join (pseudo normal join) can be retained in the chromosome, depending on the orientation of the two RSSs (Fig. 7) During deletional rearrangement, the intervening sequences are believed to most often be

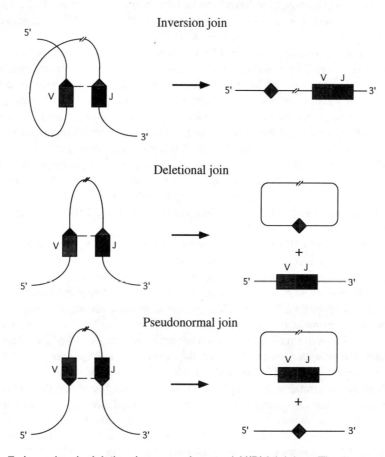

Fig. 7. Inversional, deletional or pseudonormal V(D)J joining. The sequences in between the two recombining sequences can undergo inversion or deletion, depending on the orientation of the two RSSs. When the two RSSs are in the same direction along the linear DNA, V(D)J recombination results in the inversion of intervening sequences (inversional joining). When the two RSSs face in opposite directions, the intervening sequences are deleted (deletional joining). If the recombining RSSs are oriented so that the heptamers of the RSS are facing away, then the RSS is deleted; if the heptamers of the two RSSs face toward each other, then the coding join is deleted (pseudonormal join). Black or gray squares represent coding segments, black triangles represent RSS with 12-bp spacer, and gray triangles represent RSS with 23-bp spacer sequences. (Adapted from Blackwell *et al.*, 1989.)

excised as a circle (Alt and Baltimore, 1982; Fujimoto and Yamagishi 1987; Okazaki et al., 1987). Joining of the RSS during V(D)J rearrangement is obviously important in inversional joining for maintaining the integrity of the genome (Alt and Baltimore, 1982). Furthermore, the formation of a circular excision product from deletional joining may potentially reduce the potential for reintegration of the deleted sequences into random regions of the genome. Examples of the products of these different rearrangement events have been isolated (Alt and Baltimore, 1982; Hochtl et al., 1982; Malissen et al., 1986; Fujimoto and Yamagishi, 1987; Okazaki et al., 1987; Iwashima et al., 1988; Korman et al., 1989).

The biological relevance of the different types of rearrangement events is not known. The V segments in some of the antigen receptor loci (e.g. κ) are oriented in the genome so that many of the recombination events occur by inversional joining (reviewed in Zachau, 1989). V segments of other loci (e.g. IgH), however, are oriented to predominantly undergo deletional rearrangement (reviewed in Rathbun et al., 1989; Berman and Alt, 1990). Most V segments of the human and murine TCRβ and TCRδ loci are oriented to form deletional joins, although the human and murine TCRδ loci and the murine TCRβ locus have one V segment located 3' of the constant region with its orientation resulting in the inversion of the intervening constant region upon the rearrangement of the V with a D segment (Malissen et al., 1986; Iwashima et al., 1988; Loh et al., 1988; Hata et al., 1989; Korman et al., 1989).

The configuration of the D_H segments potentially enables D_H to J_H rearrangement to occur either by deletional or inversional joining, yet D_H–J_H rearrangement of the endogenous locus predominantly undergoes deletional joining, using the J_H-proximal RSS (Meek et al., 1989). The potential factors that might cause the preference for deletional rearrangement for these D_H gene segments were investigated using extrachromosomal recombination substrates (Gauss and Lieber, 1992; Gerstein and Lieber, 1993b). Extrachromosomal recombination substrates with multiple RSSs were used to show that, at least within limited distances, the most proximal RSS is not preferentially used for rearrangement (Gauss and Lieber, 1992). Rather, analysis of extrachromosomal substrates with altered coding segments suggested that the nucleotide composition of the D_H coding region on the J_H-proximal side, as compared to the composition of the nucleotides on the J_H-distal side, may cause the predisposition for using the J_H-proximal RSS that results in the deletional joining of D_H segments (Gauss and Lieber, 1992; Boubnov et al., 1993; Gerstein and Lieber, 1993b).

MODIFICATION OF THE CODING END DURING REARRANGEMENT

N-region addition

Nucleotides that cannot be attributed to the sequence of the rearranging gene segments are often found in the junction of coding joins (Fig. 5). These nucleotides, or N-regions (Alt and Baltimore, 1982), which are often rich in G and C residues are added to the joins of TCR and IgH rearrangements of lymphocytes generated in adult animals (Elliot *et al.*, 1988; Lafaille *et al.*, 1989; Carlsson and Holmberg, 1990; Feeney, 1990, 1991, 1992; Gu *et al.*, 1990; Meek *et al.*, 1990; Aguilar and Belmont, 1991; Bangs *et al.*, 1991; Bogue *et al.*, 1991, 1992; Kyes *et al.*, 1991; Schwager *et al.*, 1991; Carlsson *et al.*, 1992; George and Schroeder, 1992). Igκ and λ rearrangements do not usually have N-region addition (Davis and Bjorkman, 1988; Rolink and Melchers, 1991). N-regions are most often associated with coding junctions, although the addition of nucleotides to the junction of RSS joins has been occasionally reported (Iwashima *et al.*, 1988; Lieber *et al.*, 1988; Korman *et al.*, 1989; Kallenbach *et al.*, 1992; Shimizu and Yamagishi, 1992).

N-region addition is notably absent in V-region gene rearrangements of lymphocytes of fetal origin (Elliot *et al.*, 1988; Lafaille *et al.*, 1989; Carlsson and Holmberg, 1990; Feeney, 1990, 1991, 1992; Gu *et al.*, 1990; Meek *et al.*, 1990; Aguilar and Belmont, 1991; Bangs *et al.*, 1991; Bogue *et al.*, 1991, 1992; Kyes *et al.*, 1991; Schwager *et al.*, 1991; Carlsson *et al.*, 1992; George and Schroeder, 1992). A close correlation between the addition of N-regions to V(D)J junctions and the expression of TdT in the lymphoid tissues of fetal versus adult animals and in various cell lines has been demonstrated (Yancopoulos *et al.*, 1986; Landau *et al.*, 1987; Bogue *et al.*, 1992; Kallenbach *et al.*, 1992). The analysis of recombination substrates introduced into cell lines expressing different levels of TdT have provided strong evidence that the expression of TdT can result in the addition of nucleotides to V-region joins (Landau *et al.*, 1987; Kallenbach *et al.*, 1992). Definitive proof that TdT is solely responsible for the physiological addition of N-regions was provided by the analysis of rearrangements from mice with lymphocytes that carried gene-targeted mutations of TdT. In adult mice with lymphocytes that do not express TdT, virtually all antigen receptor coding-join rearrangements lacked N-region addition (Gilfillan *et al.*, 1993; Komori *et al.*, 1993).

P-nucleotide addition and coding end hairpins

Palindromic (P-) nucleotides, like N-regions, are sequences that are added to the junction of rearranged gene segments. P-nucleotides were first identified in coding joins of fetal and neonatal TCRγ and δ rearrangements, which generally lack N-region addition making the P-nucleotides more conspicuous (Lafaille *et al.*, 1989). Several features distinguish P-nucleotides from N-regions. P-nucleotides are usually only one or two nucleotides and are always inserted adjacent to intact coding sequences. They differ from N-regions in that they are always palindromic to the terminal nucleotides of the coding segments (Lafaille *et al.*, 1989) (Fig. 5). Longer P-nucleotides of up to 15 bp have been reported in the joins of V(D)J recombination products isolated from *scid* mice, and from Chinese hamster ovary (CHO) cells with an analogous mutation to *scid* (Pergola et al., 1993; Taccioli et al., 1993, 1994a and b). The extraordinary length in P-nucleotides in *scid* joins appears to be a direct consequence of the defect in these genes.

The palindromic nature of P-nucleotides initially resulted in suggestions that they were encoded by the transfer of the terminal nucleotides from the opposite strand (Lafaille et al., 1989). The repeated appearance of the same nucleotides added at some junctions also resulted in speculation that a nucleotide-exchange mechanism between the joining ends and the recombination machinery was responsible for the addition of the nucleotide (McCormack et al., 1989). More recently, the origin of P-nucleotides has been attributed to the asymmetric cleavage of a hairpin structure that is an intermediate of the V(D)J recombination process, formed from the covalent ligation of the two strands of the coding segment (Lieber, 1991, 1992; Roth et al., 1992; Ferguson and Thompson, 1993), potentially mimicking recombination intermediates observed in bacterial systems. This conclusion was solidified from the cumulative observations of the palindromic nature of P-nucleotides, the longer P-nucleotides in rearrangements of *scid* lymphocytes (Kienker et al., 1991; Schuler et al., 1991), and from the isolation of DNA with covalently ligated ends adjacent to the TCRδ J segments (Roth et al., 1992). A recent study using substrates with engineered hairpin structures demonstrated that hairpin resolution can in fact result in P-nucleotide addition, and that both non-lymphoid and lymphoid cell types exhibit the ability to resolve hairpins (Lewis, 1994).

Covalently closed, hairpin DNA structures have been found adjacent to coding sequences but not RSSs of the TCRδ V-region genes in *scid* thymus (Roth et al., 1992). This is consistent with the association of P-nucleotide addition with coding but not RSS joins, as well as the influence of the *scid* defect on coding join formation more than RSS join formation (Lieber et al., 1988; Malynn et al., 1988; Blackwell et al., 1989). These findings again underscore the mechanistic differences between coding join and RSS join formation. Although hairpin intermediates have not been identified for antigen receptor loci other than the TCRδ locus, the finding of P-nucleotides in the coding junctions of most other antigen receptors suggest that the hairpin structure is a general intermediate structure that occurs during V(D)J recombination (Lieber, 1991, 1992).

Homology-mediated joining

V(D)J joins that lack N-region additions often have joins with 'overlapping' bases, whose origin cannot be unequivocally assigned to either of the two joining coding segments (Fig. 5; Gu et al., 1990; Feeney, 1992b). Overlapping nucleotides can occur when homologous nucleotides are available near the end of the joining segments, with some combinations of homologous nucleotides apparently being preferentially used over others (Boubnov et al., 1993; Gerstein and Lieber, 1993a; Komori et al.,

1993). This homology-mediated joining, however, apparently is not necessary for V(D)J rearrangement to take place, since coding sequences that frequently rearrange to form joins with overlapping nucleotides can also form joins without overlaps (Boubnov et al., 1993; Gerstein and Lieber 1993a; Komori et al., 1993), and rearrangements appear to occur equally efficiently with or without homologies (Boubnov et al., 1993; Gerstein and Lieber, 1993a). P-nucleotides are treated the same as germline-encoded V-region gene sequences and are often incorporated in these overlapping joins (Allison and Havran, 1991; Gilfillan et al., 1993; Komori et al., 1993).

The occurrence of overlapping bases suggested that short nucleotide homologies between joining segments may be used to initiate polymerization to repair the double-strand break that occurs during V(D)J rearrangement (Alt and Baltimore, 1982). As with many other features of V(D)J rearrangement, homology-mediated joining is almost always associated with coding joins, and overlapping bases are rarely detected in RSS joins. However, apparent homology-mediated joining of RSSs has been reported in the case of RSS joins of recombination substrates isolated from V(D)J recombinase-active CHO cells with a mutation in a gene that is involved in the double-strand break repair process (XR-1) (Z. Li and F. Alt, manuscript in preparation). In these cells, all resolved RSS join rearrangements were found to have 'overlapping' nucleotides, suggesting the use of a different pathway for their formation. Homologies between joining sequences may normally aid in aligning the recombining segments, perhaps by using factors involved in general DNA repair processes. To this effect, mammalian cells have been found to use short sequence homologies to join two broken double-strand DNA ends whenever such overlaps are available (Roth and Wilson, 1988).

In lymphocytes generated in the adult, N-region addition could mask the occurrence of homology-mediated joining, since N-regions provide template nucleotides that cannot be predicted (Gilfillan et al., 1993; Komori et al., 1993). The frequent occurrence of joins with overlaps in the absence of N-region addition could potentially restrict the antigen receptor repertoire that can be encoded by particular combinations of V-region gene segments (Gu et al., 1990; Feeney, 1992b). To date, such restrictions have been studied in most detail in certain T-cell subsets. The antigen receptors expressed by the murine $\gamma\delta$ dendritic epithelial cells (DEC) and vaginal intra-epithelial lymphocytes (r-IEL) exemplify striking examples of the restriction of the antigen receptor by homology-mediated joining. The DEC and r-IEL are derived from stem cells of fetal origin and express virtually identical TCR γ and δ chain V-region genes that are distinguished by the presence of overlapping sequences in their junctions (reviewed in Allison and Havran, 1991; Raulet et al., 1991). Several lines of evidence indicate that the unusual restriction of the antigen receptor repertoire on DEC and r-IEL can mostly be attributed to the influence of homologies on the mechanism of the V(D)J assembly process, rather than simply by selection for cells expressing the predominant antigen receptor (Lafaille et al., 1989; Asarnow et al., 1993; Gilfillan et al., 1993; Itohara et al., 1993; Komori et al., 1993).

LYMPHOCYTE-SPECIFIC GENES AND ACTIVITIES INVOLVED IN V(D)J RECOMBINATION

RAG-1 and *RAG*-2

The VDJ recombination reaction appears to be generated by a few lymphocyte-specific activities that confer tissue specificity, while other events (for example, polymerization and ligation) appear to be carried out by generally expressed cellular activities recruited to function in the V(D)J recombination reaction (reviewed in Alt *et al.*, 1992). The *RAG*-1 and *RAG*-2 encode phosphorylated nuclear proteins (Lin and Desiderio, 1993). *RAG*-1 and *RAG*-2 are co-expressed specifically in pre-lymphocytes, corresponding with the onset of V(D)J rearrangement (Schatz *et al.*, 1989; Oettinger *et al.*, 1990). This co-expression of *RAG*-1 and *RAG*-2 apparently provides the necessary activity to generate V(D)J recombination in all tested mammalian cells (Schatz *et al.*, 1989; Oettinger *et al.*, 1990). Other mammalian-specific factors appear to work specifically in conjunction with *RAG*-1 and *RAG*-2 to carry out V(D)J recombination as the introduction of *RAG*-1 and *RAG*-2 genes into a *Drosophila melanogaster* cell line does not appear to be sufficient for the recombination of a co-introduced extrachromosomal recombination substrate (E. Oltz, G. Taccioli and F. Alt, unpublished data). A generally expressed human factor that may potentially associate with *RAG*-1 in *vivo*, called Rch-1, has been recently identified by its ability to interact with *RAG*-1 in yeast cells (Cuomo *et al.*, 1994). Notably, Rch-1 has homology to a yeast protein SRP1 that associates with the nuclear envelope (Cuomo *et al.*, 1994).

RAG-1 and *RAG*-2 were initially identified as segments of DNA that conferred V(D)J recombination activity upon fibroblasts when introduced into their genome, as assayed by a chromosomally integrated recombination substrate (Schatz and Baltimore, 1988; Schatz *et al.*, 1989). Since both genes are required for V(D)J recombination, only the unusual proximity and compact genomic structure of *RAG*-1 and *RAG*-2 enabled the isolation of these genes by this method (Oettinger *et al.*, 1990). The protein-coding regions of the *RAG*-1 and *RAG*-2 genes are each confined to one large exon, lie within 8 kb of each other, and are transcribed towards each other (Oettinger *et al.*, 1990). This chromosomal organization of the *RAG*-1 and *RAG*-2 genes, and the conserved sequence of the genes between species (Oettinger *et al.*, 1990; Fuschiotti *et al.*, 1993), together with the conserved RSS between species, have led to speculation that the mechanism of V(D)J recombination evolved from the recombination mechanism of a virus that infected germ cells of an ancestral host (Oettinger *et al.*, 1990).

Although *RAG*-2 does not seem to have significant homology to other characterized genes, the *RAG*-1 gene has regions that are homologous to the *Saccharomyces cerevisiae HPR*-1 gene, which in turn shares some homology with the *Escherichia coli* topoisomerase I gene (Wang *et al.*, 1990). To date, neither *RAG*-1 nor *RAG*-2 proteins have been shown to bind to the RSS, which is a characteristic that might be expected of factors that are directly involved in V(D)J recombination. Kinetic studies of the expression of the *RAG* genes and the activation of V(D)J recombination strongly suggest that the products of these genes are directly involved in V(D)J recombination (Oltz, *et al.*, 1993). However, the possibility that *RAG*-1 and *RAG*-2 promote V(D)J recombination by regulating other factors directly involved in V(D)J recombination has not yet been ruled out.

The targeted mutation of *RAG*-1 or *RAG*-2 results in the generation of mice that lack mature B and T lymphocytes. Examination of the Ig and TCR loci in the lymphocytes of *RAG*-1 or *RAG*-2 deficient animals reveals a complete block in the initiation of V(D)J rearrangement (Mombaerts *et al.*, 1992; Shinkai *et al.*, 1992). The ability to undergo rearrangement can be restored in A-MuLV-transformed pre-B cell lines generated from these animals by introducing the complementing gene into the cells (G. Rathbun and F. Alt, unpublished data). The deficiency in the *RAG*-1 or *RAG*-2 gene does not appear to affect other aspects of lymphoid development in these animals, in that the cellularity of mature T and B lymphocytes is markedly restored in the offspring of *RAG*-deficient animals crossed with mice transgenic for genes that encode functional T- or B-cell antigen receptors (Shinkai *et al.*, 1993; Spanopoulou *et al.*, 1994; Young *et al.*, 1994). Together these results confirm the role of the recombinase-activating genes in providing V(D)J recombination activity during lymphocyte development.

While *RAG*-1 and *RAG*-2 genes are expressed together only in developing lymphocytes, the expression of *RAG*-2 without *RAG*-1 has been reported in B-cells of the chicken bursa (Carlson *et al.*, 1991), and *RAG*-1 without *RAG*-2 in the brain, testes and ovaries (E.M. Oltz and F.W. Alt, unpublished data; Chun *et al.*, 1991). The solitary expression patterns of *RAG*-1 or *RAG*-2 have resulted in speculation about their potential activities in these organs, such as in gene conversion, neural receptor formation or meiotic recombination (Carlson *et al.*, 1991; Chun *et al.*, 1991). However, to date, deficiency in *RAG*-1 or *RAG*-2 expression has not been shown to affect these other processes. The strictly regulated coexpression of *RAG*-1 and *RAG*-2 appears to be important, since many translocations to oncogenes that are found in lymphomas are likely due to aberrant V(D)J rearrangements (Haluska *et al.*, 1987; Finger *et al.*, 1989). In addition, in mice transgenic for the *lck* promoter-driven *RAG*-1 and *RAG*-2 genes, the deregulated expression of *RAG*-1 and *RAG*-2 seems to affect the normal development of peripheral T-cells (Wayne *et al.*, 1994).

Terminal deoxynucleotidyl tranferase

TdT is a tissue-specific but non-essential component of the V(D)J recombination machinery, since normal initiation and resolution of V(D)J rearrangement can take place in cells with or without TdT activity (Gilfillen *et al.*, 1993; Komori *et al.*, 1993). TdT was initially isolated from bovine thymus as an enzyme, and was found to have deoxynucleotide polymerizing activity that is independent of a DNA template *in vitro* (Bollum, 1974). Its specific expression in tissues and cells representing early stages of B-and T-lymphocyte development have suggested a role for TdT in early lymphocyte differentiation (Silverstone et al., 1978). This specific expression pattern of TdT together with its template-independent polymerase activity, its preference for dGTP as a substrate (Bollum, 1974), and the prevalence of G/C residues in N-regions (Alt and Baltimore, 1982) led to the suggestion that TdT may be responsible for the addition of N-regions to the junctions of antigen receptor gene rearrangements (Alt and Baltimore, 1982).

A substantial body of evidence correlating the expression of TdT and N-region addition in developing lymphocytes and cell lines has provided convincing support for the notion that TdT expression has a role in N-region addition. While lymphocytes generated in the adult express TdT and have N-region addition in more than 80% of their coding joins, both are markedly absent in fetally derived lymphocytes (Landau et al., 1987; Elliot et al., 1988; Lafaille et al., 1989; Carlsson and Holmberg, 1990; Feeney, 1990, 1991, 1992; Gu et al., 1990; Meek et al., 1990; Aguilar and Belmont, 1991; Bangs et al., 1991; Bogue et al., 1991, 1992; Kyes et al., 1991; Schwager et al., 1991; Carlsson et al., 1992; George and Schroeder, 1992). Convincing demonstration that TdT can add nucleotides to the junction of rearranging V-region segments has been provided by the introduction of *RAG*-1 and *RAG*-2 together with TdT expression vectors in fibroblast cell lines, which resulted in increased N-region addition at junctions of recombination substrates (Kallenbach et al., 1992). However, the results of these studies did not exclude the possibility of N-region addition by additional mechanisms during normal lymphoid development. The nearly exclusive role of TdT in N-region addition in normal lymphocytes was unequivocally demonstrated by the targeted gene mutation of TdT—TdT-deficient lymphocytes of the mutant adult animals did not contain N-region addition at the junction of recombined V-region coding gene segments (Gilfillan et al., 1993; Komori et al., 1993).

Analysis of antigen receptor rearrangements from the TdT-deficient lymphocytes also revealed additional potential roles for TdT expression in repertoire development beyond generating N-region diversity. Thus, the lack of N-regions in the V-region joins of TdT-mutant lymphocytes was found to dramatically increase the level of homology-mediated joins that can be detected during coding join formation, from 20% in normal animals to 80% in TdT-deficient animals (Gilfillan et al., 1993; Komori et al., 1993). In the case of some TCR V-region gene segment rearrangements, lack of N-region addition apparently resulted in the predominant expression of one antigen receptor V-region gene (Gilfillan et al., 1993; Komori et al., 1993). Thus TdT expression and N-region addition during adult lymphocyte development may prevent the formation of certain and specific antigen receptors that may only be required in the fetal antigen receptor repertoire, when TdT is absent.

The stringent regulation of TdT expression between fetal and adult animals (Bogue et al., 1992) is apparently mediated by the environment of the lymphocytes (Bogue et al., 1993; Larche and Hurwitz, 1993). When fetal-derived stem cells are transplanted into *scid* or irradiated adult mice, N-regions are added to the V-gene junctions of the fetal liver-derived lymphocytes, demonstrating that the expression pattern of

TdT *per se* is not pre-programmed in the cells (Bogue *et al.*, 1993). Disrupting the fetal environment in developing T-cells also appears to result in turning on TdT expression in fetal lymphocytes, since a high frequency of antigen receptor rearrangements with N-regions are detected in thymic organ cultures established from day 14 embryos, as compared with normal fetal thymus *in vivo* (Larche and Hurwitz, 1993). The role of TdT in effectively generating two different antigen repertoires, one during fetal and the other in adult lymphocytes, awaits further analysis of immune responses in TdT-deficient animals as well as the analysis of transgenic animals in which TdT expression is forced during fetal lymphoid development.

GENERALLY EXPRESSED ACTIVITIES INVOLVED IN V(D)J RECOMBINATION

The V(D)J recombination machinery appears to recruit several non-lymphoid-specific factors that participate in other DNA repair processes.

The *scid* gene product

The product of the murine scid gene was the first non-lymphoid specific factor implicated in normal V(D)J recombination. The murine scid mutation was initially identified as a mutation causing a general severe combined immune deficiency (SCID) (Bosma *et al.*, 1983). The phenotype of the mutation is apparently caused by the inability of *scid* lymphocytes to undergo normal V(D)J recombination (Schuler *et al.*, 1986; Hendrickson *et al.*, 1988; Lieber *et al.*, 1988a; Malynn *et al.*, 1988; Blackwell *et al.*, 1989; Ferrier *et al.*, 1990). However, unlike the *RAG*-1 and *RAG*-2 deficient lymphocytes that are unable to initiate V(D)J recombination, *scid* lymphocytes can initiate V(D)J rearrangement via the site-specific double-strand break between RSS and coding sequences; however, *scid* lymphocytes are unable to complete the normal joining process, particularly with respect to coding joins.

The *scid* mutation also affects the efficiency of double-strand break repair (DSBR) in both lymphoid and non-lymphoid cells of mutant animals. *Scid* cells are up five times more sensitive than normal cells to X-rays and other insults that induce double-strand breaks, such as bleomycin (Fulop and Phillips, 1990; Biedermann *et al.*, 1991; Hendrickson *et al.*, 1991b). The aberrant activity of cells with the single mutation in both V(D)J rearrangement and DNA repair suggested that the *scid* gene product is a generally expressed factor that normally participates in DSBR, but is recruited by the V(D)J recombinase to repair double-strand breaks that occur during the assembly of antigen receptor genes (Bosma and Carroll, 1991). In several organisms, mutations that affect DSBR processes have been found to similarly affect multiple different recombination pathways (Boyd *et al.*, 1983; Orr-Weaver and Szostak, 1988; Thompson, 1988).

The characterization of V(D)J rearrangement products from *scid* lymphocytes have provided numerous insights into the molecular mechanism of V(D)J recombination. In *scid* pre-lymphocytes, the RSS is recognized and V(D)J rearrangement is properly initiated, but the completion of rearrangement is impaired (Hendrickson *et al.*, 1988; Lieber *et al.*, 1988a; Malynn *et al.*, 1988; Blackwell *et al.*, 1989; Ferrier *et al.*, 1990). Coding join formation is highly aberrant in *scid* lymphocytes, frequently resulting in large deletions, while the RSS joins are most often normal (Hendrickson *et al.*, 1988; Malynn *et al.*, 1988; Blackwell *et al.*, 1989). Consistent with these observations, coding joins of extrachromosomal recombination substrates are rarely isolated from *scid* cell lines (Lieber *et al.*, 1988), presumably because the aberrant rearrangements delete regions of the plasmid important for its propagation or for providing drug resistance. On the other hand, precise RSS joins of extrachromosomal substrates were isolated at nearly normal frequency from *scid* pre-B cells, although some junctions were not precise (Lieber *et al.*, 1988a). These observations support the notion that the RSS and coding sequences are differentially processed by the recombination machinery, and that the *scid* mutation much more drastically affects coding, as opposed to RSS, join formation (Alt *et al.*, 1992).

The analysis of joins and recombination intermediates from *scid* lymphocytes also provided insights for understanding the mechanism of P-nucleotide addition. The isolation of extraordinarily long P-nucleotides in coding junctions isolated from *scid* lymphocytes (Kienker *et al.*, 1991; Schuler *et al.*, 1991; Harrington *et al.*, 1992) led to suggestions that the addition of P-nucleotides may result from the resolution of a covalently sealed ('hairpin') structure at the end of the coding segment (Lieber, 1991; Roth *et al.*, 1992). In *scid* but not in normal mice, an intermediate of the V(D)J recombination process that represents the step prior to P-nucleotide addition—double-strand breaks with covalently sealed ends—accumulate in the thymus (Roth *et al.*, 1992a). Together, these observations suggested that the *scid* gene product may either be necessary for resolving a covalently closed DNA hairpin that is an intermediate structure in normal V(D)J recombination (Lieber, 1991; Roth *et al.*, 1992a) or its mutant form or absence may interfere with the normal hairpin resolution process. A recent study has shown that *scid* lymphocytes are capable of efficiently resolving hairpins in extrachromosomal DNA (Lewis, 1994).

In summary, while the *scid* defect appears to lead to generation aberrant V(D)J recombination intermediates particularly with respect to coding ends, the role of the *scid* gene product in coding end hairpin resolution remains unknown. Recent evidence that the *scid* gene may encode a DNA-dependent protein kinase (see below) should provide additional insight into potential mechanisms of the molecular phenotype with respect to V(D)J recombination in *scid* mice.

Other generally expressed factors involved in V(D)J rearrangement

At least two other ubiquitously expressed factors, in addition to that affected by the *scid* mutation, appear to be components that are critical for normal V(D)J recombination. These factors have been identified in the context of DSBR mutations in CHO cells that simultaneously affect the ability of the cells to repair double-strand breaks and the ability to direct normal V(D)J recombination of recombination substrates upon the co-introduction of *RAG*-1 and *RAG*-2 into the cells (Taccioli *et al.*, 1993, 1994a). Three independent complementation groups of CHO DSBR mutants also showed defects in V(D)J recombination (Taccioli *et al.*, 1993). Notably, both the defects in DSBR and V(D)J recombination in several of these lines, xrs-6 and XR-1, could be complemented by subregions of human chromosomes 2 and 5, respectively (Taccioli *et al.*, 1993). The xrs-6 and XR-1 CHO mutants are defective in both coding and RSS join formation and are, therefore, quite distinct in their effects on V(D)J recombination from the murine *scid* mutation (Taccioli *et al.*, 1993). The third identified DSBR/V(D)J recombination mutant in CHO cells, V3, appears to be in the same complementation group as *scid* based on cell fusion studies (Taccioli *et al.*, 1994a). Correspondingly, V3 also is primarily defective in coding, as opposed to RSS, join formation, and also often exhibits long P-nucleotides in the coding junctions (Pergola *et al.*, 1993; Taccioli *et al.*, 1994a).

The observations outlined above indicate the existence of at least three separate factors that are utilized in general DSBR; these are recruited to function also in the repair of double-strand breaks generated during V(D)J recombination. Characterization of the aberrant coding and RSS joins of recombination substrates in xrs-6 and XR-1 have provided some insight into the role of the factors that are defective in these cell lines in V(D)J rearrangement. The RSS is apparently still recognized in the mutants and the V(D)J recombination process is initiated normally, since the isolated joins appear to derive from double-strand breaks that apparently occur at coding/RSS junctions. Therefore, the factors affected by these mutations, like the *scid*/V-3 factor, appear to work downstream of the (*RAG*-dependent) initiation of the V(D)J recombination process.

A previously characterized factor known as the Ku autoantigen has been implicated as being defective in the xrs-6 CHO cell line (Smider *et al.*, 1994; Taccioli *et al.*, 1994). Ku is a generally expressed and abundant nuclear factor consisting of a 70- and 86-kDa subunit that has non-specific double-strand DNA-end binding activity (de Vries *et al.*, 1989; Zhang and Yaneva, 1992), and is the DNA-binding subunit of a DNA-dependent protein kinase (Dvir *et al.*, 1992; Gottlieb and Jackson, 1993). Several independent correlations strongly support the possibility that xrs-6 mutation affects the gene encoding the 86-kDa subunit of Ku. First, the 86-kDa subunit of Ku maps to a region close to or overlapping with the region on human chromosome 2 that complements the deficiency in xrs-6 (Hafezparast *et al.*, 1993; Cai, 1994; Taccioli *et al.*, 1994b). In addition, xrs-6 apparently is deficient in the Ku DNA-end binding activity (Getts and Stamato, 1994; Rathmell and Chu, 1994; Taccioli *et al.*, 1994b). Finally, introduction of a human cDNA encoding the 86-kDa subunit of Ku restores the missing DNA end-binding activity and simultaneously restores relatively normal ability to undergo V(D)J recombination and relatively normal sensitivity to ionizing radiation. Together, these observations strongly imply that the Ku factor is a generally expressed activity involved both in DSBR and V(D)J recombination (Smider *et al.*, 1994; Taccioli *et al.*, 1994b).

Although the exact role of the Ku gene product in DNA repair and V(D)J recombination has thus far not been elucidated, the known activities of the Ku gene product offers some intriguing insights into the potential mechanistic roles of these and other generally expressed factors that participate in V(D)J recombination (Smider *et al.*, 1994; Taccioli *et al.*, 1994b). One possibility is that Ku serves a protective function in binding the free double-strand DNA ends and protecting them from nuclease activity. Another possibility that is not mutually exclusive of the first is that Ku binding may activate downstream activities which are involved in resolution of the DSBRs. In this context, the Ku complex serves as the DNA-binding subunit for the DNA-dependent protein kinase (DNA-PK), the catalytic domain of which is over 400 kDa in size (Dvir *et al.*, 1992; Gottlieb and Jackson, 1993). Upon activation through Ku binding to DNA ends, the DNA-PK phosphorylates a number of intriguing substrates *in vitro*, including p53, RNA polymerase II, topoisomerase II, several nuclear oncogenes, and Ku itself (Jackson *et al.*, 1990; Lees-Miller *et al.*, 1990; Chen *et al.*, 1991; Dvir *et al.*, 1992; Iijima *et al.*, 1992; Peterson *et al.*, 1992; Wang and Eckhart, 1992). Very recent evidence, based on studies similar to those outlined for Ku, has suggested that the DNA-PK gene may in fact be the gene affected by the murine *scid* mutation (Blunt *et al.*, 1995), further supporting the potential role of the DNA-PK complex in the DSBR/V(D)J recombination process. Clearly, further analyses of the Ku/DNA-PK complex should help elucidate unknown aspects of the mechanism of DSBR and V(D)J recombination.

CONTROL OF V(D)J RECOMBINATION

The activity of the V(D)J recombination reaction is controlled in a number of different contexts. In the most simple manifestation of control, V(D)J recombination activity is limited to cells that represent the immature stages of B- and T-lymphocyte development and not to more mature lymphoid cells or other types of cells. Such restricted expression, which is likely necessary to prevent undesired chromosomal rearrangements, appears to result solely from the regulated expression of the *RAG*-1 and *RAG*-2 gene products (Schatz *et al.*, 1989; Oettinger *et al.*, 1990).

Within developing lymphocytes, V(D)J recombination activity is also regulated in several different contexts (reviewed by Okada *et al.*, 1994).

1. *Lineage-specificity*, which refers to the fact that Ig V-region genes are completely assembled only in B-lineage cells and TCR V-region genes are completely assembled only in T-lineage cells.
2. *Developmental stage specificity*, which refers to the ordered assembly of IgH chain genes before those of IgL chain genes in developing B-cells and the assembly of TCRβ V-region genes before those of TCRα in developing T-cells.

3. *Allelic exclusion*, with respect to Ig genes, refers to regulation of the V-region gene assembly process so that a given B-cell expresses only one H chain and one L chain allele to form its expressed Ig.

As described above, all antigen receptor V-region gene segments appear to be assembled by a common V(D)J recombinase, the tissue-specific components of which are specified by the *RAG*-1 and *RAG*-2 gene products. Therefore, control of V(D)J recombination with respect to the three phenomena outlined above must be governed by mechanisms that modulate accessibility of the substrate V-region gene segments to the common V(D)J recombinase (reviewed by Okada *et al.*, 1994).

The factors that modulate accessibility are still under investigation. It has been known for some time that actively rearranging V-region gene segments, whether they be endogenous gene segments or contained within V(D)J recombination substrates, tend to be transcribed and hypomethylated (Van Ness *et al.*, 1981; Picard and Schaffner, 1984; Yancopoulos and Alt, 1985; Blackwell *et al.*, 1986; Ferrier *et al.*, 1990b; Lennon and Perry, 1990; Berman *et al.*, 1991; Engler *et al.*, 1991a; Fondell and Marcu, 1992; Goldman *et al.*, 1993; Goodhardt *et al.*, 1993; Holman *et al.*, 1993). However, examples exist in which V-region gene segments that are transcribed or hypomethylated still do not rearrange efficiently in V(D)J recombinase-expressing cells (Chen *et al.*, 1993; Okada *et al.*, 1994). Therefore, while transcription and hypomethylation are strong positive correlates of V(D)J recombinational accessibility, they may not necessarily be sufficient, at least in some cases, to confer full V(D)J recombination potential. In this regard, transcription and/or hypomethylation may simply be by-products of 'accessible' gene segments. Conversely, it is possible that transcription or hypomethylation directly contribute to the accessibility phenomenon, for example by promoting chromatin states favorable for recombinase entry or by many other potential mechanisms that have been considered in detail elsewhere (reviewed in Yancopoulos *et al.*, 1986; Blackwell *et al.*, 1989; Ferrier *et al.*, 1989; Okada *et al.*, 1994).

A variety of recent experiments have strongly implicated the transcriptional control elements associated with endogenous Ig or TCR loci with having a major role in generating tissue- or stage-specific V-region gene accessibility. For example, transgenic V-region gene mini-locus recombination substrates have been shown to require associated transcriptional enhancer elements in order for their V-region gene segments to be assembled in developing B- and T-lineage cells (Ferrier *et al.*, 1990; Kallenbach *et al.*, 1993; Lauster *et al.*, 1993; Oltz *et al.*, 1993; Lauzurica and Krangel, 1994). Additional experiments with such loci also demonstrated that TCRα and TCRβ enhancer elements can confer upon the recombination substrates the same developmental stage-specific patterns of rearrangement observed with the corresponding endogenous loci (Capone *et al.*, 1993).

Gene-targeted mutational experiments have shown that deletion or replacement mutation of the intronic IgH and Ig κ enhancer elements greatly impairs rearrangement of the corresponding endogenous loci (Chen *et al.*, 1993; Serwe and Sablitzky, 1993; Takeda *et al.*, 1993). However, in the case of the IgH intronic enhancer, rearrangement is not completely inhibited, indicating the presence of additional elements that can confer V(D)J recombinational accessibility (Chen et al., 1993; Serwe and Sablitzky, 1993). Finally, some recent transgenic recombination substrate experiments have suggested the existence of negative accessibility elements that may function analogously to silencers with respect to transcription (Lauster et al., 1993). Thus, while much progress continues to be made, it is clear that there is more to learn about the elements that regulate V(D)J recombinational accessibility.

Acknowledgments

This work was supported by the HHMI and NIH grant A.I. 20047.

REFERENCES

Aguilar, L.K. and Belmont, J.W. (1991). J. Immunol. **146**, 1348–1352.
Aguilera, R.J., Akira, S. and Sakano, H. (1987). Cell **51**, 909–917.
Akira, S., Okazaki, K. and Sakano, H. (1987). Science **238**, 1134–1138.
Allison, J.P. and Havran, W.L. (1991). Annu. Rev. Immunol. **9**, 679–705.
Alt, F.W. and Baltimore, D. (1982). Proc. Natl Acad. Sci. USA **79**, 4118–4122.
Alt, F., Rosenberg, N., Lewis, S., Thomas, E. and Baltimore, D. (1981). Cell **27**, 381–390.
Alt, F.W., Blackwell, T.K. and Yancopoulos, G.D. (1987). Science **238**, 1079–1087.
Alt, F.W., Oltz, E.M., Young, F., Gorman, J., Taccioli, G. and Chen, J. (1992). Immunol. Today **13**, 306–314.
Asarnow, D.M., Goodman, T., Lefrancois, L. and Allison, J.P. (1989). Nature **341**, 60–62.
Asarnow, D.M., Cado, D. and Raulet, D.H. (1993). Nature **362**, 158–160.
Baltimore, D. (1981). Prog. Clin. Biol. Res. **45**, 297–308.
Bangs, L., Sanz, J.M. and Teale, J.M. (1991). J. Immunol. **146**, 1996–2004.
Berman, J.E. and Alt, F.W. (1990). Int. Rev. Immunol. **5**, 203–214.
Berman, J.E., Humphries, C.G., Barth, J. and Alt, F.W. (1991). J. Exp. Med. **173**, 1529–1535.
Bernard, O., Hozumi, N. and Tonegawa, S. (1978). Cell **15**, 1133–1144.
Biedermann, K., Sun, J., Giaccia, A.J., Tosto, L.M. and Brown, J.M. (1991). Proc. Natl Acad. Sci. USA **88**, 1394–1397.
Blackwell, T.K. and Alt, F.W. (1984). Cell **37**, 105–112.
Blackwell, T.K. and Alt, F.W. (1989). Ann. Rev. Genet. **23**, 605–636.
Blackwell, T.K., Moore, M.W., Yancopoulos, G.D., Suh, H., Lutzker, S., Selsing, E. and Alt, F.W. (1986). Nature **324**, 585–589.
Blunt, T., Finnie, F.J., Taccioli, G.E., Smith, G.C.M., Demengeot, J., Gottlieb, T.M., Mizuta, R., Varghese, A.J., Alt, F.W., Jeggo, P.A., and Jackson, S.P. (1995). Cell **80**, 813–823.
Bogue, M., Candeias, S., Benoist, C. and Mathis, D. (1991). EMBO J. **10**, 3647–3654.
Bogue, M., Gilfillan, S., Benoist, C. and Mathis, D. (1992). Proc. Natl Acad. Sci. USA **89**, 11011–11015.
Bogue, M., Mossman, H., Stauffer, U., Benoist, C. and Mathis, D. (1993). Eur. J. Immunol. **23**, 1185–1188.
Bollum, F.J. (1974). In 'The Enzymes' (P. Boyer, ed.), pp.145–171. Academic Press, New York.
Bosma, G.C., Custer, R.P. and Bosma, M.J. (1983). Nature **301**, 527–530.
Bosma, M.J. and Carroll, A.M. (1991). Annu. Rev. Immunol. **9**, 323–350.
Boubnov, N., Wills, Z.P. and Weaver, D. (1993). Mol. Cell. Biol. **13**, 6957–6968.
Boyd, J.B., Harris, P.V., Presley, J.M. and Narachi, M. (1983). UCLA Symp. Mol. Cell. Biol. New Ser. **11**, 107–123.

Cai, Q.-Q. (1994). *Cytogenet. Cell Genet.* **65**, 221–227.
Capone, M., Watrin, F., Fernex, C., Horvat, B., Krippl, B., Wu, L., Scollay, R. and Ferrier, P. (1993). *EMBO J.* **12**, 4335–4346.
Carlson, L.M., Oettinger, M.A., Schatz, D.G., Masteller, E.L., Hurley, E.A., McCormack, W.T., Baltimore, D. and Thompson, C.B. (1991). *Cell* **64**, 201–208.
Carlsson, L. and Holmberg, D. (1990). *Int. Immunol.* **2**, 639–643.
Carlsson, L., Overmo, D. and Holmberg, D. (1992). *Int. Immunol.* **4**, 549–553.
Chen, J. and Alt, F.W. (1993). *Curr. Opin. Immunol.* **5**, 194–200.
Chen, Y.-R., Lees-Miller, S.P., Tegtmeyer, P. and Anderson, C.W. (1991). *J. Virol.* **65**, 5131–5140.
Chun, J.J., Schatz, D.G., Oettinger, M.A., Jaenisch, R. and Baltimore, D. (1991). *Cell* **64**, 189–200.
Covey, L.R., Ferrier, P. and Alt, F.W. (1990). *Int. Immunol.* **2**, 579–583.
Cox, M. (1988). In 'Genetic Recombination' (R. Kucherlapati and G.R. Smith, eds), pp. 429–443. American Society for Microbiology, Washington, DC.
Cuomo, C.A., Kirch, S.A., Gyuris, J., Brent, R. and Oettinger, M.A. (1994). *Proc. Natl Acad. Sci. USA* **91**, 6156–6160.
Davis, M.M. and Bjorkman, P.J. (1988). *Nature* **334**, 395–399.
de Vries, E., van Driel, W., Bergsma, W.G., Arnberg, A.C. and van der Vliet, P.C. (1989). *J. Mol. Biol.* **208**, 65–78.
Durdick, J., Moore, N.W. and Selsing, E. (1984). *Nature* **307**, 749–752.
Dvir, A., Peterson, S.R., Knuth, M.W., Lu, H. and Dynan, W.S. (1992). *Proc. Natl Acad. Sci. USA* **89**, 11920–11924.
Early, P., Huang, H., Davis, M., Calame, K. and Hood, L. (1980). *Cell* **19**, 1981–1992.
Elliot, J.F., Rock, E.P., Patten, P.A., Davis, M.M. and Chen, Y.-H. (1988). *Nature* **331**, 627–631.
Engler, P., Haasch, D., Pinkert, C.A., Doglio, L., Glymour, M., Brinster, R. and Storb, U. (1991a). *Cell* **65**, 939–947.
Engler, P., Roth, P., Kim, J.Y. and Storb, U. (1991b). *J. Immunol.* **146**, 2826–2835.
Feeney, A.J. (1990). *J. Exp. Med.* **172**, 1377–1390.
Feeney, A.J. (1991). *J. Exp. Med.* **174**, 115–124.
Feeney, A.J. (1992a). *Int. Rev. Immunol.* **8**, 113–122.
Feeney, A.J. (1992b). *J. Immunol.* **149**, 222–229.
Ferguson, S.E. and Thompson, C.B. (1993). *Curr. Biol.* **3**, 51–53.
Ferrier, P., Krippl, B., Furley, A.J., Blackwell, T.K., Suh, H.-Y., Mendelsohn, M., Winoto, A., Cook, W.D., Hood, L., Constantini, F. and Alt, F.W. (1989). In 'Cold Spring Harbor Symposia on Quantitative Biology', pp. 191–202. Cold Spring Harbor Press, Cold Spring Harbor.
Ferrier, P., Covey, L.R., Li, S.C., Suh, H., Malynn, B.A., Blackwell, T.K., Morrow, M.A. and Alt, F.W. (1990a). *J. Exp. Med.* **171**, 1909–1918.
Ferrier, P., Krippl, B., Blackwell, T.K., Furley, A.J., Suh, H., Winoto, A., Cook, W.D., Hood, L., Costantini, F. and Alt, F.W. (1990b). *EMBO J.* **9**, 117–125.
Finger, L.R., Haluska, F.G. and Croce, C.M. (1989). In 'Immunoglobulin Genes' (T. Honjo, F.W. Alt and T.H. Rabbitts, eds), pp. 221–231. Academic Press, London.
Fondell, J.D. and Marcu, K.B. (1992). *Mol. Cell. Biol.* **12**, 1480–1489.
Fujimoto, S. and Yamagishi, H. (1987). *Nature* **327**, 242–244.
Fulop, G.M. and Phillips, R.A. (1990). *Nature* **347**, 479–482.
Fuschiotti, P., Harindranath, N., Mage, R.G., McCormack, W.T., Dhanarajan, P. and Roux, K.H. (1993). *Mol. Immunol.* **30**, 1021–1032.
Gauss, G.H. and Lieber, M.R. (1992). *Genes Dev.* **6**, 1553–1561.
Gellert, M. (1992). *Trends Genet.* **8**, 408–412.
George, J.F. and Schroeder, H.W. (1992). *J. Immunol.* **148**, 1230–1239.
Gerstein, R. and Lieber, M.R. (1993a). *Nature* **363**, 625–627.
Gerstein, R.M. and Lieber, M.R. (1993b). *Genes Dev.* **7**, 1459–1469.
Getts, R.C. and Stamato, T.D. (1994). *J. Biol. Chem.* **269**, 15981–15984.
Gilfillan, S., Dierich, A., Lemeur, M., Benoist, C. and Mathis, D. (1993). *Science* **261**, 1175–1178.
Goldman, J.P., Spencer, D.M. and Raulet, D.H. (1993). *J. Exp. Med.* **177**, 729–739.

Goodhardt, M., Cavelier, P., Doyen, N., Kallenbach, S., Babinet, C. and Rougeon, F. (1993). *Eur. J. Immunol.* **23**, 1789–1795.
Gottlieb, T.M. and Jackson, S.P. (1993). *Cell* **72**, 131–142.
Gu, H., Forster, I. and Rajewsky, K. (1990). *EMBO J.* **9**, 2133–2140.
Hafezparast, M., Kaur, G.P., Zdzienicka, M., Athwal, R.S., Lehmann, A.R. and Jeggo, P.A. (1993). *Somat. Cell Mol. Genet.* **19**, 413–421.
Haluska, F., Tsujimoto, Y. and Croce, C. (1987). *Trends Genet.* **3**, 11–13.
Harrington, J., Hsieh, C.L., Gerton, J., Bosma, G. and Lieber, M.R. (1992). *Mol. Cell. Biol.* **12**, 4758–4768.
Hata, S., Clabby, M., Devlin, P., Spits, H., De Vries, J.E. and Krangel, M.S. (1989). *J. Exp. Med.* **169**, 41–57.
Hendrickson, E.A., Schatz, D.G. and Weaver, D.T. (1988). *Genes Dev.* **2**, 817–829.
Hendrickson, E.A., Liu, V.F. and Weaver, D.T. (1991a). *Mol. Cell. Biol.* **11**, 3155–3162.
Hendrickson, E.A., Qin, X.-Q., Bump, E.A., Schatz, D.G., Oettinger, M. and Weaver, D.T. (1991b). *Proc. Natl Acad. Sci. USA* **88**, 4061–4065.
Hesse, J.E., Lieber, M.R., Gellert, M. and Mizuuchi, K. (1987). *Cell* **49**, 775–783.
Hesse, J.E., Lieber, M.R., Mizuuchi, K. and Gellert, M. (1989). *Genes Dev.* **3**, 1053–1061.
Hochtl, J., and Zachau, H.G. (1983). *Nature* **302**, 260–263.
Hochtl, J., Muller, C.R. and Zachau, H.G. (1982). *Proc. Natl Acad. Sci. USA* **79**, 1383–1387.
Holman, P.O., Roth, M.E., Hunag, M. and Kranz, D.M. (1993). *J. Immunol.* **151**, 1959–1967.
Hozumi, N. and Tonegawa, S. (1976). *Proc. Natl Acad. Sci. USA* **73**, 3628–3632.
Hunkapillar, T. and Hood, L. (1986). *Nature* **323**, 15–17.
Iijima, S., Teraoka, H., Date, T. and Tsukada, K. (1992). *Eur. J. Biochem.* **205**, 595–603.
Itohara, S., Mombaerts, P., Lafaille, J., Iacomini, J., Nelson, A., Clarke, A.R., Hooper, M.L., Farr, A. and Tonegawa, S. (1993). *Cell* **72**, 337–438.
Iwashima, M., Green, A., Davis, M.M. and Chien, Y.-S. (1988). *Proc. Natl Acad. Sci. USA* **85**, 8161–8165.
Jackson, S.P., MacDonald, J.J., Lees-Miller, S. and Tjian, R. (1990). *Cell* **63**, 155–165.
Jacob, J., Kelsoe, G., Rajewsky, K., Weiss, U. (1991). *Nature* **354**, 389–392.
Kallenbach, S., Doyen, N., D'Andon, M.F. and Rougeon, F. (1992). *Proc. Natl Acad. Sci. USA* **89**, 2799–2803.
Kallenbach, S., Babinet, C., Pournin, S., Cavelier, P., Goodhardt, M. and Rougeon, F. (1993). *Eur. J. Immunol.* **23**, 1917–1921.
Kienker, L.J., Kuziel, W.A. and Tucker, P.W. (1991). *J. Exp. Med.* **174**, 769–773.
Komori, T., Okada, A., Stewart, V. and Alt, F.W. (1993). *Science* **261**, 1171–1175.
Korman, A.J., Maruyama, J. and Raulet, D.H. (1989). *Proc. Natl Acad. Sci. USA* **86**, 267–271.
Kuppers, R., Zhao, M., Hansmann, M.L., Rajewsky, K. (1993). *EMBO J.* **12**, 4955–4967.
Kyes, S., Pao, W. and Hayday, A. (1991). *Proc. Natl Acad. Sci. USA* **88**, 7830–7833.
Lafaille, J.J., DeCloux, A., Bonneville, M., Takagaki, Y. and Tonegawa, S. (1989). *Cell* **59**, 859–870.
Landau, N.R., Schatz, D.G., Rosa, M. and Baltimore, D. (1987). *Mol. Cell. Biol.* **7**, 3237–3243.
Larche, M. and Hurwitz, J.L. (1993). *Eur. J. Immunol.* **23**, 1328–1332.
Lauster, R., Reynaud, C.-A., Martensson, I.-L., Peter, A., Bucchini, D., Jami, J. and Weill, J.-C. (1993). *EMBO J.* **12**, 4615–4623.
Lauzurica, P. and Krangel, M.S. (1994). *J. Exp. Med.* **179**, 43–55.
Lees-Miller, S.P., Chen, Y.-R. and Anderson, C.W. (1990). *Mol. Cell. Biol.* **10**, 6472–6481.
Lennon, G.G. and Perry, R.P. (1990). *J. Immunol.* **144**, 1983–1987.
Lewis, S.M. (1994). *Proc. Natl Acad. Sci. USA* **91**, 1332–1336.
Lewis, S.M. and Hesse, J.E. (1991). *EMBO J.* **10**, 3631–3639.
Lewis, S., Gifford, A. and Baltimore, D. (1984). *Nature* **308**, 425–428.
Lewis, S., Gifford, A. and Baltimore, D. (1985). *Science* **228**, 677–685.
Lewis, S.M., Hesse, J.E., Mizuuchi, K. and Gellert, M. (1988). *Cell*, **55**, 1099–1107.
Li, Y.-S., Hayakawa, K. and Hardy, R.R. (1993). *J. Exp. Med.* **178**, 951–960.
Lieber, M.R. (1991). *FASEB J.* **5**, 2934–2944.

Lieber, M.R. (1992). *Cell* **70**, 873–876.
Lieber, M.R., Hesse, J.E., Lewis, S., Bosma, G.C., Rosenberg, N., Mizuuchi, K., Bosma, M.J. and Gellert, M. (1988a). *Cell* **55**, 7–16.
Lieber, M.R., Hesse, J.E., Mizuuchi, K. and Gellert, M. (1988b). *Proc. Natl Acad. Sci USA* **85**, 8588–8592.
Lin, W.C. and Desiderio, S. (1993). *Science* **260**, 953–959.
Loh, E., Cwirla, S., Serafini, A., Phillips, J.H. and Lanier, L.L. (1988). *Proc. Natl Acad. Sci. USA* **85**, 9714–9718.
McCormack, W.T., Tjoelker, L.W., Carlson, L.M., Petryniak, B., Barth, C.F., Humphries, E.H. and Thompson, C.B. (1989). *Cell* **56**, 785–791.
Malissen, M., McCoy, C., Blanc, D., Trucy, J., Devaux, C., Schmidt-Verhulst, A.-M., Fitch, F., Hood, L. and Malissen, B. (1986). *Nature* **319**, 28–33.
Malynn, B.A., Blackwell, T.K., Fulop, G.M., Rathbun, G.A., Furley, A.J., Ferrier, P., Heinke, L.B., Phillips, R.A., Yancopoulos, G.D. and Alt, F.W. (1988). *Cell* **54**, 453–460.
Matsunami, N., Hamaguchi, Y., Yamamoto, Y., Kuze, K., Kanagawa, K., Matsuo, M., Kawaichi, M. and Honjo, T. (1989). *Nature* **342**, 934–937.
Meek, K.D., Hasemann, C.A. and Capra, J.D. (1989). *J. Exp. Med.* **170**, 39–57.
Meek, K., Rathbun, G., Reininger, L., Jaton, J.C., Kofler, R., Tucker, P.W. and Capra, J.D. (1990). *Mol. Immunol.* **27**, 1073–1081.
Mombaerts, P., Iacomini, J., Johnson, R.S., Herrup, K., Tonegawa, S. and Papaioannou, V.E. (1992). *Cell* **68**, 869–877.
Morzycka-Wroblewska, E., Lee, F.E. and Desiderio, S.V. (1988). *Science* **242**, 261–263.
Oettinger, M.A., Schatz, D.G., Gorka, C. and Baltimore, D. (1990). *Science* **248**, 1517–1523.
Okada, A. and Alt, F. (1994). *Semin. Immunol.* **6**, 185–196.
Okazaki, K., Davis, D.D. and Sakano, H. (1987). *Cell* **49**, 477–485.
Okazaki, K., Nishikawa, S. and Sakano, H. (1988). *J. Immunol.* **141**, 1348–1352.
Oltz, E.M., Alt, F.W., Lin, W.-C., Chen, J., Taccioli, G., Desiderio, S. and Rathbun, G. (1993). *Mol. Cell. Biol.* **13**, 6223–6230.
Orr-Weaver, T.L. and Szostak, J.W. (1988). *Microbiol. Rev.* **49**, 33–58.
Pergola, F., Zdzienicka, M.Z. and Lieber, M.R. (1993). *Mol. Cell. Biol.* **13**, 3464–3471.
Peterson, S.R., Dvir, A., Anderson, C.W. and Dynan, W.S. (1992). *Genes Dev.* **6**, 426–438.
Picard, D. and Schaffner, W. (1984). *EMBO J.* **3**, 3031–3039.
Rathbun, G., Berman, J., Yancopoulos, G. and Alt, F.W. (1989). In 'Immunoglobulin Genes' (T. Honjo, F.W. Alt and T.H. Rabbits, eds), pp. 63–90. Academic Press, London.
Rathmell, W.K. and Chu, G. (1994). *Mol. Cell. Biol.* **14**, 4741–4748.
Raulet, D.H., Spencer, D.M., Hsiang, Y.-H., Goldman, J.P., Bix, M., Liao, N.-S., Zulstra, M., Jaenisch, R. and Correa, I. (1991). *Immunol. Rev.* **120**, 185–204.
Reth, M., Gehrmann, P., Petrac, E. and Wiese, P. (1986). *Nature* **322**, 840–842.
Rolink, A. and Melchers, F. (1991). *Cell* **66**, 1081–1094.
Roth, D. and Wilson, D. (1988). In 'Genetic Recombination' (R. Kucherlapati and G.R. Smith, eds), pp. 621–653. American Society for Microbiology, Washington, DC.
Roth, D.B., Menetski, J.P., Nakajima, P.B., Bosma, M.J. and Gellert, M. (1992a). *Cell* **70**, 983–991.
Roth, D.B., Nakajima, P.B., Menetski, J.P., Bosma, M.J. and Gellert, M. (1992b). *Cell* **69**, 41–53.
Sakano, H., Kurosawa, Y., Weigert, M. and Tonegawa, S. (1981). *Nature* **290**, 562–565.
Schatz, D.G. and Baltimore, D. (1988). *Cell* **53**, 107–115.
Schatz, D.G., Oettinger, M.A. and Baltimore, D. (1989). *Cell* **59**, 1035–1048.
Schlissel, M., Constantinescu, A., Morrow, T., Baxter, M. and Peng, A. (1993). *Genes Dev.* **7**, 2520–2532.
Schuler, W., Weiler, I.J., Schuler, A., Phillips, R.A., Rosenberg, N., Mak, T.W., Kearney, J.F., Perry, R.P. and Bosma, M.J. (1986). *Cell* **46**, 963–972.
Schuler, W., Ruetsch, N.R., Amsler, M. and Bosma, M.J. (1991). *Eur. J. Immunol.* **21**, 589–596.
Schwager, J., Burkert, N., Courtet, M. and Du Pasquier (1991). *EMBO J.* **10**, 2461–2470.
Seidman, J.G. and Leder, P. (1980). *Nature* **280**, 779–783.

Serwe, M. and Sablitzky, F. (1993). *EMBO J.* **12**, 2321–2327.
Shimizu, T. and Yamagishi, H. (1992). *EMBO J.* **11**, 4869–4875.
Shinkai, Y., Rathbun, G., Lam, K.P., Oltz, E.M., Stewart, V., Mendelsohn, M., Charron, J., Datta, M., Young, F., Stall, A.M. et al. (1992). *Cell* **68**, 855–867.
Shinkai, Y., Koyasu, S., Nakayama, K., Murphy, K.M., Loh, D.Y., Reinherz, E.L. and Alt, F.W. (1993). *Science* **259**, 822–825.
Shirakata, M., Huppi, K., Usuda, S., Okazaki, K., Yoshida, K. and Sakano, H. (1991). *Mol. Cell. Biol.* **11**, 4528–4536.
Silverstone, A.E., Rosenberg, N., Baltimore, D., Sato, V., Scheid, M.P. and Boyse, E.A. (1978). In 'Differentiation of Normal and Neoplastic Hematopoietic Cells', pp. 433–453. Cold Spring Harbor Press, Cold Spring Harbor.
Smider, V., Rathmell, W.K., Lieber, M.R. and Chu., G. (1994). *Science* **266**, 288–291.
Spanopoulou, E., Roman, C.A.J., Corcoran, L., Schilissel, M.S., Silver, D.P., Nemazee, D., Nussenzweig, M., Shinton, S.A., Hardy, R. and Baltimore, D. (1994). *Genes Dev.* **8**, 1030–1042.
Taccioli, G.E., Rathbun, G., Oltz, E., Stamato, T., Jeggo, P.A. and Alt, F.W. (1993). *Science* **260**, 207–210.
Taccioli, G., Cheng, H.-L., Varghese, A.J., Whitmore, G. and Alt, F.W. (1994a). *J. Biol. Chem.* **269**, 7439–7442.
Taccioli, G.E., Gottlieb, T.M., Blunt, T., Priestley, A., Demengoet, J., Mizuta, R., Lehmann, A.R., Alt, F.W., Jackson, S.P. and Jeggo, P.A. (1994b). *Science* **265**, 1442–1445.
Takeda, S., Zou, Y.-R., Bluethmann, H., Kitamura, D., Muller, U. and Rajewsky, K. (1993). *EMBO J.* **12**, 2329–2336.
Thompson, C.B. (1992). *Trends Genet.* **8**, 416–422.
Thompson, L. (1988). In 'Genetic Recombination' (R. Kucherlapati and G.R. Smith, eds), pp. 597–620. American Society for Microbiology, Washington, DC.
Tonegawa, S. (1983). *Nature* **302**, 575–581.
Usuda, S., Takemori, T., Matsuoka, M., Shirasawa, T., Yoshida, K., Mori, A., Ishizaka, K. and Sakano, H. (1992). *EMBO J.* **11**, 611–618.
Van Ness, B.G., Weigert, M., Colecough, C., Mather, E.L., Kelley, D.E. and Perry, R.P. (1981). *Cell* **27**, 593–600.
Wang, J.C., Caron, P.R. and Kim, R.A. (1990). *Cell* **62**, 403–406.
Wang, Y. and Eckhart, W. (1992). *Proc. Natl Acad. Sci. USA* **89**, 4231–4235.
Wayne, J., Suh, H., Misulovin, Z., Sokol, K.A., Inaba, K. and Nussensweig, M.C. (1994). *Immunity* **1**, 95–107.
Wei, Z. and Lieber, M.R. (1993). *J. Biol. Chem.* **268**, 3180–3183.
Wu, Z. and Chaconas, G. (1992). *J. Biol. Chem.* **267**, 9552–9558.
Yancopoulos, G.D. and Alt, F.W. (1985). *Cell* **40**, 271–281.
Yancopoulos, G.D. and Alt, F.W. (1986). *Annu. Rev. Immunol.* **4**, 339–368.
Young, F., Ardman, B., Shinkai, Y., Lansford, R., Blackwell, T.K., Mendelsohn, M., Rolink, A., Melchers, F. and Alt, F. (1994). *Genes Dev.* **8**, 1043–1057.
Zachau, H.G. (1989). In 'Immunoglobulin Genes' (T. Honjo, F.W. Alt and T.H. Rabbitts, eds), pp. 91–110. Academic Press, London.
Zhang, W.W. and Yaneva, M. (1992). *Biochem. Biophys. Res. Commun.* **186**, 574–579.

11

Regulation of class switch recombination of the immunoglobulin heavy chain genes

J. Zhang*, F. W. Alt† and T. Honjo‡

*The Rockefeller University, New York, NY, USA
† Howard Hughes Medical Institute Research Laboratories, The Children's Hospital, Boston, MA, USA
‡ Kyoto University Faculty of Medicine, Department of Medical Chemistry, Yoshida, Sakyo-ku, Kyoto 606, Japan

Introduction 235
Constant region genes 237
Molecular mechanism of class switch recombination 239
Role of germline transcription in regulation of class switch recombination 242
Regulation of class switch recombination by T cells 248
Models for control of class switch recombination 251
Experimental systems to study switching 257

INTRODUCTION

An immunoglobulin (Ig) molecule can be divided into two functional domains: a variable (V) region that binds specifically to antigens, and a constant (C) region that mediates the effector functions of an antibody. In most mammals, there are five classes of Ig determined by the constant region portion of the heavy chain (C_H), i.e. IgM, IgD, IgG, IgE and IgA. In mouse and human, the IgG class is further divided into four subclasses, IgG_1, IgG_{2b}, IgG_{2a} and IgG_3 for mouse and IgG_1, IgG_2, IgG_3 and IgG_4 for human, to give rise to eight isotypes.

After functional assembly of V regions of both light and heavy chains, an IgM molecule is first expressed on a B cell surface. These newly generated B cells migrate to peripheral lymphoid organs where they encounter antigens, resulting in the proliferation of a particular clone of B cells and differentiation of that clone into plasma cells that actively secrete antibodies. Activated B cells not only secrete IgM but also other isotypes of Ig by changing the C_H portion of the expressed heavy chain, permitting a clonal lineage of B cells to produce antibodies that retain V-region specificity in association with a different C_H effector function (reviewed by Lutzker and Alt, 1988a; Esser and Radbruch, 1990). This change of the C_H of the antibody is termed heavy chain class switch recombination (CSR).

The production of various heavy chain isotypes serves to direct the humoral response along different functional and anatomic pathways through the diverse effector functions conferred by the C_H portion of the antibody. It involves various interactions with components of natural immunity and inflammation, such as activation of the complement system, enhancement of phagocytosis through Fc receptor binding and antibody-dependent cell-mediated cytotoxicity.

The heavy chain class switch phenomenon was first observed when the different classes of antibodies in serum were characterized after antigen stimulation (Bauer *et al.*, 1963; Uhr and Finkelstein, 1963). In the primary response, IgM increases dramatically first and IgG appears later. In response to secondary and sequential challenges, IgM diminishes while IgG dominates. In mouse, class switching has been observed in a variety of cells including activated B cells (Nossal *et al.*, 1964; Kearney and Lawton, 1975; Kearney *et al.*, 1976; Coutinho and Forni, 1982; Radbruch and Sablitzky, 1983), Abelson murine leukemia virus (A-MuLV)-transformed pre-B cell lines (Burrows *et al.*, 1981; Alt *et al.*, 1982; Akira *et al.*, 1983), B cell lymphoma (Stavnezer *et al.*, 1985), and in myeloma and hybridoma cell lines (Preud'Homme *et al.*, 1975; Liesegang *et al.*, 1978; Radbruch, 1985). In normal pre-B and naive B cells, switching occurs very rarely. The frequencies of switched cells as detectable by immunofluorescence are below 1% (Abney *et al.*, 1978) while activated B cells switch at a much higher frequency both *in vivo* and *in vitro* (Kearney *et al.*, 1976; Andersson *et al.*, 1977; Coutinho and Forni, 1982).

Earlier experiments have demonstrated that isotype switching occurs at the antigen-dependent stage of B cell differentiation and a single clone of B cells can switch to multiple isotypes (reviewed by Lawton and Cooper, 1973; Pernis *et al.*, 1977). Experiments with mouse splenic fragment cultures also demonstrated that the V-region specificity for an antigen is retained during isotype switching (Gearhart *et al.*, 1975). This is accomplished by a recombination event distinct from V(D)J recombination that fuses the assembled V region to a downstream C_H gene.

Since the observation of CSR more than three decades ago, the structure of the whole C_H region has been characterized in detail. Much progress has been made mostly in description of the outcome of CSR and the influence of other factors on the process. However, the fundamental question of how switching occurs and how it is regulated is still unanswered. How is specificity achieved? What is the switch recombinase? Is the switch recombinase specific to the switching process? Or is it a common recombination machinery that is directed to specific and regulated switching by other factors? In this chapter, we will give an overview of what is known about the structures of these C_H genes, discuss factors that are involved in the regulation of class switch recombination and propose models for the mechanism of CSR based on recent findings.

CONSTANT REGION GENES

Organization of the heavy chain locus

The entire murine C_H locus spans approximately 200 kb and has been mapped to chromosome 12 (D'Eustachio et al., 1980). The eight C_H genes are organized as follow: $5'-J_H-(6.5\text{ kb})-C_\mu-(4.5\text{ kb})-C_\delta-(55\text{ kb})-C_\gamma 3-(34\text{ kb})-C_\gamma 1-(21\text{ kb})-C_\gamma 2b-(15\text{ kb})-C_\gamma 2a-(14\text{ kb})-C_\epsilon-(12\text{ kb})-C\alpha-3'$ (Shimizu et al., 1982). These C_H genes produce eight different isotypes of immunoglobulins: IgM, IgD, IgG_3, IgG_1, IgG_{2b}, IgG_{2a}, IgE, and IgA respectively.

The human C_H gene family is mapped to the q32 band of chromosome 14 (Kirsch et al., 1982). Its organization differs from that of mouse in that a $C_\gamma-C_\gamma-C_\epsilon-C_\alpha$ unit is duplicated downstream of the $C_\mu-C_\delta$ genes. There are also three pseudogenes with two of them located in the same locus and one translocated to chromosome 9 (Battey et al., 1982). The nine functional genes and two pseudogenes are organized as follows: $5'-J_H-(8\text{ kb})-C_\mu-(5\text{ kb})-C_\delta-(60\text{ kb})-C_\gamma 3-(26\text{ kb})-C_\gamma 1-(19\text{ kb})-\psi C_\epsilon-(13\text{ kb})-C_\alpha 1-(35\text{ kb})-\psi C_\gamma-(40\text{ kb})-C_\gamma 2-(18\text{ kb})-C_\gamma 4-(23\text{ kb})-C_\epsilon 1-(10\text{ kb})-C_\alpha 2-3'$ (Ravetch et al., 1981; Flanagan and Rabbitts, 1982; Takahashi et al., 1982; Hofker et al., 1989).

Structure of heavy chain genes

The observation of sequence homologies between these C_H genes has led to the suggestion that these C_H genes have evolved from a primordial C_H gene that is most similar to C_μ with its four exons coding for distinct protein domains (Kawakami et al., 1980; reviewed by Honjo et al., 1989). Other C_H genes are derived from gene duplication followed by functional divergence. C_ϵ resembles C_μ with four exons and C_δ contains two exons separated by a hinge exon (Tucker et al., 1980; Ishida et al., 1982; Liu et al., 1982). All the C_γ genes contain three exons and a short hinge exon between $C_H 1$ and $C_H 2$ (Sakano et al., 1979; Tucker et al., 1981). The various C_γ genes also share extensive nucleotide homology with greatest differences concentrated in the hinge exon (Miyata et al., 1980; Takahashi et al., 1982; Ollo and Rougeon, 1983; Hayashida et al., 1984). The C_α gene has three exons with hinge region encoded by the $C_H 2$ exon. The hinge region permits a flexible conformation for the $C_H 1$ and $C_H 2$ domains, allowing the antibody to bind to antigens at multiple sites to increase the strength of interaction (Oi et al., 1984). The portion containing the $C_H 2$ and $C_H 3$ region mediates the majority of the antibody effector functions (Gally, 1973).

The heavy chain proteins can be expressed as secreted forms or membrane-bound forms, which determines the final location of the complete antibody (Rogers et al., 1980, 1982). The major C_H exon nearest the 3' end encodes the carboxyl-terminal part of the secreted form. The transmembrane and cytoplasmic portions of the membrane-bound form are encoded by two separate 'mini exons' (C_α has only one exon for this part) located downstream of the other major C_H exons. The mRNAs for secreted and membrane-bound forms are generated by alternative splicing and differential usage of the polyadenylation sites 5' and 3' of the membrane exons (Alt et al., 1980; Early et al., 1980; Cushley et al., 1982; Kemp et al., 1983; Milcareck and Hall, 1985; Brown and Morrison, 1989).

The membrane-bound form has a hydrophobic transmembrane segment followed by a basic charged cytoplasmic tail. This form of Ig serves as a B cell antigen receptor and its specific interaction with antigens results in B cell activation. The 26-residue transmembrane segments of all heavy chains share extensive homology and are involved in interacting with a heterodimer of membrane proteins called Igα and Igβ, which seems to be required for efficient surface expression of all classes of transmembrane Ig (Williams et al., 1990; Venkitaraman et al., 1991). These two proteins are expressed on the cell surface non-covalently linked to Ig molecules to form a surface antigen receptor complex (Campbell and Cambier, 1990; Hombach et al., 1990). Igα and Igβ have cytoplasmic regions of 61 and 48 amino acid residues, respectively, much longer than that of the Ig molecules (2 residues for C_μ and C_δ, 14 residues for C_α, and 27 residues for C_γ and C_ϵ chains). Mutational analysis has shown that Igα and Igβ are both sufficient and necessary to mediate signal transduction by the Ig receptors in B cells (Sanchez et al., 1993).

The secreted form has a hydrophilic tail and binds to antigens to elicit various effector functions of the immune system. For IgM, IgA, and IgD, their secreted forms have additional sequences at the C termini for intermolecular interactions stabilized by a peptide called J chain, resulting in multimeric Ig molecules (reviewed by Koshland, 1975).

Effector functions of each isotype

The binding of antigen to an antibody triggers various effector functions of the antibody depending on the structure, the anatomic location and the isotype of the antibody. In immune responses against different antigen challenges and infections, distinct isotypes dominate because of their different effector functions (reviewed by Esser and Radbruch, 1990). For example, IgM can be secreted upon antigen stimulation and is very effective in complement fixation. IgD is not secreted in general and its expression is downregulated upon activation of B cells by antigens accompanied by an increase in IgM secretion and switching to other isotypes (Monroe et al., 1983). So far no obvious function has been ascribed to IgD besides its antigen-binding ability and mice lacking IgD are normal with respect to B cell development and function (Roes and Rajewsky, 1993).

IgGs are the most abundant antibodies in serum and their production is rapidly increased in B cells activated by antigens. In T cell-independent immune responses elicited by carbohydrates such as bacterial lipopolysaccharide (LPS), IgG_{2b} and IgG_3 are dramatically induced. These two classes fix complement well and have a tendency to form congregates (Grey et al., 1971; Klaus et al., 1979; Neuberger and Rajewsky, 1981). IgG_3 is hardly induced by protein antigens but plays an important role in antibacterial responses and is very efficient in promoting phagocytosis. In immune responses to viral infections, IgG_{2a} and, to a lesser degree, IgG_1 are dominant (Coutelier *et al.*, 1987, 1988). IgG_{2a} can activate complement effectively and is very efficient as a mediator of antibody-dependent cytotoxicity by binding to specific Fc receptors on macrophages (Klaus *et al.*, 1979). In T cell-dependent immune responses, IgG_1 is the dominant isotype against parasitic and viral infections (Abraham and Teale, 1987a,b). It does not activate complement well but stimulates phagocytosis most efficiently. All IgGs can readily cross the placenta through binding to a specific Fc receptor to provide neonatal immunity.

IgE production is induced by parasitic infection and mediates killing of the parasites by binding them, via Fcε receptors, to eosinophils and macrophages (Jarrett and Miller, 1982). IgE is also the isotype that triggers allergy. Mast cells and basophils express high-affinity receptors for IgE (FcεRI) that can be bound by monomeric IgE in the absence of antigens (reviewed by Beaven and Metzger, 1993). Specific antigens cause the aggregation of the IgE–FcεR complex resulting in the secretion of histamine and other chemicals that mediate the immediate hypersensitivity response.

IgA plays a key role in mucosal immunity (reviewed by Mazanec *et al.*, 1993). It provides an immune barrier by preventing the adherence and absorption of antigens, neutralizing intracellular microbial pathogens directly within epithelial cells, and eliminating locally formed immune complexes by binding to the antigens and excreting them into the lumen. IgA is transported through epithelial cells and basal membranes to extracorporeal fluid such as mucus and milk through the binding of the polymeric Ig receptor on the surface of mucosal epithelial cells.

MOLECULAR MECHANISM OF CLASS SWITCH RECOMBINATION

Switch regions

Unlike V(D)J recombination, class switch recombination does not take place at a specific site (Nikaido *et al.*, 1982; Dunnick *et al.*, 1993), but instead occurs in regions composed of tandem repetitive sequences (termed S regions) located 5' of each C_H gene except for C_δ (Davis *et al.*, 1980; Dunnick *et al.*, 1980; Kataoka *et al.*, 1980, 1981; reviewed by Lutzker and Alt, 1988a; Gritzmacher, 1989). In the mouse, these regions vary in length from 1 kb (S_ϵ) to 10 kb ($S_\gamma 1$).

Each switch region has a few sequence motifs that are repeated multiple times. Pentameric sequences GAGCT and GGGGT are the most commonly found repeats in switch regions. The S_μ region can be divided into a 3' region of sequences $[(GAGCT)_{1-7} (GGGGT)]_{150}$, and a 5' region in which these two pentamers are interspersed with the sequence of $(T/CAGGTTG)_n$ (Nikaido *et al.*, 1981; Marcu *et al.*, 1982). Other switch regions have more complex patterns of sequence motifs, but share in common with S_μ the short sequences of GAGCT and GGGGT. S_ϵ and S_α are the most homologous to S_μ with tandem repeats of 40 and 80 bp respectively (Obata *et al.*, 1981; Nikaido *et al.*, 1982; Arakawa *et al.*, 1993). The S_γ regions are composed of repeated motifs of 245 or 295 bp. These motifs themselves contain several repeating sequences 49 or 52 bp in length that share 45–82% homology with each other (Kataoka *et al.*, 1981; Nikaido *et al.*, 1982; Wu *et al.*, 1984; Szurek *et al.*, 1985; Petrini *et al.*, 1987). The degree of homology between the S_γ regions and S_μ follows the order of their positions on the chromosome, i.e. $S_\gamma 3 > S_\gamma 1 > S_\gamma 2b > S_\gamma 2a$ (Nikaido *et al.*, 1982).

With a few exceptions, all of the recombination sites in S_γ, S_ϵ, and S_α regions occur within the tandemly repeated sequences (Iwasato *et al.*, 1992; Arakawa *et al.*, 1993; Dunnick *et al.*, 1993). Switch sites in S_μ are found to cluster more in the 5' end although recombination sites are detected throughout the S_μ region and can be found outside of the S_μ region (Katzenberg and Birshtein, 1988; Dunnick *et al.*, 1993). The sequences GAGCT/GGGGT and T/CAGGTTG are frequently present near the recombination sites, but no specific sequence has been detected that can account for all the joining sites characterized so far (Petrini *et al.*, 1987; Katzenberg and Birshtein, 1988; Dunnick *et al.*, 1993).

Intrachromosomal deletion pathway

Newly formed B cells can express IgM or both IgM and IgD on their surfaces (Cooper and Burrows, 1989). The production of IgD results from alternative splicing of primary transcripts that run through the C_μ and C_δ locus resulting in the joining of the functional VDJ segments with the C_δ exons directly at RNA level (Moore *et al.*, 1981; Tucker, 1985). The expressions of other C-region genes are mediated by the class switch recombination process, which juxtaposes a downstream C_H gene such as γ, ϵ, or α to the expressed V(D)J V-region gene.

Intrachromosomal deletion has been shown to be the most common mechanism for switching. It has been found that class switch recombination occurs between S_μ and the downstream switch region with deletion of the intervening sequences including the C_μ and C_δ genes (Cory *et al.*, 1980; Rabbitts *et al.*, 1980; Yaoita and Honjo, 1980; Shimizu and Honjo, 1984). The deleted sequences are usually lost from the cells during proliferation, but occasionally have been found to be integrated back into the genome, sometimes resulting in the inversion between two switch regions (DePinho *et al.*, 1984; Jack *et al.*, 1988).

In the context of looping-out and deletion mode of class switching recombination, the intervening regions between the sites of recombination are looped out to bring the sites for recombination in close vicinity. The four free ends generated after cutting could then be religated in three possible ways to produce the original configuration (no switching), an inversion of the looped-out sequences (inversion), or deletion of the looped sequences (switching). Recent studies demonstrated that the intervening regions that are deleted after switching could be isolated as circular molecules with the joining of two switch regions (Iwasato *et al.*, 1990; Matsuoka *et al.*, 1990; Schwedler *et al.*, 1990).

Unequal exchanges between homologs or sister chromatids during mitosis also can produce progeny cells in which the downstream C_H genes are located next to a functionally assembled VDJ region on one of the chromosomes (reviewed by Harriman et al., 1993). However, there is little experimental evidence that they occur at a significant level. Interchromosomal exchange was also reported with a randomly integrated µ transgene (Gerstein et al., 1990).

Although DNA recombination has been proven to be the main mechanism for switching, RNA splicing models were proposed to explain the appearance of cells that express double isotypes either by alternative splicing of long transcripts (Perlmutter and Gilbert, 1984), or by *trans*-splicing of two transcripts (Shimizu et al., 1991; Nolan-Willard et al., 1992). However, there is no convincing demonstration that these two types of RNA splicing events contribute significantly to class switching.

Factors involved in class switching

DNA replication

The involvement of DNA replication in switching was suggested by experiments indicating that proliferation was required for switching and drugs that inhibited DNA replication also inhibited switching (Severinson-Gronowicz et al., 1979; reviewed by Severinson et al., 1982). Mutations seem to occur more frequently at the switch recombination joints suggesting that the DNA replication associated with switch recombination may be error prone (Ott et al., 1987, Winter et al., 1987; Dunnick and Stavnezer, 1990; Shapira et al., 1991; Dunnick et al., 1993).

Sequential switching

Most cells switch from µ directly to downstream C_H genes. Analyses of recombined switch regions usually have S_μ directly proceeding into the switch region of the expressed C_H gene. However, three switch region sequences were occasionally found to be linked together in the recombined region indicating that switching can also occur between two downstream switch regions (Nikaido et al., 1982; Petrini et al., 1987, Katzenberg and Birshtein, 1988; Petrini and Dunnick, 1989; Iwasato et al., 1992). This type of sequential switching enables a B cell to produce two or more isotypes during differentiation (Schultz et al., 1990; Yoshida et al., 1990). Sequential switching seems to be particularly frequent in cells that switch to IgE. Most IgE-producing cells switch to $S_\gamma 1$ first then to S_ϵ, as the expressed IgE locus and the circles of the deleted sequence often contain $S_\gamma 1$ sequence in the S_ϵ region (Yoshida et al., 1990; Mills et al., 1992; Siebenkotten et al., 1992). However, studies with mice that did not switch to γ1 (Jung et al., 1993) showed that they switched to IgE at normal level when infected with parasites, indicating that switching to γ1 is not a prerequisite for switching to ε (Jung et al., 1994).

Protein factors

The enzymatic machinery responsible for switching has not been identified. The *RAG* proteins that are essential for V(D)J recombination do not seem to be involved in class switching since very low levels of *RAG*-1 or *RAG*-2 transcripts are detected in B cells, even after stimulation with LPS (G. Rathbun, personal communication). However, more detailed analysis is needed, especially with the *RAG*-deficient mice now available, to determine conclusively whether these proteins are involved in class switch recombination. Recently, a mutant mouse strain has been generated in which a rearranged V_H segment is inserted into the heavy chain locus replacing the J_H segment (Taki *et al.*, 1993). In the homozygous mouse, all B cells produce antibodies bearing the inserted V_H region and undergo normal switching. This mouse model provides a useful background to study whether the *RAG*-1 and *RAG*-2 genes are involved in switching by crossing with the *RAG*-deficient mice and a light chain transgenic mouse.

Certain protein factors have been shown to bind to switch sequences *in vitro*. One of the factors, B cell specific activating protein (BSAP), has been shown to bind to multiple regions within the heavy chain locus (Liao *et al.*, 1992). It is coded by the *Pax*-5 gene (Adams *et al.*, 1992) and can act as a transcription factor for the CD19 gene. The same factor also seems to be essential for the ε germline promoter activity and represses the enhancer activity of the IgH 3' enhancer (Singh and Birshtein, 1993). Other protein factors have also been shown to bind to certain sequences in various switch regions, e.g. octamer-binding protein to $S_\gamma 1$; NF-κB to $S_\gamma 3$; $S_\alpha BP$ to S_α and S_μ; LPS-responsive factor LR1 to $S_\gamma 1$, $S_\gamma 3$; and S_α; and a B cell-specific factor NSF-B_1 to S_μ (Waters *et al.*, 1989; Schultz *et al.*, 1991; Williams and Maizels, 1991; Illges and Radbruch, 1992; Wuerffel *et al.*, 1992; Xu *et al.*, 1992). BSAP may be responsible for some of these binding activities (Liao *et al.*, 1992; Xu *et al.*, 1992). Genes encoding for most of these factors have not been isolated. It has yet to be demonstrated directly that these factors and their binding in the heavy chain locus are involved in the class switching process.

ROLE OF GERMLINE TRANSCRIPTION IN REGULATION OF CLASS SWITCH RECOMBINATION

Directed class switch recombination

During an immune response, different Ig isotypes are produced at different levels depending on the type of antigen, the anatomic location of the infection and the cytokines secreted by activated T helper cells and other cells involved in the process. For example, viral infection induces predominantly IgG_{2a} production while parasitic infection induces high levels of IgE production (Jarrett and Miller, 1982; Coutelier *et al.*, 1987). *In vitro* culture of B lymphocytes with certain mitogens and cytokines can also influence switching to different isotypes. Stimulation of murine splenic B cells with bacterial LPS leads to the generation of activated B cells that express IgG_{2b} or IgG_3 antibodies, whereas simultaneous treatment of splenic B cells with LPS plus interleukin (IL)-4 suppresses IgG_{2b} and IgG_3 expression but leads to production of IgG_1 and IgE (Fig. 1). Treatment of LPS-activated B cells with interferon γ (IFN-γ) leads to predominant switching to IgG_{2a} while transforming growth factor β (TGF-β) induces switching to IgA (Kearney et al., 1976; Radbruch and Sablitzky, 1983; Layton et al., 1984; Coffman et al., 1986; Snapper and Paul, 1987, Coffman et al., 1989).

Various experiments have been done to determine whether the predominant production of certain isotypes after treatment with different cytokines and mitogens is a result of directed switching or growth selection of randomly switched cells. Analyses of the status of switching in hybridomas and splenic B cells showed that often both the productive and non-productive alleles have switched to the same switch region (Radbruch et al., 1986a; Hummel et al., 1987; Winter et al., 1987; Kepron et al., 1989; Schultz et al., 1990). In general, when murine B cells producing IgG_1, IgG_3, IgG_{2a}, and IgG_{2b} have performed switch recombinations on both loci, at least 70% used the same switch region (reviewed by Radbruch et al., 1986b). Since no growth selection can be applied on the non-functional allele that does not produce protein, these results suggest strongly that a particular switch region is targeted for switching with certain types of stimulation, i.e. switching is directed by the immune response. Additional evidence supporting this concept of directed switching has also been found in cell lines that preferentially switch to certain isotypes, e.g. to γ2b in pre-B cell lines, to α in lymphoma I.29 (Burrows et al., 1983; Stavnezer et al., 1985; Lutzker and Alt, 1988b). It also helps to explain the observation that in myelomas c-*myc* is frequently translocated into the same switch region on the inactive allele that is used for switch recombination on the active allele (Tian and Faust, 1987). However, studies with human B cell lymphomas did not give a conclusive result, perhaps due to the small numbers of samples analyzed (Borzillo et al., 1987).

Germline transcription in the C_H locus

Several models have been proposed for a directed switch recombination system. The lack of homology between S regions led to the theory that isotype-specific recombinases recognize the different sequences in each switch region and the activities or expressions of these recombinases are regulated by different cytokines during an immune response (Davis et al., 1980; Jack et al., 1988). Another model proposes that a common recombinase machinery is responsible for all the switch recombinations, but that other factors are involved to direct this common recombinase to different switch regions. This specificity could be achieved by modulating the accessibility of the switch regions. There are many possible and not mutually exclusive ways to modulate the accessibility of the switch region, e.g. specific binding of protein factors, regulated transcriptional activity through the S region (Stavnezer and Sirlin, 1986; Yancopoulos et al., 1986). Although these models have not been unequivocally resolved, lines of evidence support the 'common recombinase' model.

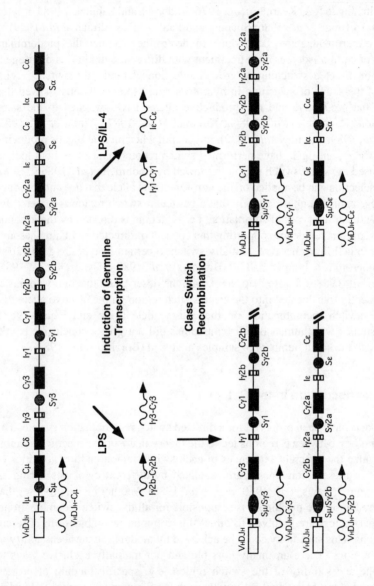

Fig. 1. Regulation of germline transcription and class switching by LPS and IL-4. Open boxes represent rearranged Ig V_H region, dark boxes represent C_H genes, circles represent switch regions, ovals represent hybrid switch regions from joining of two switch regions after switching, striped boxes represent germline exons, and wavy lines represent transcripts.

Germline transcription

The phenomenon of germline transcription was first observed in cell lines that switch spontaneously to certain isotypes in culture (Alt *et al.*, 1982; Stavnezer and Sirlin, 1986; Yancopoulos *et al.*, 1986b; Lutzker *et al.*, 1988). In several A-MuLV-transformed pre-B cell lines which switch spontaneously to γ2b, transcripts that contained the $C_\gamma2b$ gene but not any V_H sequence (termed germline transcripts) were expressed together with the switched VDJ–$C_\gamma2b$ transcript (Lutzker *et al.*, 1988). Molecular cloning and characterization of this transcript found that an exon (termed I exon) was spliced onto the $C_\gamma2b$ gene using the normal splice acceptor site (Fig. 2; Lutzker and Alt, 1988b). This germline exon is located ~700 bp 5' of the $S_\gamma2b$ region and would be deleted after switching to γ2b. It is approximately 400 bp in length and contains multiple stop codons in all reading frames. Transcription initiates heterogeneously from the 5' part (~150 bp) of the exon and no canonical promoter elements are found in the vicinity of this exon. Detailed analysis of splenic B cells stimulated with LPS showed that γ2b germline transcripts were induced within 24 hours of

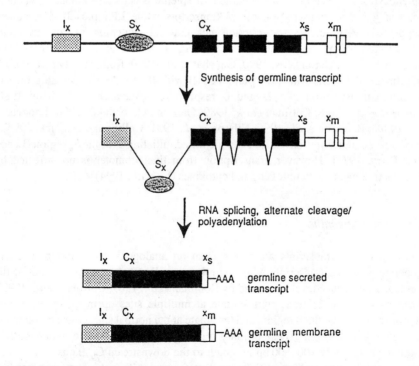

Fig. 2. Synthesis and processing of germline transcripts. Exons are depicted as boxes, switch regions as ovals, and introns as lines. (Adapted from Rothman *et al.*, 1989.)

treatment when switched VDJ–C$_\gamma$2b transcripts were barely detectable (Lutzker et al., 1988). The steady-state level of the γ2b germline transcripts reached a maximum by day 2 and remained constant thereafter. The switched VDJ–C$_\gamma$2b transcripts were not detected until day 3 of LPS treatment and increased rapidly after day 3. These results demonstrate that the germline transcription of I$_\gamma$2b precedes the actual switching to C$_\gamma$2b induced by LPS in splenocytes. Treatment of splenocytes with LPS together with IL-4 inhibited switching to γ2b and also the germline transcription of I$_\gamma$2b (Lutzker et al., 1988). These results strongly indicate a correlation between germline transcription and class switching.

Studies of a lymphoma cell line, I.29, which switches spontaneously to α, also found germline transcription from 5' of the S$_\alpha$ region (Stavnezer and Sirlin, 1986). The level of production of the α germline transcripts in subclones correlated with the ability of the subclone to switch to α. When induced, this line also switches to γ2a and ε together with induction of γ2a and ε germline transcripts. No transcripts from other regions that did not switch were detected, again indicating a correlation between germline transcription and switching (Stavnezer et al., 1988).

Similar results have been found for all the other C$_H$ genes both in cell lines and splenocytes. For example, LPS treatment of splenic B cells also induced switching to, and germline transcription of, γ3 transcripts while LPS plus IL-4 treatment suppressed germline γ3 transcription but induced switching to, and germline transcription of, γ1 and ε (summarized in Fig. 1; Berton et al., 1989; Esser and Radbruch, 1989; Gerondakis, 1990; Gauchat et al., 1990; Rothman et al., 1990a,b). Treatment of B cells with IFN-γ and TGF-β with IL-2 induced switching to, and germline transcription of, γ2a and α respectively (Stavnezer and Sirlin, 1986; Snapper et al., 1988; Coffman et al., 1989; Gaff and Gerondakis, 1990; Lebman et al., 1990a; Radcliffe et al., 1990; Nilsson et al., 1991; Collins and Dunnick, 1993).

Germline transcripts have also been identified initiating 5' of the S$_\mu$ region (Lenon and Perry, 1985). However, transcription from this promoter is not affected by various treatments with mitogens and cytokines (Li et al., 1994).

Germline transcripts

These germline transcripts all derive from an analogous transcription unit and generate germline transcripts with a similar overall structure, although not at the nucleotide sequence level (reviewed by Rothman et al., 1989; Coffman et al., 1993). Primary germline C$_H$ transcripts initiate at multiple sites upstream of a given S region, run through the S region, and terminate at the normal sites downstream of the C-region gene (Fig. 2). These primary transcripts are processed to juxtapose an I exon approximately 100–500 bp in length to the downstream C$_H$ exons. All I exons have multiple stop codons in three reading frames. Although short open reading frames that could code for small peptides are present in these exons, so far there is no evidence for the existence of these peptides.

Promoter elements regulating germline transcription

Germline transcription initiates from regions that do not have well-characterized common promoter elements such as the TATA box. Extensive analyses of several germline promoters have revealed DNA sequences that confer transcriptional activity inducible by cytokines which cause germline transcription in physiological conditions.

The I_ε promoter

One of the well-characterized promoters is the I_ε promoter (Rothman *et al.*, 1991). A 179-bp (−122 to +57 with +1 as the most 5' initiation site) DNA segment spanning the I_ε transcription initiation sites conferred LPS plus IL-4-inducible expression to a promoterless CAT construct. Electrophoretic mobility shift assays identified several nuclear proteins that bound to this region. One of the factors is constitutively expressed in normal B cells and binds in the upstream region of I_ε that is conserved between human and mouse. Analysis of the effects of point mutations in this region indicates that it contains essential sequences for I_ε promoter activity. The correlation of binding activity with the functional effect of point mutations suggests that this protein factor is directly involved in I_ε promoter activity. Two additional proteins bind near the initiation sites of I_ε and their binding activities are inducible by IL-4 treatment in splenic B cells. The transcription factor BSAP also binds to sequences directly upstream of the I_ε initiation sites that are essential for transcription activity (Coffman *et al.*, 1993).

The $I_\gamma 1$ promoter

The analyses of $I_\gamma 1$ promoter by linker-scanning mutations identified multiple independent elements required for inducible transcriptional activity by phorbol myristate acetate (PMA) and IL-4 (Xu and Stavnezer, 1992). Within a 150-bp region upstream of the initiation sites are several consensus sequences that bind known or putative transcription factors, including a C/EBP binding site as the IL-4-responsive element, four CACCC boxes, a purine-rich sequence (PU box), a TGF-β inhibitory element (TIE), an ab-IFN-α/β response element (AB-IRE) and an AP-3 site. This fragment is sufficient to confer PMA and IL-4 responsiveness to a heterologous promoter and all of these elements are required for this inducibility. An IL-4-inducible nucleoprotein complex was identified to bind to the region around the initiation sites of $I_\gamma 1$ (Illges and Radbruch, 1992). Although the exact nucleotide sequences that bind to this complex have not been characterized, the DNA fragment used for the assay contains the same sequence elements that are present in the I_ε region which binds IL-4-inducible protein factors (Rothman *et al.*, 1991).

The I_α promoter

Analysis of mouse I_α region identified at least two distinct elements that were involved in constitutive and TGF-β inducible transcription activity (Lin and Stavnezer, 1992). An ATF/CRE site in the region from −1 to −106 is involved in basal activity and a tandemly repeated sequence between −127 and −105 can confer partly the TGF-β response. Human $I_\alpha 1$ and $I_\alpha 2$ germline promoters showed a more complex organization of regulatory elements (Lars and Sideras, 1993). Two clusters of elements have been identified: proximal positive elements (−248 to +79) required for basal and TGF-β inducible expression, and distal negative elements (−731 to −352) contributing to the B cell-specific expression.

The LPS inducible promoters, $I_\gamma 2b$ and $I_\gamma 3$

Germline transcription from γ2b and γ3 are induced dramatically by LPS both in normal B cells and in A-MuLV-transformed pre-B cells (Lutzker et al., 1988; Rothman et al., 1990b). No obvious promoter elements are present in the vicinity of the transcription initiation sites of γ2b and γ3 although there is a short stretch (~100 bp) of scattered homology between mouse $I_\gamma 2b$, $I_\gamma 3$, and human $I_\gamma s$ spanning the initiation sites (Sideras et al., 1989; Rothman et al., 1990b). Unlike the germline promoters for the other isotypes, DNA fragments surrounding the $I_\gamma 2b$ region did not confer a highly LPS-inducible transcriptional activity that is comparable to the endogenous level, although basal promoter activity can be located within a few hundred base pairs of the initiation sites (J. Zhang and F. Alt, unpublished data).

REGULATION OF CLASS SWITCH RECOMBINATION BY T CELLS

During an immune response, another major lymphocyte population, T cells, are also activated through the binding of T cell antigen receptors (TCR) by processed antigens presented by antigen-presenting cells including B cells (Lanzavecchia, 1985; reviewed by Hedrick, 1989; Parker, 1993). One of the results of T cell activation is the secretion of cytokines, which play important roles in lymphocyte proliferation, differentiation and regulation of lymphocyte functions (reviewed by Paul, 1989). Some of the cytokines secreted by T helper (T_H) cells are directly involved in regulation of class switching in B cells (reviewed by Finkelman et al., 1990). Interactions between B and T cells through various surface molecules also provide signals necessary for activation (reviewed by Parker, 1993).

Role of cytokines in directed switching

Two subclasses of T_H cells have been identified: $T_H 1$ and $T_H 2$ (reviewed by Mosmann and Coffman, 1989; Powrie and Coffman, 1993). T_H1 cells secret IL-2 and IFN-γ while T_H2 cells produce IL-4 and IL-5 but not IFN-γ.

Interleukin-4

The first cytokine found to exert an influence on the outcome of switching is IL-4 (reviewed by Coffman et al., 1993). It has been shown that IL-4 treatment induces dramatically the production of IgG_1 and IgE, and at the same time inhibits the production of IgG_3 and IgG_{2b} both in vivo and in vitro (Isakson et al., 1982; Bergstedt-Lindqvist et al., 1984; Sideras et al., 1985; Vitetta et al., 1985; Coffman and Carty, 1986). The germline transcription of these C_H genes are affected by IL-4 in a similar fashion (Lutzker et al., 1988; Berton et al., 1989; Esser and Radbruch, 1989; Gerondakis, 1990; Rothman et al., 1990a,b). The gene encoding for IL-4 has been isolated and bacterial-expressed recombinant IL-4 has been shown to have the same effect on regulating switching as the purified factor (Coffman et al., 1986; Lee et al., 1986; Noma et al., 1986; Snapper and Paul, 1987).

Mice with an overexpressed or deleted IL-4 gene have been generated. In IL-4 transgenic mice, the serum Ig isotype repertoire is dramatically changed: IgG_1 and IgE levels are elevated; IgG_3, IgG_{2a} and IgG_{2b} levels are depressed (Burstein et al., 1991). In IL-4-deficient mice, the serum levels of IgG_1 and IgE are markedly reduced. The IgG_1 dominance in a T cell-dependent immune response is lost and IgE is not detectable upon nematode infection (Kuhn et al., 1991). However, the development of B and T cells in mutant mice is normal. These results suggest that IL-4 is critical for generating certain Ig isotypes in immune responses, but not essential for the development of the immune system.

Recently it has been found that another cytokine, IL-13, has similar effects on switching as IL-4. Recombinant human IL-13 induced switching to IgE and IgG_4 and induced germline ε transcripts in human B cells independently of IL-4 (McKenzie et al., 1993; Punnonen et al., 1993). However, given the finding that little IgE and IgG_1 is present in mice deficient for IL-4 (Kuhn et al., 1991), it is not clear whether IL-13 induces IgE and IgG_1 synthesis in murine B cells.

Interleukin-5

IL-5 was originally identified as a factor that induces antigen-primed B cells to differentiate into antibody-producing cells or induces growth of BCL1 B-cell tumor cells. Molecular cloning of murine and human IL-5 cDNA has shown that IL-5 has the activities of an eosinophil differentiation factor (Azuma et al., 1986; Kinashi et al., 1986; Campbell et al., 1988).

The effect of IL-5 on class switching is still controversial. IL-5 was shown to increase IgA production by murine B cells that carry IgA on their surface. No IgA secretion was induced in the surface IgA-negative cells, suggesting that IL-5 does not direct switching to IgA (Murray et al., 1987; Beagley et al., 1988; Harriman et al., 1988). IL-5 was also reported to enhance switching to IgG_1 in dextran–anti-IgD activated B cells cultured with IL-4 (Purkerson and Isakson, 1992; Mandler et al., 1993). Few effects on human B cells by human IL-5 have been reported. The role of IL-5 in class switching *in vivo* remains to be clarified.

Interferon-γ

During an immune response to viral infection, IgG_{2a} is the predominant isotype produced (Coutelier et al., 1987). The cytokine that causes the selective expression of IgG_{2a} is IFN-γ. Studies with purified IFN-γ and anti-IFN-γ antibodies showed that IFN-γ stimulated switching to IgG_{2a} and inhibited IL-4 induction of IgG_1 and IgE both *in vivo* and *in vitro* (Snapper and Paul, 1987; Finkelman et al., 1988; Snapper et al., 1988). IFN-γ also increases germline transcription from the γ2a locus (Collins and Dunnick, 1993). The effect of IFN-γ on switching to IgG_3 seems to depend on the mode of B cell activation. IFN-γ inhibits switching to IgG_3 by B cells activated by LPS but induces switching to IgG_3 in B cells activated with anti-IgD conjugated to dextran (Snapper and Paul, 1987; Snapper et al., 1992). In the latter case, germline γ3 transcription is also induced by IFN-γ.

Transforming growth factor β

In vitro studies indicate that TGF-β alone can induce switching to IgA in LPS-stimulated B cells and the production of IgA is enhanced when IL-2 is present at the same time (Coffman et al., 1989). A similar effect has also been observed with IL-5 (Coffman et al., 1989; Sonoda et al., 1989). Treatment of TGF-β alone induces germline transcription of α in LPS-stimulated B cells while IL-2 does not affect the level of germline transcription (Lebman et al., 1990a,b; Radcliffe et al., 1990). The DNA regions deleted upon switching to α induced by TGF-β have been characterized further, supporting the notion that TGF-β alone induces switching to IgA (Matsuoka et al., 1990; Iwasato et al., 1992). However, other studies suggested that TGF-β is a partial switch signal or secondary signal that causes switching to IgA in a population that has already committed to switching to IgA (Whitmore et al., 1991; Ehrhardt et al., 1992). It has yet to be demonstrated that TGF-β induces switching to IgA *in vivo*. A recently cloned isoform of TGF-β, TGF-β1, has also been shown to stimulate IgG_{2b} secretion by LPS-activated B cells *in vitro* (McIntyre et al., 1993).

Interactions between T and B cells

The activation of B cells requires not only the signals delivered through the Ig antigen receptor, but also co-stimulatory signals derived from T_H cells (reviewed by Noelle and Snow, 1992). T_H cells provide help for B cell activation by secreting cytokines and by taking part in cell–cell interactions with B cells.

Until recently, *in vitro* studies of switching have all relied on activation of B cells by anti-IgM, LPS or anti-Ig conjugated to dextran. LPS stimulates both proliferation and differentiation including switching to IgG_{2b} and IgG_3. In *in vivo* T cell-dependent isotype switching, the activation signal to B cells is delivered through direct contact between B and T cells (reviewed by Noelle and Snow, 1991, 1992; Parker, 1993). This contact-mediated signal does not require living T cells, as plasma membranes of activated T_H cells can induce B cells to switch to IgE and IgG_1 when incubated together with IL-4 (Hodgkin *et al.*, 1990).

A newly identified surface molecule on activated T_H cells, a ligand for the B cell differentiation antigen CD40, is a key component of B cell activation (Armitage *et al.*, 1992; Spriggs *et al.*, 1992). Recently, antigen-induced activation of B cells has been shown to require a second signal through the CD40 molecule, which interacts with CD40 ligand (CD40L) expressed on activated T_H cells (reviewed by Banchereau *et al.*, 1994). When B cells recognize self antigens by their surface IgM without the second signal through CD40, because of the absence of specific T_H cells, B cells generally die by apoptosis (Valentine and Licciardi, 1992; Tsubata *et al.*, 1993, 1994; Parry *et al.*, 1994). B cells activated by two signals are ready to undergo specific S–S recombination directed by lymphokines. A small fraction of spleen cells can be induced to switch by the addition of anti-CD40 antibody (or CD40L) and IL-4. However, this is probably due to preactivation of spleen cells by antigen. For example, studies with anti-CD40 antibodies showed that IL-4 alone induced germline ε transcription but not switching to IgE, while cross-linking of CD40 increased the level of $I_ε$ and induced switching to IgE (Gauchat *et al.*, 1992; Shapira *et al.*, 1992). In X-linked hyper-IgM immunodeficiency characterized by the absence of serum IgG, IgA and IgE, mutations were found in the CD40L gene resulting in non-functional or defective expression of CD40L on the surface of T cells (reviewed by Callard *et al.*, 1993). These results indicate that the interaction between CD40 and CD40L provide necessary signals in addition to those from cytokines for T cell-dependent heavy chain class switching and from surface IgM by antigen binding. Class switching appears to require at least three signals: antigen, CD40 and lymphokine.

MODELS FOR CONTROL OF CLASS SWITCH RECOMBINATION

The accessibility model

The correlation of germline transcription of a specific I region and switching to that particular locus has led to the hypothesis that germline transcription of a C_H locus directs class switch recombination to that particular locus (Stavnezer and Sirlin, 1986; Yancopoulos *et al.*, 1986; reviewed by Lutzker and Alt, 1988a; Rothman *et al.*, 1989). It has been proposed that genetic recombination is initiated by a break in the DNA, such as a single-stranded nick or a double-stranded break (Holliday, 1964; Meselson and Radding, 1975; Szostak *et al.*, 1983). However, DNA in cells is associated with histones and various other proteins to form a tight DNA–protein complex that is not readily accessible. The observation that a transcriptionally active gene is often more sensitive to nucleases than when it is silent has led to the suggestion that the transcription process could serve to disrupt the chromatin structure and render the DNA accessible to the recombinase machinery. Studies in yeast have demonstrated that transcription activation of a region enhanced its ability to be the target for recombination (Thomas and Rothstein, 1989).

Recently, gene targeting techniques were used to investigate the role of germline transcription and/or transcripts in heavy chain class switching. Jung et al. (1993) generated mice in which the $I_\gamma 1$ promoter and exon were deleted. These mutant mice exhibit a selective agammaglobulinemia with respect to IgG_1 production due to a block of switching to the $S_\gamma 1$ region. The block of switching to $S_\gamma 1$ was specific on the mutated allele indicating a *cis*-acting defect caused by the deletion of the γ1 germline transcription unit. Zhang et al. (1993) replaced the germline $I_\gamma 2b$ promoter and I exon with an expressed neomycin-resistance gene (*neo*^{r}) in embryonic stem (ES) cells, which were then analyzed in somatic chimeras using the *RAG*-2-deficient blastocyst complementation system (Chen et al., 1993). B lymphocytes from chimeras derived by injection of homozygous mutant ES cells were deficient in IgG_{2b} production both *in vivo* and *in vitro*, but normal with respect to production of other Ig heavy chain isotypes. The lack of ability to produce IgG_{2b} by the mutant B cells correlated with lack of germline transcription and resulted from a *cis* defect in class switch recombination to $S_\gamma 2b$ on the mutated chromosome. Together, these studies demonstrate that the I region is an important regulatory element for control of class switch recombination.

The precise role(s) of germline transcription and/or transcripts in the class switch recombination process is unknown. The importance of the I region and the germline transcription it generates can be explained by possibilities that are not mutually exclusive. Several aspects of the germline transcription process can be involved in the regulation of switching, e.g. the transcription itself, the germline transcripts or protein factors that bind to the I region could all play important roles in class switch recombination.

Transcription

The question of whether transcription *per se* is sufficient to direct switching to a particular locus was also studied by gene targeting experiments. In an A-MuLV-transformed pre-B cell line, the I_ε region on one chromosome was replaced by an E_μ–V_H transcription cassette that conferred constitutive transcription through the ε locus (Xu et al., 1993). Normally, germline transcription of ε and switching to ε requires stimulation with both LPS and IL-4 (Rothman et al., 1990). In contrast, the mutant cell line had constitutive transcription through the S_ε region and switched to ε in the absence of IL-4, by PCR analyses. This result suggests that constitutive transcription from the E_μ–V_H promoter in the absence of IL-4 can direct switching to ε in this pre-B cell line.

The same mutation was introduced into ES cells to generate somatic chimeras using the *RAG*-2-deficient blastocyst complementation system (Bottaro et al., 1994). In B cells heterozygous or homozygous for this mutation, a substantial level of transcription through the ε locus was present after treatment with LPS alone or LPS plus IL-4. However, no switching comparable to the normal level was detected in either LPS or LPS plus IL-4 treated homozygous mutant cells and LPS-treated heterozygous mutant cells. Low levels of switching to ε (1–10% of normal) were detected in heterozygous and homozygous mutant cells treated with LPS, in contrast to no switching at all in LPS-treated normal B cells, indicating transcription *per se* did cause switching albeit at a very low level. When stimulated with LPS plus IL-4, heterozygous mutant cells switched to ε at wild-type levels, while homozygous mutants still switched to IgE at very low levels. These results indicate that transcription *per se* can direct switching to the transcribed locus but is not sufficient to confer efficient switching at a normal level.

Regulation of class switching was also analyzed in B cells that had the J_H segments and E_μ deleted by gene targeting (Gu et al., 1993). On the mutant chromosome, switch recombination at the S_μ region is substantially suppressed correlating with a lack of transcription through the S_μ region. However, rearrangement of the $S_\gamma 1$ region occurred efficiently in the form of intra-S region deletion after LPS plus IL-4 treatment, suggesting that switch recombination is controlled independently at individual switch regions.

Germline transcripts

Another possibility for the function of the I region is that the germline transcripts encoded by the I exon could be an integral part of the switch recombination process.

One of the possible functions of these mature germline transcripts is that they could interact with double-stranded (ds) DNA containing the same sequence (i.e. the I exon and the C_H exons) causing the looping-out of the switch region in between and exposing it to the switch recombinase. Such interactions between dsDNA and RNA could also serve to disrupt the chromatin structure and render the DNA accessible to the switch recombinase. A hybrid structure of dsDNA and RNA has been reported for the S_α region (Reaban and Griffin, 1990). The nascent transcript of a supercoiled plasmid containing the S_α region was found to remain base-paired to its DNA templates resulting in the relaxation of supercoils of the plasmid. A triplex structure using conventional Watson–Crick base-paring and Hoogstein base pairs was proposed for the transcript and its dsDNA template. However, the S_α region used in this study contains the purine-rich AGGAG repeats that are unique to S_α and similar studies have not been done with other switch region sequences.

It is also conceivable that germline transcripts could interact with the switch recombinase machinery and their interactions serve to direct recombination to the accessible DNA. RNA–protein interactions have been shown to be essential in many biological processes such as protein translation (ribosome) and RNA splicing (spliceosomes) in which the RNAs themselves are often directly involved in the activities. However, it is not known whether the germline transcripts, particularly the I exon part can interact with any protein factors.

The role of germline transcripts was studied using antisense transcripts or oligonucleotides. In the case of germline α transcripts, an expression construct that could be induced to produce antisense transcripts to the I_α exon was transfected into a B cell line (Wakatsuki and Strober, 1993). The steady-state level of α germline transcripts was downregulated in these cells together with a decrease in IgA production. Experiments with antisense oligonucleotides to I_α gave similar results suggesting that the germline α transcripts are important for switching to α. However, results from studies using antisense oligonucleotides to the $I_\gamma 2b$ region are more complex (Tanaka et al., 1992). Incubation of splenic B cells stimulated with LPS or LPS plus IL-4 with antisense oligonucleotides to the $I_\gamma 2b$ exon inhibited secretion of all Ig isotypes. It also stimulated DNA synthesis by resting B cells. Most strikingly, it caused an increase in the steady-state level of $I_\gamma 2b$ germline transcripts by 10–20 fold. It is not clear whether this increase in $I_\gamma 2b$ transcripts is due to an increase in transcription rate or message stability. Incubation of normal B cells with large amounts of oligonucleotides may also elicit other responses besides its intended antisense function, as indicated by the stimulant effect on DNA synthesis. Expression of antisense transcripts inside the cell by transfection in cell lines, transgenic mice or the *RAG*-2 chimera system would be useful to determine whether germline transcripts are necessary component(s) in the class switching process.

Binding of protein factors in the I region

The I region could also provide binding site(s) for protein factors that are involved in switch recombination. Protein factors have been identified to bind to various I regions. Some of these factors could be part of the basal transcription complex responsible for germline transcription. They could also be involved in the actual switching process, e.g. as part of the switch recombinase machinery. Their binding to the DNA could serve to direct the switch recombinase to the locus. In the latter case, their binding to the DNA would be regulated in the same fashion as switching is regulated for a particular locus.

A locus control mechanism for LPS-responsive class switch recombination

The accessibility model provides a mechanism by which different factors, e.g. cytokines, mitogens, DNA-binding factors, could influence class switch recombination through regulation of germline transcription. It would require the existence of *cis* regulatory elements that control germline transcription in response to the various agents.

Local promoter elements

Germline promoters do not have the canonical transcription elements, such as TATA boxes, that are normally present in other promoters. Studies on some of the germline promoters have identified promoter elements that respond to certain cytokines, such as IL-4 and TGF-β (Rothman et al., 1991; Lin and Stavnezer, 1992; Xu and Stavnezer, 1992; Lars and Sideras, 1993). However, efforts to identify LPS-responsive elements have not been successful. One interesting finding from the $I_\gamma 2b$ replacement experiment by Zhang et al. (1993) showed that the expression of the neo^r gene replacing the $I_\gamma 2b$ promoter was also induced by LPS treatment of mutant B cells. In contrast, when the whole targeting construct was randomly integrated into the genome, the expression of the neo^r gene was no longer LPS inducible (J. Zhang and F. Alt, unpublished data). These results indicated that the sequences deleted by the mutation or the region surrounding $I_\gamma 2b$ (up to 6 kb 5′ and 2 kb 3′ of $I_\gamma 2b$) did not have all the necessary elements to confer LPS-inducible transcription activity. In the I_ε gene targeting experiment, the same neo^r gene is brought into the I_ε locus together with the E_μ–V_H cassette that replaced I_ε (Bottaro et al., 1994). Similarly, the expression of the neo^r gene in the I_ε locus was also induced by LPS while the random integrated construct had a very low level of non-LPS-inducible neo^r gene expression (A. Bottaro and F. Alt, unpublished data). These results imply that the whole IgC_H region can be activated transcriptionally by LPS through a more general-acting locus control element. This element is probably not the IgH intron enhancer (E_μ) since the randomly integrated I_ε gene targeting construct contains the E_μ enhancer.

The IgH 3′ enhancer

Another B cell-specific enhancer element has been identified 3′ of C_α in the rat (Pettersson et al., 1990). A mouse sequence 16 kb 3′ of C_α showed 80% homology to the rat sequence (Dariavach et al., 1991; Lieberson et al., 1991). This 3′ enhancer has a minimal core sequence of 700 bp flanked by repetitive sequences and contains some enhancer elements found in the E_μ enhancer (Grant et al., 1992). Transient transfection analyses showed that the mouse 3′ enhancer was a weak enhancer for transcriptional activity in B cell lines at a later stage of differentiation (Dariavach et al., 1991; Lieberson et al., 1991). Studies of the methylation status and DNase hypersensitivity of this 3′ enhancer also indicated that it was relatively inactive during early stages of B cell differentiation and became increasingly accessible at later stages, particularly in plasma cells in which this region was completely demethylated (Giannini et al., 1993). It is postulated that this enhancer is operative at the B cell to plasma cell stage of differentiation, regulating the transcription of Ig genes in actively secreting cells. It could be important not only for the increase in antibody production by activated B cells after encountering antigen, but also for class switching and somatic mutation.

Recently, mice lacking the 3' enhancer and its flanking region were generated (Cogne et al., 1994). B cells bearing this mutation had normal VDJ rearrangement and produced IgM at normal levels both *in vivo* and *in vitro*. Strikingly, their abilities to switch were affected dramatically and differently with respect to each isotypes. *In vitro* cultured splenic B cells derived from homozygous mutant ES cells ($E_{3'}$−/−) had a combined deficiency in IgG_{2a}, IgG_{2b}, IgG_3, IgE and, at least partially, IgA production. For example, no switching to IgG_{2b} or IgG_3 was detectable in $E_{3'}$−/− splenic B cells treated with LPS, and in $E_{3'}$−/− B cells treated with LPS plus IL-4 switching to IgG_1 occurred at close to normal level while switching to IgE was drastically reduced. Interestingly, the status of germline transcription corresponded to that of switching, i.e. no γ2b or γ3 germline transcripts were present in LPS-treated B cells and very low levels of ε germline transcripts were detected in cells treated with LPS plus IL-4 (J. Zhang, M. Cogne and F. Alt, unpublished data). In line with the evidence for a functional role of germline transcription in the class switching process, these results suggest that the 3' enhancer might act as a locus control element in response to LPS treatment. The mutation introduced could block germline transcription, especially those that are LPS responsive, such as γ2b and γ3, resulting in the shutdown of switching to these isotypes. It also could explain the fact that so far no LPS-responsive transcriptional element(s) for each individual isotype has been found. However, these mutant mice produced normal levels of IgG_{2b} in serum, suggesting that alternative pathways independent of $3'E_H$ may exist for switching to different isotypes.

A competition model for control of class switch recombination

A locus control region (LCR) that functions over a large area has been characterized for other gene clusters, such as the α- and β-globin locus (Talbot et al., 1989; Higgs et al., 1990). The β-globin LCR regulates chromatin structure, replication and transcription of five β-like globin genes that lie from 11 to 60 kb downstream (Crossley and Orkin, 1993). A competition model has been proposed in which expression of a particular gene in the β-globin locus is a result of stronger interaction between its promoter and the LCR enhancer.

The IgH 3' enhancer region could also act as an LCR that regulates the transcriptional activation of germline promoters following some forms of B cell activation. The 3' E_H mutation may delete critical sequences of this control element and/or block its activity through the insertion of a competing promoter associated with the neo^r gene. Consistent with a competition mechanism, the expression of the neo^r gene inserted in the 3' E_H region is LPS inducible in splenic B cells (J. Zhang, M. Cogne and F. Alt, unpublished data). In the context of such a mechanism, the C_H genes induced by LPS alone (γ2b and γ3) would have the weakest interaction with the 3' E_H LCR, while the C_H genes induced by LPS plus other cytokines would have stronger interactions that allow them to compete better for the LCR.

EXPERIMENTAL SYSTEMS TO STUDY SWITCHING

In vitro switch substrate constructs

Most studies of isotype switching and its regulation are done *in vitro* with polyclonal populations of normal B cells or with transformed cell lines. Analyses of switching at DNA level in normal B cells have been difficult due to the imprecise nature of the joints occurring over a large region. Tumor cell lines provide a clonal or partial clonal system and give insight to the mechanism of switching. However, due to their restricted switching patterns and possible non-physiological environment, analysis of switching to the whole C_H locus is severely hindered.

To distinguish between the various possibilities of how germline transcription regulates class switching, more defined and detailed analyses of the promoter, transcripts and the protein factors associated with them are needed.

One potentially very useful way to analyze the mechanism of class switching is by using switch recombination substrate constructs similar to those used for VDJ recombination analyses. In such substrate constructs, various mutations could be introduced into all the elements that are possibly involved in switching and analyzed rapidly. However, attempts to use switch substrate constructs to study the mechanism of switching have generated limited success so far (reviewed by Coffman *et al.*, 1993). A retroviral switch construct was developed by flanking a thymidine kinase gene (*tk*) with S_μ and $S_\gamma 2b$ sequences (Ott *et al.*, 1987; Ott and Marcu, 1989). Switch recombinants which have the *tk* gene deleted were selected by growing cells in BrdU. Analyses of stable transfectants of the construct in pre-B and non-lymphoid cells showed that certain components of S region recombination are B cell specific and switch sequences are important in the recombination process. However, analyses were only done with portions of S_μ and $S_\gamma 2b$. Another *in vitro* assay was developed in which the recombinant construct (containing S_μ and $S_\gamma 3$) was transiently transfected into LPS-stimulated normal B cells and subsequently analyzed for recombination utilizing a bacterial selection system (Leung and Maizels, 1992). A positive effect of transcription on switching was observed by adding the E_μ enhancer into the construct. However, most of the recombination occurred outside of the switch region in this construct.

These switch substrates all have limited switch region sequences and do not contain proper regulatory sequences such as I region in the construct. It is therefore difficult to prove that the recombinations observed with these constructs are relevant to physiological class switching. It may be necessary to use large DNA fragments that include not only the switch regions but also the flanking sequences to establish a substrate construct that can perform specific and regulated class switching.

In vivo studies of mutant mice generated by gene targeting techniques

With recent developments in gene targeting techniques, it is now possible to introduce small mutations into ES cells without leaving the *neo*^{r} gene in place of the

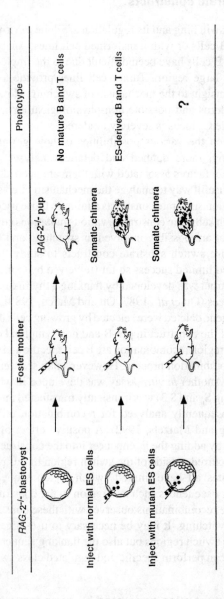

Fig. 3. The *RAG-2*-deficient blastocyst complementation system. (Adapted from Chen and Alt, 1993.)

mutation, which could interfere with the interpretation of the results. In particular, the Cre-loxP site-specific recombination system of bacteriophage P1 has been successfully utilized in mammalian cells to delete the neo^r gene after selection for homologous recombination (Sauer and Henderson, 1989; Gu et al., 1993). For example, small portions of the germline transcripts could be deleted without affecting the germline promoter to see if germline transcripts are necessary for switching. The *RAG*-2-deficient blastocyst complementation system (Chen et al., 1993; Fig. 3) allows for rapid analyses of different mutations through normal B cell development without the need for time-consuming germline transmission. Furthermore, the highly reproducible complementation in the resulting chimeras will allow analysis of many mutations in a short time period. Therefore, various mutations could be introduced into the C_H region to define how different parts are involved in the class switch recombination process.

Acknowledgements

This work was supported by the Howard Hughes Medical Institute and grants from the National Institutes of Health and by a grant from the Ministry of Education, Science and Culture of Japan. This work was done while TH was a Fogarty scholar-in-residence at National Institute of Health, USA.

REFERENCES

Abney, E.R., Cooper, M.D., Kearney, J.F., Lawton, A.R. and Parkhouse, R.M.E. (1978). *J. Immunol.* **120**, 2041–1987.
Abraham, K.M. and Teale, J.M. (1987a). *J. Immunol.* **138**, 1699–1704.
Abraham, K.M. and Teale, J.M. (1987b). *J. Immunol.* **139**, 2530–2537.
Adams, B., Dorfler, P., Aguzzi, A., Kozmik, Z., Urbanek, P., Maurer-Fogy, I. and Busslinger, M. (1992). *Genes Dev.* **6**, 1589–1607.
Akira, S., Sugiyama, H., Yoshida, N., Kikutani, H., Yamamura, Y. and Kishimoto, T. (1983). *Cell* **34**, 545–556.
Alt, F.W., Bothwell, M., Knapp, M., Siden, E., Mather, E., Koshland, M. and Baltimore, D. (1980). *Cell* **29**, 293–310.
Alt, F.W., Rosenberg, N., Casanova, R.J., Thomas, E. and Baltimore, D. (1982). *Nature* **296**, 325–331.
Andersson, J., Coutinho, A., Lernhardt, W. and Melchers, F. (1977). *Cell* **10**, 27–34.
Arakawa, H., Iwasato, T., Hayashida, H., Shimizu, A., Honjo, T. and Yamagishi, G. (1993). *J. Biol. Chem.* **268**, 4651–4655.
Armitage, R., Fanslow, W., Strockbine, L., Sato, T., Clifford, K., Macduff, B., Anderson, D., Gimpel, S., Davissmith, T., Maliszewski, C., Clark, E., Smith, C., Grabstein, K., Cosman, D. and Spriggs, M. (1992). *Nature* **357**, 80–82.
Azuma, C., Tanabe, T., Konishi, M., Kinashi, T., Noma, T., Matsuda, F., Yaoita, Y., Takatsu, K., Hammarström, L., Smith, E., Severinson, E. and Honjo, T. (1986). *Nucleic Acids Res.* **14**, 9149–9158.
Banchereau, J., Bazan, F., Blanchard, D., Briére, F., Galizzi, J.-P., Kooten, K.-V., Liy, Y.-J., Rousset, F. and Saeland, S. (1994). *Annu. Rev. Immunol.* **12**, 881–922.
Battey, J., Max, E.E., McBride, W.O., Swan, D. and Leder, P. (1982). *Proc. Natl Acad. Sci. USA* **79**, 5956–5960.
Bauer, D.C., Maties, M.J. and Stavitsky, A.B. (1963). *J. Exp. Med.* **117**, 889–907.

Beagley, K.W., Eldridge, H., Kiyono, H., Everson, M.P., Koopman, W.J., Honjo, T. and McGhee, J.R. (1988). *J. Immunol.* **141**, 2035–2042.
Beaven, M.A. and Metzger, H. (1993). *Immunol. Today* **14**, 222–226.
Bergstedt-Lindqvist, S., Sideras, P., MacDonald, H.R. and Severinson, E. (1984). *Immunol. Rev.* **78**, 25–50.
Berton, M.T., Uhr, J.W. and Vitetta, E.S. (1989). *Proc. Natl Acad. Sci. USA* **86**, 2829–2833.
Borzillo, G.V., Cooper, M.D., Kubagawa, H., Landay, A. and Burrows, P.D. (1987). *J. Immunol.* **139**, 1326–1335.
Bottaro, A., Lansford, R., Xu, L., Zhang, J., Rothman, P. and Alt, F.W. (1994). *EMBO J.*, **3**, 665–674.
Brown, S.L., and Morrison, S.L. (1989). *J. Immunol.* **142**, 2072–2080.
Burrows, P.D., Beck, G.B. and Wabl, M.R. (1981). *Proc. Natl Acad. Sci. USA* **78**, 564–568.
Burrows, P.D., Beck-Engeser, G.B. and Wabl, M.R. (1983). *Nature* **306**, 243–246.
Burstein, H.J., Tepper, R.T., Leder, P. and Abbas, A.K. (1991). *J. Immunol.* **147**, 2950–2956.
Callard, R.E., Armitage, R.J., Fanslow, W.C. and Spriggs, M.K. (1993). *Immunol. Today* **14**, 559–564.
Campbell, K.S. and Cambier, J.C. (1990). *EMBO J.* **9**, 441–448.
Canbell, H.D., Sanderson, C.J., Wang, Y., Hort, Y., Martinson, M.E., Tucker, W.Q., Stellwagen, A., Strath, M. and Young, I.G. (1988). *Eur. J. Biochem.* **174**, 345–352.
Chen, J. and Alt, F.W. (1994). In 'Analysis of the Immune System Utilizing Transgenesis and Targeted Mutagenesis' (H. Bluthman and P. Ohashi, eds) pp. 35–49. Academic Press, New York.
Chen, J., Lansford, R., Stewart, V., Young, F. and Alt, F.W. (1993). *Proc. Natl Acad. Sci. USA* **90**, 4528–4532.
Coffman, R.L. and Carty, J. (1986). *J. Immunol.* **136**, 949–954.
Coffman, R.L., Ohara, J., Bond, M.W., Carty, J., Zlotnick, A. and Paul, W.E. (1986). *J. Immunol.* **136**, 4538–4541.
Coffman, R.L., Lebman, D.A. and Shrader, B. (1989). *J. Exp. Med.* **170**, 1039–1044.
Coffman, R.L., Lebman, D.A. and Rothman, P.B. (1993). *Adv. Immunol.* **54**, 229–270.
Cogne, M., Bottaro, A., Lansford, R., Zhang, J., Gorman, J., Young, F., Cheng, H.-L. and Alt, F.W. (1994). *Cell*, **77**, 737–747.
Collins, J.T. and Dunnick, W.A. (1993). *Int. Immunol.* **5**, 885–891.
Cooper, M.D. and Burrows, P. (1989). In 'Immunoglobulin Genes' (T. Honjo, F.W. Alt and T.H. Rabbitts, eds), pp. 1–21. Academic Press, London.
Cory, S., Jackson, J. and Adams, J.M. (1980). *Nature* **285**, 450–456.
Coutelier, J.-P., van der Logt, J.T.M., Heesen, F.W.A., Warner, G. and van Snick, J. (1987). *J. Exp. Med.* **165**, 64–69.
Coutelier, J.-P., van der Logt, J.T.M., Heesen, F.W., Vink, A. and van Snick, J. (1988). *J. Exp. Med.* **168**, 2373–2378.
Coutinho, A. and Forni, L. (1982). *EMBO J.* **1**, 1251–1257.
Crossley, M. and Orkin, S.H. (1993). *Curr. Opin. Genet. Dev.* **3**, 232–237.
Cushley, W., Coupar, B.E., Nichelson, C.A. and Williamson, A.R. (1982). *Nature* **298**, 77–79.
Dariavach, P., Williams, G.T., Campbell, K., Pettersson, S. and Neuberger, M.S. (1991). *Eur. J. Immunol.* **21**, 1499–1504.
Davis, M., Kim, S. and Hood, L. (1980). *Science* **209**, 1360–1365.
DePinho, R., Kruger, K., Andrews, N., Lutzker, S., Baltimore, D. and Alt, F.W. (1984). *Mol. Cell. Biol.* **4**, 2905–2910.
D'Eustachio, P., Pravtcheva, D., Marcu, K. and Ruddle, F. (1980). *J. Exp. Med.* **151**, 1545–1550.
Dunnick, W.A. and Stavnezer, J. (1990). *Mol. Cell. Biol.* **10**, 397–400.
Dunnick, W., Rabbitts, T.H. and Milstein, C. (1980). *Nature* **286**, 669–675.
Dunnick, W.A., Hertz, G.Z., Scappino, L. and Gritzmacher, C. (1993). *Nucleic Acids Res.* **21**, 365–372.
Early, P., Rogers, J., Davis, M., Calame, K., Bond, M., Wall, R. and Hood, L. (1980). *Cell* **20**, 313–319.
Ehrhardt, R.O., Strober, W. and Harriman, G.R. (1992). *J. Immunol.* **148**, 3830–3836.
Esser, C. and Radbruch, A. (1989). *EMBO J.* **8**, 483–488.
Esser, C. and Radbruch, A. (1990). *Annu. Rev. Immunol.* **8**, 717–735.

Finkelman, F.D., Katona, I.M., Mosmann, T.R. and Coffman, R.L. (1988). *J. Immunol.* **140**, 1022–1027.
Finkelman, F.D., Holmes, J., Katona, I.M., Urban, J.F., Beckman, M.P., Park, L.S., Schooley, K.A., Coffman, R.L., Mosmann, T.R. and Paul, W.E. (1990). *Annu. Rev. Immunol.* **8**, 303–333.
Flanagan, J.G. and Rabbitts, T.H. (1982). *Nature* **300**, 709–713.
Gaff, C. and Gerondakis, S. (1990). *Int. Immunol.* **12**, 1143–1148.
Gally, J.A. (1973). In 'The antigens' (M. Sela, ed.), vol. 1, pp. 161–289. Academic Press, New York.
Gauchat, J.F., Lebman, D.A., Coffman, R.L., Gascan H. and de Vries, J.E. (1990). *J. Exp. Med.* **172**, 463–473.
Gauchat, J.F., Aversa, G., Gascan, H. and de Vries, J.E. (1992). *Int. Immunol.* **4**, 397–406.
Gearhart, P.J., Sigal, N.H. and Klinman, N.R. (1975). *Proc. Natl Acad. Sci. USA* **72**, 1707–1711.
Gerondakis, S. (1990). *Proc. Natl Acad. Sci. USA* **87**, 1581–1585.
Gerstein, R.M., Frankel, W.M., Hsieh, C.L., Durdik, J., Rath, S., Collin, J.M., Nissonof, A. and Selsing, E. (1990). *Cell* **63**, 537–548.
Giannini, S.L., Singh, M., Calvo, C.F., Ding, G. and Birshtein, B.K. (1993). *J. Immunol.* **150**, 1772–1780.
Grant, P.A., Arulampalam, V., Ahrlund-Richter, L. and Pettersson, S. (1992). *Nucleic Acids Res.* **20**, 4401–4408.
Grey, H.M., Hirst, J.W. and Cohn, M. (1971). *J. Exp. Med.* **133**, 289–304.
Gritzmacher, C.A. (1989). *Crit. Rev. Immunol.* **9**, 173–200.
Gu, H., Zou, Y. and Rajewsky, K. (1993). *Cell* **73**, 1155–1164.
Harriman, G.R., Kunimoto, D.Y., Elliott, J.F., Paetkav, V. and Strober, W. (1988). *J. Immunol.* **140**, 3033–3039.
Harriman, W., Volk, H., Defranoux, N. and Wabl, M. (1993). *Annu. Rev. Immunol.* **11**, 361–384.
Hayashida, H., Miyata, T., Yamawaki-Katoaka, Y., Honjo, T., Wels, J. and Blattner, F. (1984). *EMBO J.* **3**, 2047–2053.
Hedrick, S.M. (1989). In 'Fundamental Immunology', 2nd edn (W. Paul, ed.), pp. 291–314. Raven Press, New York.
Higgs, D.R., Wood, W.G., Jarman, A.P., Sharpe, J., Lida, J., Pretorius, I.M. and Ayyub, H. (1990). *Genes Dev.* **4**, 1588–1601.
Hodgkin, P.D., Yamashita, L.C., Coffman, R.L. and Kehry, M.R. (1990). *J. Immunol.* **145**, 2025–2034.
Hofker, M.H., Walter, M.A. and Cox, D.W. (1989). *Proc. Natl Acad. Sci. USA* **86**, 5567–5571.
Holliday, R. (1964). *Genet. Res.* **5**, 282–304.
Hombach, J., Takishi, T., Leclerco, L., Stappert, H. and Reth, M. (1990). *Nature* **343**, 760–762.
Honjo, T., Shimizu, A. and Yaoita, Y. (1989). In 'Immunoglobulin Genes' (F. Alt, T. Honjo and T.H. Rabbitts, eds), pp. 123–149. Academic Press, London.
Hummel, M., Berry, J.K. and Dunnick, W. (1987). *J. Immunol.* **138**, 3539–3548.
Illges, H. and Radbruch, A. (1992). *Mol. Immunol.* **29**, 1265–1272.
Isakson, P.C., Pure, E., Vitetta, E.S. and Krammer, P.H. (1982). *J. Exp. Med.* **155**, 734–748.
Ishida, N., Ueda, S., Hayahida, H., Miyata, T. and Honjo, T. (1982). *EMBO J.* **1**, 1117–1123.
Inwasato, T., Shimizu, A., Honjo, T. and Yamagishi, H. (1990). *Cell* **62**, 143–149.
Iwasato, T., Arakawa, H., Shimizu, A., Honjo, T. and Yamagishi, H. (1992). *J. Exp. Med.* **175**, 1539–1546.
Jack, H.M., McDowell, M., Steinberg, C. and Wabl, M. (1988). *Proc. Natl Acad. Sci. USA* **85**, 1591–1581.
Jarrett, E.E.E. and Miller, H.R.P. (1982). *Prog. Allergy* **31**, 178–233.
Jung, S., Rajewsky, K. and Radbruch, A. (1993). *Science* **259**, 984–987.
Jung, S., Sienbenkotten, G. and Radbruch, A. 1994. *J. Exp. Med.* **179**, 2023–2026.
Kataoka, T., Kawakami, T., Takahashi, N. and Honjo, T. (1980). *Proc. Natl Acad. Sci. USA* **77**, 919–923.
Kataoka, T., Miyata, T. and Honjo, T. (1981). *Cell* **23**, 357–368.
Katzenberg, D. and Birshtein, B. (1988). *J. Immunol.* **140**, 3219–3227.
Kawakami, T., Takahashi, N. and Honjo, T. (1980). *Nucleic Acids Res.* **8**, 3933–3945.
Kearney, J.F. and Lawton, A.R., (1975). Differentiation induced by lipopolysaccharide. *J. Immunol.* **115**, 671–676.
Kearney, J.F., Cooper, M.D. and Lawton, A.R. (1976). *Immunol.* **117**, 1567–1572.

Kemp, D., Morahan, G., Cowman, A. and Harris, A. (1983). *Nature* **301**, 84–86.
Kepron, M.R., Chen, Y.-W., Uhr, J.W. and Vitetta, E.S. (1989). *J. Immunol.* **143**, 334–339.
Kinashi, T., Harada, N., Severinson, E., Tanabe, T., Sideras, P., Konishi, M., Azuma, C., Tominaga, A., Bergstedt-Lindqvist, S., Takahashi, M., Matsuda, F., Yaoita, Y., Takatsu, K. and Honjo, T. (1986). *Nature* **324**, 70–73.
Kirsch, J.R., Morton, C.C., Nakahara, K. and Leder, P. (1982). *Science* **216**, 301–303.
Klaus, G.G.B., Pepys, M.B., Kitajima, K., Askonas, B.A. (1979). *Immunology* **38**, 687–695.
Koshland, M.E. (1975). *Adv. Immunol.* **1**, 1–96.
Kuhn, R., Rajewsky, K. and Muller, W. (1991). *Science* **254**, 707–710.
Lanzavecchia, A. (1985). *Nature* **314**, 537–539.
Lars, N. and Sideras, P. (1993). *Int. Immunol.* **5**, 271–282.
Lawton, A.R. and Cooper, M.D. (1973). *Contemp. Top. Immunobiol.* **3**, 193–225.
Layton, J.E., Vitetta, E.S., Uhr, J.W. and Krammer, P.H. (1984). *J. Exp. Med.* **160**, 1850–1863.
Lebman, D.A., Lee, F.D. and Coffman, R.L. (1990a). *J. Immunol.* **144**, 952–959.
Lebman, D.A., Nomura, D.Y., Coffman, R.L. and Lee, F.D. (1990b). *Proc. Natl Acad. Sci. USA* **87**, 3962–3966.
Lee, F., Yokata, T., Otsuka, T., Meyerson, P., Villaret, D., Coffman, R., Mosmann, T.R., Rennick, D., Roehm, N., Smith, C., Zlotnik, A. and Arai, K. (1986). *Proc. Natl Acad. Sci. USA* **83**, 2061–2065.
Lenon, G.G. and Perry, R.P. (1985). *Nature* **318**, 475–478.
Leung, H. and Maizels, N. (1992). *Proc. Natl Acad. Sci. USA* **89**, 4154–4158.
Li, S.C., Rothman, P.B., Zhang, J., Chan, C., Hirsh, D. and Alt, F. (1994). *Int. Immunol.*, **6**, 491–497.
Liao, F., Giannini, S.L. and Birshtein, B.K. (1992). *J. Immunol.* **148**, 2909–2917.
Lieberson, R., Giannini, S.L., Birshtein, B.K. and Eckhardt, L.A. (1991). *Nucleic Acids Res.* **19**, 933–937.
Liesegang, B., Radbruch, A. and Rajewsky, K. (1978). *Proc. Natl Acad. Sci. USA* **75**, 3901–3905.
Lin, Y.C. and Stavnezer, J. (1992). *J. Immunol.* **149**, 2914–2925.
Liu, F.-T., Alband, K., Sutchliffe, J. and Katz, D. (1982). *Proc. Natl Acad. Sci. USA* **79**, 7852–7856.
Lutzker, S. and Alt, F.W. (1988a). In 'Mobile DNA' (Berg D.E. and Howe M.M. eds), pp. 693–714. American Society for Microbiology Washington, DC.
Lutzker, S. and Alt, F.W. (1988b). *Mol. Cell. Biol.* **8**, 1849–1852.
Lutzker, S., Rothman, P., Pollock, R., Coffman, R. and Alt, F.W. (1988). *Cell* **53**, 177–184.
McIntyre, T.M., Klinman, D.R., Rothman, P.B., Lugo, M., Dasch, J.R., Mond, J.J. and Snapper, C.M. (1993). *J. Exp. Med.* **177**, 1031–1037.
McKenzie, A.N.J., Culpepper, J.A., Malefyt, J., Briere, R.W., Punnenen, F., Aversa, J., Sato, G., Dang, A., Cocks, W., Menon, S., De Vries, J.E., Banchereau, J. and Zurawski, G. (1993). *Proc. Natl Acad. Sci. USA* **90**, 3735–3739.
Mandler, R., Chu, C.C., Paul, W.E., Max, E.E. and Snapper, C.M. (1993). *J. Exp. Med.* **178**, 1577–1586.
Marcu, K., Lang, R., Stanton, L. and Harris, L. (1982). *Nature* **298**, 87–89.
Matsuoka, M., Yoshida, K., Maeda, T., Usuda, S. and Sakano, H. (1990). *Cell* **62**, 135–142.
Mazanec, M.B., Nedrud, J.G., Kaetzel, C.S. and Lamm, M.E. (1993). *Immunol. Today* **14**, 430–435.
Meselson, M.S. and Radding, C.M. (1975). *Proc. Natl Acad. Sci. USA* **72**, 358–361.
Milcareck, C. and Hall, B. (1985). *Mol. Cell. Biol.* **5**, 2514–2520.
Mills, F.C., Thyphronitis, G., Finkelman, F.D. and Max, E.E. (1992). *J. Immunol.* **149**, 1075–1085.
Miyata, T., Yasunaga, T., Yamawaki-Kataoka, Y., Obata, M. and Honjo, T. (1980). *Proc. Natl Acad. Sci. USA* **77**, 2143–2147.
Monroe, J.G., Harvan, W.L. and Cambier, J.C. (1983). *Eur. J. Immunol.* **13**, 208–213.
Moore, K.W., Rogers, J., Hunkerpillar, T., Early, P., Nattengerg, C., Weissmann, I., Bazin, H., Wall, R. and Hood, L.E. (1981). *Proc. Natl Acad. Sci. USA* **78**, 1800–1804.
Mosmann, T.R. and Coffman, R.L. (1989). *Annu. Rev. Immunol.* **7**, 145–173.
Murray, P., McKenzie, D., Swain, S. and Kagnoff, M. (1987). *J. Immunol.* **139**, 2669–2674.
Neuberger, M.S. and Rajewsky, K. (1981). *Eur. J. Immunol.* **11**, 1012–1016.
Nikaido, T., Nakai, T. and Honjo, T. (1981). *Nature* **292**, 845–848.
Nikaido, T., Yamawaki-Kataoka, Y. and Honjo, T. (1982). *J. Biol. Chem.* **257**, 7322–7329.

Nilsson, L., Islam, K.B., Olafsson, O., Zalcberg-Quintana, I., Smakovlis, C., Hammarstrom, L., Smith, C.I.E. and Sideras, P. (1991). *Int. Immunol.* **3**, 1107–1115.
Noelle, R.J. and Snow, E.C. (1991). *FASEB J.* **5**, 2770–2776.
Noelle, R.J. and Snow, E.C. (1992). *Curr. Opin. Immunol.* **4**, 333–337.
Nolan-Willard, M., Berton, M.T. and Tucker, P. (1992). *Proc. Natl Acad. Sci. USA* **89**, 1234–1238.
Noma, Y., Sideras, P., Naito, T., Bergstedt-Lindqvist, S., Azuma, C., Severinson, E., Tanabe, T., Kinashi, T., Matsuda, F., Yaoita, Y. and Honjo, T. (1986). *Nature* **319**, 640–646.
Nossal, G.J.V., Szengerg, A., Ada, G.L. and Austin, C.M. (1964). *J. Exp. Med.* **119**, 485–502.
Obata, M., Kataoka, T., Nakai, S., Yamagishi, H., Takahashi, N., Yamawaki-Kataoka, Y., Nikaido, T., Shimizu, A. and Honjo, T. (1981). *Proc. Natl Acad. Sci. USA* **78**, 2437–2441.
Oi, V.T., Vuong, T.M., Hardy, R., Reidler, J., Dangl, J., Herzenberg, L.A. and Stryer, L. (1984). *Nature* **307**, 136–140.
Ollo, R. and Rougeon, F. (1983). *Cell* **32**, 515–523.
Ott, D.E. and Marcu, K.B. (1989). *Int. Immunol.* **1**, 582–591.
Ott, D.E., Alt, F.W. and Marcu, K.B. (1987). *EMBO J.* **6**, 577–584.
Parker, D.C. (1993). *Annu. Rev. Immunol.* **11**, 331–360.
Parry, S.L., Hasbold, J., Holman, M. and Klaus, G.G.B. (1994). *J. Immunol.* **152**, 2821–2829.
Paul, W.E. (1989). *Cell* **57**, 521–524.
Perlmutter, A.P. and Gilbert, W. (1984). *Proc. Natl Acad. Sci. USA* **81**, 7189–7193.
Pernis, B., Forni, L. and Luzzati, A. (1977). *Cold Spring Harbor Symp. Quant. Biol.* **41**, 175–183.
Petrini, J. and Dunnick, W.A. (1989). *J. Immunol.* **142**, 2932–2935.
Petrini, J., Shell, B., Hummel, M. and Dunnick, W. (1987). *J. Immunol.* **138**, 1940–1946.
Pettersson, S., Cook, G.P., Bruggemann, M., Williams, G.T. and Neuberger, M.S. (1990). *Nature* **344**, 165–168.
Powrie, F. and Coffman, R.L. (1993). *Immunol. Today* **14**, 270–274.
Preud'Homme, J.-L., Birshtein, B. and Scharf, M.D. (1975). *Proc. Natl Acad. Sci. USA* **72**, 1427–1430.
Punnonen, J., Aversa, G., Cocks, B.G., McKenzie, A.N.J., Menon, S., Zurawski, G., Malefyt, R.W. and De Vries, J.E. (1993). *Proc. Natl Acad. Sci. USA* **90**, 3730–3734.
Purkerson, J.M. and Isakson, P.C. (1992). *J. Exp. Med.* **175**, 973–982.
Rabbitts, T.H., Forster, A., Dunnick, W. and Bentley. D.L. (1980). *Nature* **283**, 351–356.
Radbruch, A. (1985). In 'Handbook of Experimental Immunology' (D.M. Weir and L.A. Herzenberg, eds), pp. 110.1–110.12. Blackwell Scientific Publications, Oxford.
Radbruch, A. and Sablitzky, F. (1983). *EMBO J.* **2**, 1929–1935.
Radbruch, A., Muller, W. and Rajewsky, K. (1986a). *Proc. Natl Acad. Sci. USA* **83**, 3954–3957.
Radbruch, A., Burger, C., Klein, S. and Muller, W. (1986b). *Immunol. Rev.* **89**, 69–83.
Radcliffe, G., Lin, Y.-C., Julius, M., Marcu, K.B. and Stavnezer, J. (1990). *Mol. Cell. Biol.* **10**, 382–386.
Ravetch, J.V., Sienbenlist, U., Korsmeyer, S., Waldmann, T. and Leder, P. (1981). *Cell* **27**, 583–591.
Reaban, M. and Griffin, J.A. (1990). *Nature* **348**, 342–344.
Roes, J. and Rajewsky, K. (1993). *J. Exp. Med.* **177**, 45–55.
Rogers, J., Early, P., Carter, C., Calame, K., Bond, M., Hood, L. and Wall, R. (1980). *Cell* **20**, 303–312.
Rogers, J., Choi, E., Souza, L., Carter, C., Word, C., Kuehl, M., Eisenberg, D. and Wall, R. (1982). *Cell* **26**, 19–27.
Rothman, P., Li, S.C. and Alt, F.W. (1989). *Semin. Immunol.* **1**, 65–77.
Rothman, P., Chen, Y.-Y., Lutzker, S., Li, S.C., Stewart, V., Coffman, R. and Alt, F.W. (1990a). *Mol. Cell. Biol.* **10**, 1672–1679.
Rothman, P., Lutzker, S., Gorham, B., Stewart, V., Coffman, R. and Alt, F.W. (1990b). *Int. Immunol.* **2**, 621–627.
Rothman, P., Li, S.C., Gorham, B., Glimcher, L., Alt, F.W. and Boothby, M. (1991). *Mol. Cell. Biol.* **11**, 5551–5561.
Sakano, H., Rogers, J., Huppi, K., Brack, C., Traunecker, A., Maki, R., Wall, R. and Tonegawa, S. (1979). *Nature* **277**, 627–633.
Sanchez, M., Misulovin, Z., Burkhardt, A.L., Mahajan, S., Costa, T., Franke, R., Bolen, J.B. and Nussenzweig, M. (1993). *J. Exp. Med.* **178**, 1049–1055.

Sauer, B. and Henderson, N. (1989). *Nucleic Acids Res.* **17**, 147–161.
Schultz, C., Petrini, J., Collins, J., Claflin, J.L., Dennis, K.A., Gearhart, P., Gritzmacher, C., Manser, T., Shulman, M. and Dunnick, W. (1990). *J. Immunol.* **144**, 363–370.
Schultz, C.L., Elenich, L.A. and Dunnick, W.A. (1991). *Int. Immunol.* **3**, 109–116.
Schwedler, U., Jack, H.-M. and Wabl, M. (1990). *Nature* **345**, 452–456.
Severinson, E., Bergstedt-Lindqvist, S., van der Loo, W. and Fernandez, C. (1982). *Immunol. Rev.* **67**, 73–85.
Severinson-Gronowicz, E., Doss, C. and Schroder, J. (1979). *J. Immunol.* **123**, 2057–2062.
Shapira, S.K., Jabara, H.H., Thiennes, C.P., Ahern, D.J., Vercelli, D., Gould, H.J. and Geha, R.S. (1991). *Proc. Natl Acad. Sci. USA* **88**, 7528–7532.
Shapira, S.K., Vercelli, D., Jabara, H.H., Fu, S.M. and Geha, R.S. (1992). *J. Exp. Med.* **175**, 289–292.
Shimizu, A. and Honjo, T. (1984). *Cell* **36**, 801–803.
Shimizu, A., Takahashi, N., Yaoita, Y. and Honjo, T. (1982). *Cell* **28**, 499–506.
Shimizu, A., Nuzzenzweig, M.C., Han, H., Sanchez, M. and Honjo, T. (1991). *J. Exp. Med.* **173**, 1385–1393.
Sideras, P., Bergstedt-Lindqvist, S. and Severinson, E. (1985). *Eur. J. Immunol.* **15**, 593–598.
Sideras, P., Mizuta, T.-R., Kanamori, H., Suzuki, N., Okamoto, M., Kuze, K., Ohno, H., Doi, S., Fukuhara, S., Hassan, M.S., Hammastrom, L., Smith, E., Shimizu, A. and Honjo, T. (1989). *Int. Immunol.* **1**, 631–642.
Siebenkotten, G., Esser, C., Wabl, M. and Radbruch, A. (1992). *Eur J. Immunol.* **22**, 1827–1834.
Singh, M. and Birshtein, B.K. (1993). *Mol. Cell. Biol.* **13**, 3611–3622.
Snapper, C.M. and Paul, W.E. (1987). *Science* **236**, 944–947.
Snapper, C., Peschel, C. and Paul, W. (1988). *J. Immunol.* **140**, 2121–2127.
Snapper, C.M., Mcintyre, T.M., Mandler, R., Pecanha, L.M., Finkelman, F.D., Lees, A. and Mond, J.J. (1992). *J. Exp. Med.* **175**, 1367–1371.
Sonoda, E., Matsumoto, R., Hitoshi, Y., Ishii, T., Sugimoto, M., Araki, S., Tominaga, A., Yamaguchi, N. and Takatsu, K. (1989). *J. Exp. Med.* **170**, 1415–1420.
Spriggs, M.K., Armitage, R.J., Strokbine, L., Clifford, K.N., Macduff, B.M., Sato, T.A., Maliszewski, C.R. and Fanslow, W.C. (1992) *J. Exp. Med.* **176**, 1543–1550.
Stavnezer, J. and Sirlin, S. (1986). *EMBO J.* **5**, 95–102.
Stavnezer, J., Sirlin, S. and Abbott, J. (1985). *J. Exp. Med.* **161**, 577–601.
Stavnezer, J., Radcliffe, G., Lin, Y.-C., Nietupski, J., Berggren, L., Sitia, R. and Severinson, E. (1988). *Proc. Natl Acad. Sci. USA* **85**, 7704–7708.
Szostak, J.W., Orr-Weaver, T.L., Rothstein, R.J. and Stahl, F.W. (1983). *Cell* **33**, 25–35.
Szurek, P., Petrini, J. and Dunnick, W. (1985). *J. Immunol.* **135**, 620–626.
Takahashi, N., Ueda, S., Obata, M., Nikaido, T., Nakai, S. and Honjo, T. (1982). *Cell* **29**, 671–679.
Taki, S., Meiering, M. and Rajewsky, K. (1993). *Science* **262**, 1268–1271.
Talbot, D., Collis, P., Antoniou, M., Vidal, M., Grosveld, F. and Greaves, D. (1989). *Nature* **338**, 352–255.
Tanaka, T., Chu, C.C. and Paul, W.E. (1992). *J. Exp. Med.* **175**, 597–607.
Thomas, B.J. and Rothstein, R. (1989). *Cell* **56**, 619–630.
Tian, S.-S. and Faust, C. (1987). *Mol. Cell. Biol.* **7**, 2614–2619.
Tsubata, T., Wu, J. and Honjo, T. (1993). *Nature* **364**, 645–648.
Tsubata, T., Murakami, M. and Honjo, T. (1994). *Curr. Biol.* **4**, 8–17.
Tucker, P.W. (1985). *Immunol. Today* **6**, 181–182.
Tucker, P., Liu, C.P., Muskinski, J. and Blattner, F. (1980). *Science* **209**, 1353–1360.
Tucker, P., Slightom, J. and Blattner, F. (1981). *Proc. Natl Acad. Sci. USA* **78**, 7684–7688.
Uhr, J.W. and Finkelstein, M.S. (1963). *J. Exp. Med.* **117**, 457–477.
Valentine, M.A. and Licciardi, K.A. (1992). *Eur. J. Immunol.* **22**, 3141–3148.
Venkitaraman, A.R., Williams, G.T. and Dariavac, P. (1991). *Nature* **352**, 777–781.
Vitetta, E.S., Ohara, J., Myers, C.D., Layton, J.E., Krammer, P. and Paul, W.E. (1985). *J. Exp. Med.* **162**, 1726–1731.
Wakatsuki, Y. and Strober, W. (1993). *J. Exp. Med.* **178**, 129–138.
Waters, S.H., Saikh, K.U. and Stavnezer, J. (1989). *Mol. Cell. Biol.* **9**, 5594–5601.
Whitmore, A.C., Prowse, D.M., Haughton, G. and Arnold, L.W. (1991). *Int. Immunol.* **3**, 95–103.

Williams, G.T., Venkitaraman, A.R., Gilmore, D.J. and Neuberger, M.S. (1990). *J. Exp. Med.* **171**, 947–952.
Williams, M. and Maizels, N. (1991). *Genes Dev.* **5**, 2353–2361.
Winter, E., Krawinkel, U. and Radbruch, A. (1987). *EMBO J.* **6**, 1663–1671.
Wu, T.T., Reid-Miller, M., Perry, H.M. and Kabat, E.A. (1984). *EMBO J.* **3**, 2033–2040.
Wuerffel, R., Jamieson, C.E., Morgan, L., Merkolov, G.V., Sen, R. and Kenter, A.L. (1992). *J. Exp. Med.*, **176**, 339–349.
Xu, L., Kim, M.G. and Marcu, K.B. (1992). *Int. Immunol.* **4**, 875–887.
Xu, L., Gorman, B., Li, S.C., Bottaro, A., Alt, F.W. and Rothman, P. (1993). *Proc. Natl Acad. Sci. USA* **90**, 3705–3709.
Xu, M. and Stavnezer, J. (1992). *EMBO J.* **11**, 145–155.
Yancopoulos, G.D., DePinho, R., Zimmerman, K., Lutzker, S., Rosenberg, N. and Alt, F.W. (1986). *EMBO J.* **5**, 3259–3266.
Yaoita, Y. and Honjo, T. (1980). *Nature* **286**, 850–853.
Yoshida, K., Matsuoka, M., Usuda, S., Mori, A., Ishizaka, K. and Sakano, H. (1990). *Proc. Natl Acad. Sci. USA* **87**, 7829–7833.
Zhang, J., Bottaro, A., Li, S., Stewart, V. and Alt, F.W. (1993). *EMBO J.* **12**, 3529–3537.

12

Generation of diversity by post-rearrangement diversification mechanisms: the chicken and the sheep antibody repertoires

Jean-Claude Weill and Claude-Agnès Reynaud

Institut Necker, Université René Descartes—Paris V, Paris, France

Introduction 267
Gene conversion in chicken B cells..269
Hypermutation in sheep B cells..279
Conclusions 286

INTRODUCTION

The immune system has evolved different molecular mechanisms directed towards the same goal: the production of a large number of different antigen-recognizing units. These mechanisms are site-specific rearrangement, homologous recombination, gene conversion and hypermutation. Some of them are part of the general machinery of the cell, being used at a low rate on the whole genome during evolution. In the immune system they are used at a high rate in a developmentally regulated mode on a few specific genes. Strikingly, when activated in a deregulated fashion in somatic cells, some of these molecular mechanisms may also, depending upon the gene involved, lead to neoplastic processes.

In this chapter we describe two different molecular mechanisms used by two different species to generate their pre-immune repertoire: gene conversion used by chickens (and by rabbits as well), and hypermutation used by sheep. Activation of hypermutation during early B cell development in the sheep represents a new function for this mechanism, known so far to be responsible for affinity maturation during T-dependent immune responses.

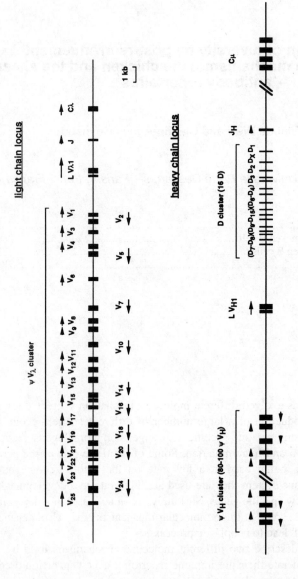

Fig. 1. Organization of the chicken light and heavy chain loci. The chicken light chain locus contains a single functional V gene ($V_\lambda 1$), a single J–C unit and a cluster of 25 pseudogenes (Reynaud et al., 1985, 1987). The heavy chain locus contains single C_μ, J_H, and functional V_H ($V_H 1$) elements, a cluster of 16 D, and a group of pseudogenes (80–100 ψV_H in 60–80 kb of DNA) (Reynaud et al., 1989, 1991b). The respective order of the D elements in parentheses has not been determined. Horizontal arrows indicate transcriptional polarities.

GENE CONVERSION IN CHICKEN B CELLS

Organization of Ig genes in the chicken

Light and heavy chain loci

Both heavy and light chain loci share a similar gene organization with unique functional V and J elements, indicative of a striking co-evolution (Fig. 1).

For the light chain locus, which is a λ isotype, the unique $V_\lambda 1$ gene is 1.8 kb upstream of a single J–C unit. This V gene has all the characteristics (promoter with octamer sequence, leader peptide, leader intron, recombination signals) of mouse V elements. Upstream of $V_\lambda 1$ are 25 V pseudogenes clustered in 19 kb of DNA with alternate polarities (Reynaud et al., 1985, 1987). Most of these pseudogenes (but not all) lack recombination signal sequences. The homology with the $V_\lambda 1$ gene never extends further in 5' than approximately 40 bp in the leader intron. None of them have upstream regulatory elements, i.e. transcription signals and leader sequence, some of them being even truncated genes with only part of the V coding sequence.

For the heavy chain, the unique $V_H 1$ and J_H elements are 15 kb apart, and J_H is approximately 15 kb upstream of the C_μ gene. Between $V_H 1$ and J_H are 16 D elements (Reynaud et al., 1989, 1991b). A pool of V_H pseudogenes, larger than for the light chain locus, has been analysed but not completely sequenced: it covers 60–80 kb of DNA, with a total of 80–100 pseudogenes. The alternance of polarities is quasi-systematic. Like V_λ pseudogenes, homology with $V_H 1$ is restricted in 5' to ca. 80 bp in the leader intron, with no leader sequence or transcription signals. In 3', they have no recombination signals but, surprisingly, a 'D-like' segment fused to the V element; this 'D-like' sequence is even terminated in some pseudogenes by a few nucleotide's homology with J_H. The presence of leader intronic sequence in some of these pseudogenes argues against a processed gene origin for these fused V–D structures.

D elements

D elements are the only functional V-encoding elements present in multiple copies in the chicken genome: 16 D elements exist, with 15 of them extremely homologous (some even in several identical copies) and one rather different (Dx) (Fig. 2). However this last is poorly functional, since it has a low rearrangement frequency and is counterselected during B cell expansion in the bursa (Reynaud et al., 1991b).

The amino acid composition of the three reading frames is strikingly different: Gly-Ser-Ala-Tyr-Cys, i.e. hydrophilic and aromatic residues, in reading frame 1; hydrophobic amino acids (Leu, Val) in reading frame 2; and one or two stop codons in reading frame 3. This amino acid composition resembles that of the mouse Dfl16 and Dsp families, and a bias for usage of reading frame 1 is observed in the chicken as in the mouse.

D6	GGATTTTGG	TCAACGTTGTGT	CACCGTG	GGT AGT GGT TAC TGT GGT AGT GGT GCT TAT ▼TGT	CACCGTG	CTCCATCCCATA	ACAAAAACC
D1	---------	--------A-	---T---	GGT▼AGT GCT TAC GGT TGT ( ) GGT GCT TAT	-------	------------	---------
D2	---------	------C----	-------	GGT AGT GCT ( ) TGT TGT ( ) GGT CCT TAT	-------	------------	---------
D3	---------	-----------	-------	GGT AGT GCT TAC TGT TGT AGT GCT TAT	-------	A-----------	---------
D4	---------	-----------	-------	GGT AGT GCT TAC TGT ( ) TGG GAT GCT GAT	---AA--	------------	---------
D5	---------	--------A-	---T---	GGT AGT GCT TAC TGT GGT AGT GGT GCT TAT	-------	------------	---------
D7	---------	-----------	-------	GGT AGC GCT TAC TGT ( ) TGG TAT GCT GAT	-------	------------	---------
D8	---------	-----------	-------	GGT AGT GCT TAC TGT ( ) TGG TAT GCT GAT	---AA--	------------	---------
D9	---------	----A----A-	-------	GGT AGT GGT TAC TGT GGT AGT GCT GCT TAT	-------	------------	---------
D10	---------	----------A-	-------	GGT AGT GGT TAC TGT GGT TGG GCT GCT TAT	-------	------------	---------
D11	---------	------C----	-------	GGT AGT GCT TAC TGT ( ) TGG GAT GCT GAT	---AA--	------------	---------
D12	---------	----A----A-	-------	GGT AGT GGT TAC TGT GGT AGT GCT GCT TAT	-------	------------	---------
D13	---------	--------A-	-------	GGT AGT GGT TAC TGT GGT AGT GCT GCT TAT	-------	------------	---------
D14	---------	--------AC	-------	GGT AGT GGT TAC TGT GGT TGG AGT GCT TAT	-------	------------	---------
D15	---------	--------A-	---T---	GGT AGT GGT TAC TGT GGT AGT GGT GCT GAT	-------	------------	---------
Dx	---------	-----C---A-	---T---	GGT ACT TCT GGT GCC TGC ACC TTT TTC TAT	↓CCT TCC TGC CCT TAT ⎣------A⎦	-----C-----	

Fig. 2. Genomic D sequences, including recombination signal sequences (Reynaud et al., 1991b). D6 has been chosen arbitrarily to be the comparison, with dashes for nucleotide identity, and parentheses and arrows for, respectively, gaps and insertions introduced to maximize homology. The D coding part is written in a triplet mode in reading frame 1, and indicated in full to facilitate comparison. Heptamer–nonamer signal sequences are boxed.

The germline-encoded D elements are only minor contributors to the overall heavy chain repertoire, which results mainly from gene conversion (see Fig. 5). What is thus the function of this pool of genomic D elements? Its main contribution could be the formation of D–D junctions: these amount to 25% among DJ alleles and are maintained at similar percentages in 'functional' VDJ sequences (cf. Table II). Such junctions may represent an alternative way of generating a large length heterogeneity of the third complementarity-determining region (CDR3) in the absence of N additions. Surprisingly, this D–D joining requires rearrangement between two signals with the same 12-bp spacing, a mechanism that would be expected from model systems to be extremely inefficient (Hesse et al., 1989).

Generation of B cell precursors

The generation of the primary antibody repertoire occurs in each species in a specific organ, starting from a population of specific B cell progenitors. In mouse and human the primary B cell organ is the bone marrow in which as yet uncharacterized population of multipotent and committed stem cells generate the B lineage. These bone marrow precursors have their Ig loci in germline configuration and generate a diversified B cell population by ongoing rearrangement of their Ig coding elements.

In the chicken, approximately 10 000–30 000 B cell precursors (two to three precursors per bursal follicle) from which the repertoire will be derived accumulate at the beginning of bursal growth (day 8–14 of embryonic development). Strikingly, these precursors have already undergone Ig gene rearrangement once they start to proliferate in the bursa, and for each cell the rearrangement involves the same functional $V_\lambda 1$ and $V_H 1$ gene, as already suggested from the structure of the Ig loci. The most striking feature of B cell development in the chicken is thus the existence of a unique wave of Ig gene rearrangement in B cell progenitors, the entire B cell compartment being derived from the 2–3 × 10⁴ productively rearranged bursal stem cells that develop within bursal follicles (Pink and Lassila, 1987; Weill and Reynaud, 1987).

Because of this unique feature of the precursors, their formation during embryonic development can be traced by studying the status of their Ig loci at the single cell sensitivity level.

DJ_H rearrangement can be detected first in the yolk sac at day 5–6. At this stage the B lineage segregates from the other haematopoietic lineages (DJ_H rearrangement does not occur in chicken T cell precursors, in contrast to the mouse). D to J_H rearrangement is detected a few days later in the spleen (day 6–7), and then in the bursa, the bone marrow and the thymus (day 9–10). One or two days later, whether in the blood or in a lymphoid organ, B cell progenitors undergo heavy and light chain V gene rearrangement simultaneously. Quantitatively, we have estimated that there are approximately 1000 DJ_H progenitors in the yolk sac at day 8; 125 000–250 000 precursors can be detected in the blood at day 10 until day 13, at which time comparable amounts of DJ_H progenitors can be found also in the spleen, the bone marrow and start to accumulate in the bursa. At day 17, these B lymphoid populations decline in blood, spleen and bone marrow and B cells expand in the bursa (Table I) (Reynaud et al., 1992).

The ontogenetic appearance of B cells, taking into account ours and other's results (Moore and Owen, 1967; Lassila et al., 1978, 1982; Dieterlen-Lièvre and Martin, 1981; Ratcliffe et al., 1986; Houssaint et al., 1991), can be described as follows (Fig. 3). Multipotent haematopoietic precursors are detected first at day 3–4 in the intraembryonic aortic region. Then they migrate to the yolk sac at day 5–6 where DJ_H rearrangement starts. DJ_H joining continues, and V_H and V_λ rearrangement starts as cells go back to the embryo and are retained in the various lymphoid organs. The role of the various regions of the embryo in turning on rearrangement is not clear at the moment, this cascade of events being probably part of a cellular programme that could have been triggered at the very beginning of the process. Several millions of cells are thus engaged in this differentiation pathway, most of them ending in sites where they cannot expand and further differentiate. Based on the absence of

Table I. Number of DJ_H-committed progenitors in various embryonic compartments*

	Day of development			
	Day 8	Day 10	Day 13	Day 17
Yolk sac	**1000–2000**			
Blood		**125 000–250 000**	**250 000–500 000**	500–1000
Spleen		2000–5000	**500 000**	20 000–40 000
Bone marrow		10–20	**100 000**	2000–10 000
Bursa		50–200	30 000	**>1–2 × 10⁶**
Thymus		100	400–1000	1000–4000

* The total number of DJ_H-committed progenitors in various compartments at four developmental stages was estimated from a quantitative PCR analysis, reported to the size of each organ at each embryonic stage (Reynaud et al., 1992). Figures in bold represent the major DJ_H progenitor compartment(s) at a given stage.

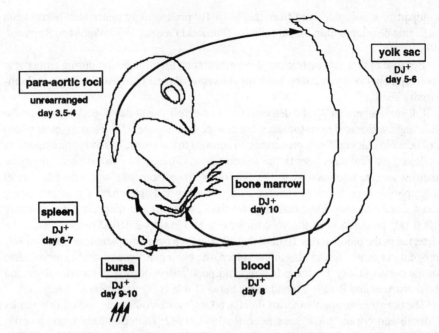

Fig. 3. A proposed scheme for the emergence of B cell progenitors in the chicken embryo (Reynaud et al., 1992). Haematopoietic progenitors are first detected within the embryo in the para-aortic region (Lassila et al., 1978; Dieterlen-Lièvre and Martin, 1981); they further migrate through the yolk sac, where DJ_H-committed progenitors start to segregate from the other cell lineages and seed, via the general circulation, the various lymphoid organs. These progenitor populations regress thereafter in spleen and bone marrow and only expand in the bursa (↑↑↑) (cf. Table I). The first day of detection of DJ_H rearrangement is indicated for each compartment.

selection for functional rearrangement in these extra-bursal sites and also on the absence of proliferation, we favour the hypothesis that these cells slowly disappear from the system and therefore do not migrate back to the bursa, although they can restore B cell function if injected into a B cell-depleted animal (Ratcliffe et al., 1986; Houssaint et al., 1989, 1991).

Regulation of allelic exclusion in chicken B cells

Most B cell precursors that have colonized bursal follicles have undergone heavy and light chain gene rearrangement on one allele while the other allele is maintained in germline configuration. As mentioned previously, rearrangement of heavy and light chain genes occurs simultaneously during a short period of embryonic development, with no heavy to light chain transition. The short time window during which rearrangement is performed precludes a second trial on the other allele in the case of an abortive joining.

Such a situation is different from the one observed in mouse B cell precursors, where heavy and light chain gene rearrangement can proceed from one allele to the other. For the light chain κ locus for example, about 40% of the κ-positive cells carry an abortive light chain rearrangement (Coleclough *et al.*, 1981). In a model in which rearrangement continues, feedback mechanisms become necessary to achieve allelic exclusion: if the first attempt on the heavy chain locus is productive, this event probably inhibits rearrangement on the other heavy chain allele and enhances rearrangement of the light chain locus (Alt *et al.*, 1981, 1984; Reth *et al.*, 1987; Ehlich *et al.*, 1993). It is not known how these different steps are precisely regulated (reviewed in Melchers *et al.*, 1993). Surprisingly, despite the differences between the chicken and mouse models, analysis of rearrangement of the chicken light chain locus may suggest a clue to this problem.

We investigated which regulatory elements are necessary for rearrangement of the chicken light chain locus in pre-B cells from transgenic mice. A transgene of 11.5 kb containing the chicken light chain locus in its natural configuration experienced rearrangement in mouse B cells (Bucchini *et al.*, 1987). Mutation and deletion of this transgene allowed us to define four regions necessary for rearrangement. Two of these are the promoter region, which must include for efficient rearrangement a segment larger than the octamer motif, and the enhancer element located 3' of the C_λ region. Such a role of enhancer elements on rearrangement had been previously reported for mouse Ig loci by transgenic and knock-out experiments (reviewed in Okada and Alt, 1994). Surprisingly, a negative regulatory element showing a strong silencing activity in an *in vitro* chloramphenicol acetyl transferase (CAT) assay was located in the V–J intervening sequence which is deleted upon rearrangement. Flanking the silencing element, there were one or two positive elements that counteracted the negative effect of the silencer. When these sites were mutated the mutation abolished rearrangement, but had no effect when the intervening segment was replaced by a neutral fragment of DNA (Lauster *et al.*, 1993).

We have proposed that this set of positive and negative elements could control rearrangement during chicken embryonic development (Fig. 4). During a short period of time the silencing of the locus will be relieved by the presence of anti-silencing factors. The transient access of the recombinase to the locus will allow rearrangement to occur productively on one allele in a certain fraction of cells. The silencer will remain active thereafter in order to exclude the other allele from the rearrangement machinery. However, the negative regulation is dispensable once the recombination machinery is shut off.

Our results, which have been derived from experiments in transgenic mice, suggest that this mode of regulation could be utilized in the mouse to regulate allelic exclusion of the Ig loci. As for chicken B cells, limited accessibility through silencer/anti-silencer regulation would reduce the chances of rearrangement occurring on both chromosomes. Moreover, in mouse pre-B cells carrying a VDJ⁺/DJ configuration, silencing of the unrearranged heavy chain allele may protect it from the recombinase machinery, while rearrangement continues at the light chain locus. Such a regulation could also be utilized at the κ light chain locus in a VJ⁺/germline

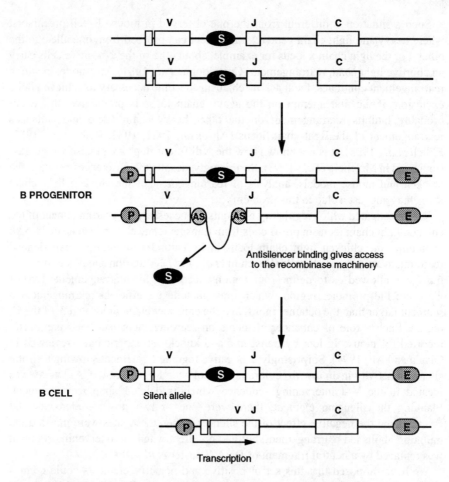

Fig. 4. Silencer/anti-silencer regulation of chicken light chain gene rearrangement. Four DNA elements regulating rearrangement of the chicken light chain locus in transgenic mice have been described (Lauster et al., 1993): two positive regulatory elements, the promoter and the enhancer regions (the enhancer is located 3' of C_λ, as described by Hagman et al. (1990) for the mouse λ locus); one negative control element, corresponding to a strong transcriptional silencer, located in the V–J intervening sequence excised upon rearrangement; one (or two) putative elements located on one (or both) side(s) of the V–J segment, antagonizing the effect of the silencer. It is proposed that the anti-silencer factors would be present transiently in chicken B cell progenitors, removing the silencer and allowing rearrangement to be performed on one allele. In the mature B cell, the remaining silencing element maintains the unrearranged allele in a silent configuration, the other allele being actively transcribed. Binding of the corresponding promoter (P), enhancer (E), silencer (S) and anti-silencer (AS) factors is represented.

configuration, in order to maintain one allele germline and to allow possible rearrangement processes such as editing to continue on the productively rearranged allele. We have in fact noticed a homology between the anti-silencer sequence present in the chicken, 3' of V_λ, and an equivalent region upstream of the mouse J_κ genes (Weaver and Baltimore, 1987). Recent results obtained from mice in which this region has been mutated support the idea that a similar type of regulation may be used in the two species (Ferradini *et al.*, in preparation).

Functional Ig sequences are selected during B cell proliferation in the bursa

After bursal colonization by B cell progenitors, there is a defined period (day 10–18) during which selection for 'functional' Ig sequences takes place, before and independently of the occurrence of gene conversion. This selection affects both in-frame sequences and the D reading frame, and can be easily monitored on the total bursal population since only one Ig allele is rearranged per cell (McCormack *et al.*, 1989; Reynaud *et al.*, 1991b).

Heavy and light chain gene rearrangement in the bursa, analysed at the time of DJ progenitor colonization (day 10–12), represents the outcome of the joining process, with characteristics similar to those described for the mouse fetal repertoire: absence of N additions, biased joining at both V–D and D–J junctions due to terminal homology, favouring the D reading frame 1 (Table II). The proportion of the three reading frames on the DJ allele does not change with time, which suggests that selection for the D reading frame is produced on the assembled VDJ structure presented at the cell surface (Reynaud *et al.*, 1991b). For the light chain, joining appears to be roughly random (McCormack *et al.*, 1989).

Most Ig sequences are already in-frame (with a D in reading frame 1) at day 15–18 of development, before their modification by gene conversion (Table II). What is the

Table II. Selection of heavy chain rearrangements in bursal vs. non-bursal sites*

	Number of sequences	In-frame sequences (%)	D reading frame			D–D junctions (%)
			Frame 1 (%)	Frame 2 (%)	Frame 3 (%)	
DJ	48		50	31	19	25
VDJ bursa day 13	28	78	67	30	3	21
VDJ bursa day 15	12	92	92	8	0	17
VDJ bursa day 18	45	96	95	5	0	15
VDJ spleen day 15	22	50	59	27	13	18

* Heavy chain rearranged sequences have been analysed with regard to the overall reading frame, the frame of the D elements, and the incidence of D–D junctions, at days 13, 15 and 18 of development in the bursa and at day 15 in the spleen. The reading frame of the D is referred to J for DJ sequences and to V for VDJ sequences (from Reynaud *et al.*, 1991b).

structure that transmits the signal for cell proliferation, and thus selects for 'functional' Ig molecules? It has been proposed that the germline-encoded specificity (i.e. the $V_H 1$–$V_\lambda 1$ pair) could recognize some bursal determinant that would induce further development (Reynaud et al., 1989; McCormack et al., 1989; see also the discussion in Langman and Cohn, 1993). Alternatively, the mere presence of an IgM molecule at the cell surface could transmit the signal for proliferation, and would result in selection of in-frame sequences with a D region ensuring the proper folding of the Ig molecule, i.e. in reading frame 1.

Ig diversification by gene conversion

Based on the homology observed between segments of spleen light chain cDNA sequences and the sequence of the first three pseudogenes present on the light chain locus, we proposed that a gene conversion process between genomic V elements could generate diversity in the chicken (Reynaud et al., 1985). The analysis of the complete genomic locus and rearranged sequences at different stages of development confirmed that a gene conversion process generated the chicken antibody repertoire (Fig. 5) (Reynaud et al., 1987). Ongoing diversification of the $V_\lambda 1$ gene was also demonstrated by using restriction sites located in the V gene sequence (Thompson and Neiman, 1987).

What do we know almost 10 years after this initial observation? Gene conversion occurs in *cis* and is directional, from the pseudogenes to the $V_\lambda 1$ acceptor sequence. This was observed by Carlson et al. (1990) by studying cell lines with restriction enzyme polymorphisms. It occurs at a high rate (one event every 10–20 cell divisions). It is homology dependent, the pseudogenes showing the highest homology to $V_\lambda 1$ being the most frequently used (Reynaud et al., 1987; McCormack and Thompson, 1990). This dependence upon homology will also influence successive gene conversion events, each event favouring the use of its more homologous pseudogene. Gene conversion includes untemplated events, mainly point mutations, occurring frequently at the border of the converted segment. It is rather precise. In the DT40 cell line, which undergoes gene conversion in culture, few mistakes, in the percent range, have been detected (its rate of gene conversion is however much lower than that of the bursa) (Buerstedde et al., 1990; Kim et al., 1990). It involves DNA tracks of various lengths, from 10 to more than 100 bp (Reynaud et al., 1987; McCormack and Thompson, 1990).

What could be the mechanism controlling gene conversion in chicken B cells? The closest model is the budding yeast (*Saccharomyces cerevisiae*) mating type switch. The mating type (MAT) switch is a regulated, non-reciprocal recombination event in which MAT-specific sequences are replaced by alternative DNA sequences copied from silent donor genes. The silent loci contain a complete copy of MAT genes including their promoter but remain unexpressed. Silencing results most probably from a specific chromatin structure preventing access to the genes. The maintenance of silencing depends on two sites surrounding each locus and on several genes

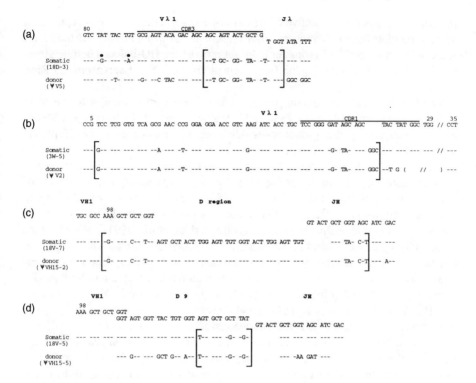

Fig. 5. Examples of gene conversion events on light and heavy chain V sequences. Four examples of gene conversion events in bursal V sequences are depicted. (a) A light chain sequence isolated from a day 18 embryo with a conversion at the V–J junction, the donor pseudogene sequence ending with a three base homology with the J segment. Dots indicate untemplated mutations. (b) A large gene conversion event in a light chain sequence at 3 weeks after hatching. The donor has a 20-bp deletion (from codons 27 to 35) on the 3′ side of the conversion that created a shifted alignment resulting in a one-codon insertion in CDR1. (c) A heavy chain sequence at day 18 of embryonic development showing conversion of the whole D region up to the D–J junction, due to a small homology of the donor with the J_H sequence. (d) A small gene conversion at the D–J junction of a heavy chain V sequence at the same stage, with part of the original D genomic sequence (D9) retained.

participating in the process. The analysis of DNA has shown different steps in the gene conversion of MATa to MATα. First, a double-strand break is induced by the HO endonuclease in the MATa gene that is going to be modified (the acceptor gene). The break is processed by a 5′ to 3′ exonuclease, which produces a long single-strand end that initiates strand invasion from the donor gene and new DNA synthesis. The whole process is completed in 30 min and is supposed to involve Holliday junctions. These junctions are rarely resolved by crossing over and it has been suggested that the branched structures could be unwound by topoisomerases rather than by cleavage, therefore allowing the two DNA segments after copy of the donor to return to their initial position. Selective expression of HO and access to the acceptor gene seem to regulate the whole process. Transcription *per se* is not involved in the access since deletion of the MAT promoter does not prevent the HO-induced double-strand break of the corresponding gene (reviewed in Haber, 1992; Laurenson and Rine, 1992).

The chicken gene conversion process, which appears very similar to the MAT switch, could be controlled also by specific double-strand cuts in the acceptor V gene and a silencing of the pseudogenes allowing for unidirectionality. Heptamer sequences specific for the rearrangement process exist in two locations within the $V_\lambda 1$ gene (and in $V_H 1$ and half of the D elements as well) and it was tempting to propose that they could be recognized by some lymphoid, heptamer-specific endonuclease activity (Reynaud et al., 1987). Nevertheless it was difficult to envision how two entry sites could generate gene conversion tracks with so diverse endpoints. This proposal was raised again by Carlson et al. (1991), who observed high *RAG*-2 gene expression in the bursa and proposed that *RAG*-2 might represent this heptamer-nicking activity. However, correlation of *RAG*-2 expression with gene conversion was put in question due to the observation that disruption of both *RAG*-2 alleles in the DT-40 cell line did not modify its gene conversion activity (Takeda et al., 1992). The way specific DNA breaks promoting gene conversion in the V sequence could be generated is thus an open issue.

Based on the observation that longer stretches of homology flank gene conversion tracks in 5′ rather than in 3′, McCormack and Thompson (1990) have proposed a model of recombination that involves strand displacement following the direction of transcription and resolution by strand dissociation, instead of Holliday junction formation. As for MAT switching, and more generally for intrachromosomal gene conversion, the formation and the mode of resolution of Holliday junctions are complex issues, and we have discussed elsewhere whether specific bursal circular DNA structures isolated recently (Kondo et al., 1993) might represent gene conversion intermediates after resolution of Holliday structures (Reynaud et al., 1994).

In conclusion, we would like to propose the following hypothetical scheme for the early events of chicken B cell development. Heamatopoietic precursors of the B lymphoid lineage initially undergo rearrangement of their DJ_H segments on both heavy chain alleles. This event may be triggered in the yolk sac or may result from the relief of a block as B cell progenitors leave the intra-embryonic para-aortic region (Reynaud et al., 1992). Thereafter, they will express *trans*-acting factors that counteract the silencing element present upstream of the J–C unit. This event, which occurs in the circulation or in a lymphoid tissue, will lead to rearrangement of the $V_H 1$ and $V_\lambda 1$ genes on one allele and deletion of the silencer element in *cis* (Lauster et al. 1993). At this stage, if the cells are in the proper bursal environment and can express a functionally rearranged H and L gene, they will receive the appropriate signal promoting further differentiation and triggering the expression of a V-specific endonuclease. The DNA breaks induced in the rearranged V genes by the endonuclease will initiate the gene conversion process on the rearranged V genes of each Ig locus. An upstream silencing of the pseudogene pool will ensure the unidirectionality of the process and protect the donor genes (the pseudogenes) from the endonuclease. By analogy with the yeast model, donor genes that are not accessible to the endonuclease could nevertheless remain active for recombination. In such a scheme, the shut-down of the endonuclease could also signal the end of the gene conversion programme and allow the cells to migrate out of the bursa to the periphery.

HYPERMUTATION IN SHEEP B CELLS

Ig rearrangement in sheep ileal Peyer's patches

We wanted to investigate whether the chicken model of B cell production was unique or whether mammalian gut-associated lymphoid tissue (GALT) could behave also as a primary organ of B cell diversification during ontogeny. Such a proposition had been made 25 years ago after the initial description of the role of the chicken bursa (Archer *et al.* 1963; Cooper and Lawton, 1972; Good, 1973), but was then rapidly abandoned as more and more data arising from the mouse underlined the role of the bone marrow as the primary site of B cell production.

The most appropriate model for such an investigation is, in our opinion, sheep ileal Peyer's patches (IPP). IPP in the sheep, as opposed to jejunal Peyer's patches, possess many histological and physiological similarities with the avian bursa. They are organized in segregated follicles that are colonized by progenitors around 110 days of fetal life (birth is at 150 days). These progenitors thereafter divide extensively within the follicles and generate a large population of surface IgM B cells (95% of follicle cells are IgM positive). This active lymphopoiesis (more than 10^9 B cells produced per hour) goes on until 3 months after birth at which time IPP, contrary to jejunal Peyer's patches, start to involute, and by 18 months their follicular structure has disappeared. During this massive proliferation, most cells die *in situ* (95%) and a small proportion migrate to the periphery and generate the B cell compartment of the animal. Based on these features, Reynolds and colleagues proposed that sheep IPP had the property of a primary B cell organ and therefore could be considered as a mammalian bursa equivalent (Reynolds and Morris, 1983; Reynolds, 1987).

When looking at this model at the molecular level, we first observed that progenitors which accumulate in IPP follicles during fetal life (two to three per follicle, 10^5 follicles) had already undergone Ig gene rearrangement (Fig. 6). Moreover, in most of these cells, one Ig light chain allele was rearranged, while the other allele was in germline configuration (this was further confirmed by the low frequency of abortive rearrangements among VJ sequences). A similar picture was obtained when looking at the progenitors that colonize bursal follicles during embryonic development, indicating that rearrangement is not ongoing in the chicken bursa, but probably occurs during a short window of development (cf. previous section). We therefore expected that, as in the bursa, a post-rearrangement diversification process was probably active in sheep B cells during their proliferation in the IPP. However, while chicken B cells diversify their Ig receptors by gene conversion, sheep B cells use another mechanism to generate their antibody repertoire—hypermutation (Reynaud et al., 1991a).

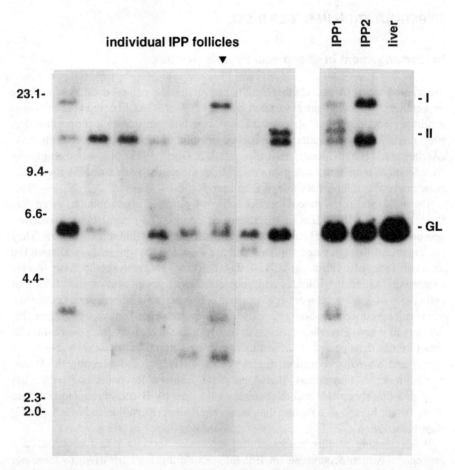

Fig. 6. Restricted rearrangement pattern of the sheep λ light chain locus. λ light chain gene configuration was analysed in isolated IPP follicles by Southern blotting after *Eco*RI digestion using a J_λ probe (the lane marked ▼ represents a mixture of two follicles). 'IPP2' and 'liver' represents controls from the same animal from which follicles were isolated, while 'IPP1' is an ileal Peyer's patch DNA sample from a different sheep, showing differences in the rearrangement pattern. The two major rearrangements are marked I and II, and the germline Jλ fragment 'GL'. Sizes (in kb) of λ *Hin*dIII restriction fragments are indicated.

Ig diversification in IPP B cells

The major light chain isotype of the sheep is λ. The λ light chain locus has been analysed and is composed of a J–C unit that is mainly used (other C_λ-hybridizing elements have been detected, but their use is probably minor, if any). A family of approximately 100 V_λ genes exist in the sheep genome, of which about one-fifth are pseudogenes (27 V_λ genes have been analysed so far from two different libraries; Reynaud et al. 1991a, 1995). This genomic V_λ pool may not be functionally as large, because about one-third of the genes have variant recombination signals and may not rearrange efficiently, and also because many of them share identical CDRs. A κ isotype has also been described in approximately 25% of sheep B cells (Griebel et al., 1992).

The analysis by Southern blot of total IPP DNA or of DNA prepared from single follicles showed a restricted number of rearrangement events (Fig. 6). Among them, two rearrangements corresponding to specific genomic V genes were present in a higher proportion (50%), and these were analysed at different stages of development. This analysis showed that rearranged V_λ genes accumulated sequence modifications with time in IPP. There was on average one modification per sequence at the end of fetal life, and more than 11 at 4 months after birth. These modifications were mainly point mutations, but they also occurred in doublets. There were very few cases of insertions or deletions. Mutations were highly clustered in CDRs but were also present in the framework regions, the leader, the leader intron and the J–C intron. There was a high number of replacement mutations in CDRs and a lower proportion of silent mutations, suggesting that a selection process was exerted on the binding sites of the mutated antibody molecules. On the other hand, framework regions (FR) where the Ig structure must be maintained carried fewer replacement mutations. When compared to the genomic V_λ sequences, we could never correlate a stretch of somatic modifications with any sequence present in the V_λ genomic genes. Overall, these results suggested that an untemplated hypermutation process that started in the antigen-free environment of the fetus was active in IPP B cells for several months after birth and that a positive selection was probably acting on the mutant B cells. Assuming a division cycle of 20 hours, the rate of mutation was about $0.5–1.2 \times 10^{-4}$ bp/cell division, such a rate being comparable to the hypermutation process acting during immune responses to T-dependent antigens (Reynaud et al., 1991a).

In a second set of experiments, we wanted to find out whether food and bacterial antigens present in the gut and able to reach the IPP lymphoid follicles through specialized epithelial cells could be responsible for the selection observed. We therefore analysed rearranged V_λ sequences in experimental conditions where IPP were never in contact with foreign antigens during development. This analysis was performed either in sterile isolated segments of IPP detached from the gut during fetal life and analysed several weeks after birth ('ileal loops') or in sheep maintained in gnotobiotic conditions for 3–4 weeks after birth. In these different experimental conditions, the growth of the lymphoid follicles was impaired and showed about 50% reduction of size several weeks after their isolation from gut constituents.

However, the analysis of approximately 100 sequences carrying about 500 modifications showed that, in spite of this reduced growth, the pattern of mutation was qualitatively and quantitatively identical to the normal situation (Table III). Mutations were clustered in CDRs with a high number of replacement mutations; silent mutations were again favoured in FR sequences and modifications were also found in the J segment and the leader intron (Reynaud et al., 1995).

Table III. Somatic mutations in loop, germ-free and thymectomy experiments

	Mutations per sequence	R/S CDR	R/S FR	CDR/FR clustering ratio*
Germ-free 5 weeks no. 1	5.2	6.1	2.4	9.0
Germ-free 5 weeks no. 2	4.3	8.3	1.7	8.3
Control 5 weeks	4.7	5.8	1.0	10.7
Thymectomy 5 weeks	4.1	12.0	1.7	8.3
Control 5 weeks	4.9	>45†	1.0	11.5
Loop 8 weeks	7.8	11.2	5.5	11.2
Control 8 weeks	7.5	8.3	1.3	11.1

* Mutation rate per nucleotide of complementarity-determining regions (CDR) vs. framework regions (FR).
† 45 replacement mutations (R) and no silent mutations (S) observed.

The hypermutation process in sheep IPP

When analysing all the rearranged VJ sequences obtained from normal IPP and from IPP obtained from antigen-free animals (loop and germ-free conditions), we observed the same distribution of mutations with a trend for transitions (60%) over transversions (40%), the theoretical values being 33% and 67%, respectively. G↔A transitions were much more frequent than C↔T and, overall, mutations were twice as frequent on purines than on pyrimidines (Table IV). This ratio was also observed when analysing mutations occurring in non-coding intronic sequences (on a total of 55 mutations from the two sets of studies).

Table IV. Nucleotide substitutions (%) in sheep ileal Peyer's patches V sequences

To:	G	A	T	C	Total	
From:						
G	–	25.3	4.3	8.6	38.2 ⎫	
A	22.5	–	4.0	3.8	30.4 ⎭	68.6
T	8.4	7.3	–	5.6	21.3 ⎫	
C	2.6	1.4	6.1	–	10.1 ⎭	31.4
Transitions	59.5	G/A	47.8	T/C	11.7	
Transversions	40.5					

Most of the mutations were targeted to CDRs of the rearranged VJ sequences, and within the CDRs to a subset of the CDR codons. These mutated subsets remained very similar in the normal and the antigen-free situation (Fig. 7), suggesting that they were not antigen selected.

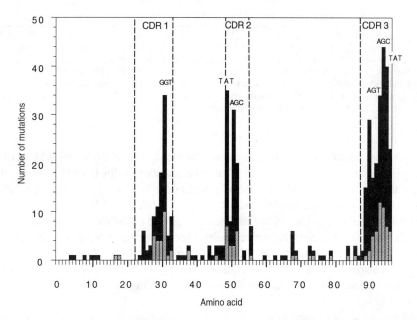

Fig. 7. Pattern of mutation of the 5.1 gene in 'loop' and normal conditions. Mutations obtained for the V 5.1 coding sequence (rearrangement type I) are represented with discrimination of the mutations from the 'loop' experiments (in grey, 116 mutations) vs. the ones obtained from normal animals (in black, 405 mutations) (Reynaud et al., 1991a, 1995). The ordinate represents the number of mutations at each amino acid position. Limits of CDRs are delineated. The particular amino acid positions with more than 30 mutations are indicated.

When looking at the base pair composition of the V_λ genes analysed in this study, which represent the most utilized V_λ genes, there is a striking bias for G or A in the first and second position of each CDR triplet and a bias against G or A in the third base. This bias is further strengthened when looking at the CDR codons that are hotspots of mutations (Table V). The other functional V_λ genomic genes, although we do not know their exact usage frequency, show the same trend for G or A in the first two bases of their CDR triplets.

One must therefore re-estimate a 'mechanistic' replacement over silent substitution ratio (R/S) value, taking into account the two-fold higher mutation rate of purines compared to pyrimidines. Moreover, if only the subset of CDR codons in

Table V. Percentage of purines in codons of sheep V_λ 5.1 and 16.1 genes

Nucleotide position in the codon	FR		CDR total		CDR 'mutated subset'*	
	First + second	Third	First + second	Third	First + second	Third
V5.1†	47	39	75	18	88	0
V16.1†	48	39	77	23	81	19

* CDR mutated subsets represent 17 out of 27 amino acids for the 5.1 gene, accounting for 95% of all CDR mutations (CDR1, positions 28–31, 33; CDR2, 49–52, CDR3, 89–96). For gene 16.1, it represents 16 out of 29 amino acids, accounting for 92% of total CDR mutations (CDR1, positions 29, 31–35; CDR2, 51–54; CDR3, 90–92, 94–98). Mutations are compiled from Reynaud et al., (1991a, 1995).

† 5.1 and 16.1 are the genomic genes to which somatic sequences from rearrangements type I and type II, respectively, are compared (5.1 being the closest homologous gene, with two differences in the coding sequence, and 16.1 the precise genomic counterpart).

which most mutations are clustered is considered (17 out of 27 amino acids for the 5.1 gene), a theoretical R/S value of 8.7 is expected, close to the one observed experimentally. As far as the concentration of mutations in the CDRs is concerned, a threefold clustering per nucleotide in CDRs vs. FRs is observed for silent mutations. When extrapolated to the ratio of total mutations (see calculation in Reynaud et al., 1995), one can estimate that the mechanism itself can generate 5.5 more mutations in CDRs in the absence of antigen selection. Therefore, taking advantage of the G/A nucleotide specificity and of the bias in codon usage of the V_λ genes, the mechanism per se generates preferentially replacement mutations in sheep IPP B cells (Reynaud et al., 1995).

The molecular mechanism of hypermutation

Our results suggest that the molecular process which operates in sheep IPP B cells during the generation of the primary antibody repertoire is similar to the one acting in mouse B cells undergoing an immune response.

Effectively, a similar bias for mutations on purines and for concentration in CDRs was observed on passenger Ig transgenes that are mutated but not submitted to antigen-induced selection during secondary immune responses in the mouse (Betz et al., 1993). Such an 'unselected' pattern reflects therefore the DNA specificity of the process. When mouse Ig V genes undergoing mutation are submitted to antigenic selection, some specific antigen-driven mutations are superimposed on this basic mechanistic pattern, and they may sometimes obscure the molecular trend. As noted in the mouse, the imbalance between purine and pyrimidine mutations indicates that the mutation mechanism is targeted to one strand of DNA.

It was proposed 25 years ago that V genes undergoing hypermutation were submitted to nucleolytic attacks and repaired by an error-prone polymerase thus introducing mutations in the sequence (Brenner and Milstein, 1966). However no data have yet been produced that could explain the mutational process. When looking at the error frequencies of most polymerases they are in the range of 10^{-3}–10^{-5} (Kunkel, 1991). Therefore inhibition of proofreading during and after replication (3'–5' exonuclease and mismatch repair) may lead to the rate of mutation observed during V gene hypermutation. If such inhibition takes place in B cells, it has to be specifically targeted to the mutating V gene and to affect no other region of the genome.

It is believed that the specific microenvironment of the germinal centre that arises during a T-dependent immune response triggers B cell hypermutation. IPP follicles contain 95% of surface IgM B cells but also many different cell types such as reticular cells, stromal cells, dendritic cells, tingible body macrophages, T cells. Whether any of these constituents plays a role in inducing the diversification process remains unknown at the moment. We have studied sheep thymectomized at 70 days of fetal development. When analysed at 5 weeks after birth, at which stage circulating T cells are below 10% of the normal value (Hein et al., 1990), the distribution of mutations on rearranged V_λ sequences in IPP B cells was strictly comparable to the normal situation (Table III) (Reynaud et al., 1995).

Selection of B cells

What does this model tell us about the selection events taking place in a primary B cell organ? It has been shown that most IPP B cells have an immature phenotype and that, similarly to immature mouse B cells, they are negatively signalled when cross-linked through their surface Ig receptor *in vitro* (Griebel et al., 1991). Moreover the nature of the diversification process used in sheep B cells, which like gene conversion introduces modifications in the three CDRs, suggests that some B cells carrying antibodies with high affinity against self constituents must be constantly produced in IPP. These anti-self B cells will most probably be silenced or deleted in the tissue as they are produced, but it is difficult to estimate what could be the contribution of negative selection to the high cell death rate (95%) taking place in IPP during development.

As for positive selection, we thought at first that sheep IPP would represent an ideal model to measure, through the R/S ratios, the contribution of external antigens to this process. The detailed analysis of the hypermutation process occuring in sheep IPP in different experimental situations has shown that:

1. patterns of mutations remain unchanged in antigen-free conditions;
2. the accumulation of replacement substitutions in CDRs is due for the most part to the specific action of the mutator and the particular base pair composition of the genomic V_λ genes.

Nevertheless our estimations, based on R/S values and CDR clustering, remain imprecise and we cannot exclude that internal ligands or external antigens may exert a selection on newly formed B cells within IPP in the normal situation. Moreover as

B cells migrate to the periphery they could, as proposed for mouse B cells (Gu *et al.*, 1991), be signalled through their Ig receptors and be driven to different peripheral pools. What these interactions could be and whether they involve external antigens remains totally open at the moment.

CONCLUSIONS

There are different cellular and molecular strategies operating in order to generate an antibody repertoire but overall these strategies have evolved towards the same goal: the production of a large number of different antibody molecules carrying different antigen-recognizing domains.

In mouse and human, this is accomplished by maintaining in the genome a rather large number of different V gene families, each gene being theoretically able to produce a new antibody molecule by rearrangement. In chicken and rabbit (Becker and Knight, 1990), it is accomplished by selecting for a single family of homologous genomic V genes and pseudogenes, which will be transferred in segments by ongoing gene conversion to a rearranged acceptor V gene. In sheep, the mechanism involves a pool of functional V genes that harbour a specific base pair composition in their CDRs, diversity being generated by a mutator acting more specifically and at a high rate on these particular regions.

Why certain species use GALT or bone marrow to generate their primary B cell repertoire remains an open issue. The molecular strategies have evolved along with the structure of the primary B cell organs. Ongoing rearrangement takes place in bone marrow and involves a few cellular divisions corresponding to the B cell lifetime in the organ. Post-rearrangement diversification processes take place in lymphoid follicles, in which B cells can proliferate extensively over several weeks and thus accumulate modifications on their antigen receptors. Our results do not point to a role for external antigens in the specificity of the B cell repertoire in animals using GALT. However the non-specific potent mitotic activity provided by gut bacterial antigens could be the necessary stimuli that allows B cells to proliferate for several months and thus to accumulate modifications on their Ig receptors by gene conversion or hypermutation.

Some differences also exist in the physiology of the B cell system when comparing species using GALT instead of bone marrow. Primary GALT involutes at the end of development and it is not clear how a diversified repertoire at this stage can be maintained in the periphery in the absence of a life-long production of diversified B cells. The adult animal will have to rely on a population of naive B cells and on a pool of memory cells that have been produced during development. It then becomes necessary to postulate the existence of a post-GALT stem cell in the periphery (Toivanen and Toivanen, 1973) in order to self-maintain the naive B cell population.

REFERENCES

Alt, F.W., Rosenberg, N., Lewis, S., Thomas, E. and Baltimore, D. (1981). *Cell* **27**, 381–390.
Alt, F.W., Yancopoulos, G.D., Blackwell, T.K., Wood, C., Thomas, E., Boss, M., Coffman, R., Rosenberg, N., Tonegawa, S. and Baltimore, D. (1984). *EMBO J.* **3**, 1209–1219.
Archer, O.K., Sutherland, D.E.R. and Good, R.A. (1963). *Nature* **200**, 337–339.
Becker, R.S. and Knight, K.L. (1990). *Cell* **63**, 987–997.
Betz, A.G., Rada, C., Pannell, R., Milstein, C. and Neuberger, M.S. (1993). *Proc. Natl Acad. Sci. USA* **90**, 2385–2388.
Brenner, S. and Milstein, C. (1966). *Nature* **211**, 242–243.
Bucchini, D., Reynaud, C.-A., Ripoche, M.A., Grimal, H., Jami, J. and Weill, J.-C. (1987). *Nature* **326**, 409–411.
Buerstedde, J.M., Reynaud, C.A., Humphries, H.E., Olson, W., Ewert, D.L. and Weill, J.C. (1990). *EMBO J.* **9**, 921–927.
Carlson, L.M., McCormack, W.T., Postema, C.E., Humphries, E.H. and Thompson, C.B. (1990). *Genes Dev.* **4**, 536–547.
Carlson, L.M., Oettinger, M.A., Schatz, D.G., Masteller, E.L., Hurley, E.A., McCormack, W.T., Baltimore, D. and Thompson, C.B. (1991). *Cell* **64**, 201–208.
Coleclough, C., Perry, R.P., Karjalainen, K. and Weigert, M. (1981). *Nature* **290**, 372–378.
Cooper, M.D. and Lawton, A.R. (1972). *Contemp. Top. Immunobiol.* **1**, 49–68.
Dieterlen-Lièvre, F. and Martin, C. (1981). *Dev. Biol.* **88**, 180–191.
Ehlich, A., Schaal, S., Gu, H., Kitamura, D., Müller, W. and Rajewsky, K. (1993). *Cell* **72**, 695–704.
Good, R.A. (1973). *Harvey Lect.* **67**, 1–107.
Griebel, P.J., Davis, W.C. and Reynolds, J.D. (1991). *Eur. J. Immunol.* **21**, 2281–2284.
Griebel, P.J., Kennedy, L., Graham, T., Davis, W.C. and Reynolds, J.D. (1992). *Immunology* **77**, 564–570.
Gu, H., Tarlinton, D., Müller, W., Rajewsky, K. and Förster, I. (1991). *J. Exp. Med.* **173**, 1357–1371.
Haber, J.E. (1992). *Trends Genet.* **8**, 446–452.
Hagman, J., Rudin, C.M., Haasch, D., Chaplin, D. and Storb, U. (1990). *Genes Dev.* **4**, 978–992.
Hein, W.R., Dudler, L. and Morris, B. (1990). *Eur. J. Immunol.* **20**, 1805–1813.
Hesse, J.E., Lieber, M.R., Mizuuchi, K. and Gellert, M. (1989). *Gene Dev.* **3**, 1053–1061.
Houssaint, E., Lassila, O. and Vainio, O. (1989). *Eur. J. Immunol.* **19**, 239–243.
Houssaint, E., Mansikka, A. and Vainio, O. (1991). *J. Exp. Med.* **174**, 397–406.
Kim, S., Humphries, E.H., Tjoelker, L., Carlson, L. and Thompson, C.B. (1990). *Mol. Cell. Biol.* **10**, 3224–3231.
Kondo, T., Arakawa, H., Kitao, H., Hirota, Y. and Yamagishi, H. (1993). *Eur. J. Immunol.* **23**, 245–249.
Kunkel, T.A. (1991). In 'Somatic Hypermutation in V Regions' (E.J. Steele, ed.), pp. 159–178. CRC Press, Boca Raton, FL.
Langman, R.E. and Cohn, M. (1993). *Res. Immunol.* **144**, 422–446.
Lassila, O., Eskola, J., Toivanen, P., Martin, C. and Diterlen-Lièvre, F. (1978). *Nature* **272**, 353–354.
Lassila, O., Martin, C., Dieterlen-Lièvre, F., Gilmour, D., Eskola, J. and Toivanen, P. (1982). *Scand. J. Immunol.* **16**, 265–268.
Laurenson, P. and Rine, J. (1992). *Microbiol. Rev.* **56**, 543–560.
Lauster, R., Reynaud, C.-A., Mårtensson, L., Peter, A., Bucchini, D., Jami, J. and Weill, J.-C. (1993). *EMBO J.* **12**, 4615–4623.
McCormack, W.T. and Thompson, C.B. (1990). *Genes Dev.* **4**, 548–558.
McCormack, W.T., Tjoelker, L.W., Barth, C.F., Carlson, L.M., Petryniak, B., Humphries, E.H. and Thompson, C.B. (1989). *Genes Dev.* **3**, 838–847.
Melchers, F., Karasuyama, H., Haasner, D., Bauer, S., Kudo, A., Sakaguchi, N., Jameson, B. and Rolink, A. (1993). *Immunol. Today* **14**, 60–68.
Moore, M.A. and Owen, J.J.T. (1967). *Nature* **215**, 1081–1082.
Okada, A. and Alt, F.W. (1994). *Semin. Immunol.* **6**, 185–196.
Pink, J.R.L. and Lassila, O. (1987). *Curr. Top. Microbiol. Immunol.* **135**, 57–64.

Ratcliffe, M.J.H., Lassila, O., Pink, J.R.L. and Vainio, O. (1986). *Eur. J. Immunol.* **16**, 129–133.
Reth, M., Petrac, E., Wiese, P., Lobel, L. and Alt, F.W. (1987). *EMBO J.* **6**, 3299–3305.
Reynaud, C.A., Anquez, V., Dahan, A. and Weill, J.C. (1985). *Cell* **40**, 283–291.
Reynaud, C.A., Anquez, V., Grimal, H. and Weill, J.C. (1987). *Cell* **48**, 379–388.
Reynaud, C.A., Dahan, A., Anquez, V. and Weill, J.C. (1989). *Cell* **59**, 171–183.
Reynaud, C.A., Mackay, C.R., Müller, R.G. and Weill, J.C. (1991a). *Cell* **64**, 995–1005.
Reynaud, C.A., Anquez, V. and Weill, J.C. (1991b). *Eur. J. Immunol.* **21**, 2661–2670.
Reynaud C.A., Imhof, B.A., Anquez, V. and Weill, J.C. (1992). *EMBO J.* **12**, 4349–4358.
Reynaud, C.A., Bertocci, B., Dahan, A. and Weill, J.C. (1994). *Adv. Immunol.* **57**, 353–378.
Reynaud, C.A., Garcia, C., Hein, W.R. and Weill, J.C. (1995). *Cell* **80**, 115–125.
Reynolds, J.D. (1987). *Curr. Top. Microbiol. Immunol.* **135**, 43–56.
Reynolds, J.D. and Morris, B. (1983). *Eur. J. Immunol.* **13**, 627–635.
Takeda, S., Masteller, E.L., Thompson, C.B. and Buerstedde, J.M. (1992). *Proc. Natl Acad. Sci. USA* **89**, 4023–4027.
Thompson, C.B. and Neiman, P. (1987). *Cell* **48**, 369–378.
Toivanen, P. and Toivanen, A. (1973). *Eur. J. Immunol.* **3**, 585–595.
Weaver, D. and Baltimore, D. (1987). *Proc. Natl. Acad. Sci. USA* **84**, 1516–1520.
Weill, J.-C. and Reynaud, C.-A. (1987). *Science* **238**, 1094–1098.

13

Immunoglobulin heavy chain genes of rabbit

Katherine L. Knight and Chainarong Tunyaplin

Department of Microbiology and Immunology, Loyola University Chicago, Stritch School of Medicine, Maywood, IL, USA

Introduction	289
V_H genes	290
C_H genes	304
Concluding remarks	311

INTRODUCTION

Rabbit immunoglobulins (Ig) are unique in that they have allotypic specificities not only in the constant (C) regions but also in the variable (V) regions of the heavy chains. Studies of rabbit Ig allotypes have contributed significantly to our knowledge of Ig structure and genetics. In the 1960s, Todd (1963) and Feinstein (1963) studied the inheritance and molecular localization of these allotypic specificities, and their studies led to the revolutionary idea that two genes encode one polypeptide chain (Dreyer and Bennett, 1965; Lennox and Cohn, 1967). Allelic exclusion was first discovered in rabbit when Pernis *et al.* (1965) and Cebra *et al.* (1966) examined the expression of the allotypic specificities and showed that in individual plasma cells, only one heavy chain/light chain allele was expressed. Other studies of the rabbit V_H and C_H allotypes demonstrated that genes encoding the V and C regions were linked (Dubiski, 1969; Mandy and Todd, 1970), that V and C region genes are expressed in *cis* (Kindt *et al.*, 1970; Landucci-Tosi *et al.*, 1970) and that germline recombination occurs between V_H and C_H genes at a high frequency (Mage *et al.*, 1971, 1982; Kindt and Mandy, 1972; Kelus and Steinberg, 1991).

For many years the allelic inheritance of V_Ha allotypes, a1, a2 and a3, was the most perplexing aspect of rabbit Ig. The rabbit genome contains more than 100 V_H genes, and because 80–90% of serum Ig molecules bear the V_Ha allotypes, one would assume that most germline V_H genes encode the V_Ha allotypic specificities. The question that researchers faced was, if many of these V_H genes are used in the antibody repertoire, how then was the rabbit V_H chromosomal region protected from meiotic recombination, which would shuffle the V_H genes encoding the different allotypes, such that the allelic inheritance of the allotypes would be lost.

Other intriguing findings regarding rabbit Ig and the genes that encode them are that (1) the germline contains 13 C_α genes but has only one C_γ gene, (2) expression of the 13 C_α genes is differentially regulated in various lymphoid tissues, and (3) *trans*-association of V_H and C_H regions occurs in a significant number of IgG and IgA heavy chains. In this chapter we summarize our current knowledge of the organization and expression of V_H and C_H genes and identify the problems yet to be solved.

V_H GENES

V_H gene organization

The V_H locus of rabbits is organized in a manner similar to that of other species in that the locus contains multiple V_H, D and J_H gene segments (Fig. 1) (Gallarda *et al.*, 1985; Currier *et al.*, 1988; Becker *et al.*, 1989). By Southern analysis using a pan-V_H probe, Gallarda *et al.* (1985) estimated that the rabbit genome contains approximately 100 V_H genes. Currier *et al.* (1988) cloned a total of 750 kb of the V_H locus from a cosmid library and estimated that these clones contained approximately 100

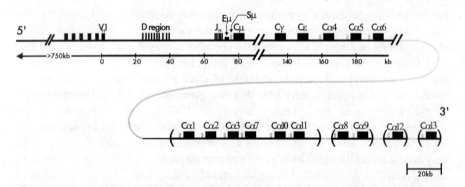

Fig. 1. Rabbit Ig heavy chain locus. The order of D gene segments from 5' to 3' is D3, D1a, D4, D1b, D6, D7, D2a, D1c, D2b, D1d and D5. Of the 13 C_α genes, the germline location of only $C_\alpha 4$, $C_\alpha 5$ and $C_\alpha 6$ is known; the other C_α genes are shown as clusters in parentheses. The order of C_α genes in each cluster is as shown but the relative position of each cluster is unknown. Each C_H gene is associated with switch regions shown as gray boxes 5' of each C_H gene. The germline clones that span the heavy chain locus are from rabbits of the G haplotype and are described in Currier *et al.*, (1988); Burnett *et al.* (1989); Becker *et al.* (1990); Raman *et al.* (1994); Knight *et al.* (1985).

V_H genes. Many of these V_H containing cosmid clones are overlapping and can be grouped into clusters, but many of these clusters are not linked to each other, and hence they do not span the entire V_H region. Therefore, it is likely that the rabbit V_H locus spans over 750 kb and that there are more than 100 germline V_H genes.

All rabbit V_H genes belong to a single family because, based on the nucleotide sequences of about 50 published rabbit germline V_H genes, they share more than 80% similarity (Bernstein et al., 1985; McCormack et al., 1985; Currier et al., 1988; Fitts and Metzger, 1990; Knight and Becker, 1990; Roux et al., 1991; Short et al., 1991). This is in contrast to the situation in mouse and human in which the genome contains 7–13 V_H gene families (Kirkham et al., 1992). On the basis of cross-species hybridization with family-specific V_H probes, the rabbit V_H genes are analogous to the human V_H3 family and the mouse V_H7183 and S107 families (Tutter and Riblet, 1989). These families of V_H genes are thought to represent the primordial V_H gene because they are the only V_H families found in every species examined so far (Tutter and Riblet, 1989). In fact, the rabbit V_H genes are so similar to the primordial V_H gene that the antibodies to V_Ha allotypes cross-react with V_H regions of Ig from several species, including human, shark and toad, which are evolutionarily distant from rabbit (Knight et al., 1975; Rosenshein and Marchalonis, 1985).

The V_H genes that have been studied most extensively are the V_H genes nearest the 3' end on each of three heavy chain alleles, a^1, a^2 and a^3. We cloned from 4 to 11 of the V_H genes closest to the 3' end from DNA of rabbits homozygous for one of the three alleles and determined their nucleotide sequences (Knight and Becker, 1990; Raman et al., 1994). These V_H genes were assigned a sequential number, with the gene nearest the 3' end being V_H1 (Fig. 2). On the basis of the nucleotide sequence, the pool of V_H genes from each allele contains both functional and non-functional V_H genes, which are present in a ratio of approximately 1 : 1 (Knight and Becker, 1990;

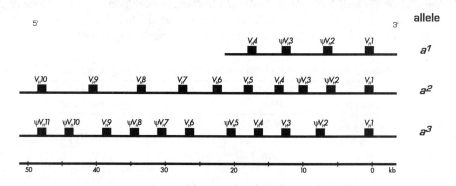

Fig. 2. The V_H genes nearest the 3' end of a^1/a^1, a^2/a^2 and a^3/a^3 rabbits. Pseudogenes are indicated with ψ. No designation of functionality is made for V_H5 to V_H10 of the a^2 allele because only partial nucleotide sequences are available. The V_H genes of the a^1 allele are from a rabbit of the C haplotype, V_H genes from the a^2 allele are from rabbits of the E haplotype and V_H genes of the a^3 allele are from rabbits of the G haplotype.

Raman et al., 1994). While the location of functional and non-functional V_H genes varies between alleles, the V_H gene nearest the 3' end, V_H1, is invariably a functional gene.

V_H gene usage

The V_H regions of nearly all rabbit Ig can be placed serologically into one of three groups—**a**, **x** and **y**—on the basis of the allotypic specificities (Dray et al., 1963; Kim and Dray, 1972). These three groups appear genetically as if they are controlled by three separate but closely linked V_H loci (Dray et al., 1974b). Each group is polymorphic, and allelic forms of each are described. Since the germline contains over 100 V_H genes within several hundred kilobases of DNA, it was difficult to understand how the V_H genes mapped as only three, rather than as over 100, V_H loci. We now know that such an inheritance pattern of the rabbit V_H region is due to the fact that essentially the entire antibody repertoire is developed from rearrangements of only three to four V_H genes. Evidence for such limited usage of V_H genes is described below.

V_H gene usage in $V_H a$ molecules

$V_H a$ molecules account for approximately 80% of the serum Ig molecules and are encoded by the *a* locus (Dray et al., 1963b). As mentioned above, this group of **a** molecules is polymorphic: three allelic forms—a1, a2 and a3—are found in laboratory rabbits and react with anti-allotype antisera, anti-a1, anti-a2 and anti-a3, respectively (Table I). According to amino acid sequence analysis of pooled Ig and of specifically purified antibodies, the a1, a2 and a3 allotypic specificities differ by multiple amino acid differences distributed in framework regions 1 and 3 (Mage et al., 1984) (Fig. 3). Ten additional alleles of the *a* locus, a^{100} to a^{109}, are found in wild rabbits (Haouas et al., 1989); these allotypes will not be discussed further.

Table I. V_H allotypes and the utilized allotype-encoding V_H genes

	Allotype	Serum Ig (%)	Alleles	Gene (germline location)
$V_H a$	a1, a2, a3	80–90	a^1, a^2, a^3	$V_H 1$ (3' most V_H gene)
$V_H a^-$	x32, x⁻	5–10	x^{32}, x^-	$V_H x^*$ (>50 kb 5' of $V_H 1$)
	y33, y⁻	5–10	y^{33}, y^-	$V_H y$† (>50 kb 5' of $V_H 1$)

* $V_H x$ has only been cloned as cDNA and it is assumed to encode x32 molecules; however, this has not been shown experimentally.
† $V_H y$ encodes y33 molecules as shown by *in vitro* expression and reactivity of the expressed product with anti-y33 antibody.

FR1

```
a1   V Q C / Q S V E E S G G R / / / / L V T P G T P L T L T C
a2   - - - / - - - K - - E - - G / / / / - F K - T D T - - - - -
a3   - - - / - - L - - - - - D / / / / - - K - - A S - - - - -
x32  - - Q / E Q L K - - - - G / / / / - - K - - G S - K - S -
y33  - - - / - - L E Q - - - G A G G G - - K - - G S - E - C -
z    - - - / R - - - H - - - G / / / / - - Q P R G S - K - C -
```

CDR1 FR2

```
a1   V S G F S L S S / Y A M S W V R Q A P G K G L E W I G I I
a2   - - - - - - - - / N - I - - - - - - - - N - - - - - A -
a3   A - - - - F - - S - Y - C - - - - - - - - - - - - - A C -
x32  A - - - D F - - / - G V - - - - - - - - - - - - - - - Y -
y33  A - - - - - - - S - W I C - - - - - - - - - - - - - - C -
z    A - - - T F - - / - Y - C - - - - - - - - - - - - - - C -
```

CDR2 FR3

```
a1   / S S S G S T Y Y A S W A K G R F T I S K T S / / T T V D
a2   / G - - - - A - - - - - - S - S - - T R N T N L N - - T
a3   A G - - - T - - - - - - - - - - - - - - - - S / - - - T
x32  P V F / - - T - - - - - V N D - - - - - S H N A Q N - L Y
y33  A G - / - - T H - - - - V N - - - - L - R D I D P S - G C
z    A G - - - A - - - - - - V N - - - - L - R D I D Q S - G C
```

```
a1   K I T S P T T E D T A T Y F C A R
a2   - M - - L - A A - - - - - - - - -
a3   Q M - - L - A A - - - - - - - - -
x32  Q L N - L - P A - - - - - - - - -
y33  Q L N - L - A A - - - M - Y - - -
z    Q L N - L - A A - - - M - Y - - -
```

Fig. 3. Amino acid sequences of V$_H$a allotype molecules, a1, a2, a3 and of V$_H$a⁻ allotype molecule, x32, y33 and z. The a1, a2 and a3 sequences are deduced from the allelic V$_H$1 genes, V$_H$1-a1, V$_H$1-a2 and V$_H$1-a3 (Knight and Becker, 1990). The x32 sequence is deduced from cDNA (Freidman et al., 1994); the y33 sequence is deduced from germline V$_H$y33 (Short et al., 1991); the z sequence is deduced from cDNA (Freidman et al., 1994) and is presumed to encode a V$_H$a⁻ molecule but of unknown allotype. The a1, a2 and a3-allotype associated residues are in bold. Dashes represent identity to a1.

Several hypotheses have been proposed to explain the allelic inheritance phenomenon of $V_H a$ allotypes; but until recently, all of them remained hypothetical. The simplest explanation was that only one $V_H a$ allotype-encoding gene was used in the VDJ gene rearrangements and that in the various heavy chain haplotypes these utilized V_H genes were allelic. Although it was difficult to envision that most of the VDJ repertoire was derived from one V_H gene, several lines of evidence, as discussed below, indicate that, in fact, most of the antibody repertoire is derived from $V_H 1$, a V_H gene known to encode prototypic $V_H a$ molecules (Knight and Becker, 1990).

V_H gene usage in B cell tumors Some years ago we reasoned that studies of V_H gene usage in rabbits would be enhanced if B cell tumors and/or B cell lines were available. Because Adams *et al.* (1985) showed that B-lineage tumors developed in transgenic mice carrying the c-*myc* oncogene driven by the heavy chain enhancer, E_μ, we introduced the c-*myc* oncogene driven by either E_μ or by the κ chain enhancer (E_κ) into rabbit zygotes, with the goal of generating transgenic rabbits that would develop B cell tumors. Such transgenic rabbits were generated and, indeed, these transgenic rabbits developed B cell malignancies (Knight *et al.*, 1988; Sethupathi *et al.*, 1994).

The transgenic rabbits with E_μ–*myc* spontaneously developed B cell leukemias at a young age (Knight *et al.*, 1988). From these leukemic B cells, Becker *et al.*, (1990) cloned VDJ genes from genomic phage libraries. To their surprise, they found that the nucleotide sequences of the V_H regions of some of these VDJ genes were identical to each other. This observation suggested that these leukemic B cells used a limited number of V_H genes. Further, the V_H regions of the VDJ genes obtained from leukemic B cells of a^1/a^1 rabbits were identical to germline $V_H 1$ from the a^1 allele, $V_H 1$-$a1$; the V_H regions of the VDJ genes obtained from a^2/a^2 rabbits were identical to germline $V_H 1$ from the a^2 allele, $V_H 1$-$a2$; and one of the VDJ genes from an a^3/a^3 rabbit was identical to germline $V_H 1$ from the a^3 allele, $V_H 1$-$a3$. The utilized germline V_H gene can be unmistakably determined by directly comparing the regions 5' of the VDJ genes with the region 5' of the germline V_H genes, a region that is unique among the V_H genes. By analyzing the 5' regions, Knight and Becker (1990) showed that these clones did indeed use $V_H 1$. Of eight VDJ genes cloned from leukemic rabbits, only one did not utilize $V_H 1$; this one used a gene we now know as the y33-encoding gene.

The E_κ–*myc* transgenic rabbits developed several types of tumors, including hepatocarcinoma, ovarian carcinoma, basal cell carcinoma, embryonic carcinoma and, most importantly to us, B cell lymphomas (Sethupathi *et al.*, 1994). We developed a cell line from one of the B lymphomas and cloned the VDJ genes from this cell line. We found VDJ genes on both alleles and showed that both genes utilized $V_H 1$ in their VDJ gene rearrangements. Taken together, the data showed that $V_H 1$ was preferentially used in the VDJ genes of the B cell tumors from both E_μ–*myc* and E_κ–*myc* transgenic rabbits.

$V_H 1$ gene usage in normal rabbit B cells To extend the study of V_H gene usage from B cell tumors to normal rabbit B cells, Raman *et al.* (1994) generated several

stable V_Ha allotype-secreting rabbit X mouse heterohybridomas. They then cloned the VDJ genes from genomic phage libraries and determined whether the heterohybridomas used V_H1 in their VDJ genes. Because these hybridomas were derived from B cells of adult rabbits in which the VDJ genes were presumably diversified, they determined whether V_H1 was the utilized gene by analyzing the 5' regions. Raman et al. (1994) found that, indeed, the 5' region of eight of nine VDJ clones had a restriction map identical to the 5' region of V_H1 of the appropriate allele. This study showed that most V_Ha allotype Ig-secreting B cells from adult rabbits use V_H1 in their VDJ genes.

Deletion of V_H1 correlates with loss of V_Ha allotype expression in a mutant rabbit Kelus and Weiss (1986) contributed to the understanding of V_Ha allotype inheritance by identifying the *ali* mutation on the a^2 heavy chain allele. The *ali/ali* homozygous mutant rabbit, Alicia, was genetically a^2/a^2, but it expressed only a low level of Ig molecules of the a2 allotype. The Ig molecules in Alicia rabbits are designated V_Ha⁻ molecules. When this mutation was first described, it was difficult to understand what mutation could so dramatically render Alicia unable to express the V_Ha2 allotype, especially when the germline contained more than 100 V_H genes, most of which presumably encoded the V_Ha allotype. Because most V_Ha allotype Ig is derived from the V_H1 gene, it was hypothesized that the *ali* mutation could be explained easily by a mutation in V_H1. The evidence to support this hypothesis was reported by Allegrucci et al. (1991) who, by Southern analysis, showed a 10–15 kb deletion of the 3' end of the V_H locus in Alicia rabbits, whereas the rest of the V_H locus appeared similar to that of the normal a2 allele. Knight and Becker (1990) cloned the V_H genes nearest the 3' end from a splenic cosmid library of a homozygous *ali/ali* rabbit and showed that the *ali* mutation correlated to a deletion of 10 kb of V_H region that included V_H1–a2. They concluded that the deletion of V_H1–a2 rendered Alicia rabbits unable to express Ig molecules with the a2 allotype.

V_H genes encoding V_Ha⁻ molecules

Ten to twenty percent of the serum Ig of normal rabbits do not have the V_Ha allotypic specificities and are designated V_Ha⁻ (Dray et al., 1963b). In general, these V_Ha⁻ molecules comprise **x** and **y** groups, each representing approximately 5–10% of the serum Ig molecules (Kim and Dray, 1973) (Table I). The **x** molecules are genetically controlled by the *x* locus, and from classical genetic mapping studies two alleles were described, x^{32} and x^-. The x^{32} allele encodes the Ig molecules that bear the x32 allotypic specificity, the only known allotypic marker of the **x** group. The other allelic form, x^-, encodes the x⁻ molecules, which do not have the x32 or any other known allotypic specificity. The Ig of the **y** group are genetically controlled by the *y* locus. Two alleles of the *y* locus are known, y^{33} and y^-. The y^{33} allele encodes Ig that carry the y33 allotypic specificity, and the y^- allele encodes y⁻ molecules that do not have the y33 or any other known allotypic specificity. The *a*, *x* and *y* loci are closely linked and are inherited as a group, known as an allogroup or haplotype (Table II). Roux (1981) described another group of $V_H a^-$ molecules, w, which represent less than 1% of total Ig of normal rabbits. Although we assume that the amino acid sequences of the various $V_H a^-$ allotype Ig molecules differ from each other, the extent of these differences is not known.

The $V_H a^-$ molecules were studied most extensively in a2 allotype rabbits in which the $V_H a^-$ Ig molecules have the x32 or y33 allotypic specificities. To study the $V_H a^-$-encoding genes in a2 rabbits, Short *et al.* (1991) suppressed the production of a2 allotype Ig by injecting a2 rabbits at birth with anti-a2 antiserum. These a2-suppressed rabbits produced normal levels of serum Ig, but these Ig molecules were essentially all $V_H a^-$. The VDJ genes encoding these $V_H a^-$ Ig were amplified by polymerase chain reaction (PCR) from lymphoid tissue of the allotype-suppressed rabbits and cloned. On the basis of the similarity of the nucleotide sequences in the V_H region, these VDJ genes were divided into two groups, **x** and **y**, with each group having an approximately equal number of clones. The **y** group of the VDJ genes had V regions similar to a non-$V_H 1$-utilizing VDJ gene that was cloned from a B cell leukemia and was shown by *in vitro* expression studies to encode the y33 allotype. It was concluded that the VDJ genes in this group encoded y33 molecules. Short *et al.* (1991) also cloned a germline $V_H y33$ gene from an a2 genomic phage library and showed that it is at least 50 kb 5′ of $V_H 1$ (Table I). The V_H genes in the other group, group **x**, are also similar to each other but are different from $V_H 1$-a2 and $V_H y33$. Because the VDJ genes in this group were found at approximately the same frequency as those of the y33 group and because immunochemical analysis showed that x32 and y33 molecules occur at approximately equal levels (Kim and Dray, 1973), Short *et al.* (1991) suggested that these VDJ genes encoded x32 molecules and that their V regions represent a consensus sequence of the germline $V_H x32$. This idea needs to be tested

Table II. V_H and C_H allotypic specificities in rabbits of various haplotypes

Haplotype	V_H			C_μ		C_γ		C_α	
	a	x	y	Ms	n	d	e	f	g
A	1	–	–	16	81	12	15	73	74
B	1	–	33	17	80	12	15	71	75
C	1	–	33	17	80	11	15	72	74
I	1	–	33	17	80	12	14	69	77
J	1	–	–	16	81	12	15	70	76
C–G	1	–	33	17	80	12	15	71	75
E	2	32	33	17	80	12	15	71	75
F	2	32	33	17	80	12	15	69	77
F–I	2	32	33	17	80	12	14	69	77
M	2	32	33	16	81	12	15	73	74
P	2	32	33	16	81	12	14	69	77
G	3	32	–	17	80	12	15	71	75
H	3	32	–	16	81	11	15	72	74
H–F	3	32	–	17	80	12	15	69	77

by cloning the germline $V_H x32$ and by expressing the presumed x32-encoding VDJ genes *in vitro* and testing the reactivity of the expressed protein with anti-x32 antibody.

Recently, a third group of VDJ genes encoding $V_H a^-$ Ig has been described in a2 rabbits. Friedman *et al.* (1994) PCR-amplified VDJ genes from newborn a2 rabbits and found that, of 68 VDJ genes analyzed, two had V_H regions that were nearly identical to each other but were different from $V_H 1$, $V_H x32$ and $V_H y33$. These genes were designated $V_H z$. $V_H z$ presumably encodes $V_H a^-$ molecules because it does not encode any of the $V_H a2$ allotype-associated residues. We do not know which allotypic specificity, if any, is encoded by $V_H z$. Although neither germline $V_H x$ nor germline $V_H z$ genes have been cloned, we do know that they must be at least 50 kb 5' of $V_H 1$ because they are not represented in any of the 10 V_H genes nearest the 3' end on the a^2 allele. In total, the results from studies of $V_H a^-$-encoding VDJ genes in a2-suppressed rabbits indicate that $V_H a^-$ molecules are encoded by as few as three V_H genes.

The $V_H a^-$ molecules in Alicia rabbits have also been studied. Because Alicia rabbits are genetically a2, one would expect the $V_H a^-$ molecules in Alicia rabbits to be similar to the $V_H a^-$ molecules of normal a2 rabbits. This prediction was confirmed by data obtained by Chen *et al.* (1993), who analyzed the V_H gene usage in PCR-amplified VDJ genes from spleen of 2–8-week-old Alicia rabbits. They found that these $V_H a^-$-encoding VDJ genes were similar to the x and y groups previously identified in the a2-suppressed rabbits (Short *et al.*, 1991). At 2 weeks of age, Alicia rabbits utilized $V_H x32$ and $V_H y33$ at approximately the same frequency, but as the rabbits aged Chen *et al.* found that the Alicia rabbits preferentially used $V_H y33$ rather than $V_H x32$ and that, by 8 weeks of age, nearly all of the $V_H a^-$ molecules were derived from $V_H y33$. This finding is similar to the finding of DiPietro *et al.* (1990) who studied V_H gene usage in an 8-week-old Alicia rabbit and showed that six of seven VDJ genes cloned from a splenic cDNA library were y33-like molecules and none were x32-like molecules. These results indicated that, indeed, the $V_H a^-$ molecules in young Alicia rabbits are similar to those in normal a2 rabbits. However, unlike normal a2 rabbits, in which x32 molecules are found throughout life (Kim and Dray, 1973), Alicia rabbits seem to lose x32 allotype-encoding mRNA as they get older. No $V_H z$-utilizing VDJ genes were found in these studies.

To study the $V_H a^-$ molecules, x32 and y-, in a3 allotype rabbits (Table II), DiPietro *et al.* (1992) analyzed V_H gene usage in VDJ genes cloned from a $V_H a3$-allotype-suppressed rabbit. They injected a neonatal a^3/a^3 rabbit with anti-a3 antiserum and showed that, by 5 weeks of age, the rabbit produced mainly $V_H a^-$ molecules. The VDJ genes were reverse-transcribed, PCR-amplified and cloned from splenic mRNA. On the basis of nucleotide sequence analyses of the VDJ genes, DiPietro *et al.* (1992) concluded that, again, a limited number of V_H genes had been used. The VDJ genes could be arranged into three groups. One group corresponded to the consensus $V_H x32$, and these presumably encode the serum x32 Ig molecules. The V_H regions of a second group were very similar to $V_H 3$–$a3$, indicating that $V_H 3$–$a3$ or a very similar gene was used in the VDJ gene rearrangement. We obtained direct evidence that $V_H 3$

was used in at least some VDJ genes when we analyzed the region 5' of a VDJ gene cloned from a genomic phage library of the a3 allotype-suppressed rabbit and showed that it was identical to the region 5' of V_H3–$a3$ (K.L. Knight, unpublished data). The V regions of the third group of the VDJ genes were not similar to any of the known germline V_H genes from the a^3 allele. We do not know which allotypic specificity is encoded by either V_H3 or the unidentified germline V_H gene. We suspect that one or both of them may encode the so-called y⁻ molecules. In summary, as the $V_H a^-$ molecules in a2 rabbits seem to be encoded by three germline V_H genes, $V_H x32$, $V_H y33$ and $V_H z$, the $V_H a^-$ molecules in a3 rabbits also seem to be encoded primarily by three germline V_H genes—$V_H x32$, $V_H 3$ and an unidentified V_H gene.

V_H gene usage in the total VDJ gene repertoire

All of the studies described above specifically investigated V_H gene usage for either $V_H a$ or $V_H a^-$ molecules, but how these genes contribute to the total VDJ gene repertoire was not investigated directly. To study the complete VDJ gene repertoire in normal rabbits, Friedman *et al.* (1994) examined PCR-amplified VDJ genes cloned from bone marrow and spleen of 7–10-day-old a2 rabbits. Neonatal rabbits were used for this study because at this age the VDJ genes are undiversified (Short *et al.*, 1991), and the utilized V_H gene can be unequivocally identified. In a2 rabbits, 80% of serum Ig are a2, 5–10% are x32 and 5–10% are y33 (Table I). According to these numbers, most of the VDJ genes from a2 rabbits would utilize $V_H 1$, a small number would utilize $V_H x32$ or $V_H y33$, and $V_H z$ would be used infrequently. Friedman *et al.* (1994) found that, regardless of which tissue the VDJ genes were derived from, $V_H 1$ was used in 75% of the VDJ genes. Of the remaining ones, 7% used $V_H x32$, 7% used $V_H y33$ and 3% used $V_H z$. These data indicated that the VDJ gene repertoire in a2 rabbits used a total of four V_H genes: $V_H 1$, $V_H x32$, $V_H y33$ and $V_H z$.

The hierachy of V_H gene utilization, particularly the utilization of $V_H 1$ in approximately 80% of VDJ genes, can be explained either by preferential rearrangement or by selective expansion of B cells. To test these possibilities, it would be important to examine VDJ genes in B cell precursors that express no surface Ig and, therefore, would not be selected by antigen. If the preferential usage occurs by preferential rearrangement, then we would expect to find VDJ gene rearrangements that utilized one of a limited number of V_H genes. In fact, Friedman *et al.* (1994) analyzed VDJ genes in neonatal bone marrow, which presumably contained B cell precursors, and found only four V_H genes used. We think, therefore, that the preferential usage of these V_H genes is because of preferential rearrangement rather than selection by antigen. However, this finding of Friedman *et al.* (1994) needs to be confirmed in a purified pre-B or pro-B cell population.

Preferential rearrangement of V_H genes could be due to variations in the recombination signal sequences (RSS) (Ramsden and Wu, 1991). We think that differences in RSS do not explain the preferential rearrangement of $V_H 1$, because RSS are conserved among the rabbit germline V_H genes examined thus far and many of them have an RSS that is identical to that of V_H1. Another mechanism by which to explain preferential rearrangement of V_H1 is that V_H1 is readily accessible to the recombination machinery because it is the gene closest to the 3' end (Yancopoulos et al., 1984). However, the accessibility model does not fully explain the preferential use of V_H1, because Alicia rabbits do not rearrange the functional V_H gene nearest the 3' end, V_H4; instead they primarily rearrange V_Hx32 and V_Hy33, genes which are at least 50 kb further upstream. Preferential rearrangement could be due to other factors such as cis-regulatory elements and the chromosomal topology of the V_H locus.

D and J_H gene usage

The entire germline D locus was cloned from a cosmid library of an a3-allotype rabbit and was shown to span approximately 25 kb of DNA between V_H and J_H gene segments (Becker et al., 1990) (Fig. 1). Eleven D gene segments have been identified in this region (Becker et al., 1990; Friedman et al., 1994), and they are grouped into seven families—D1 to D7 (Fig. 4). Each D gene segment is not equally represented in the VDJ gene repertoire. For example, while D2a, D2b, D3 and D5 are frequently used in VDJ gene rearrangements, D1 is used much less frequently, and D6 has not yet been found to be used in a VDJ gene (Friedman et al., 1994). Upon analyzing VDJ genes, we found several in which the V regions were undiversified and the D regions did not correspond to any of the known germline D gene segments. Because some of these D regions were found in more than one VDJ clone, we think that they represent germline D gene segments that have not yet been identified.

One open reading frame (RF) of every D gene is rich in Gly and Tyr amino acid residues. In mouse, the RF of D gene segments that is rich in these two amino acids is preferentially utilized in VDJ genes (Kaartinen and Mäkelä, 1985; Ichihara et al.,

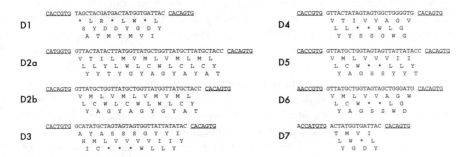

Fig. 4. D gene segments and the amino acid sequences encoded by each reading frame. The underlined sequences are the conserved heptamer sequence of the RSS. All three RFs are shown with RF1 on the top line and RF3 on the bottom line. The nucleotide sequences of D1a, D1b, D1c and D1d are identical and are indicated as D1.

1989). We examined the cloned VDJ genes to determine whether rabbits also used D gene segments in a preferred RF. Indeed, we found that each rabbit D gene segment is preferentially utilized in one RF, the RF rich in Gly and Tyr residues (Friedman et al., 1994; Knight and Crane, 1994). Currently, we do not know whether the molecular basis for this preferential utilization of the D RF is preferential rearrangement or selection by antigen. We examined a limited number of available DJ gene rearrangements and found that each RF of the D gene segments is used at approximately the same frequency. This result indicates that the preferential usage of the D RF is probably not due to preferential rearrangement but rather is due to selection, presumably antigen-driven selection.

We studied the germline J_H region in three rabbit haplotypes that have Ig of the a1, a2 and a3 allotypes (Becker et al., 1989; Knight et al., 1995). The J_H locus is contained within 2 kb of DNA located 63 kb downstream of V_H1, and the nucleotide sequence of the entire region was determined (Fig. 5). This region has greater than 95% similarity at the nucleotide sequence level and contains six J_H gene segments, J_H1 to J_H6, all of which appear functional, except for J_H1, which lacks the RSS. The J_H locus appears to have resulted from gene duplication during evolution (Becker et al., 1989), so it is not surprising that the coding regions of the six J_H gene segments are very similar to one another. This gene duplication must have occurred before the divergence of V_Ha allotypes, because the entire 2-kb J_H region is nearly identical between the C, E and G haplotypes. Despite the similarity between each J_H gene segment, J_H4 is used in more than 80% of the VDJ genes (Knight and Crane, 1994; Fitts and Mage, 1995). The basis for this preferential usage of J_H4, like the preferential usage of V_H and D genes, is not known.

Generation of antibody diversity

Mechanisms for generating antibody diversity include combinatorial joining, N-segment addition, somatic hypermutation and somatic gene conversion. Although rabbits use all of these mechanisms to diversify their VDJ gene repertoire, they use combinatorial joining to a very limited extent. Because of the limited number of V_H genes used in VDJ gene rearrangements, rabbits have a rather limited VDJ gene repertoire until they are a few weeks of age, at which time they undergo diversification by somatic gene conversion and somatic hypermutation. Before the VDJ genes somatically diversify, rabbits partially compensate for the limited VDJ repertoire by the addition of N segments, which are found in essentially all VDJ genes cloned from neonatal rabbits. Clearly, N-segment addition begins before birth.

The evidence that rabbits diversify their VDJ genes by somatic gene conversion was obtained when Becker and Knight (1990) analyzed the diversified V_H1-utilizing VDJ genes and found that much of the diversification occurred as clusters of nucleotide changes that often included codon insertion(s) and/or deletion(s). Such codon insertions and deletions cannot occur by simple point mutation but they can occur by replacement of the V_H region in the VDJ genes with a new V_H gene or by

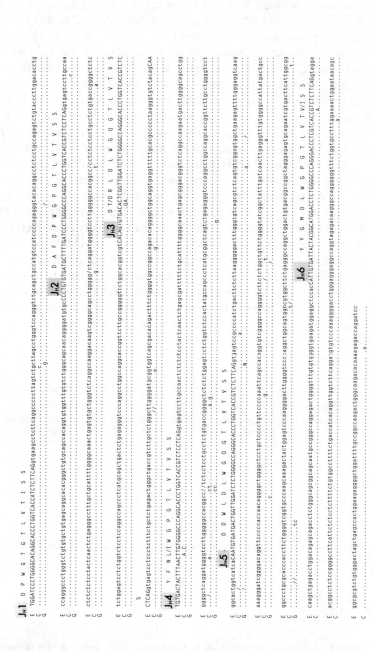

Fig. 5. Comparison of nucleotide sequences of the JH locus of E (a2), C (a1) and G (a3) haplotype rabbits.

somatic gene conversion using upstream V_H genes as donors. These authors showed that the diversification in these VDJ genes was not due to V gene replacement, because the DNA 5' of the diversified VDJ genes remained identical to the region 5' of V_H1. Rather, they searched for and found V_H genes 5' of V_H1, which had nucleotide sequences identical to those of the diversified region of the V_H1-utilizing VDJ genes and could have been used as donors in gene conversion events (Becker and Knight, 1990). Two striking examples of such diversification that extended over a 100-bp stretch of DNA from the leader intron through most of framework region 1 are shown in Fig. 6. In one of these examples, the nucleotide sequence of the upstream gene V_H8 was identical to the diversified region of a VDJ clone, 15–23, and the nucleotide sequence of V_H3 was nearly identical to the diversified region of the other VDJ clone, 30–10 (Raman et al., 1994). The data indicated that V_H8, V_H3 or V_H genes similar to them were used as donor genes to convert the sequences of the V_H1-utilizing VDJ genes. Several such examples of gene conversion-like events were found in each of two studies (Becker and Knight, 1990; Raman et al., 1994). We conclude that rabbit VDJ genes diversify extensively by a somatic gene conversion-like process. We cannot, however, rule out the possibility that the so-called gene conversion events are actually double recombination events between an upstream V_H gene and the VDJ gene. To distinguish between these possibilities, we need to identify the upstream donor gene and to determine whether it was unchanged, as would be expected for a gene conversion event, or whether it had taken on characteristics of the V_H gene used in the VDJ gene rearrangement, as would be expected for a recombination event.

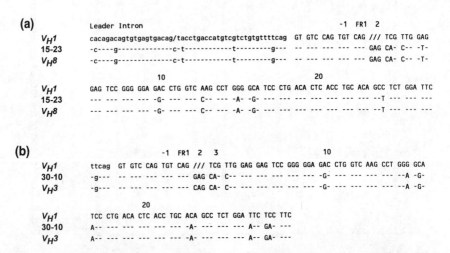

Fig. 6. Clustered nucleotide changes in V_H1-utilizing VDJ genes and nucleotide sequences of potential donor V_H genes. Codon numbers are shown: (a) clone 15–23, (b) clone 30–10. (From Raman et al., 1994.)

The D regions of VDJ genes seem to diversify by somatic hypermutation rather than by somatic gene conversion (Short et al., 1991). With regard to the diversification of D regions by gene conversion, no potential donor segments for the D region have been found in the germline. In particular, no VD-like genes that serve as donor sequences for the diversification in the D regions of chicken VDJ genes (Reynaud et al., 1989) have been found in rabbit germline DNA. However, Short et al. (1991) found that the D regions of VDJ genes start to diversify when the rabbits are 3–6 weeks of age, and by 9 weeks of age the D regions are highly diversified. Diversification of the D regions appears to be due to an accumulation of point mutations rather than to gene conversion. We think that somatic hypermutation probably also contributes to diversification of V regions in VDJ genes because it is unlikely that somatic hypermutation would be limited to the few nucleotides comprising the D region. To determine whether the V_H regions are diversified by somatic hypermutation, we can search for evidence of somatic mutation in DNA 5' and/or 3' of the VDJ gene.

Although most species use the same mechanisms to generate antibody diversity, the stage of development at which these mechanisms are used varies. For example, species such as mice and humans develop their pre-immune repertoire by combinatorial joining, junctional diversity and/or N-segment addition (Early et al., 1980; Tonegawa, 1983; Alt et al., 1987), whereas other species such as chickens, sheep and rabbits rely primarily on somatic diversification for development of their primary antibody repertoire (Table III). In the case of chickens and sheep, the primary antibody repertoire develops during embryonic life by somatic gene conversion and somatic hypermutation, respectively (Reynaud et al., 1987, 1989, 1991). Rabbits develop their primary antibody repertoire at a few weeks of age, probably by both somatic gene conversion and somatic mutation (Short et al., 1991; Knight and Crane, 1994; Weinstein et al., 1994). Before the primary antibody repertoire develops, however, neonatal rabbits compensate for their limited repertoire by extensive N-segment addition. In summary, the primary, or pre-immune, antibody repertoire develops somewhat differently in various species; therefore, it will be important to determine the mechanisms that regulate the ontogenic timing of processes such as N-segment

Table III. Pre-immune antibody repertoire in selected species

Species	Contribution by:		Age developed	Mechanism of somatic diversification
	Combinatorial* joining	Somatic diversification		
Rabbit	Limited†	Yes	4–6 weeks	Gene conversion and somatic hypermutation
Chicken	None	Yes	18-day gestation	Gene conversion
Sheep	Limited	Yes	Embryonic	Somatic hypermutation
Human	Extensive	No	Embryonic	N/A
Mouse	Extensive	No	Embryonic	N/A

* Only V_H gene usage (not D or J_H genes) is considered here.
† 3–4 V_H genes are used.

addition, preferential V_H gene usage and somatic diversification, all of which are involved in developing the antibody repertoire of various species.

C_H GENES

Ig isotypes and allotypes

Unlike V_H genes, which encode the region of the Ig molecules responsible for antigen binding, the C_H genes encode the region of the molecules responsible for the effector function. Like other mammals, rabbits have immunoglobulin classes IgM, IgG, IgA and IgE. However, unlike other mammals, rabbits have only one IgG isotype and they have as many as 13 IgA isotypes (Knight *et al.*, 1985; Burnett *et al.*, 1989). The C genes that encode each of the isotypes are closely linked to each other and to the V region genes (Becker *et al.*, 1989) (Fig. 1). The C_μ, C_γ and C_α chains are associated with allotypic specificities that are inherited as a group (Dray *et al.*, 1974) (Table II). The C_γ locus encodes two sets of allotypic specificities, **d** and **e** (Dubiski, 1969; Prahl *et al.*, 1969). The **d** allotypic specificities, d11 and d12, are inherited in an allelic fashion, and they correlate with a single amino acid difference in the hinge region (Kindt, 1975). The **e** allotypic specificities, e14 and e15, are also inherited in an allelic fashion, and they correlate with a single amino acid difference in the C_H3 domain (Kindt, 1975). The **d** and **e** allotypic specificities are inherited together in one of three haplotypes, $d^{11}e^{15}$, $d^{12}e^{14}$ and $d^{12}e^{15}$ (Table II).

For IgM, several allotypic specificities have been described, but because they were identified by different laboratories two different nomenclatures, **Ms** and **n**, are used (Table II) (Gilman-Sachs and Dray, 1972; Naessens *et al.*, 1978; Gilman-Sachs *et al.*, 1982). Although the structural correlates of these allotypic specificities are not yet known, it is reasonable to assume that the Ms and n anti-allotype antibodies identify similar epitopes.

For IgA, two sets of allotypic specificities, **f** and **g**, were described, and they appear to be controlled by two closely linked loci, $C_\alpha f$ and $C_\alpha g$ (Knight and Hanly, 1975) (Table II). The IgA-f and IgA-g allotypes are complex allotypes with several antigenic determinants that are distributed on both the Fd and Fc portions of the α-chains (Knight and Hanly, 1975). Muth *et al.* (1983) analyzed IgA-f molecules by quantitative immunoprecipitation techniques and showed that the IgA-f molecules comprised a mixture of IgA types. They concluded that rabbits have at least three, but likely more, subclasses of IgA. As we describe below, Burnett *et al.* (1989) found, on cloning the C_α genes, that the germline contains 13 C_α genes.

Rabbit IgE was first discovered by Lindqvist (1968) who immunized rabbits with tetanus toxoid and identified an antibody that was capable of eliciting an anaphylactic reaction. No allotypic variants of C_ε were described.

IgD has not been conclusively identified in rabbit even though experiments with rabbit B cells indicated that sIg other than IgM, IgA and IgG is present on most B cells. A search for the IgD class of Ig was undertaken by analyzing surface Ig of rabbit B lymphocytes. By membrane immunofluorescence, Eskinazi et al. (1979) showed that molecules with V_H and κ-chain allotypes were expressed on the surface of B cells even after IgM was removed from the surface. Similarly, Wilder et al. (1979) immunoprecipitated lysates of surface-labeled spleen cells with anti-V_Ha allotype antibodies, after the removal of IgM molecules, and showed that a heavy chain-like molecule was precipitated. In both experiments, the residual Ig-like molecules were neither IgA nor IgG. Taken together, the data indicated that rabbit B cells express surface IgD. However, as we describe below, a gene encoding C_δ has not been identified.

C_H gene organization and expression

C_γ, C_μ and C_ε genes

The rabbit C_H chromosomal region was cloned from cosmid and phage libraries and was shown to span a distance of over 200 kb (Knight et al., 1985; Burnett et al., 1989) (Fig. 1). The C_μ gene resides 8 kb downstream from the J_H gene segments; a single C_γ gene is found 55 kb 3' of C_μ; and a single C_ε gene resides 13 kb 3' of C_γ.

The genomic organization and nucleotide sequence of the germline C_γ gene was shown to be similar to that of other species, with separate exons encoding C_H1, hinge, C_H2 and C_H3 domains, each separated by 102–135 bp (Martens et al., 1984). The rabbit C_γ gene is approximately 70% similar to mouse C_γ genes and approximately 80% similar to human C_γ genes. The possibility that the germline contained two C_γ genes was proposed by Wolf et al. (1983), who reported finding so-called latent IgG molecules that have **d** and **e** allotypes other than those expected from the family pedigree. In 1980 we identified two C_γ-hybridizing fragments in germline DNA of some rabbits and suggested that one of these may encode the latent C_γ allotype (Martens et al., 1982). After further analysis, however, we showed that these two C_γ-hybridizing fragments represented polymorphic forms of a single germline C_γ gene (Knight et al., 1985). To date, we have no evidence to suggest that the haploid genome contains more than one C_γ gene.

The nucleotide sequence of C_μ was determined from a cDNA clone (Bernstein et al., 1984). It is approximately 70 and 75% similar to mouse and human C_μ genes, respectively. The intron/exon structure of the C_μ gene has not been studied; however, a switch region has been identified 1.5 kb 5' of C_μ. Further, the heavy chain enhancer region, E_μ, was found 4 kb 5' of C_μ. E_μ was cloned from the J_H–C_μ intron (Knight et al., 1985), and its nucleotide sequence was shown to be highly similar to that of mouse and human (Mage et al., 1989). Evidence that this region has enhancer activity *in vivo* was obtained from transgenic rabbits (Knight et al. 1988). In these experiments, E_μ was used to drive the tissue-specific expression of c-*myc* and the rabbits developed B cell leukemia. Interestingly, the fragment of E_μ used in the E_μ–*myc* transgene did not have the ATTTGCAT octamer sequence or its inversion, which are associated with enhancer activity of mouse E_μ (Staudt and Lenardo, 1991), indicating that the octamer motif is not required for the enhancer activity of rabbit E_μ.

C_α genes

Organization Serologic studies of IgA with anti-IgA-f and anti-IgA-g allotype antisera indicated that there are at least three C_α genes in the germline. To our surprise, Southern analysis with a rabbit C_α probe showed many more than three hybridizing fragments (Burnett *et al.*, 1989). We cloned the germline C_α genes from cosmid and phage genomic libraries and identified 13 non-allelic rabbit C_α genes, which indicated that extensive duplication of the C_α genes had occurred (Burnett *et al.*, 1989) (Fig. 1). Each C_α gene is separated by 8–18 kb. Altogether the C_α genes span at least 160 kb of the heavy chain chromosomal region. Each of the C_α genes is associated with a switch region that was identified by hybridization with a human S_μ probe. Although it appears that we have cloned all C_α genes identified by Southern analysis, not all of the C_α genes have been linked by overlapping cosmid or phage clones (Fig. 1). In attempts to clone the entire C_α chromosomal region, we have screened eight genomic libraries with C_α probes. However, we have not been able to clone all of the DNA between the 13 C_α genes, and we suggest that the DNA in this region is highly recombinogenic and is readily deleted during cloning.

The nucleotide sequences of all 13 C_α genes have been determined, and they are organized in a manner similar to that of the C_α genes in mouse and human, with separate exons for C_H1, hinge and C_H2, C_H3 and tail piece, and membrane regions (Burnett *et al.*, 1989). On the basis of nucleotide sequence analyses, each of the 13 C_α genes appears functional. The C_H2 and C_H3 domains of these genes are approximately 65–95% similar to each other, whereas the C_H1 domains and hinge regions are highly diverse (Burnett *et al.*, 1989).

In vitro *expression of IgA genes and functional studies of the IgA isotypes*

Because all 13 germline C_α genes appeared functional by nucleotide sequence analysis, Schneiderman *et al.*, (1989) tested whether all of them could be expressed *in vitro*. They cloned each of the C_α genes into a eukaryotic expression vector that contained a murine VDJ gene and transfected them into a murine plasmacytoma cell line. After analyzing supernatant from each transfectoma for the presence of IgA, they found that at least 12 of the C_α genes were expressed, and 11 of them reacted with anti-IgA-f antisera. Only one, IgA13 (derived from the $C_\alpha 13$ gene), reacted with anti-IgA-g antiserum. Thus, most IgA isotypes are of the IgA-f type, while only one, IgA13, is of the IgA-g type.

One possible explanation why rabbits have 13 C_α genes is that the different IgA isotypes may have different *in vivo* functions such as activation of complement or binding with polymeric Ig receptor. Using an *in vitro* complement fixation assay, Schneiderman *et al.* (1990) found that each of the IgA isotypes activated complement by the alternative pathway and with approximately the same efficiency. None of the

IgA isotypes activated complement through the classical pathway. To examine the association of the IgA isotypes with the polymeric Ig receptor, they cocultured the IgA-producing transfectomas with a Madin–Darby canine kidney cell line that expressed the rabbit polymeric Ig receptor, and they tested whether the IgA associated with polymeric Ig receptor then was transcytosed (Schneiderman et al., 1989). They found that each IgA isotype was transcytosed and that all of the IgA-f molecules covalently bound to secretory component by a disulfide linkage, whereas the IgA-g isotype (IgA13) bound non-covalently. This observation confirmed results of earlier immunochemical studies (Muth et al., 1983). The physiologic significance of the covalent versus the non-covalent association of secretory component to IgA is not known. Although we assume that the 13 different IgA isotypes are functionally distinct, these differences have not yet been identified. Two other functional properties that could be investigated are (1) whether organisms of the normal microbial flora of rabbits produce many different IgA proteases, and that, like the human IgA_1 and IgA_2 isotypes, the rabbit IgA isotypes differ in their sensitivity to these proteases (Knight et al., 1973; Plaut et al., 1975), and (2) whether the divergence in transmembrane and cytoplasmic regions of the 13 α-chains cause differences in B cell activation as a result of different cytoplasmic signals.

Differential in vivo expression of IgA isotypes   Although each of the 13 C_α genes can be expressed *in vitro*, it was important to determine whether they were differentially expressed *in vivo*. To examine the expression of each C_α gene in various lymphoid tissues, Spieker-Polet et al., (1993) performed RNase protection analysis by using probes specific for each of the 13 C_α genes. They found that at least 10 of the C_α genes are expressed in gut, appendix, mesenteric lymph node and mammary tissue (Fig. 7). It was surprising that only one C_α gene, $C_\alpha 4$, was expressed in lung and tonsil. Further, $C_\alpha 4$ was the only C_α gene expressed in some Peyer's patches. The mechanism by which C_α genes are differentially regulated is not yet known. One possibility is that each C_α gene is individually regulated by different tissue-specific factors. To address this, it would be useful to analyze the I-regulatory regions 5' of each C_α gene and to identify transcription factors that are associated with these regions. It is striking that the only C_α gene expressed in tonsil and lung is $C_\alpha 4$, the C_α gene nearest the 5' end that resides downstream of the C_α gene. Spieker-Polet et al. (1993) hypothesized that, during isotype switching to IgA, the B cells first switch to the $C_\alpha 4$ gene and subsequently switch to other C_α genes. The idea of sequential isotype switching can be explored by inducing isotype switching in activated B cells and examining the kinetics of IgA isotype synthesis. We do not yet know whether this isotype switching is developmentally regulated or is antigen driven.

IgA genes in other lagomorphs   Because the genomes of mouse and human have only one and two C_α genes, respectively, we wondered about the phylogenetic origin of duplicated C_α genes and whether the gene duplication events took place before or after diversification of lagomorphs. Burnett et al. (1989) obtained genomic

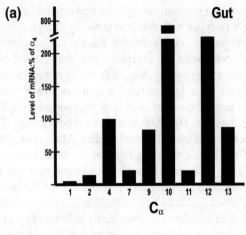
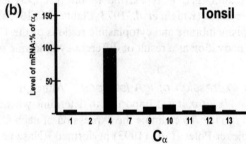
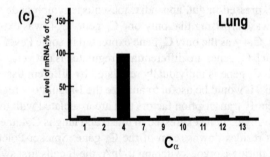

Fig. 7. RNase protection assay of nine C_α genes using $C_\alpha 1$, $C_\alpha 2$, $C_\alpha 4$, $C_\alpha 7$, $C_\alpha 9$, $C_\alpha 10$, $C_\alpha 11$, $C_\alpha 12$ and $C_\alpha 13$ probes, and RNA from (a) gut, (b) tonsil and (c) lung. The radioactivity of each sample was normalized to that of $C_\alpha 4$; the radioactivity of $C_\alpha 4$ was designated as 100%. (From Spieker-Polet et al., 1993.)

DNA from other members of the two families of lagomorphs, Leporidae and Ochotonidae, and performed Southern analysis using a rabbit C_α probe. From the Leporidae family, DNA was obtained from four genera, *Oryctolagus* (domestic rabbit), *Silvilagus* (cotton-tail rabbit), *Lepus* (jackrabbit) and *Pentalagus* (Amami rabbit); from the other family, Ochotonidae, DNA from the single living genus *O. princeps* (pika) was obtained. Multiple C_α hybridizing bands were observed in each sample of DNA (Fig. 8). This study indicated that all lagomorphs have multiple C_α genes and that the duplication of C_α genes occurred in an ancestor of all lagomorphs. Because lagomorphs are not clearly placed in the phylogenetic tree of mammals, we do not know the closest ancestor to lagomorphs and therefore cannot test where in evolution the duplication of C_α genes occurred. We also do not know whether the duplication of C_α genes confers increased survival advantage to lagomorphs or whether it was simply a genetic event with no selective advantage.

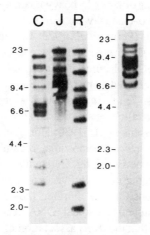

Fig. 8. Southern analysis of *Bam*HI-restricted lagomorph genomic DNA with a rabbit C_α probe. DNA samples: domestic rabbit (R), cotton-tail rabbit (C), jackrabbit (J) and pika (P). (From Burnett *et al.*, 1989.)

The C_δ puzzle

As we indicated above, investigators (Wilder *et al.*, 1979; Eskinazi *et al.*, 1979) have described an IgD-like molecule on the surface of B lymphocytes. However, a C_δ gene has not yet been found. We extensively searched the 55 kb of DNA between the C_μ and C_γ genes for hybridization with any one of several mouse and human C_δ probes (Cheng *et al.*, 1982; White *et al.*, 1985), including one for the most conserved exon between murine and human C_δ genes, the membrane exon. We especially focused on the region 8–15 kb 3′ of the C_μ gene, the region in which C_δ is found in the mouse and human genome (Cheng *et al.*, 1982; White *et al.*, 1985). We have not identified any segment of the DNA that specifically hybridized with C_δ probes, and we conclude that either the rabbit genome does not contain a C_δ gene or the rabbit C_δ gene has so little similarity with mouse and human C_δ genes that they did not cross-hybridize.

V_H and C_H recombinants

Trans-*association of V_H and C_H regions in Ig heavy chains* The presence of genetic markers in both the V_H and C_H regions of rabbit heavy chains allowed immunologists to examine the contribution that V_H and C_H genes make to the synthesis of individual heavy chains. By using rabbits heterozygous for V_Ha and C_γ allotypes, Landucci-Tosi et al. (1970) and Kindt et al. (1970) showed that the V_H and C_H genes were expressed in *cis*, such that most IgG heavy chains in heterozygous $a^1e^{14}/a^2 e^{15}$ rabbits were either a1e14 or a2e15. Similarly, Knight et al. (1974) showed that, in heterozygous a^1g^{74}/a^2g^{75} rabbits; most IgA-g molecules were a1g74 or a2g75. Unexpectedly, a small number of *trans* Ig heavy chains, such as a1e15 and a2e14 γ-chains, and a1g75 and a2g74 α-chains, were identified by analyzing Ig molecules as well as individual plasma cells (Pernis et al., 1973; Tosi and Tosi, 1973; Knight et al., 1974; Martens et al., 1981). The *trans* IgG heavy chains occurred at a frequency of approximately 1% of both IgG molecules and IgG-producing plasma cells, whereas the *trans* IgA molecules represented as much as 8% of both total IgA molecules and IgA-producing plasma cells. *Trans* Ig molecules were also identified in V_H/C_H heterozygous rabbits that were allotype suppressed at birth by injections with anti-V_H and/or anti-C_H antisera (Naessens et al., 1980; Horng et al., 1982). *Trans* Ig heavy chains could occur through *trans*-chromosomal recombination during VDJ gene rearrangement or during isotype switching. Gerstein et al. (1990) identified *trans* Ig molecules in VDJ–C_μ transgenic mice. In that case, they found the VDJ transgene associated with an endogenous C_γ gene. Further, they showed that such *trans* IgG molecules resulted from interchromosomal DNA recombination during isotype switching. It is also possible that the *trans* association of V_H and C_H occurs by *trans*-RNA splicing as has been proposed by Shimizu et al. (1991). In rabbits, because essentially no *trans* IgM heavy chains are found and *trans* IgA molecules are found in higher numbers than *trans* IgG molecules, we think that the *trans* Ig molecules in rabbits occur by interchromosomal DNA recombination during isotype switching rather than during VDJ gene recombination or *trans*-splicing.

Recombinant haplotypes Despite the fact that V_H and C_H loci are closely linked, investigators studying the inheritance of V_H and C_H allotypes in families of laboratory rabbits have identified 11 recombinant events that occurred within the heavy chain chromosomal region (Mage et al., 1982; Kelus and Steinberg, 1991). Most of the recombinations occurred between the genes encoding the V_Ha allotype, presumably V_H1, and either C_μ or C_γ. In an exhaustive study that took place over several decades, Kelus and Steinberg (1991) analyzed 6142 offspring from 1138 litters of informative crosses of rabbits. They found seven V_H/C_H recombinants and estimated that the frequency of recombination was 0.1%. This is a remarkably high recombination frequency, because the V_Ha-encoding gene, V_H1, is separated from C_μ and C_γ by only 75 kb and 130 kb, respectively. The molecular basis of this high frequency of recombination is not known but may be due to the presence of a recombination hotspot at this locus. Each of the recombination events took place in the male parent, and, for this reason, Kelus and Steinberg (1991) suggested that the recombinations took place in meiotic events during spermatogenesis. Further, they speculated that the enzymes involved in these recombinations may be expressed in gonial cells and be responsible for the high frequency of germline recombination observed between V_H and C_H genes.

CONCLUDING REMARKS

The field of immunoglobulin genetics began when Oudin described allotypic variants of rabbit immunoglobulins nearly four decades ago (Oudin, 1956). Since that time the immunoglobulin genes of mice, chickens, humans and rabbits have been studied extensively. Although the basic organization and structure of these genes are similar among the species, each species has its unique characteristics, many of which are described in various chapters in this book. For rabbit, these unique characteristics include:

1. the germline has only one C_γ gene but has 13 C_α genes;
2. the 13 C_α genes are differentially expressed in various lymphoid tissues;
3. as much as 80% of the VDJ gene repertoire derives from utilization of one V_H gene, V_H1, two to three D gene segments, and one J_H gene segment, J_H4;
4. the V_H genes are all approximately 80% similar and therefore are members of a single large V_H gene family; and
5. VDJ genes are extensively diversified by a somatic gene conversion-like mechanism.

From the results of studies on the organization and expression of rabbit Ig genes, several intriguing problems have developed that need to be addressed. These include (1) the biologic significance of 13 IgA isotypes as well as their differential expression in mucosal tissues, (2) the mechanism by which a small number of V_H, D and J_H gene segments are used in VDJ genes, (3) the mechanisms for somatic gene conversion of V_H regions of the VDJ genes and for somatic mutation of D regions of the VDJ genes, (4) the order of V_H, D and J_H gene rearrangements, and (5) identification and location of the germline V_H genes that encode the $V_H a^-$ molecules.

Another intriguing problem of the rabbit immune system that needs to be addressed is how B cells develop. We know that the neonatal VDJ repertoire is limited because at birth most VDJ gene rearrangements utilize one V_H gene, V_H1, and the VDJ gene rearrangements are undiversified. Beginning at about 1 month of age, the VDJ genes undergo somatic diversification and give rise to the primary, pre-immune antibody repertoire (Knight and Crane, 1994). By approximately 2 months of age, essentially all VDJ genes are diversified. Although it is clear that the neonatal antibody repertoire develops into the pre-immune antibody repertoire within a short period of time, we do not know how or where this process is initiated. Knight and

Crane (1994) proposed a model for B cell development in which nearly all B cells in rabbits develop early in ontogeny, migrate to the gut-associated lymphoid tissue (GALT) within the first few weeks of life and there are somatically diversified. Weinstein et al., 1994 have recently shown that indeed the VDJ genes can be diversified in GALT, and now we need to know if GALT is the primary site of antibody diversification. If so, it will be important to identify the signals that are responsible for the migration of B cells to GALT as well as those that initiate somatic diversification. In the model of Knight and Crane, the somatically diversified B cells are maintained by self-renewing and little or no B lymphopoiesis occurs in adult rabbits. Clearly, this idea needs to be tested experimentally. Whatever the mechanism for B cell development and generation of the antibody repertoire is, the rabbit provides us with another example of how various species perform similar functions in a different manner.

Acknowledgements

Supported by NIH grants AI16611 and AI11234.

REFERENCES

Adams, J.M., Harris, A.W., Pinkert, C.A., Corcoran, L.M., Alexander, W.S., Corex, S., Palmiter, R.D. and Brinster, R.L. (1985). *Nature* **318**, 533–538.
Allegrucci, M., Young-Cooper, G.O., Alexander, C.B., Newman, B.A. and Mage, R.G. (1991). *Eur. J. Immunol.* **21**, 411–417.
Alt, F.W., Blackwell, T.K. and Yancopoulos, G.D. (1987). *Science* **238**, 1079–1087.
Becker, R.S. and Knight, K.L. (1990). *Cell* **63**, 987–997.
Becker, R.S., Zhai, S.K., Currier, S.J. and K.L. Knight, (1989). *J. Immunol.* **142**, 1351–1355.
Becker, R.S., Suter, M. and Knight, K.L. (1990). *J. Immunol.* **20**, 397–402.
Bernstein, K.E., Alexander, C.B., Reddy, E.P. and Mage, R.G. (1984). *J. Immunol.* **132**, 490–495.
Bernstein, K.E., Alexander, C.B. and Mage, R.G. (1985). *J. Immunol.* **134**, 3480–3488.
Burnett, R.C., Hanly, W.C., Zhai, S.K. and Knight, K.L. (1989). *EMBO J.* **8**, 4041–4047.
Cebra, J.J., Colberg, J.E. and Dray, S. (1966). *J. Exp. Med.* **123**, 547–557.
Chen, H.T., Alexander, C.B., Young-Cooper, G.O. and Mage, R.G. (1993). *J. Immunol.* **150**, 2783–2793.
Cheng, H., Blattner, F.R., Fitzmaurice, L., Mushinski, J.F. and Tucker, P.W. (1982). *Nature* **296**, 410–415.
Currier, S.J., Gallarda, J.L. and Knight, K.L. (1988). *J. Immunol.* **140**, 1651–1659.
DiPietro, L.A., Short, J.A., Zhai, S.K., Kelus, A.S., Meier, D. and Knight, K.L. (1990). *Eur. J. Immunol.* **20**, 1401–1404.
DiPietro, L.A., Sethupathi, P., Kingzette, M., Zhai, S.K., Suter, M., and Knight, K.L. (1992). *J. Immunol.* **4**, 555–561.
Dray, S., Young, G.O. and Gerald, L. (1963a). *J. Immunol.* **91**, 403–415.
Dray, S., Young, G.O. and Nisonoff, A. (1963b). *Nature* **199**, 52–55.
Dray, S., Kim, B.S. and Gilman-Sachs, A. (1974). *Ann. Inst. Pasteur* **125**, 41–47.
Dreyer, W.J. and Bennett, J.C. (1965). *Proc. Natl Acad. Sci. USA* **54**, 864–869.
Dubiski, S. (1969). *J. Immunol.* **103**, 120–128.
Early, P., Huang, H., Davis, M., Calame, K. and Hood, L. (1980). *Cell* **19**, 981–992.

Eskinazi, D.P., Bessinger, B.A., McNicholas, J.M., Leary, A.L. and Knight, K.L. (1979). *J. Immunol.* **122**, 469–474.
Feinstein, A. (1963). *Nature* **199**, 1197–1199.
Fitts, M.G., and Mage, R.G. (1995) *Eur. J. Immunol.* **25**, 700–707.
Fitts, M.G. and Metzger, D.W. (1990). *J. Immunol.* **145**, 2713–2717.
Friedman, M.L., Tunyaplin, C., Zhai, S.K. and Knight, K.L. (1994). *J. Immunol.* **152**, 632–641.
Gallarda, J.L., Gleason, K.S. and Knight, K.L. (1985). *J. Immunol.* **135**, 4222–4228.
Gerstein, R.M., Frankel, W.N., Hsieh, C.-L., Durdik, J.M., Rath, S., Nisonoff, M.A. and Selsing, E. (1990). *Cell* **63**, 537–548.
Gilman-Sachs, A. and Dray, S. (1972). *Eur. J. Immunol.* **2**, 505–509.
Gilman-Sachs, A., Roux, K.H., Horng, W.J. and Dray, S. (1982). *J. Immunol.* **128**, 451–456.
Haouas, H., El Gaaied, A. and Cazenave, P.-A. (1989). *Res. Immunol.* **140**, 265–273.
Horng, W.J., Roux, K.H., Gilman-Sachs, A. and Dray, S. (1982). *Mol. Immunol.* **19**, 151–158.
Ichihara, Y., Hayashida, H., Miyazawa, S. and Kurosawa, Y. (1989). *Eur. J. Immunol.* **19**, 1849–1854.
Kaartinen, M. and Mäkelä, O. (1985). *Immunol. Today* **6**, 324–327.
Kelus, A.S. and Steinberg, C.M. (1991). *Immunogenetics* **33**, 255–259.
Kelus, A.S. and Weiss, S. (1986). *Proc. Natl Acad. Sci. USA* **83**, 4883–4886.
Kim, B.S. and Dray, S. (1972). *Eur. J. Immunol.* **2**, 509–514.
Kim, B.S. and Dray, S. (1973). *J. Immunol.* **111**, 750–760.
Kindt, T.J. (1975). *Adv. Immunol.* **21**, 35–86.
Kindt, T.J. and Mandy, W.J. (1972). *J. Immunol.* **108**, 1110–1113.
Kindt, T.J., Mandy, W.J. and Todd, C.W. (1970). *Biochemistry* **9**, 2028–2032.
Kirkham, P.M., Mortari, F., Newton, J.A. and Schroeder, H.W. Jr (1992). *EMBO J.* **11**, 603–609.
Knight, K.L. and Becker, R.S. (1990). *Cell* **60**, 963–970.
Knight, K.L. and Crane, M.A. (1994). *Adv. Immunol.* **56**, 179–218.
Knight, K.L. and Hanly, W.C. (1975). *Contemp. Top. Mol. Immunol.* **4**, 55–88.
Knight, K.L., Kingzette, M., Crane, M.A. and Zhai, S.-K. (1995). *J. Immunol.* **155**, 684–691.
Knight, K.L., Lichter, E.A. and Hanly, W.C. (1973). *Biochemistry* **12**, 3197–3203.
Knight, K.L., Malek, T.R. and Hanly, W.C. (1974). *Proc. Natl Acad. Sci. USA* **71**, 1169–1173.
Knight, K.L., Malek, T.R. and Dray, S. (1975). *Nature* **253**, 216–217.
Knight, K.L., Burnett, R.C. and McNicholas, J.M. (1985). *J. Immunol.* **134**, 1245–1250.
Knight, K.L., Spieker-Polet, H., Kazdin, D.S. and Oi, V.T. (1988). *Proc. Natl Acad. Sci. USA* **85**, 3130–3134.
Landucci-Tosi, S., Mage, R.G. and Dubiski, S. (1970). *J. Immunol.* **104**, 641–647.
Lennox, E.S. and Cohn, M. (1967). *Annu. Rev. Biochem.* **36**, 365–406.
Lindqvist, K.J. (1968). *Immunochemistry* **5**, 525–542.
Mage, R.G., Young-Cooper, G.O. and Alexander, C.B. (1971). *Nature New Biol.* **230**, 63–64.
Mage, R.G., Dray, S., Gilman-Sachs, A., Hamers-Casterman, C., Hamers, R., Hanly, W.C., Kindt, T.J., Knight, K.L., Mandy, W.J. and Naessems, J. (1982). *Immunogenetics* **15**, 287–297.
Mage, R.G., Bernstein, K.E., McCartney-Fransis, N., Alexander, C.B., Young-Cooper, G.O., Padlan, E.A. and Cohen, G.H. (1984). *Mol. Immunol.* **21**, 1067–1081.
Mage, R.G., Newman, B.A., Harindranath, N., Bernstein, K.E., Becker, R.S. and Knight, K.L. *Mol. Immunol.* **26**, 1007–1010.
Mandy, W.J. and Todd, C.W. (1970). *Biochem. Genet.* **4**, 59–71.
McCormack, W.T., Laster, S.M., Marzluff, W.F. and Roux, K.H. (1985). *Nucleic Acids Res.* **13**, 7041–7055.
Martens, C.L., Gilman-Sachs, A. and Knight, K.L. (1981). In 'The Immune System 1. Past and Future' (C.M. Steinburg and I. Lefkovits, eds), pp. 291–298. Karger, Basel.
Martens, C.L., Moore, K.W., Steinmetz, M., Hood, L. and Knight, K.L. (1982). *Proc. Natl Acad. Sci. USA* **79**, 6018–6022.
Martens, C.L., Currier, S.J. and Knight, K.L. (1984). *J. Immunol.* **133**, 1022–1027.
Muth, K.L., Hanly, W.C. and Knight, K.L. (1983). *Mol. Immunol.* **20**, 989–999.
Naessens, J., Hamers-Casterman, C., Hamers, R. and Okerman, F. (1978). *Immunogenetics* **6**, 17–27.

Naessens, J., Hamers-Casterman, C. and Hamers, R. (1980). *Eur. J. Immunol.* **10**, 776–781.
Oudin, J. (1956). *C.R. Acad Sci (D) (Paris)* **242**, 2489–2490.
Pernis, B., Chiappino, M.B., Kelus, A.S. and Gell, P.G.H. (1965). *J. Exp. Med.* **122**, 853–876.
Pernis, B., Forni, L., Dubiski, S., Kelus, A.S., Mandy, W.J. and Todd, C.W. (1973). *Immunochemistry* **10**, 281–285.
Plaut, A.G., Gilbert, J.V., Artenstein, M.S. and Capra, J.D. (1975). *Science* **190**, 1103–1105.
Prahl, J.W., Mandy, W.J. and Todd, C.W. (1969). *Biochemistry* **8**, 4935–4940.
Raman, C., Spieker-Polet, H., Yam, P.-C. and Knight, K.L. (1994). *J. Immunol.* **152**, 3935–3945.
Ramsden, D.A. and Wu, G.E. (1991). *Proc. Natl Acad. Sci. USA* **88**, 10721–10725.
Reynaud, C.-A., Anquez, V., Grimal, H. and Weill, J.-C. (1987). *Cell* **48**, 379–378.
Reynaud, C.-A., Dahan, A., Anquez, V. and Weill, J.-C. (1989). *Cell* **59**, 171–183.
Reynaud, C.-A., Mackay, C.R., Muller, R.G. and Weill, J.-C. (1991). *Cell* **64**, 995–1005.
Rosenshein, I.L. and Marchalonis, J.J. (1985). *Mol. Immunol.* **22**, 1177–1183.
Roux, K.H. (1981). *J. Immunol.* **127**, 626–632.
Roux, K.H., Dhanarajan, P., Gottschalk, P., McCormack, W.T. and Renshaw, R.W. (1991). *J. Immunol.* **146**, 2027–2036.
Schneiderman, R.D., Hanly, W.C. and Knight, K.L. (1989). *Proc. Natl Acad. Sci. USA* **86**, 7561–7565.
Schneiderman, R.D., Lint, T.F. and Knight, K.L. (1990). *J. Immunol.* **145**, 233–237.
Sethupathi, P., Spieker-Polet, H., Polet, H., Yam, P., Tumyaplin, C. and Knight, K. (1994). *Leukemia* **8**, 2144–2155.
Shimizu, A., Nussenzweig, M.C., Han, H., Sanchez, M. and Honjo, T. (1991). *J. Exp. Med.* **173**, 1385–1393.
Short, J.A., Sethupathi, P., Zhai, S.K. and Knight, K.L. (1991). *J. Immunol.* **147**, 4014–4018.
Spieker-Polet, H., Yam, P.-C. and Knight, K.L. (1993). *J. Immunol.* **150**, 5457–5465.
Staudt, L.M. and Lenardo, M.J. (1991). *Annu. Rev. Immunol.* **9**, 373–398.
Todd, C.W. (1963). *Biochem. Biophys. Res. Commun.* **11**, 170–175.
Tonegawa, S. (1983). *Nature* **302**, 575–581.
Tosi, S.L. and Tosi, R.M. (1973). *Immunochemistry* **10**, 65–71.
Tutter, A. and Riblet, R. (1989). *Proc. Natl Acad. Sci. USA* **86**, 7460–7464.
Weinstein, P.D., Anderson, A.O. and Mage, R.G. (1994). *Immunity* **1**, 647–659.
White, M.B., Shen, A.L., Word, C.J., Tucker, P.W. and Blattner, F.R. (1985). *Science* **228**, 733–737.
Wilder, R.L., Yuen, C.C., Coyle, S.A. and Mage, R.G. (1979). *J. Immunol.* **122**, 464–468.
Wolf, B., Yee, F.J. and Ikeda, M.T. (1983). *J. Immunol.* **131**, 1860–1864.
Yancopoulos, G.D., Desiderio, S.V., Paskind, M., Kearney, J.F. and Baltimore, D. (1984). *Nature* **311**, 727–733.

14

The structure and organization of immunoglobulin genes in lower vertebrates

Jonathan P. Rast, Michele K. Anderson and Gary W. Litman

University of South Florida, All Children's Hospital, St Petersburg, Florida, USA

Introduction 315
Immunoglobulin heavy chains 317
Immunoglobulin light chains...327
Evolution of immunoglobulin variable region genes 332
Agnathans: the jawless vertebrates 333
Lower vertebrate immunoglobulin and the immunoglobulin gene superfamily........334
Immune system organs and tissues 335
Conclusions 337

INTRODUCTION

The focus of this chapter will be on the structure and organization of immunoglobulin (Ig) genes in non-mammalian vertebrate species, with emphasis on the cold-blooded vertebrates. Where possible, inferences will be drawn regarding the relationships of the properties of these genes to antibody function and the generation of diversity. In considering the origins of gene structure and organization in lower vertebrates, it is important to recognize the vast amounts of information regarding mammalian systems as well as the unique case of avian Ig genes, detailed elsewhere in this volume. The discussion of lower vertebrate Ig genes will proceed from the most divergent representative forms of jawed vertebrates, relative to mammals, through the various vertebrate radiations; immunity in jawless vertebrates will be discussed separately as will the relationships between Ig and T cell antigen receptors (TCRs).

The accumulated information regarding lower vertebrate Ig genes is extensive; however, many questions regarding the extended chromosomal organization of these genes and various aspects of antibody diversification are not yet resolved. While studies of lower vertebrate systems carried out to date have been highly informative, it is critical to recognize that many routine approaches that have facilitated our understanding of the generation of antibody diversity in mammals and avians are not applicable to the more phylogenetically distant species, e.g. cell lines, monoclonal antibodies for the detection of specific cell types and inbred species are not yet available. As will become apparent, the Ig gene systems in many vertebrate species are highly complex, consisting of hundreds of presumably independently functioning gene loci, complicating assignment of specific functions to individual gene sets. Despite these complications, it is possible to describe the nature of Ig gene organization and structure in representatives of the most significant vertebrate radiations with extant species and formulate general hypotheses as to the primary mechanisms that have governed their evolutionary diversification.

Ig genes are easily recognizable as light and heavy chains in all the jawed vertebrates and these diverged from one another before the divergence of the chondrichthyans and bony fish. Therefore, we will discuss the phylogeny of each separately. Until recently, Ig heavy chains have received the greatest attention, thus enabling a more detailed discussion of their evolution. Recent work on both chondrichthyan and osteichthyan light chain genes have shown that both their isotypic diversity and organizational plasticity are comparable to that of the heavy chains. The evolutionary relationships among heavy and light chain genes are unknown as there is little context in which to develop hypotheses (i.e. intermediates are presently unknown). Although comparison to TCRs may be invaluable, it is difficult to align these sequences with Igs and even among themselves with confidence that evolutionary relationships are reflected in the comparisons. It is notable that all variable region-containing proteins (TCRs and Igs) are made up of dimers in which one chain possesses diversity (D) elements and the other does not. It presently is unclear whether this represents a vertical evolutionary relationship between quaternary structure in these proteins, the ease of acquisition of additional recombining segments, and/or functional limits in the extent of diversity that can be carried by a single antigen receptor. Within Igs, light chain constant regions show the greatest sequence identity to the fourth exon of the μ-type heavy chain constant region, which may reflect the origin of Ig from a monovalent TCR-like molecule where duplication of internal chain domains (exons) within one chain resulted in its elongation into a heavy chain and a concurrent transition from a monovalent to divalent state.

Although the peptide chains that comprise the Igs and the genetic elements that constitute them are similar in all jawed vertebrates, the diversity of their specific sequences, chromosomal arrangement and the mechanisms in which they are employed to create diversity are extraordinarily variable phylogenetically. In a sense, this is fortuitous as one vertebrate group may serve as a natural experiment and help to explain function in another.

IMMUNOGLOBULIN HEAVY CHAINS

The *Heterodontus* prototype

Relative to the mammals, the most divergent extant jawed vertebrates are the chondrichthyans, which include the elasmobranchs (sharks, skates and rays) and the holocephalans (chimeras; see summary of vertebrate interrelationships below). As chondrichthyans comprise the most phylogenetically distant taxa to humans in which specific humoral immune responses to a variety of defined antigenic determinants have been characterized, these species have been the focus of significant efforts on our part (Litman *et al.*, 1985a). The most intensively studied chondrichthyan, *Heterodontus francisci* (horned shark), fails to exhibit affinity maturation and lacks fine specificity differences in the hapten-specific antibody responses among genetically unrelated animals (Mäkelä and Litman, 1980). This lack of affinity maturation led to the hypothesis that somatic mutation may not take place in the Ig loci of *Heterodontus* in the same manner observed in higher vertebrate immune responses. However, recent genetic evidence indicates that somatic mutation plays an important role in diversifying specific Ig genes (Hinds-Frey *et al.*, 1993), although it may not be associated with affinity maturation. One possible explanation for this apparent disparity is that the selection mechanisms that are responsible for affinity maturation in mammals are inefficient or absent in the chondrichthyans (Hinds-Frey *et al.*, 1993).

The single heavy chain type described in *Heterodontus* corresponds most closely in its physicochemical characteristics to mammalian IgM, and is designated as µ-type. Like the mammalian counterpart, *Heterodontus* IgM is expressed in its secreted form as a pentamer, but also is found in high concentration in the monomeric configuration. We originally detected V_H gene homologs in *Heterodontus* using heterologous cross-hybridization with a mouse V_H probe, which enabled us to subsequently define the genomic structure of the *Heterodontus* heavy chain loci. The *Heterodontus* µ-type heavy chain genes exist as closely linked $V_H-D_H1-D_H2-J_H-C_H$ clusters dispersed throughout the genome, at multiple loci and on different chromosomes (Amemiya and Litman, 1991; C.T. Amemiya and G.W. Litman, unpublished data). The V_H, D_H and J_H segments are separated by ~300 bp and the overall length of a cluster is estimated at ~18 kb. An organizational model is shown in Fig. 1A. The V_H, D_H and J_H elements share significant sequence identity to the corresponding mammalian elements (Litman *et al.*, 1990), e.g. V_H genes are ~60% identical at the DNA level. All of the V_H genes belong to a single gene family ($V_H I$) in which the individual members are ~90% related at the nucleotide level, with the exception of one cluster that contains the monotypic representative of a second V_H gene family ($V_H II$), which is only 60% related to $V_H I$. Efforts to detect the association of other V_H families by negative selection strategies employing C_H specific probes have been uniformly unsuccessful, as have been efforts to detect additional C_H families. The D_H and J_H elements are remarkably similar between different clusters (Litman et al., 1993a). The recombination signal sequences (RSSs) in the *Heterodontus* $V_H I$ gene cluster are arranged so that either $D_H 2$, or both $D_H 1$ and $D_H 2$ (but not only $D_H 1$) may participate in VDJ joining during recombination (Litman et al., 1993b). In the $V_H II$ gene cluster, the intervening sequence (including the RSSs) between $D_H 1$ and $D_H 2$ is inverted, permitting either $D_H 1$ or $D_H 1-D_H 2$ joining but not $D_H 2$ joining alone.

Using cluster-specific probing techniques, which permit the identification of transcription products of specific gene clusters, we have established that V_H rearrangement occurs only within and not between clusters, eliminating combinatorial diversity as a mechanism in the generation of diversity. The high level of homology between $V_H I$ family members in different clusters severely restricts this potential source of diversity. Comparisons of genomic parent clusters ($V_H I$ and $V_H II$) sequences to their corresponding cDNA sequences indicate that both extensive

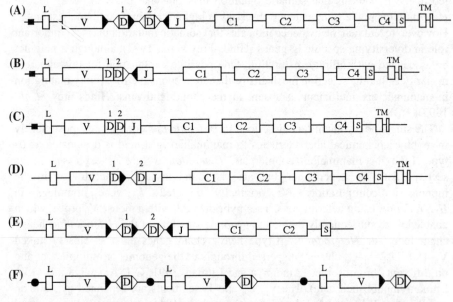

Fig.1. Non-mammalian Ig heavy chain gene organization. (A) Unrearranged V–D₁–D₂–J–C cluster arrangement found in *Hydrolagus*, *Heterodontus* and *Raja*. (B) Partially rearranged VDD–J–C cluster found in *Heterodontus* and *Raja*. (C) Fully rearranged VDD–C arrangement found in *Heterodontus* and *Hydrolagus*. (D) Partially joined VD–DJ–C cluster found in *Raja*. (E) IgX cluster found in *Raja*. (F) Tandem repeating V–D organization found in the living coelacanth *Latimeria chalumnae*. L, hydrophobic leader sequence; V, variable segment; D, diversity segment; J, joining segment; C1–C4, constant region exons; S, secretory exon; TM, transmembrane exons; ■, decamer–nonamer TCR CRE promoter element (found only in *Heterodontus*); ●, Ig octamer promoter element. RSS spacer lengths are indicated by triangles: ◁▷, 12 bp; ▶◀, 23 bp.

deletions and N (non-templated)-additions occur in these animals during the joining process (Hinds-Frey et al., 1993). The introduction of a joining site at D_H1/D_H2, also seen in TCRδ (Chien et al., 1987), provides an additional site for variation that may compensate in part for the reduced levels of V_H variability and limited V_H family diversity in the elasmobranchs (Hinds-Frey et al., 1993).

In terms of isotypic complexity all constant region genes are μ-type, consisting of six exons: C_H1, C_H2, C_H3, C_H4/SEC, TM1 and TM2 (Litman et al., 1993a) (Fig. 1A–D). There are significant degrees of sequence difference between the C_H exons identified in individual clusters, particularly in C_H1 (Kokubu et al., 1987). Although the entire J_H–C_H intron has been sequenced (M. Anderson, and G. Litman, unpublished observations), there is no evidence for the presence of isotype switching sequences associated with these clusters, consistent with the existence of only one heavy chain isotype (Kokubu et al., 1988a).

A second characteristic unique to the chondrichthyans is the presence of germline-joined Ig genes. Approximately half of the germline (non-lymphoid) heavy chain genes examined thus far are partially (VD–J) or fully (VDJ) joined, lacking intervening sequences between some or all of the V region elements, respectively (Litman et al., 1990; Fig. 1B, C). The V_H coding sequences of these genes closely resemble non-joined genes at the nucleotide and predicted peptide levels, with the exception of the D regions (Litman et al., 1993a), which exhibit distant relatedness to typical, unjoined D regions. These differences cannot be explained by the typical variation that accompanies V–D–J joining; instead it appears that these D segments either have sustained very high levels of regionalized somatic variation, were formed from a different type of D segment(s), or have diversified extensively by typical evolutionary mechanisms. The intact leader and J_H–C_H intervening sequences (see below), typical J_H–C_H splice sites (correct reading frame in VDJ-joined genes), upstream (5') putative regulatory sequences (see below), absence of internal termination codons and presence of additional 5' and 3' sequence identity with unjoined cluster-type genes suggests that these genes are not pseudogenes (Kokubu et al., 1988b). It is unlikely that these genes arise from B cell contamination in the tissues used to construct the genomic libraries as the incidence of such genes is high, and because identical germline-joined genes have been recovered from both liver and gonadal DNA of unrelated animals (Litman et al., 1993b). As will be described below, joined light chain genes also have been described.

Joined germline genes have several important implications.

1. The specificity of these genes (unlike the unjoined clusters) may be selected for entirely by conventional evolutionary processes.
2. The inherited specificities of VDJ-joined genes (or to a lesser extent VD–J joined genes), which are compromised in terms of junctional diversity, could be modified by somatic mutation and/or provide antigen-binding motifs serving specialized needs.
3. These genes may be substrates for gene conversion, or serve as substrates for secondary recombination events (Kokubu et al., 1988b).

Although the joined genes appear to be transcriptionally competent in *Heterodontus*, extensive searches for transcripts arising from joined heavy chain genes using a variety of direct screening and polymerase chain reaction (PCR)-based approaches have been unsuccessful. However, transcription of at least one completely joined heavy chain gene in the holocephalan, *Hydrolagus colliei*, has been demonstrated (see below).

Several characteristics of *Heterodontus* Ig gene loci are more reminiscent of mammalian TCR loci than mammalian Ig loci, supporting the hypothesis that the cartilaginous fishes possess Igs that retain characteristics of the putative Ig/TCR common ancestor. Shared properties include: (1) close proximity of D_H, J_H and C_H segments (TCRβ and γ), (2) closely linked V_H, J_H, and C_H (murine TCRγ), (3) two D_H segments closely linked to J_H (murine TCRδ), (4) absence of hyperconserved regulatory octamer in V_H (all TCR), (5) presence of decamer–nonamer cyclic AMP reactive element (CRE; TCRβ; Anderson *et al.*, 1989), (6) low affinity of antigen binding (Matsui *et al.*, 1991). However, it must be emphasized that the exon organization of the μ-type heavy chain, predicted amino acid sequence and capacity for somatic mutation(s) are distinctive Ig properties. The finding of TCR genes in the cartilaginous fish (see below) provides a basis for defining genetic mechanisms involved in the divergence of Ig and TCR systems.

Raja as a developmental model

Most of the characteristics of *Heterodontus* heavy chain loci that distinguish them from the osteichthyan heavy chain locus, including cluster organization, germline-joined genes and restriction(s) in V_H family diversity, extend to the other cartilaginous fishes, including the skate, *Raja*. *Raja* possesses an advantage over *Heterodontus* in that staged embryos are available for developmental studies of Ig gene expression (Luer, 1989). In addition, the skates and rays provide an important group for assessing diversification of the chondrichthyan Ig gene system over an extended evolutionary time period as they diverged from the lineage leading to *Heterodontus* ~220 million years ago. Studies of the μ-type gene loci in the little-skate, *Raja erinacea*, show that these clusters consist of one V_H element, two D_H elements, a J_H element and six C_H constant region exons, including the two TM exons, paralleling the Ig exon arrangement in *Heterodontus* (Fig. 1A) (Amemiya and Litman, 1991). *Raja* V_H elements are related closely to *Heterodontus* V_H elements, with 78% nucleotide identity in framework regions (FR) and 60% overall amino acid identity. *Raja* V_H elements, like those of the *Heterodontus*, exhibit a very high degree of nucleotide identity (~90%) between different clusters. The minimal variability of D_H segments in *Raja* is similar to that observed in *Heterodontus*, but *Raja* J_H segments are considerably more variable (Harding *et al.*, 1990a). Preliminary studies indicate that the μ-type heavy chain clusters are present at different chromosomal locations in *Raja* (Anderson *et al.*, 1994) as well as in *Heterodontus* (Litman *et al.*, 1993a). In *Raja*, germline joining occurs in this locus in ~50% of the clones characterized, although the completely joined VDJ form has not been detected (Harding et al., 1990a). The most common pattern of joining in Raja is $V_H D_H 1 - D_H 2 J_H$ (Fig. 1D) (Litman et al., 1993a).

As indicated above, analyses of *Heterodontus* Ig genes have established that only one constant region type is found in association with the V_H type described in this shark species. Other unrelated heavy chains may be present, but there is currently no DNA or protein evidence to suggest this in *Heterodontus*. However, both protein and nucleic acid analyses have identified a second heavy chain type, IgX, in *Raja*. There is also evidence from protein studies for a second H chain isotype in the frill shark (*Chlamydoselachus anguineus*), a member of a basal elasmobranch lineage (Kobayashi et al., 1992), suggesting that multiple H chain types may occur throughout the elasmobranchs. The *Raja* V_x elements are ~60% related in nucleotide sequence to the V_H elements of *Raja* IgM (Harding et al., 1990b). Comprehensive analyses of both nucleotide and predicted peptide structure suggest that the IgX heavy chain gene is unique to the Chondrichthyes and unrelated to isotypes found in higher vertebrates. The IgX heavy chain loci are in the cluster configuration consisting of one V_x, two D_x ($D_x 1$ and $D_x 2$), one J_x element and two C_x exons ($C_x 1$ and $C_x 2$) and a cysteine-rich exon that may correspond to a secretory (SEC) exon (Fig. 1E) (Anderson et al., 1994). A putative promoter sequence (TATTTAAA) is located 18 bp 5' of the start codon. The octamer (ATGCAAAT), which is found invariably 50–70 bp upstream of all higher vertebrate Ig V_H element transcriptional start sites (Parslow et al., 1984; Atchison et al., 1990; Litman et al., 1993a), does not occur within 1500 bp upstream of the IgX start codon. The V_H region elements (V_x, $D_x 1$, $D_x 2$ and J_x) of the IgX heavy chain locus are flanked by typical RSSs. The chromosomal distribution of Ig genes, as revealed by fluorescence *in situ* hybridization (FISH), suggests that, as with IgM, there are several dispersed IgX loci in the genome of *Raja* and that the clusters for the two isotypes are not contiguous (Anderson et al., 1994).

The presence of two Ig isotypes, both in multiple cluster form and on different chromosomes, raises questions about the regulation of Ig gene expression and the nature of allelic/isotypic exclusion in this species. IgX and IgM are co-expressed on individual embryonic spleen cells in the skate *Bathyraja aleutica* (Aleutian skate), but not on adult spleen cells, indicating that these loci are able to escape isotypic exclusion during at least one stage of development (Kobayashi et al., 1985). As indicated elsewhere in this book, isotypic exclusion in mammals involves a combination of isotype switching and alternative mRNA processing, which is dependent on a linear-arrangement of different C_H isotype exons. The lack of DNA sequences resembling higher vertebrate isotype switch sequences in the *Raja* $J_H - C_H 1$ intervening sequence and the failure to detect both IgX heavy chain and µ-type cluster elements at the same chromosomal loci is consistent with the absence of conventional class switching between IgM and IgX (Harding et al., 1990b). *Trans*-acting factors may inhibit rearrangement of other clusters in a manner analogous to allelic exclusion, perhaps in a developmental stage-specific manner.

In order to identify potential *cis*-acting regulatory elements, the $J_x - C_x 1$ intervening sequence (IVS) of one genomic clone was subjected to rigorous sequence comparison searches using a panel of Ig- and TCR-related transcriptional control sequence motifs. The results indicate that there are many regions that show high identity with sequences known to be important in transcriptional control in both mammalian Ig and TCR gene loci. Sequences related to the mammalian Ig J_H–C_H regulatory sequences that have been identified in the *Raja eglanteria* J_x–C_x IVS include (fraction of nucleotide identity is shown in parentheses) the octamer (ATGCAAAT) (7/8) (Landolfi *et al.*, 1986), the Ig heavy chain enhancer (9/10) (Calame and Eaton, 1988), B/µE1 (11/16) (Calame and Eaton, 1988), and IgHC.12 (8/8) (Ephrussi *et al.*, 1985). In addition, two copies of the inverse octamer (ATTTGCAT) are present. Identities also were found to several Ig light chain regulatory motifs, including 3'κF4 (11/15; 12/15) (Meyer *et al.*, 1990) and κE2 (8/9) (Murre *et al.*, 1989). Several TCR regulatory motifs also are present, including TCR-NFα4 (10/12; 9/12) (Winoto and Baltimore, 1989), TCR-NFα3 (11/14) (Winoto and Baltimore, 1989), TCRβ706 (9/12) (Krimpenfort *et al.*, 1988) and TCRγ3' (10/12) (Hsiang *et al.*, 1993). The arrangement of these sequence motifs differs considerably from the arrangement of regulatory elements in both the mammalian Ig J_H–C_H IVS and the *Heterodontus* IgM J_H–C_H IVS (M. Anderson, unpublished data). Although the presence of these sequences suggests the possibility of a functional role, there is no direct evidence that they regulate transcription of the IgX locus.

The control of Ig gene transcription in mammals is linked directly to rearrangement of the V_H elements, which brings the octamer promoter in close proximity to the enhancer elements present in the J_H–C_H1 IVS, resulting in transcriptional activation of the locus. Certain characteristics of the cartilaginous fish Ig loci differ from mammalian Ig loci, including cluster-type organization, multiple dispersed loci, absence of octamer and germline-joined genes, suggesting a different mechanism of transcriptional regulation. For example, germline-joined gene expression does not appear to be dependent on somatic V region rearrangement, since the V region is already in a rearranged configuration. It is possible, however, that germline-joined genes are under transcriptional regulation that is independent of the unjoined clusters. Further study is necessary to determine the role of isotypic exclusion in the expression of IgM and IgX, and its implications for specificity and autoimmunity.

Like mammalian Ig, *Raja* IgM gene loci generate either SEC or transmembrane (TM) forms of heavy chains by alternative splicing and polyadenylation (Harding *et al.*, 1990a). IgX gene loci, however, give rise to at least three different types of transcripts, none of which contain traditional TM exons (Anderson *et al.*, 1994). The putative SEC mRNAs ($V_xD_x1D_x2J_xC_x1C_x2SEC$) are the fully spliced transcription products of fully rearranged gene loci. There are other IgX cDNAs, however, that represent transcription products of loci which are not fully rearranged in the V_x region. These transcripts: (1) contain germline sequence surrounding the V_H elements, including intact RSSs, (2) are correctly J_x–C_x1 spliced, and (3) terminate in the C_x1–C_x2 IVS, without an apparent polyadenylation signal sequence. One of these transcripts exhibits apparent elimination of the D_x2 element as well as the RSS 5' of D_x2 and most of the RSS 3' of D_x2. This transcript may arise from an IgX heavy chain cluster lacking a D_x2 element, or may be the result of an aberrant joining event, possibly an open–shut joint that has been subjected to exonuclease activity. A third class of IgX mRNAs contain V_x regions from either rearranged or incompletely joined loci, correct $J_x-C_x1-C_x2$ splicing and ~1250 bp of 3′ sequence, downstream of C_x2, which does not include SEC. This 'tail' does not contain traditional TM exons, but may instead consist of additional Ig-type domains that have not yet been defined at the genomic level (M. Anderson, unpublished data). Because Ig transcripts containing the SEC exon instead of the TM exon(s) are not expressed in high abundance until activated B cells differentiate into plasma cells (Desiderio, 1993), the absence of SEC in the incompletely joined transcripts of genes may indicate a relatively early stage of B-cell development.

Hydrolagus, an independent phylogenetic lineage of the cartilaginous fish

The holocephalans, or chimeras and ratfishes, are a group of predominantly deep-sea chondrichthyans that are thought to form an outgroup to the extant elasmobranchs (Maisey, 1984) and to have diverged from them about 350 million years ago. The heavy chain gene arrangement in the spotted ratfish, *Hydrolagus colliei*, is generally of the elasmobranch cluster type. Thus far three types of clusters have been identified. The unjoined $V_H-D_1-D_2-J_H$ type is the most numerous (about 70 copies per genome estimated from genomic library screening; Fig. 1A). This gene type is closely related to the most common cDNAs identified by spleen library screening. Although each cDNA of this type has a unique constant region, no λ genomic clones of this type are linked to a constant region exon, consistent with a greater linkage distance between V_H and C_H than in *Heterodontus* or *Raja*.

A monotypic VDJ-joined gene (Fig. 1C) also has been identified in which the variable region and first exon of the constant region are only 80% and 75% related, respectively, to the $V_H-D_1-D_2-J_H$-unjoined type. Linkage distances are equivalent to those found in *Heterodontus* heavy chain clusters. Transcripts of both the TM and SEC forms of this gene have been identified using reverse transcriptase PCR (RT–PCR) with both second complementarity determining region (CDR2) and 5′ untranslated region specific primers (J. Rast and C. Amemiya, unpublished data).

A third cluster type has been identified in cosmid clones. V_H regions in this cluster type form a family that is about 85% related in nucleotide sequence to those of the first type. A C_H1-type exon is missing in these clusters and is replaced by a second C_H2-like exon, raising questions about heavy and light chain interactions. Although transcripts from this cluster type have not been identified by library screening, they have been detected using RT-PCR of spleen mRNA. As with *Heterodontus* and *Raja*, V_H genes in *Hydrolagus* lack a 5′ regulatory octamer. *Hydrolagus* genomic library screening indicates that ~10% of V_H^+ clones also are J_H^+, a number approximately equivalent to the number of C_H4^+ plaques, accounting for all of the cluster-type genes described above. Sequencing and PCR analyses have demonstrated that the remaining 90% of V_H^+ genomic clones represent truncated V_H pseudogenes, which are nearly identical to each other and the V_H element of the monotypic, joined heavy chain cluster. Whether these genes play any role in the generation of variable region diversity is unknown, but their near identity makes this possibility unlikely. The presence of a cluster-type gene arrangement in a holocephalan extends this organization to the entire vertebrate class Chondrichthyes and in both *Raja* and *Hydrolagus* extensive diversification of heavy chain gene clusters has taken place.

Teleosts: bony fishes

Ig heavy chain gene organization has been investigated in a number of teleost species and appears to be generally similar to that found in mammals. In *Elops saurus*, a member of a primitive teleost lineage, closely linked V_H genes and pseudogenes have been identified in individual λ genomic clones (Amemiya and Litman, 1990). V_H elements contain typical RSSs with 23-bp spacers and are associated with typical, mammalian-like upstream octamers. Similar findings are described for the V_H elements of the trout, *Oncorhynchus mykiss* (Matsunaga et al., 1990). In *Elops*, several J_H elements each having RSSs with 23-bp spacing are < 3.6 kb upstream of the first C_H exon. Studies of the channel catfish (*Ictalurus punctatus*) genomic J_H region revealed nine J_H elements in a 2.2 kb region that is ~1.8 kb upstream of C_H1. These J_H genes are similar to each other except in their 5' regions, where sequence variation can contribute to CDR3 diversity, unless it is eliminated during the joining process. Each catfish J_H has a typical RSS with 22–24 bp spacing (Hayman et al., 1993). Field inversion gel electrophoresis (FIGE) mapping in *Elops* indicates that V_H elements are found within ~100 kb of C_H. Only a single, μ-type C_H region isotype has been found in all teleost species thus far investigated with the exception of the Atlantic salmon, *Salmo salar*, in which two very similar (98% nucleic acid identity) C_μ type constant regions are present, consistent with the quasi-tetraploid genome of this species (Hordvik et al., 1992).

In *Elops saurus*, two V_H gene families have been identified that share about 50% nucleotide identity and more families may be present (Amemiya and Litman, 1990). Five V_H families have been identified in *Ictalurus punctatus* (Ghaffari and Lobb, 1991). Southern blot analyses using V_H family-specific probes derived from *Ictalurus* indicate that V_H region family diversity is present throughout the teleosts (Jones et al., 1993). DNA sequencing of isolated clones has established the existence of at least three V_H families in the pufferfish, *Spheroides nephelus* (M. Margittai and J. Rast, unpublished data), which possesses a relatively small genome. It is reasonable to conclude that extensive V_H family diversity is present throughout the teleosts.

The splicing of the TM exons to the third rather than the fourth C_H exon is a feature of IgM gene expression that appears to be unique to the teleosts. In all of the teleosts studied thus far, including channel catfish (Wilson et al., 1990), rainbow trout (Andersson and Matsunaga, 1993), Atlantic cod (Bengten et al., 1991) and Atlantic salmon (Hordvik et al., 1992), the typical consensus 5' splice site embedded in the vertebrate C_H4 exon is absent, consistent with the sequences of the TM cDNAs which demonstrate direct splicing of C_H3 to TM. Genomic Southern blot and gene titration experiments indicate that there is only one copy of the IgM-type heavy chain gene locus in teleost fish, suggesting that this type of transcript must result from alternative splicing events, rather than from the product of a related gene locus (Ghaffari and Lobb, 1989). Although C_H4 is absent from the TM, the transcript is longer than the SEC owing to the presence of a long 3' untranslated region (Bengten et al., 1991). The loss of the C_H4 exon in membrane-bound IgM does not appear to affect function, suggesting that C_H4 function is specific to the SEC form of IgM, possibly in the formation of complex covalent structures (Wilson et al., 1990).

Crossopterygians: *Latimeria*, the living coelacanth

Among the living sarcopterygians, V_H gene structure has been investigated most extensively in the coelacanth, *Latimeria chalumnae*, and closely linked V_H elements have been identified, as in *Elops*. A significant portion of the V_H gene population is comprised of pseudogenes, also reminiscent of those described originally in *Elops* (Amemiya and Litman, 1990). Each non-pseudogene V_H element is associated with a typical upstream octamer and RSS with a 23-bp spacer. However, in contrast to the typical osteichthyan arrangement, a D_H element is located ~190 bp downstream of each non-pseudogene V_H segment. The D_H segments are flanked by RSSs with 12-bp spacers (Fig. 1F). Obtaining further information on the heavy chain locus in this species has been problematic owing to the difficulty in procuring lymphoid tissue that is suitable for the extraction of mRNA. The atypical, somewhat chondrichthyan, arrangement of the V_H gene elements is consistent with a number of other chondrichthyan-like characters found in *Latimeria* (Amemiya et al., 1993).

Dipnoi: the lungfishes

Studies at the protein level have shown that the African lungfish (*Protopterus aethiopicus*), a sarcopterygian, possess three different C_H isotypes (Litman et al., 1971). cDNAs encoding two different isotypes have been sequenced (J. Rast, unpublished data) and the V_H region associated with each C_H is representative of a different family. Northern blot analyses of spleen mRNA indicate that each of these V_H family-specific probes and their respective constant region-specific probes hybridizes to different size bands. These preliminary findings contrast with observations in other tetrapods in which different V_H families associate with different constant region isotypes through class switching. This interpretation suggests that the heavy chain locus structure in this species may be complex. Germline arrangements of heavy chain genes in *Protopterus* presently are unknown, as such analyses are complicated by the extraordinarily large genome of this species (~37 times that of humans). This group is of great interest in terms of future study, owing to these preliminary observations as well as its unique phylogenetic position, high haploid cellular DNA content and evolutionary relationship to *Latimeria*.

Amphibians

Like the teleosts, amphibians possess the single extended locus form of Ig chromosomal organization, including multiple V_H gene families. However, the amphibians as well as Dipnoi and reptiles possess multiple C_H isotypes. The most extensively studied representative of this group, *Xenopus laevis* (South African clawed toad), an anuran, possesses three heavy chain isotypes: IgM (Schwager *et al.*, 1988), IgX (Haire *et al.*, 1989) and IgY (Amemiya *et al.*, 1989). At least two Ig isotypes that are homologs of *Xenopus* IgM and IgY (Fellah *et al.*, 1992, 1993) are found in a urodele, the Mexican axolotl (*Amblystoma mexicanum*). *Xenopus* IgM appears to be homologous to mammalian IgM, which is found throughout the radiations of jawed vertebrates; however, IgX and IgY may lack mammalian counterparts. The observations that IgM and IgY share common J_H sequences (Kokubu *et al.*, 1987) and the isolation of cDNAs containing μ- or υ-type constant regions (Wilson *et al.*, 1992) and identical V_H regions are consistent with a class switch occurring between these isotypes. However, the IgM to IgY class switch differs from mammalian class switching (which typically occurs after antigen stimulation) in that IgM levels remain high after the appearance of IgY. There is a weak secondary response to antigen in *Xenopus*, associated with the IgM to IgY class switch and limited affinity maturation, which is intermediate between chondrichthyan and mammalian levels.

The limitations in affinity maturation in *Xenopus* are not due to a paucity of genetic elements available for rearrangement. *Xenopus* possesses 11 highly diversified V_H families, at least 17 D_H elements and at least 10 different J_H elements, which may exceed the level of genomic complexity found in mammalian systems (Haire *et al.*, 1990). The overall relatedness of the V_H regions (FR1–FR3) belonging to different families ranges from 35 to 70% nucleotide identity. The *Xenopus* V_H family members, with some exceptions, appear to be interspersed throughout the V_H locus (Haire *et al.*, 1991). Some of these families do not contain certain amino acids that are highly conserved in FR1 and FR2 of human, mouse, *Heterodontus*, caiman, chicken and *Elops* V_H genes, further expanding the diversity of the *Xenopus* V_H repertoire (Haire *et al.*, 1990). In addition, *Xenopus* D_H elements can be fused, inverted or truncated, and are used in different reading frames. There is ample evidence that the number of elements available for combinatorial diversity, the level of junctional diversity and the rate of somatic mutation are comparable to that found in mammalian systems. One explanation for the limited affinity maturation of specificities may be the lack of highly developed germinal centres in these animals (Wilson *et al.*, 1992).

Reptilians

Studies of the caiman (*Caiman crocodylus*) also are consistent with an extended locus form of Ig gene organization. Caiman heavy chain gene coding segments show 65–70% nucleotide identity and 60–65% amino acid identity to mammalian V_H genes, most of which is in the framework regions (Litman et al., 1983). One of the caiman genes examined has several unusual features including an RSS located in the split leader intron (Litman et al., 1985b). The presence of the RSS in an atypical site in this gene suggests the possibility that at one point in evolution the RSS may have been mobile, which would have facilitated the duplication and divergence of the primordial exon into the currently existing families of rearranging genes, i.e. Igs and TCRs. In addition, this gene exhibits an RSS spacer that potentially could allow joining of J_H directly to V_H, bypassing D_H, providing a novel recombination mode. Notably, *in situ* hybridization using heavy chain probes in the snapping turtle, *Chelydra serpentina*, has demonstrated the presence of multiple, chromosomally dispersed loci (C.T. Amemiya and G.W. Litman, unpublished data).

Avians

While detailed elsewhere in this book, it is essential to describe the Ig gene system of birds, which possess an entirely unique form of Ig heavy and light chain gene organization. The chicken heavy chain locus consists of a 60–80 kb cluster of V_H pseudogenes and a single functional V_H gene, $V_H 1$. Several D_H elements and a single J_H element are located downstream. The pseudogenes lack RSSs, exhibit 5' truncation, lack leader sequences and are in different relative transcriptional polarity. Some of the V_H pseudogenes appear to be fused to closely related potential D_H regions, and some of these contain some putative J_H codons (Reynaud et al., 1989).

VDJ rearrangement involves only the $V_H 1$ gene, one of the nearly identical D_H elements and the single J_H gene, providing only limited combinatorial diversity and little apparent junctional diversity. Diversification is achieved primarily when gene conversion occurs between the rearranged allele and upstream pseudogenes, which donate blocks of sequence to $V_H 1$. This sequence transfer may be partially driven by the pseudogene homology to D_H and, in some cases, J_H. The use of different pseudogenes as donors is not completely random, with some preferred exchanges. Multiple exchange events result in a gradual increase in modifications with increasing cell divisions; the greatest degree of diversity accumulates in the D_H region (Reynaud et al., 1989). Three constant region isotypes, IgM, IgG (IgY) and IgA, occur downstream of J_H. The evolutionary relationship between these isotypes and those denoted similarly in other vertebrates is unclear, although the functional relationships appear equivalent.

IMMUNOGLOBULIN LIGHT CHAINS

In higher vertebrates, light chains may not participate in antigen binding to the same extent as heavy chains; however, their class diversity and organizational complexity exceed that of the heavy chains. For example, the presence of two loci (κ and λ) and the organizational variability of the λ locus within mammals surpasses heavy chain variation within the entire osteichthyan/tetrapod lineage. As additional light chain genes in lower vertebrates are investigated, it appears that complex structural and organizational variation exceed that found in mammals. Continuing analysis of patterns of light chain variation in lower vertebrates may shed important new light on the overall mechanisms of evolution of antigen-binding receptors.

Chondrichthyans

In the cartilaginous fishes, Ig light chain genes are arranged in clusters of closely linked V_L, J_L and C_L elements. At least three families of light chain clusters are present in the chondrichthyans (Fig. 2A, B). The type-I family has been identified in both *Heterodontus* and *Raja*. Type I genes exhibit high identity at the nucleotide level to mammalian TCR β chains. However, amino acid sequence and gross constant region structure unequivocally ally these with the Ig light chains (Shamblott and Litman, 1989a). Notably, when V_H and $C_H 4$ exon genes are used as outgroups in phylogenetic analyses, the elasmobranch type I light chains form a

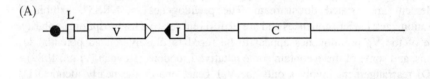

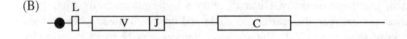

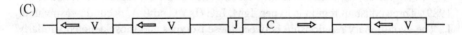

Fig. 2. Non-mammalian Ig light chain gene organizations. (A) Unjoined cluster found in *Heterodontus* type I, and nurse shark (*Ginglymostoma*) and possibly *Heterodontus* type III (κ) clusters. (B) Germline rearranged cluster arrangement of *Raja* type I clusters and all known chondrichthyan type II clusters. (C) Catfish (*Ictalurus*) light chain genes cluster organization. Arrows indicate transcriptional orientation (Ghaffari and Lobb, 1993). L, hydrophobic leader sequence; V, variable segment; J, joining segment; C, constant region exon; ●, Ig octamer promoter element, which may be present in catfish clusters. RSS spacer lengths are indicated by triangles:, ▷, 12 bp; ◀, 23 bp. Presumably RSSs are present in *Ictalurus* but have not been defined.

basal assemblage relative to all other vertebrate light chain types, including the other two known chondrichthyan light chains. There are at least 40 unique type I light chain clusters in the *Heterodontus* genome. The multigene, cluster-type arrangement of the type I light chain genes parallels the *Heterodontus* heavy chain genes. However, unlike shark heavy chain genes, the type I genes possess the invariant light chain octamer (ATTTGCAT) 5' of their V regions. No germline-joined type I genes have been detected in *Heterodontus* (Shamblott and Litman, 1989b). Thus far, homologs of the elasmobranch type I light chains have not been identified in *Hydrolagus*.

Type II light chain genes were described first in the sandbar shark (*Carcharhinus plumbeus*) (Hohman *et al.*, 1992). The discovery of homologous light chains in the holocephalan, *Hydrolagus colliei*, suggests that this light chain family is distributed throughout the cartilaginous fishes and, in fact, type II genes have also been characterized in *Heterodontus* and *Raja* (Rast *et al.*, 1994). Type II genes are organized similarly to type I genes, but differ significantly at the amino acid sequence level. By identity with peptide sequence determined for another species of holocephalan (De Ioannes and Aguila, 1989), they appear to be a major expressed light chain form in the holocephalans. All type II genes thus far characterized from the chondrichthyans are germline-joined and possess an upstream 5' regulatory octamer (Hohman *et al.*, 1993; Rast *et al.*, 1994).

A third light chain type (type III) was described originally in the nurse shark (*Ginglymostoma cirratum*) (Greenberg *et al.*, 1993) and later in *Heterodontus* (Rast *et al.*, 1994). This light chain is in a cluster organization similar to other shark light chains and the predicted amino acid sequence of its V region is most similar to the mammalian κ-type light chains. The relationship between the type III light chain and mammalian κ light chains also implies that a κ-type light chain was present in the ancestor of all extant jawed vertebrates and probably is widely distributed, although it has not been detected by cross-hybridization experiments in *Raja* or *Hydrolagus* (M. Anderson, unpublished data). These light chains appear to display independent evolution of amino acid sequence and gene organization as exemplified by the tandem arrangement of mammalian V_κ vs. the multiple cluster arrangement of the elasmobranch type III genes.

Two types of light chain genes have been detected in *Raja* that share homology with the type I and type II light chain genes of other cartilaginous fishes (Rast *et al.*, 1994). *Raja* type I and type II light chain genes, which both exist in cluster-type loci, show ~50% nucleotide identity. Extensive studies of the V_L–J_L regions in over 50 different clusters have established that all of the type I light chain genes are V_L–J_L germline joined (Anderson *et al.*, 1995), in contrast to *Heterodontus* in which there is no apparent V_L–J_L germline joining (Shamblott and Litman, 1989b). Furthermore, there is strong evidence that at least some of these germline-joined genes are expressed at the mRNA level, suggesting that these are physiologically relevant. The type II loci that have been characterized are germline joined as well (Rast *et al.*, 1994). The existence of potentially all of the light chain genes in a germline-joined form severely limits the genetic mechanisms available for generating diversity in the antigen-combining sites. However, there are additional indications that at least some of the germline-joined type I loci are targets of extensive somatic hypermutation. Mammalian Ig V region hypermutation is normally restricted to B cell Ig loci that have undergone complete V region rearrangement (Roes et al., 1989). It is possible, however, that in *Raja* germline-joined hypermutation is not subject to the same developmental restrictions, due to the prejoined state of the gene clusters. Therefore, the light chain clusters in *Raja* may represent a completely unique Ig gene system, in which the lack of combinatorial and junctional diversity achieved through gene reorganization in other systems may be compensated for through germline commitment of many distinct specificities, and the ability to modify this 'committed' repertoire in an efficient manner through hypermutation mechanisms.

V_L region divergence into distinct gene families (isotypes) has been well documented in mammals (Kabat et al., 1991) and amphibians (Schwager et al., 1991). An equivalent level of diversification is found in *Hydrolagus*, in which type II light chain cDNAs have been identified that share nearly identical constant regions but exhibit only ~70% nucleotide identity in variable regions (Rast et al., 1994). Light chain genomic clones that hybridize with a type II C_L probe hybridize with either of the two V_L family probes, but no clones hybridize with all three probes. Thus, different V_L types appear to be associated with the same C_L type, in characteristic cluster organization (i.e. at least one variable region upstream of a constant region). In addition, the 3' untranslated sequences associated with cDNAs encoding the two different V_L families are almost identical, suggesting that constant region identity is the result of recent duplications, unequal crossing over or gene conversion, and not merely purifying selection. This represents the only case where a chondrichthyan Ig gene cluster has variable segments representing different gene families in association with the same C_L type.

Teleosts

Genomic Southern blot and cDNA sequence data for the Ig light chains in two teleost species, the cod (*Gadus morhua*) and rainbow trout (*Oncorhynchus mykiss*), indicate multiple copies of both V_L and C_L region genes. Sequence analyses of light chain cDNAs and Northern blot analyses revealed several different types of aberrant light chain transcripts: (1) a rearranged and spliced cDNA(s) lacking the V_L sequence upstream of the CDR3, (2) an unspliced cDNA containing C_L preceded by what is presumably J_L–C_L intervening sequence, and (3) a J_L–C_L spliced cDNA containing some J_L–C_L intervening sequence (Daggfeldt et al., 1993). The presence of these transcripts is consistent with transcription patterns that are found in mammalian pre-B cells. Transcription of unrearranged Ig loci may be an essential component in promoting the accessibility of the locus to recombination machinery.

Genomic analyses of the catfish, *Ictalurus punctatus*, have shown multiple V_L regions upstream and downstream of single J_L and C_L elements (Fig. 2C). This pattern is found in multiple copies (>15) in the catfish genome. V_L elements are in opposite transcriptional orientation to J_L regions and presumably are rearranged by inversion rather than deletion. As in the cases described above, a number of aberrantly processed cDNAs were detected (Ghaffari and Lobb, 1993).

Amphibians

Xenopus possesses at least three types of light chains. The σ light chains associate preferentially with IgM heavy chains, and the ρ light chains associate with IgY heavy chains. This type of preferential association has not been described in any other species. The expression of both σ and ρ light chains in spleen, in the absence of high levels of IgY heavy chain mRNA, suggests that Ig heavy chain gene activity does not regulate Ig light chain expression in this system. The genes that encode these light chains are present at two separate loci, one comprised of V_σ and C_σ segments, the other comprised of V_ρ and C_ρ segments. The σ locus contains multiple $V_\sigma 1$ segments, a few $V_\sigma 2$ segments, and $C_\sigma 1$ and $C_\sigma 2$ segments. The V_σ elements are unusual in several regards and are comprised of 115 residues rather than the typical 110. In addition, the two cysteines that form the intradomain disulfide bond are separated by 73 residues rather than the usual 65. The otherwise invariant Trp36 and Trp47 are substituted by Ile or Leu. The two C_σ regions also differ by 26 nucleotides in the coding region, resulting in a nine amino acid substitution located outside the four-stranded β-pleated sheet that (in mammalian Ig) participates in C_L–C_H interactions. C_σ shares 29% residue identity with C_ρ and 29–33% identity with the C_L residues of shark, chicken or mammals, suggesting that these isotypes diverged early. However, the expressed V_σ elements are highly conserved, with little divergence in the CDR regions in outbred animals (Schwager *et al.*, 1991). If the V_ρ elements are similarly limited in diversity, the Ig light chains may be a limiting factor for antibody diversity in *Xenopus*. However, as the light chain appears to be less crucial than the heavy chain in direct antibody–antigen interactions, the lack of Ig V_L region diversity may not limit antigen-binding capabilities. A third light chain with sequence that is more similar to typical vertebrate light chains has been cloned. The existence of this light chain was anticipated in protein studies (Hsu *et al.*, 1991). This isotype appears to have multiple constant region genes (possibly in cluster form) that are associated with a number of variable region families (R. Haire, unpublished data).

Avians

The avian light chain locus is described in detail elsewhere in this book and therefore is described here in a limited manner, for comparison with other vertebrate light chains. Avians possess a single light chain locus, which shows greatest sequence identity to the mammalian λ-type gene. The chicken Ig light gene locus is arranged similarly to the heavy chain locus, as the light chain locus contains 25 V_L pseudogenes upstream of one functional V_L, one J_L and one C_L (Reynaud et al., 1987; Thompson and Nieman, 1987). Like the heavy chain genes, chicken light chain genes are diversified by gene conversion, but there also is evidence indicating that imprecise joining and somatic point mutations are involved in generation of diversity in this system. The avian light chain and heavy chain gene loci represent a clear case of parallel evolution where a unique diversity-generating mechanism acts at two loci that diverged before the mechanism became operative.

Isotypic classification of light chains

Traditionally lower vertebrate light chains have been categorized into grouping determined originally in mammalian systems. While κ and λ classifications are potentially valid, it must be realized that some light chain types may have diverged prior to the divergence of these mammalian gene classes. Phylogenetic analyses using a wide variety of vertebrate light chain sequences (Rast et al. 1994) show that while 'V_κ' genes from a wide variety of vertebrates cluster into a seemingly meaningful evolutionary group, variable regions of 'λ' genes and the constant regions of the κ-like genes of elasmobranchs tend to form a number of separate clusters that cannot be classified as either κ or λ relative to mammalian prototypes.

EVOLUTION OF IMMUNOGLOBULIN VARIABLE REGION GENES

The evolution of multigene families is often characterized by non-Mendelian processes collectively termed molecular drive (Dover, 1982). These processes include unequal crossing over and gene conversion, both of which tend to homogenize the members of gene families, although in special cases they may enhance diversity. The variable region genes in the mammalian V_H locus appear to undergo homogenizing processes at a very low rate, if at all, compared to other gene families such as ribosomal RNA genes. In fact, evolution at the CDR regions of these genes appears to proceed in the absence of purifying selection at a rate among the highest for proteins (Gojobori and Nei, 1984). Tanaka and Nei (1989) demonstrate that positive Darwinian selection occurs at the CDR regions of mammalian V_H genes and invoke a diversity-enhancing selection mechanism such as over-dominant selection as the major driving force in their evolution. Ota and Nei (1994) also found little evidence for gene conversion in mammalian V_H genes and described the evolution of these genes as a process of birth by gene duplication, divergence by diversifying selection, and death by dysfunctional mutation. Whether these same processes apply also to the cluster-type genes of the chondrichthyans is not yet known as comparisons of the divergence of orthologous (vertically related) gene pairs in closely related species have not been made. It is possible that because these species have an entirely different Ig gene organization, gene conversion may play a larger role in their evolution.

AGNATHANS: THE JAWLESS VERTEBRATES

This chapter has described in detail the remarkable conservation of recombining segmental elements that typifies Ig gene structure and organization in the extant jawed vertebrates. This group was derived at least 450 million years ago from the jawless vertebrates (Forey and Janvier, 1993), which formed a vast assemblage represented today only by hagfish and lampreys. Little is known about Igs and antibodies in the surviving jawless vertebrates. First it is essential to state that these two species are highly divergent forms, although they may share a common ancestor since their divergence from the ancestors of the jawed vertebrates (Stock and Whitt, 1992). Historically, two views have been presented as regards the nature of humoral immune recognition molecules in these species: (1) that the recognition molecules, such as found in the lamprey, were heterodimers resembling labile forms of 'modern-type' Ig (Litman et al., 1993a) and (2) that inducible, specific recognition is mediated by a very labile molecule of atypical (lacking typical heavy and light chains) structure. These possibilities are not yet resolved; however, it recently was shown that the hagfish molecule (which has been equated with the 'conventional' lamprey recognition protein) possesses a heterodimeric structure with very limited peptide identity to Ig. Molecular genetic studies show that the structural resemblance is coincidental and that the Ig-like molecule is a complement component homolog (Ishiguro et al., 1992). Extensive screening of both lamprey and hagfish genomic DNA and cDNA libraries with a wide variety of Ig gene probes and various degenerate PCR amplification strategies have failed as yet to yield products exhibiting a significant degree of sequence identity with the highly conserved segments of higher vertebrate Ig. These findings suggest that authentic homologs of Ig genes do not exist in these species or that they are highly divergent.

Recent studies have resolved considerable amounts of peptide sequence information for the putative lamprey inducible recognition molecule (anti-blood group 'O' specificity) and shown there to be little appreciable sequence identity with higher vertebrate Ig or TCR genes (A. Zilch, unpublished data). Inducible humoral recognition of foreign antigens in jawless vertebrates may involve genes that are only distantly related, if at all, to Ig.

As indicated elsewhere in this chapter, the recent development of a minimal degeneracy PCR amplification approach may facilitate elucidation of true Ig or TCR gene homologs. Based on the general assumption that cellular immunity precedes humoral immunity, it is possible that these species may possess TCR and not Ig or may possess a TCR-like gene that can be expressed in an extracellular form.

LOWER VERTEBRATE IMMUNOGLOBULIN AND THE IMMUNOGLOBULIN GENE SUPERFAMILY

As should be apparent from the foregoing discussion, Igs from all species are similar at the protein level, which to a certain degree limits what can be inferred about their evolution from comparison of gross structure. However, at the most fundamental level, the characteristic domain structure of Igs allies them, along with a wide variety of both functionally related and unrelated proteins, in an assembly termed the Ig gene superfamily (Williams and Barclay, 1988). The hallmark of this family, the 'immunoglobulin fold', is composed of seven β strands arranged into two β sheets and is stabilized by an intrachain disulfide bond. Many of the proteins that contain this fold are cell surface adhesion molecules and a number of these are involved in various facets of immune recognition and the nervous system. Ig superfamily domains can be classed as either variable (V) or constant (C) type by analogy to Ig V and C region domains. This division appears to represent an early split in this gene family where the V-type domains acquired an additional pair of β strands. One theory of the evolution of specific immune recognition is that a self-oriented cell surface receptor evolved into a non-self-directed cell surface receptor and ultimately acquired a diversification mechanism(s) that would result in directed specificity against a variety of molecules (Williams and Barclay, 1988). Because mammalian CD8 shares a number of similarities with Ig and TCR variable region genes and binds major histocompatibility complex (MHC) class I molecules, as do TCRs, it has been proposed that this protein may resemble the non-rearranging ancestor of both TCRs and Ig (Davis and Bjorkman, 1988).

Outside of the vertebrates, proposed members of the Ig supergene family have been isolated from a number of invertebrates, including both molluscs (Williams et al., 1988) and insects (Sun et al., 1990). A bacterial chaperon protein also appears to possess this characteristic fold (Holmgren et al., 1992). While it remains debatable whether many of these examples are actual homologs of the vertebrate Ig domains or represent convergent phenomena, it is probable that the protein domain that is the central feature of Ig structure is considerably older than the role of Igs and TCRs as inducible, genetically rearranging, antigen-recognition molecules. Detection of invertebrate analogs to Ig and their corresponding genes represent a formidable challenge and may require functional definition before structural characterization. However, recent studies in our laboratory have shown that gene amplification requiring only a pair of three to four amino acid identities within 200 bp may afford a means for isolating highly diverged variable region-type genes. This approach has proven highly successful in the isolation of lower vertebrate TCR homologs and may lead to the discovery of new rearranging variable region-containing gene families (possibly TCR–Ig intermediates) (Rast and Litman, 1994). In any event, clarification of the phylogeny of TCRs may itself lead to new insights into the origins of Igs.

IMMUNE SYSTEM ORGANS AND TISSUES

Based on the foregoing descriptions, it is apparent that a considerable body of knowledge regarding the Ig and TCR genes now exists for lower vertebrate species. However, in terms of fully understanding cellular mechanisms that generate and select recognition diversity, as well as the nature of alternative forms of antigen receptor genes, it is instructive to consider what is known regarding immune system organs and lymphopoiesis in these species. The phylogenetic distribution of the tissues and cells associated with B cell development and Ig production potentially can provide insight into the evolution of the immune system. Because these analyses are based mainly on histological considerations, where function is inferred by analogy to mammalian systems, care must be taken to avoid interpreting a lack of certain aspects of mammalian immunity in a divergent vertebrate group as equivalent to the absence of immune system complexity. Unique adaptations of the non-mammalian immune systems may only become evident when these systems are investigated using tools of similar sophistication as have been used in mammalian studies. These considerations have been reviewed extensively (Du Pasquier, 1989).

In the agnathans, immune system tissue is limited to that associated with the digestive system and is termed gut-associated lymphoid tissue (GALT), which is present in all vertebrates. The thymus and spleen of the jawed vertebrates can be viewed as specialized forms of GALT. The most detailed hematopoietic studies carried out on the agnathans have employed lampreys. In the ammocoete lamprey larvae, blood production begins at isolated blood islands, but later moves to typhlosole (a structure of the gut considered by some to represent a spleen homolog), and then to intertubular and fat regions of the nephric fold, which represent the primary regions of blood formation and lymphopoiesis throughout larval life. Involution of these tissues takes place at metamorphosis and hematopoiesis moves to the protovertebral arch in the adult (Percey and Potter, 1976). No definite thymus homolog has been described in either hagfish or lamprey.

A number of cell types that are morphologically similar to the immune cells of the jawed vertebrates have been described. Zapata *et al.* (1981) have reported cells that morphologically resemble plasma cells in the ammocoete larvae of *Petromyzon marinus*. In the hagfish, *Eptatretus stoutii*, Raison *et al.* (1987) describe two cell types that appear to be involved in mixed leukocyte reactions. As no gene homologs of the jawed vertebrate Ig or TCRs have been identified in the agnathans, it is difficult to move beyond speculation with regard to the relations, if any, of agnathan tissues and cells to those of the higher vertebrate immune system.

All jawed vertebrates possess a well-defined thymus and spleen. The elasmobranch thymus has been described from a number of species and is delineated by a connective tissue capsule. Typical lobate structure with cortical and medullary zonation is exhibited (Zapata, 1980). Similar thymic structure has been observed in a number of phylogenetically diverse elasmobranch species and distinct age-dependent involution appears to occur (C. Luer, personal communication). Thymic lymphocytes in the shark *Heterodontus japonicus* appear to lack surface Ig (Tomonaga et al., 1985).

As with mammals, the elasmobranch spleen contains both erythropoietic red pulp and lymphopoietic white pulp, which appears to be the major site of lymphopoiesis in these taxa. The splenic white pulp lymphatic tissue is organized around blood vessels in a fashion similar to that observed in the mammals and is rich in Ig-positive lymphocytes (Tomonaga et al., 1985). Plasma cells are apparent in elasmobranch splenic tissue but are rare in the peripheral circulation (Zapata, 1980).

In addition to the thymus and spleen, the elasmobranchs possess two unique organs that histologically appear to be associated with immune function: the Leydig's organ (specialized GALT) and the epigonal organs (Zapata, 1981; Mattison and Fänge, 1982). Both of these organs are similar in structure and contain granulocytes, lymphocytes and plasma cells. Some elasmobranch species lack either the epigonal or Leydig's organ, but at least one is always present. The holocephalan cartilaginous fishes (e.g. ratfish) lack these organs, but similar lymphomyeloid tissue can be found in the ocular orbit and the roof of the mouth. Major aggregations of lymphocytes are associated with the spiral valve of the elasmobranch intestine (Tomonaga et al., 1985).

In the bony fishes, the kidney provides both hematopoietic and lymphopoietic function. GALT is present and probably has important immunological function (Rombout et al., 1993). Both the thymus and spleen of the bony fishes are similar in structure to those found in mammals.

Lymphopoietic bone marrow is present only in the amniotes and some anuran amphibians. Interestingly, even among the anurans the use of bone marrow for hematopoiesis is both seasonally and phylogenetically variable.

While it is increasingly evident that all of the jawed vertebrates possess Ig, both classes of MHC and TCR genes, lymph nodes and germinal centers, which serve as sites for immune cell interactions, appear to be absent in all but the mammals and avians. The lack of significant affinity maturation in vertebrates, other than the mammals and avians, has been attributed to an absence of germinal centers (Wilson et al., 1992; Hinds-Frey et al., 1993). An analysis of reptilian lymphoid tissue may indicate whether these structures have arisen independently in these two lines or from the avian–mammalian common ancestor.

Inferring immune function from similarities among vertebrate tissues can be problematic; however, two points emerge.

1. The invariance of certain splenic and thymic characteristics in all jawed vertebrates indicates that the immune processes that take place within these structures, and at least some of their associated functions, were well established prior to the radiation of these taxa.

2. The ubiquity of GALT throughout the agnathans and jawed vertebrates and the possibility that the spleen and thymus of higher vertebrates is a derivation of this tissue may contain clues as to the origins of the vertebrate immune system.

At the same time, there is a certain flexibility with regard to the anatomical sites of lymphopoiesis. If this flexibility extends to sites of immune cell interactions, then some immune functions, analogous to those already well defined in the mammals, or other unique interactions, may go unrecognized with currently available methods. With the identification of the genes encoding Ig, TCR and MHC proteins in all the major jawed vertebrate groups, as well as the establishment of a method for detecting unique types of antigen receptor genes (Rast and Litman, 1994), it should be possible to enhance our understanding of the tissues and cellular interactions involved in the immune systems of these animals.

CONCLUSIONS

Although there is major variation in the organization of Ig gene loci between different taxa, the genetic elements and recombination mechanisms that promote the formation of an antigen-combining site capable of expressing multiple specificities are highly conserved. The widespread presence of the Ig fold domain, not only in the Igs, but in a diverse array of additional proteins, suggests that a common ancestral domain duplicated and subsequently diverged. The emergence of V- and C-type domains appears to predate the emergence of the recombining, antigen-binding proteins themselves. The evolution of the Ig gene system may have involved the segmentation of a primordial exon encoding a single V-type Ig domain into separate V and J elements to give a light chain type protein. This may have been followed by a second segmentation event resulting in D regions. This secondary segmentation may have occurred only once in the ancestor of both TCR β and δ chains and the Ig heavy chains or more than once in the lines leading to each. This initial event may have been mediated by a transposon containing an RSS-like sequence or by intragenic recombination; one outcome of such events would be a genetic system capable of rearrangement. The introduction of enzymatic deletion and non-templated nucleotide addition processes would regionally diversify the recombined gene. Further rounds of duplication and divergence ultimately resulted in the emergence of Ig heavy chain, Ig light chain and TCR gene systems, although their order of evolution is unclear. Recent findings of TCR gene homologs in cartilaginous fishes have the potential for resolving this longstanding issue. The acquisition of additional domains in the constant region of one member of a heterodimer could transform a monovalent TCR-type protein into a divalent, Y-shaped Ig-type protein. The emergence of a SEC mechanism could lead to a system of humoral immunity like that of the Igs either before or after the transition to a divalent state. Throughout the evolution of this system, processes that allow the selection of advantageous and the elimination of detrimental specificities also were evolving.

Studies of divergent vertebrate species demonstrate that there are at least four major Ig heavy chain gene organizational motifs: (1) the multiple cluster type in the Chondrichthyes, (2) the single, pseudogene block-linked locus type in the chicken, (3) the osteichthyan/tetrapod single extended locus type, and (4) the repeated V–D pattern of *Latimeria*. All appear to share homologous V and C region elements and highly conserved RSSs. Somatic mutation of the rearranged V region elements (including both point mutations and gene conversions) occurs throughout vertebrate phylogeny. Junctional diversity also occurs in all species studied, although the germline-joined genes in the chondrichthyans are not, or only partially, subject to junctional diversity and the chicken Ig loci appear to derive little of their diversity through this mechanism. While the study of light chain phylogeny lags behind that of the heavy chain, a complexity both in gene type and organizational diversity is beginning to emerge that may rival that of the heavy chain. Figure 3 summarizes the distribution of Ig-associated characteristics in the context of vertebrate phylogeny.

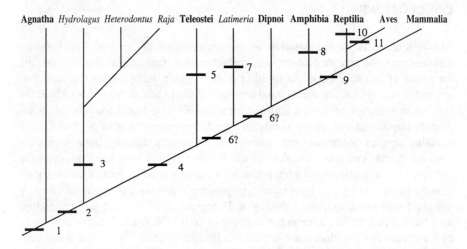

Fig. 3. Summary of vertebrate Ig-associated characteristics superimposed on a hypothetical vertebrate phylogeny. 1, Humoral immunity (?); 2, MHC class I and II proteins, rearranging TCR, μ-type IgH and IgL, somatic mutation of Ig V region genes, isotypic diversification of light chains and divergence of κ-type variable regions; 3, cluster-type Ig gene arrangement, some intercluster diversification and emergence of new constant region isotypes; 4, tandem IgH chain gene organization, V_H family diversification, significant affinity maturation; 5, transmembrane exons spliced to $C_H 3$; 6, constant region isotype diversification (class switching); 7, V–D–tandem heavy chain organization; 8, extensive V_H family diversification and interspersion; 9, enhanced affinity maturation; 10, multiple heavy chain loci dispersed on separate chromosomes; 11, single copy (limited copy number) V_H and V_L, pseudogene gene conversion diversification system. Placement of traits is inferred from the phylogenetic distribution of where they have been established and in some cases (e.g. MHC and TCR) may be present in more primitive taxa. The position of character 6 is ambiguous in the absence of a complete data set. Placement of the sarcopterygians (*Latimeria* and the Dipnoi) relative to the tetrapods is uncertain.

Unlike the other mechanisms that depend upon the most conserved characteristics of V region rearrangement, combinatorial diversity is the least universal mechanism because it is dependent on the chromosomal organization of the loci, which as stated before is highly variable. Recent findings regarding a TCR homolog in *Heterodontus* demonstrates that a highly diversified, multiple element V_T and J_T locus is probably associated with a single C_T region, establishing that combinatorial systems either were present prior to the emergence of the Igs and/or have evolved multiple times (as is probably the case with mammalian V_κ and V_H systems). These comparisons illustrate the various pathways by which the common goal of antibody diversity can be achieved.

This chapter has outlined the similarities and differences between Ig genes of species representing divergent points in vertebrate evolution. It is remarkable that the gene elements encoding Ig heavy and light chains, which have nearly parallel structures and functions, exhibit such varied chromosomal arrangements between different taxa. In a recent survey of the molecular sequence databases, a comparison of divergence between homologous mouse and human proteins revealed that those associated with the immune system evolve at a higher rate than other protein categories and it is suggested that this is in avoidance of molecular mimicry and subsequent subversion of the immune system by pathogenic organisms (Murphy, 1993). Thus, the immune system may evolve especially quickly because it is reacting to organisms that, unlike the typical factors to which living systems must adapt (e.g. changes in the physical environment), are able to counter-adapt in response. Knowledge of the molecular genetics of antigen-binding molecules such as Igs and TCRs from widely divergent species has important implications in understanding the overall nature and developmental regulation of adaptive immunity. Ongoing studies will attempt to define the earliest forms of Ig genes and examine the relationships between gene organization and the developmental regulation of the generation of antibody diversity.

REFERENCES

Amemiya, C.T., Haire, R.N. and Litman, G.W. (1989). *Nucleic Acids Res.* **17**, 5388.
Amemiya, C.T. and Litman, G.W. (1990). *Proc. Natl Acad. Sci. USA* **87**, 811–815.
Amemiya, C.T. and Litman, G.W. (1991). *Am. Zool.* **31**, 558–564.
Amemiya, C.T., Ohta, Y., Litman, R.T., Rast, J.P., Haire, R.N. and Litman, G.W. (1993). *Proc. Natl Acad. Sci. USA* **90**, 6661–6665.
Anderson, M.K., Amemiya, C.T., Luer, C.A., Litman, R.T., Rast, J.P., Niimura, Y. and Litman, G.W. (1994). *Int. Immunol.* **6**, 1661–1670.
Anderson, M.K., Shamblott, M.J., Litman, R.T. and Litman, G.W. (1995). *J. Exp. Med.* **182**, 109–119.
Anderson, S.J., Miyake, S. and Loh, D.Y. (1989). *Mol. Cell. Biol.* **9**, 4835–4845.
Andersson, E. and Matsunaga, T. (1993). *Immunogenetics* **38**, 243–250.
Atchison, M.L., Delmas, V. and Perry, R.P. (1990). *EMBO J.* **9**, 3109–3117.
Bengten, E., Leanderson, T. and Pilstrom, L. (1991). *Eur. J. Immunol.* **21**, 3027–3033.
Calame, K. and Eaton, S. (1988). *Adv. Immunol.* **43**, 235–275.
Chien, Y.-H., Iwashima, M., Wettstein, D.A., Kaplan, K.B., Elliott, J.F., Born, W. and Davis, M.M. (1987). *Nature* **330**, 722–727.

Daggfeldt, A., Bengten, E. and Pilström, L. (1993). *Immunogenetics* **38**, 199–209.
Davis, M.M. and Bjorkman, P.J. (1988). *Nature* **334**, 395–402.
De Ioannes, A.E. and Aguila, H.L. (1989). *Immunogenetics* **30**, 175–180.
Desiderio, S.V. (1993). In, 'Developmental Immunology' (E. Cooper and E. Nisbet-Brown, eds pp. 129–152. Oxford University Press, New York.
Dover, G. (1982). *Nature* **299**, 111–117.
Du Pasquier, L. (1989). In 'Fundamental Immunology' (W.E. Paul, ed.) pp. 139–168. Raven Press, New York.
Ephrussi, A., Church, G.M., Tonegawa, S. and Gilbert, W. (1985). *Science* **227**, 134–140.
Fellah, J.S., Wiles, M.V., Charlemagne, J. and Schwager, J. (1992). *Eur. J. Immunol.* **22**, 2595–2601.
Fellah, J.S., Kerfourn, F., Wiles, M.V., Schwager, J. and Charlemagne, J. (1993). *Immunogenetics* **38**, 311–317.
Forey, P. and Janvier, P. (1993). *Nature* **361**, 129–134.
Ghaffari, S.H. and Lobb, C.J. (1989). *J. Immunol.* **143**, 2730–2739.
Ghaffari, S.H. and Lobb, C.J. (1991). *J. Immunol.* **146**, 1037–1046.
Ghaffari, S.H. and Lobb, C.J. (1993). *J. Immunol.* **151**, 6900–6912.
Gojobori, T. and Nei, M. (1984). *Mol. Biol. Evol.* **1**, 195–212.
Greenberg, A.S., Steiner, L., Kasahara, M. and Flajnik, M.F. (1993). *Proc. Natl Acad. Sci. USA* **90**, 10603–10607.
Haire, R., Shamblott, M.J., Amemiya, C.T. and Litman, G.W. (1989). *Nucleic Acids Res.* **17**, 1776.
Haire, R.N., Amemiya, C.T., Suzuki, D. and Litman, G.W. (1990). *J. Exp. Med.* **171**, 1721–1737.
Haire, R.N., Ohta, Y., Litman, R.T., Amemiya, C.T. and Litman, G.W. (1991). *Nucleic Acids Res.* **19**, 3061–3066.
Harding, F.A., Cohen, N. and Litman, G.W. (1990a). *Nucleic Acids Res.* **18**, 1015–1020.
Harding, F.A., Amemiya, C.T., Litman, R.T., Cohen, N. and Litman, G.W. (1990b). *Nucleic Acids Res.* **18**, 6369–6376.
Hayman, J.R., Ghaffari, S.H. and Lobb, C.J. (1993). *J. Immunol.* **151**, 3587–3596.
Hinds-Frey, K.R., Nishikata, H., Litman, R.T. and Litman, G.W. (1993). *J. Exp. Med.* **178**, 825–834.
Hohman, V.S., Schluter, S.F. and Marchalonis, J.J. (1992). *Proc. Natl Acad. Sci. USA* **89**, 276–280.
Hohman, V.S., Schuchman, D.B., Schluter, S.F. and Marchalonis, J.J. (1993). *Proc. Natl Acad. Sci. USA* **90**, 9882–9886.
Holmgren, A., Kuehn, M.J., Bränden, C.-I. and Hultgren, S.J. (1992). *EMBO J.* **11**, 1617–1622.
Hordvik, I., Voie, A.M., Glette, J., Male, R. and Endresen, C. (1992). *Eur. J. Immunol.* **22**, 2957–2962.
Hsiang, Y.-H., Spencer, D., Wang, S., Speck, N.A. and Raulet, D.H. (1993). *J. Immunol.* **150**, 3905–3916.
Hsu, E., Lefkovits, I., Flajnik, M. and Du Pasquier, L. (1991). *Mol. Immunol.* **28**, 985–994.
Ishiguro, H., Kobayashi, K., Suzuki, M., Titani, K., Tomonaga, S. and Kurosawa, Y. (1992). *EMBO J.* **11**, 829–837.
Jones, J.C., Ghaffari, S.H. and Lobb, C.J. (1993). *J. Mol. Evol.* **36**, 417–428.
Kabat, E.A., Wu, T.T., Foeller, C., Perry, H.M. and Gottesman, K. (1991). 'Sequences of Proteins of Immunological Interest' US. Dept. Health and Human Services, Washington, DC.
Kobayashi, K., Tomonaga, S., Teshima, K. and Kajii, T. (1985). *Eur. J. Immunol.* **15**, 952–956.
Kobayashi, K., Tomonaga, S. and Tanaka, S. (1992). *Dev. Comp. Immunol.* **16**, 295–299.
Kokubu, F., Hinds, K., Litman, R., Shamblott, M.J. and Litman, G.W. (1987). *Proc. Natl Acad. Sci. USA* **84**, 5868–5872.
Kokubu, F., Hinds, K., Litman, R., Shamblott, M.J. and Litman, G.W. (1988a). *EMBO J.* **7**, 1979–1988.
Kokubu, F., Litman, R., Shamblott, M.J., Hinds, K. and Litman, G.W. (1988b). *EMBO J.* **7**, 3413–3422.
Krimpenfort, P., de Jong, R., Uematsu, Y., Dembic, Z., Ryser, S., von Boehmer, H., Steinmetz, M. and Berns, A. (1988). *EMBO J.* **7**, 745–750.
Landolfi, N.F., Capra, J.D. and Tucker, P.W. (1986). *Nature* **323**, 548–551.
Litman, G.W., Wang, A.C., Fudenberg, H.H. and Good, R.A. (1971). *Proc. Natl Acad. Sci. USA* **68**, 2321–2324.
Litman, G.W., Berger, L., Murphy, K., Litman, R., Hinds, K.R. and Erickson, B.W. (1983). *Nature* **303**, 349–352.

Litman, G.W., Berger, L., Murphy, K., Litman, R., Hinds, K.R. and Erickson, B.W. (1985a). *Proc. Natl Acad. Sci. USA* **82**, 2082–2086.
Litman, G.W., Murphy, K., Berger, L., Litman, R.T., Hinds, K.R. and Erickson, B.W. (1985b). *Proc. Natl Acad. Sci. USA* **82**, 844–848.
Litman, G.W., Amemiya, C.T., Haire, R.N. and Shamblott, M.J. (1990). *Bioscience* **40**, 751–757.
Litman, G.W., Rast, J.P., Shamblott, M.J., Haire, R.N., Hulst, M., Roess, W., Litman, R.T., Hinds-Frey, K.R., Zilch, A. and Amemiya, C.T. (1993a). *Mol. Biol. Evol.* **10**, 60–72.
Litman, G.W., Amemiya, C.T., Hinds-Frey, K.R., Litman, R.T., Kokubu, F., Suzuki, D., Shamblott, M.J., Harding, F.A. and Haire, R.N. (1993b). In 'Developmental Immunology' (E.L. Cooper and E. Nisbet-Brown, eds), pp. 108–128. Oxford University Press, New York.
Luer, C.A. (1989). In 'Nonmammalian Animal Models for Biomedical Research' (A.D. Woodhead, ed.), pp. 121–147. CRC Press, Boca Raton, FL.
Maisey, J.G. (1984). *J. Vert. Paleont.* **4**, 359–371.
Mäkelä, O. and Litman, G.W. (1980). *Nature* **287**, 639–640.
Matsui, K., Boniface, J.J., Reay, P.A., Schild, H., Fazekas de St. Groth, B. and Davis, M.M. (1991). *Science* **254**, 1788–1791.
Matsunaga, T., Chen, T. and Tormanen, V. (1990). *Proc. Natl Acad. Sci. USA* **87**, 7767–7771.
Mattison, A. and Fänge, R. (1982). *Biol. Bull.* **162**, 182–194.
Meyer, K.B., Sharpe, M.J., Surani, M.A. and Neuberger, M.S. (1990). *Nucleic Acids Res.* **18**, 5609–5615.
Murphy, P.M. (1993). *Cell* **72**, 823–826.
Murre, C., McCaw, P.S. and Baltimore, D. (1989). *Cell* **56**, 777–783.
Ota, T. and Nei, M. (1994). *Mol. Biol. Evol.* **11**, 469–482.
Parslow, T.G., Blair, D.L., Murphy, W.J. and Granner, D.K. (1984). *Proc. Natl Acad. Sci. USA* **81**, 2650–2654.
Percey, R. and Potter, I.C. (1976). *J. Zool. Lond.* **178**, 319–340.
Raison, R.L., Gilbertson, P. and Wotherspoon, J. (1987). *Immunol. Cell Biol.* **65**, 183–188.
Rast, J.P. and Litman, G.W. (1994). *Proc. Natl Acad. Sci. USA* **91**, 9248–9252.
Rast, J.P., Anderson, M.K., Ota, T., Litman, R.T., Margittai, M., Shamblott, M.J. and Litman, G.W. (1994). *Immunogenetics* **40**, 83–99.
Reynaud, C.-A., Anquez, V., Grimal, H. and Weill, J.-C. (1987) *Cell* **48**, 379–388.
Reynaud, C.-A., Dahan, A., Anquez, V. and Weill, J.-C. (1989). *Cell* **59**, 171–183.
Roes, J., Huppi, K., Rajewsky, K. and Sablitzky, F. (1989). *J. Immunol.* **142**, 1022–1026.
Rombout, J.H.W.M., Taverne-Thiele, A.J. and Villena, M.I. (1993). *Dev. Comp. Immunol.* **17**, 55–66
Schwager, J., Mikoryak, C.A. and Steiner, L.A. (1988). *Proc. Natl Acad. Sci. USA* **85**, 2245–2249.
Schwager, J., Burckert, N., Schwager, M. and Wilson, M. (1991). *EMBO J.* **10**, 505–511.
Shamblott, M.J. and Litman, G.W. (1989a). *Proc. Natl Acad. Sci. USA* **86**, 4684–4688.
Shamblott, M.J. and Litman, G.W. (1989b). *EMBO J.* **8**, 3733–3739.
Stock, D.W. and Whitt, G.S. (1992). *Science* **257**, 787–789.
Sun, S.-C., Lindström, I., Boman, H.G., Faye, I. and Schmidt, O. (1990). *Science* **250**, 1729–1732.
Tanaka, T. and Nei, M. (1989). *Mol. Biol. Evol.* **6**, 447–459.
Thompson, C.B. and Nieman, P.E. (1987). *Cell* **48**, 369–378.
Tomonaga, S., Kobayashi, K. and Hagiwara, K. (1985). *Dev. Comp. Immunol.* **9**, 617–626.
Williams, A.F. and Barclay, A.N. (1988). *Annu. Rev. Immunol.* **6**, 381–405.
Williams, A.F., Tse, A.G.D. and Gagnon, J. (1988). *Immunogenetics* **27**, 265–272.
Wilson, M.R., Marcuz, A., van Ginkel, F., Miller, N.W., Clem, L.W., Middleton, D. and Warr, G.W. (1990). *Nucleic Acids Res.* **18**, 5227.
Wilson, M., Hsu, E., Marcuz, A., Courtet, M., Du Pasquier, L. and Steinberg, C. (1992). *EMBO J.* **11**, 4337–4347.
Winoto, A. and Baltimore, D. (1989). *EMBO J.* **8**, 729–733.
Zapata, A. (1980). *Dev. Comp. Immunol.* **4**, 459–472.
Zapata, A. (1981). *Dev. Comp. Immunol.* **5**, 43–52.
Zapata, A., Ardavin, C.F., Gomariz, R.P. and Leceta, J. (1981). *Cell Tissue Res.* **221**, 203–208.

PART III: IMMUNOGLOBULIN GENE EXPRESSION

15

Ig gene expression and regulation in Ig transgenic mice

Ursula Storb

Department of Molecular Genetics and Cell Biology, University of Chicago, Chicago, IL, USA

Introduction 345
Ig transgenes 346
Influence of Ig transgenes on the expression of endogenous Ig genes and the
   development of B cells 350
Rearrangement test transgenes 356
Somatic hypermutation of Ig transgenes 358
Conclusions 360

INTRODUCTION

Transgenic mice can be produced in several ways, the most popular being microinjection of DNA into the male pronucleus of a zygote (Brinster et al., 1985). (Alternatively, transgenic mice may be produced via transfection or microinjection of DNA into embryonic stem (ES) cells and transfer of selected ES cell clones into mouse blastocysts.) The injected embryos are developed to term in pseudopregnant foster mothers. Between 0 and 60% of the resulting pups contain the injected gene and generally have the foreign DNA stably integrated in all (the integration occurred before the first embryonic cell division) or a large proportion of cells. Most often, the DNA is present in germ cells.

Most introduced genes can be expressed correctly and at near-normal levels under the direction of their own control regions, provided the plasmid or phage cloning vectors have been eliminated or restricted to a few hundred base pairs (Palmiter and Brinster, 1986).

Compared with other structural genes, immunoglobulin (Ig) genes (and T cell receptor (TCR) genes) are subject to a variety of regulatory controls besides transcription that are specific for these genes. Thus, beyond allowing determination of the regulation of tissue-specific expression, which itself may be different in the transgenic mice from transfected cells, the use of transgenic mice permits the study of the specific expression of antibody genes: (1) turning Ig gene rearrangement on and off; (2) B cell development; (3) allelic and isotypic exclusion of Ig gene expression; (4) somatic hypermutation; (5) B and T cell interactions; (6) B cell activation; (7) B cell tolerance. The first four of these as well as antibody engineering in transgenic mice will be the topics of this chapter.

Data with Ig transgenic mice published up to 1987 are only briefly summarized in the text and the reader is referred to the first edition of this book for details. Especially, some interesting but odd observations that remain unexplained are not repeated here.

IG TRANSGENES

Many rearranged H and L genes or substrates for rearrangement have been introduced into the germline of mice, thereby creating new mouse strains. The genes are generally integrated in multiple tandem copies at one or rarely at two or three sites in the genome. The integration sites differ between founder mice and there is no evidence for integration into the homologous H or L locus (Storb *et al.*, 1985). The transgenes are, in general, stably integrated and inherited. Maintenance of the transgenic lines is mostly done by breeding with non-transgenic inbred mouse lines. Attempts at homozygosing the transgene loci have been successfully undertaken a few times.

κ Transgenes

The first Ig transgenic mice were made with a κ transgene. The transgene was derived from the myeloma MOPC-21 (Brinster *et al.*, 1983) and showed B cell specific expression and high levels of serum κ chains with the mobility and isoelectric focusing pattern of MOPC-21 κ chains (Brinster *et al.*, 1983; Storb *et al.*, 1985).

Other κ transgenic mice were produced by several laboratories (Storb *et al.*, 1986; Petterson *et al.*, 1989; Carmack *et al.*, 1991). While the κ transgenes are always expressed, mostly in a tissue-specific fashion (but see Storb *et al.*, 1986), the effect on endogenous Ig gene rearrangement varies (see below).

An interesting finding was made with a transgenic strain that carried both a κ transgene and a skeletal muscle myosin light chain gene at the same chromosomal integration site (Einat *et al.*, 1987). Each gene appeared to be expressed in its specific tissue, lymphoid and muscle respectively, despite the proximity of the genes. This may suggest that an expressed domain in DNA can be very small and, furthermore, that suppression of gene expression (for example of the myosin gene in B cells) does not suppress activation of a co-integrated gene, and vice versa.

λ Transgenes

The first λ transgenic mice were produced before the λ enhancers were discovered (Hagman et al., 1990). The transgenes were under the control of the λ promoter and the heavy chain intron enhancer (Hagman et al., 1989; Neuberger et al., 1989) and were expressed in B and T cells. Transcription in T cells is presumably a consequence of the presence of the heavy chain enhancer. Heavy chain transgenes with the intron enhancer likewise are expressed in T cells (Weaver et al., 1985; Storb et al., 1986). While T cells normally express and rearrange D/J_H genes, they do not transcribe or rearrange V_H genes. Thus, normally T cells do not contain rearranged VDJ_H genes. Presumably, the proximity of the V_H or V_λ promoters to the heavy chain enhancer permits transcription in T cells.

A curious observation was made with a human λ transgene under control of its own promoter and the mouse heavy chain enhancer (Vasicek et al., 1992). The transgenic mice had a depression in B cell numbers of varying severity in different lines. The effect was perhaps caused by the early expression of the human λ gene in certain of the mouse lines and may be due to incompatibility of the λ protein with the pre-B cell receptor. Since this finding was not made with mouse λ1 (Neuberger et al., 1989) or λ2 (Hagman et al., 1989; Bogen and Weiss, 1991) transgenes, it is possible that the mouse proteins can substitute for the $\lambda 5/V_{pre-B}$ component of the surrogate light chain/µ pre-B cell receptor, but that human λ cannot do so.

Recently, λ1 (Eccles et al., 1990) and λ2 (Doglio et al., 1994) transgenes have been expressed via the λ-specific enhancers. The transgenes are transcribed only in B cells. A high proportion of splenic B cells and all hybridomas retaining the λ2 transgene express the λ2 protein (Doglio et al., 1994). Thus, it appears likely that all B cells have the potential to produce λ, and there is no evidence for a κ-only B cell population (Gollahon et al., 1988).

µ Transgenes

A number of transgenic mice have been produced with µ transgenes alone or with µ in combination with other heavy chain genes and/or κ genes (Grosschedl et al., 1984; Rusconi and Kohler, 1985; Storb et al., 1986; Nussenzweig et al., 1987; Durdik et al., 1989; Goodnow et al., 1989; Muller et al., 1989; Nemazee and Buerki, 1989; Brombacher et al., 1991). Both, the secreted and membrane forms of µ mRNA are found in the spleen of µ transgenic mice (Storb et al., 1986). There is more of the secreted form than of the membrane form, which suggests that a large proportion of the µ transcripts in the spleen are derived from plasma cells.

Generally, non-lymphoid organs do not transcribe the μ transgenes (Grosschedl et al., 1984; Storb et al., 1986). However, as mentioned above, in contrast to κ and λ transgenes with their own promoters and enhancers, the μ transgenes are expressed in T cells. Lyt-2-positive T cells from lymph nodes and thymus are enriched for transgenic μ RNA (Grosschedl et al., 1984). The μ RNA in the thymus is of the size of secreted and membrane mRNAs, with a slight preponderance of the membrane form (Storb et al., 1986). About 60% of thymus cells are μ protein positive in the cytoplasm, compared with less than 1% in normal thymus (Storb et al., 1986). A similar proportion of thymocytes express TCR genes (Roehm et al., 1984). In μ transgenic thymus the μ protein seems to remain in the cytoplasm (Storb et al., 1986). No μ is seen on the T cell surface by immunofluorescence and therefore, as expected, μ does not seem to associate with the TCR.

In one study μ transgenes were found to be expressed in skeletal muscle (Jenuwein and Grosschedl, 1991). This was independent of the intron enhancer and V_H promoter.

Transgenic mice with a μ gene whose membrane exons were deleted (μ del mem) have also been produced (Storb et al., 1986; Nussenzweig et al., 1988). The mice express transgenic μ RNA for secreted μ protein at high levels in spleen and thymus.

In transgenic mice with both μ and κ transgenes, the two genes are co-expressed in most cases. This seems to be independent of the mode of combining the two genes, i.e. whether both on one plasmid (Rusconi and Kohler, 1985), or μ and κ on separate plasmids but co-injected (μ + κ), or by mating a μ transgenic mouse with a κ transgenic mouse (μ × κ) (Storb et al., 1986). In the first, and most likely the second, case the μ and κ transgenes are inserted at the same chromosomal site, whereas in the last case they are at different sites. Interestingly, in the μ + κ mice, transgenic μ and κ mRNAs are expressed at a high level in the spleen, whereas in the thymus only μ mRNA is found (Storb et al., 1986). Thus, similar to the κ/myosin transgenic mice cited above, the close proximity of the suppressed κ gene does not interfere with μ transcription, nor does the active μ gene induce κ gene transcription. Presumably this is due to the tissue specificities of the promoters.

Details of the transcriptional control of Ig genes are discussed in Chapter 18.

Other heavy-chain transgenes

Transgenic mice were produced that carry δ (Iglesias et al., 1987; Goodnow et al., 1989), γ (Yamamura et al., 1986; Tsang et al., 1988; Gram et al., 1992; Offen et al., 1992; Tsao et al., 1992) and α (Lo et al., 1991) heavy chain transgenes.

The transgenes are expressed in B cells. Expression was also tested in T cells in γ2b transgenic mice and found to be positive (Tsang et al., 1988; Offen et al., 1992). Surprisingly, γ2b transcript levels in thymus were higher than in spleen in the mice produced by Tsang et al. (1988). In peripheral lymphatic organs, both B and T cells contain the γ2b protein (D. Lo., K. Gollahon, R. Brinster and U. Storb, unpublished data).

Human antibody production in mice

Human heavy chains were produced in mice carrying a human transgene minilocus of unrearranged heavy chain genes (Brueggemann et al., 1989). Recently, transgenic mice have been engineered whose endogenous heavy and κ genes were knocked out by homologous recombination and which carry unrearranged human heavy and κ genes (Green et al., 1994; Lonberg, 1994; Lonberg et al., 1994). The mice produce human Igs with a varied repertoire based on mixing and matching of the several human V, (D) and J sequences present in the transgenes. The mice created by Lonberg et al. (1994) contain a C_γ transgene in addition to the C_μ transgene, and switching to γ heavy chains occurs readily. These investigators have also shown that somatically mutated human antibodies are produced. All the mice show a depletion of B lymphocytes by at least 50% compared with normal mice. This may be a consequence of lower transcriptional efficiency of the human promoters/enhancers or defective cues by the human μ chains in B cell development.

Another mouse was produced by replacing the mouse C_κ gene with human C_κ (Zou et al., 1993). These mice show normal levels of serum Igs, most of which contain chimeric κ light chains.

Summary of Ig transgene expression

Ig transgenic mice show that Ig genes, with relatively small additions of upstream and downstream flanking sequences, are sufficient for expression in B (and T) lymphocytes. Thus, specific expression in B cells is not a direct consequence of rearrangement.

Furthermore, there do not seem to be strong position effects. Random insertion into the genome is compatible with correct regulation. The high success rate of Ig transgene expression suggests that strong tissue-specific control elements present in the transgenes may override negative effects of certain chromosomal sites. In the case of globin transgenes, consistent expression was only obtained when the locus control region was included (Palmiter and Brinster, 1986; Grosveld et al., 1987). Perhaps, the heavy and light chain gene enhancers act as locus control regions.

The heavy chain transgenes cited above all contain only the intron enhancer, which is apparently sufficient for chromatin activation in the B cell lineage in the absence of the 3′ heavy chain enhancer (the role of the heavy chain intron enhancer in somatic hypermutation will be discussed below). The κ transgenes most often contained the 3′ κ enhancer as well as the intron enhancer. We found that DNA sequences located beyond 6 kb 3′ of C_κ are required for high expression of κ transgenes (U. Storb, S. McKnight, C. Pinkert and R. Brinster, unpublished data). Similarly, Rusconi (1984) found that a κ gene that contains only 1 kb 3′ of C_κ is inactive in transgenic mice. Also, κ transgenes with about 7 kb 3′ of C_κ, but lacking the 3′ enhancer showed only low levels of κ transcripts in spleen B cell hybridomas (Betz et al., 1994). In contrast, other κ transgenic mice with only about 1.5 kb of

DNA 3' of C_κ strongly transcribed the κ gene (Carmack *et al.*, 1991). These findings are unresolved. There could perhaps be an unusual V_κ promoter which, in contrast to the promoters present in other κ transgenes, may be lacking a silencing sequence.

It can be concluded that Ig transgenes with their own control regions are correctly expressed in the B lymphocytes of transgenic mice. These mice therefore represent an excellent model for the study of the control of Ig gene expression and the interplay of Igs with B cell developmental cues.

INFLUENCE OF IG TRANSGENES ON THE EXPRESSION OF ENDOGENOUS IG GENES AND THE DEVELOPMENT OF B CELLS

Apparently, Ig transgenic mice are healthy and of normal longevity, implying that the high expression of the transgenes does not eliminate immunity against pathogens. However, the transgenes do affect endogenous Ig gene expression and B cell development as discussed in the following sections.

Effects of heavy chain transgenes on endogenous Ig genes and B cell development

The synthesis of μ heavy chains has been shown to be essential for B cell development. Inactivation of the membrane domains of μ by gene targeting results in B cell-deficient mice (Kitamura *et al.*, 1991). It appears that the μ protein may have at least four functions during B cell development: the feedback inhibition of heavy chain gene rearrangement, the upregulation of the rate of light chain gene rearrangement, the provision of a maturation or survival signal and, in combination with light chains, the feedback inhibition of light chain and DJ_H gene rearrangements by terminating V(D)J recombinase production.

The feedback inhibition of H gene rearrangement has been postulated to be required for allelic exclusion, namely the synthesis of only one heavy chain per B cell. Allelic exclusion was first observed with allotype heterozygous individuals whose plasma cells produce antibodies of only one or the other allotype, but not both (Bernier and Cebra, 1964; Pernis *et al.*, 1965; Weiler, 1965). On the genetic level, allelic exclusion is due to the presence of only one functionally rearranged allele (Perry *et al.*, 1980; Coleclough *et al.*, 1981; Walfield *et al.*, 1981). Since both alleles are transcriptionally active, allelic exclusion cannot be due to a priori inactivation of one allele (Perry *et al.*, 1981). Two basic hypotheses were formulated to explain allelic exclusion. The *regulated* model proposed that when a correct H- or L-gene rearrangement had occurred the cell would receive a specific signal to stop further rearrangement (Alt *et al.*, 1981). The *stochastic* model was based on the finding of a high proportion of incorrectly rearranged Ig genes. It proposed that the probability of two productive rearrangements in both alleles was too low to result in the frequent occurrence of two different functional H or L chains (Coleclough *et al.*, 1981; Walfield *et al.*, 1981).

The sequence of Ig gene rearrangements in mouse during B cell development follows a scheme of first DJH rearrangement in pro-B cells, then V to DJ rearrangement in immature pre-B cells, and finally light chain rearrangement in mature pre-B cells (Hardy *et al.*, 1991). The sequential activation of first heavy then light chain genes may not be rigorously followed, however; at least there is now evidence that there is no strict regulation of light chain gene rearrangement by heavy chains (Ehlich *et al.*, 1993).

Evidence from μ transgenic mice in favor of a positive role of μ chains in feedback inhibition of heavy chain gene rearrangement has not been completely convincing. This question was studied in Abelson murine leukemia virus (A-MuLV)-transformed pre-B cell lines and hybridomas of spleen B cells from μ transgenic mice (Rusconi and Kohler, 1985; Weaver *et al.*, 1985; Manz *et al.*, 1988). A-MuLV lines from the bone marrow of μ-transgenic mice all produced the transgenic μ protein (Weaver *et al.*, 1985). When their endogenous H genes were analyzed, it was found that 40% (10 of 25) of the cell lines had at least one H allele in the germline J_H configuration. In contrast, in A-MuLV cell lines from the bone marrow of normal mice no germline J_H genes were found. Furthermore, in μ-transgenic splenic hybridomas representing mainly the mature B cell stage, 9% (Rusconi and Koehler, 1985), 10% (Weaver *et al.*, 1985) and 20% (Manz *et al.*, 1988) of the cells had at least one unrearranged J_H gene. In hybridomas from normal mice unrearranged H genes are rarely seen. In addition, the μ transgenic hybridomas contained a higher proportion of DJ rearrangements without VD joining than hybridomas normally show (Rusconi and Koehler, 1985; Weaver *et al.*, 1985; Manz *et al.*, 1988). Thus, apparently, the presence of a transgenic μ gene suppresses the rearrangement of endogenous μ genes. This effect may reflect a physiological response and not simply the competition by multiple transgene copies for *trans*-acting factors because of the following result: no germline H genes were observed in A-MuLV-transformed pre-B cell lines or in hybridomas from mice which combined κ and μ transgenes where the μ gene lacked the exons encoding the membrane terminus (Manz *et al.*, 1988).

This result also indicates that secreted H chains cannot exert a feedback inhibition of H gene rearrangement. In support of this notion, human μ transgenes that encode only the secreted protein do not suppress mouse μ synthesis (Nussenzweig *et al.*, 1988). On the other hand, transgenic mice with a human μ gene encoding only the membrane form of μ show suppression of mouse μ mRNA in the spleen (Nussenzweig *et al.*, 1987). Also, 25–35% of the spleen cells carry human μ and the proportion of mouse μ-positive B cells is reduced to 5–10% from normally 50%.

While the results with μ transgenic mice support the notion of feedback inhibition of heavy chain gene rearrangement, the partial nature of the effect raised the concern that the apparent feedback was due to unknown unphysiological effects of the transgenes. Recently, however, stronger support for heavy chain gene feedback has been obtained with γ2b transgenic mice (Roth *et al.*, 1993). Such mice have greatly depleted numbers of B cells in the periphery due to a block in B cell development.

This block occurs in the large B220^{lo} pre-B cells. Old γ2b mice will eventually achieve almost normal numbers of peripheral B cells; however, all B cells express μ with or without the γ2b transgene. Apparently, γ2b strongly inhibits the rearrangement of endogenous μ genes, but γ2b cannot replace μ in the maturation/survival function, thus leading to the B cell depletion.

Further evidence for inhibition of heavy chain gene rearrangement was obtained in crosses between γ2b transgenic and μ transgenic mice (Roth et al., 1993). While the μ transgenic mice showed only little suppression of endogenous μ in splenic B cells, the μ × γ2b crosses expressed transgenic μ, but essentially no endogenous μ, suggesting that γ2b strongly enhances the feedback effect.

A caveat with the transgenic experiments is the concern that the expression of the transgenic heavy chain may upregulate the rearrangement of endogenous light chain genes, so that the pre-B cells may progress at an accelerated rate to a stage at which both heavy and light chain genes are produced. Once a complete mature Ig receptor is present, the cells are ready to terminate rearrangement altogether by downregulating the production of V(D)J recombinase. Thus, the presence of unrearranged or DJ-arrested B cells in a μ transgenic mouse may be caused by feedback inhibition of recombinase rather than specific H gene feedback inhibition. In this context it is useful to ask what the target for the postulated H gene feedback may be. It would have to be very specific, because the feedback does not block the transcription of rearranged heavy chain genes. Rearrangement and transcriptional competence are closely coupled (Perry et al., 1980; Storb et al., 1981; Yancopoulos and Alt, 1985) and may involve the same *trans*-acting factors. The H locus continues to be transcribed at the functionally rearranged VDJ$^+$ allele, as well as aberrantly rearranged VDJ$^-$ and DJ alleles. Thus, as a consequence of the postulated heavy chain feedback, a change must occur in pre B cells that interferes with further heavy chain gene rearrangement, but not with transcription of rearranged H genes. It was found that unrearranged V$_H$ genes are DNaseI sensitive and transcribed in early pre-B cell lines, but not in more mature B cells (Yancopoulos and Alt, 1985). It has not been investigated if a similar change in transcription occurs at the D genes.

The results to date are compatible with the following scheme. After a correct heavy chain gene has been produced, heavy chain gene rearrangement is actively stopped by inactivation of unrearranged V$_H$ genes as targets for the recombinase, thus preventing V to DJ joining. D to J$_H$ joining continues until the V(D)J recombinase is eliminated. The increased proportion of endogenous germline heavy chain genes seen in B cells of H gene transgenic mice may be due to the accelerated shutoff of the recombinase. The increased proportion of endogenous heavy chain genes arrested at the DJ step may be due to the absence of further VD rearrangements as a consequence of heavy chain feedback by the transgene. The latter can be exerted by μ and γ2b proteins. It is not known which other pre-B cell proteins may be required for this feedback.

The second function of μ appears to be the *upregulation* of light chain gene rearrangement (Ehlich et al., 1993). This function is shared by γ2b (Roth et al., 1993). However, induction of light chain gene rearrangement *per se* does not seem to require a signal from heavy chains (Ehlich et al. 1993).

Thirdly, μ appears to deliver a survival or maturation signal. Possibly, bcl-2 or a similar anti-apoptosis protein is induced by the μ-containing pre-B cell receptor, since it was shown that the *scid* phenotype of B cell deficiency can be rescued by expressing a bcl-2 transgene in pre-B cells of *scid* mice (Strasser et al., 1994). The γ2b transgenes cannot provide the survival/maturation signal (Roth et al., 1993). Since γ2b is apparently able to deliver a heavy chain feedback signal but not a signal for B cell maturation, either different regions of the heavy chains are responsible for the different functions, or a certain region of μ can deliver both signals but γ2b can deliver only one. A major structural difference between μ and γ2b is the cytoplasmic tail, which consists of three amino acids in μ but 27 in γ2b (Kabat et al., 1991). However, transgenic mice with a γ2b hybrid gene whose transmembrane and cytoplasmic portions have been replaced by those of μ show the same B cell defect as γ2b transgenic mice (Roth et al., 1993). Possibly, the transmembrane domain is involved in the heavy chain feedback, since μ and γ proteins differ no more in this domain than do μ and δ (Kabat et al., 1991), which both provide a feedback as well as a maturation signal (Iglesias et al., 1987). γ2b can interact with the surrogate light chains, λ5/V_{pre-B} (Roth et al., 1995).

Finally, in cooperation with a bona fide light chain, κ or λ, μ is responsible for the feedback inhibition of the V(D)J recombinase production (see below).

While it is clear that γ2b transgenes prevent B cell maturation, because of the strong feedback effect by γ2b coupled with an inability of γ2b to deliver a maturation signal, one unique γ2b transgenic mouse was found in which the γ2b transgene is permissive for B cell development (Roth et al., 1995). This mouse line, named the C line, shows strong suppression of μ, as do the other γ2b lines. However, it does not have the B cell defect and its peripheral B cells are mostly γ2b only. When crossed with a μ-knockout mouse, B cell development proceeds in the absence of μ. The C-line phenotype is dominant in crosses with other γ2b transgenic lines. There is no evidence for a higher or otherwise altered expression of the γ2b transgene in the C line indicative of a position effect on transgene expression, and the transgene copies were found to be unmutated. It was therefore concluded that in this line the transgene must be integrated in a chromosomal site at which its presence induces a gene whose expression can overcome the need for μ in B cell development. Interestingly, the B cell rescue apparently requires the γ2b expression on the pre-B cell membrane, since deletion of the λ5 genes prevents B cell development in the C line γ2b transgenic mouse.

Interesting results have been obtained with Ig transgenic mice with respect to tolerance and autoimmunity.

Effects of light chain transgenes on endogenous Ig gene rearrangement

The wish to distinguish between the regulated and stochastic models of allelic exclusion of κ genes led to the first production of Ig transgenic mice (Brinster et al.,

1983). Hybridomas from the spleens of κ transgenic mice were analyzed for the effect on endogenous κ gene rearrangement (Ritchie et al., 1984). A clear effect in several different founder mice was seen in hybridomas that produced endogenous heavy chains together with the transgenic light chain; endogenous κ genes were in germline conformation.

Thus, the presence of κ chains together with H chains apparently causes feedback inhibition of κ gene rearrangement. In these mice it seemed that any heavy chain in combination with the κ transgene could exert the effect. However, this κ chain may be unusual, in that it may be particularly capable of combining with any H chain. It is the same κ chain as the one produced by the myeloma NS-1 (a derivative of MOPC-21). This myeloma has been widely used as a fusing line for hybridomas (Koehler and Milstein, 1976). In such hybridomas mixed Igs are produced in which the NS-1 κ chain is paired with H chains from the fusion partner.

Various degrees of feedback inhibition have been obtained with subsequent κ transgenes. In transgenic mice carrying μ and κ genes isolated from an anti-TNP (trinitrophenol) hybridoma, feedback inhibition of endogenous κ genes could not be demonstrated (Rusconi and Koehler, 1985). However, in these mice μ and κ synthesis was unbalanced: the level of the κ chains was only about one-tenth that of the μ chains. In transgenic mice with the MOPC-167 anti-PC (phosphorylcholine) κ transgene, only a few of the B cells immortalized in hybridomas had not rearranged their endogenous κ genes (Manz et al., 1988). Another κ transgene encoding the light chain of an anti-influenza hemagglutinin showed very strong feedback, with 90% of endogenous κ genes in germline configuration in one transgenic line, but only 0–18% in another line (Carmack et al., 1991). This clearly cannot be due to the variable region alone, since the two lines carried identical transgenes. Furthermore, they were both backcrossed to the same mouse strain, Balb/c, thus eliminating the possibility of strain-specific tolerance/autoimmunity. Perhaps the levels of expression of these transgenes at a time critical for feedback inhibition were different due to differential genomic position effects.

Presumably, the inhibition of κ gene rearrangement is caused by the regulated shutoff of the V(D)J recombinase. A clue as to the molecular events that lead to this feedback inhibition has been obtained with transformed pre-B cell lines from N-*myc* transgenic mice (Ma et al., 1992). These cells express a B cell receptor of μ/κ on the cell surface, but are positive for the expression of the *RAG* genes. When treated with anti-μ antibodies, the synthesis of *RAG* mRNAs is downregulated. While *RAG* mRNAs reappear in these transformed cells after removal of the anti-μ, this experimental system may nevertheless be a paradigm for permanent shutoff of the *RAG* genes and related genes in normal pre-B cells. Presumably, pre-B cells producing a membrane-bound complete Ig molecule need to encounter a specific, as yet unidentified, ligand for the induction of the recombinase shutoff. It was found in hybrids between pre-B cells and mature B cells that B cells contain an inhibitor which prevents Ig gene rearrangements in rearrangement substrates (Engler et al., 1991a). The loss of certain B cell chromosomes in the pre-B × B cell hybrids allows the recombinase to persist and rearrangement to occur, suggesting that one of the lost chromosomes carried the inhibitor gene. The inhibitor appears to be acting on the level of transcription, since introduction of the *RAG* genes under control of a ubiquitous promoter into B cells permits their expression and the rearrangement of transfected rearrangement substrates (P. Roth, J. Zhao, K. Fuller and U. Storb, unpublished data). Perhaps crosslinking of the Ig receptor in pre-B cells by the unidentified ligand leads to the activation of this inhibitor.

The λ light chains appear to have the same feedback effect as κ chains as first demonstrated with λ1 and λ2 transgenes transcribed under the control of the λ promoter and the heavy chain intron enhancer (Hagman *et al.*, 1989; Neuberger *et al.*, 1989). The λ transgenes under control of a λ enhancer also cause feedback inhibition of endogenous κ gene rearrangement (Doglio *et al.*, 1994). Compared with a λ transgene with the heavy chain enhancer, the λ enhancer-driven transgene is expressed about a day later in fetal liver (Doglio *et al.*, 1994). However, feedback inhibition seems to occur at about the same stage of B cell development, regardless of the enhancer present in the λ transgene. Thus, the feedback is not necessarily coincident with the assembly of a heavy/light chain complex in pre-B cells in analogy to the findings with the transformed cell lines described above (Ma *et al.*, 1992).

In splenic B cell hybridomas from λ transgenic mice, germline H genes are found relatively frequently (Hagman *et al.*, 1989; Doglio *et al.*, 1994). These may be the result of early transgenic light chain synthesis in some pre-B cells in which the second heavy chain gene allele has not yet had a chance for rearrangement. Based on the proposed mechanism of the feedback, these cells would then have had to have an early encounter with the proposed B cell receptor feedback ligand.

It has been postulated that the escape of endogenous Ig genes from feedback inhibition of heavy chain gene rearrangement by μ transgenes may be restricted to B-1 (CD5) B cells (Herzenberg and Stall, 1989). The ability of B-1 and B-2 (conventional B) cells to resist or escape feedback inhibition of light chain gene rearrangement (i.e. presumably the inhibition of the V(D)J recombinase) was investigated in λ2 transgenic mice (Rudin *et al.*, 1991). In these mice, the λ gene was under the control of the heavy chain intron enhancer. B-1 and B-2 cells were distinguished in these mice by surface markers in the spleen and peritoneal exudate cells. Furthermore, bone marrow from λ transgenic mice more than 2 months old was used as a source of conventional B cells to reconstitute sublethally irradiated mice homozygous for the *scid* mutation. Levels of serum κ chains and of splenic B cells expressing κ chains were analyzed 1–3 months after transfer. The results clearly showed that both types of B cells are subject to feedback inhibition of endogenous light chain gene rearrangement and that both can escape feedback inhibition in about equal proportions (Rudin *et al.*, 1991). Thus, as alluded to above, in transgenic mice a certain percentage of B cells are not allelically excluded, in contrast to normal mice in which most B cells display strict allelic exclusion. Most likely, κ$^+$ B cells in λ transgenic mice represent the result of strong selection of B cells that express a specificity different from the transgenic light chain. B cells with endogenous light chain gene rearrangements must arise at varying rates, depending on the strength of the feedback, presumably due in part on the chromosomal position of the transgene. Cells that express 'useful' endogenous specificities may then become preferentially expanded.

Isotypic exclusion of κ and λ genes

Transgenic experiments relating to κ/λ isotypic exclusion are discussed in the context of λ genes in Chapter 9.

REARRANGEMENT TEST TRANSGENES

Transgenic mice with rearrangement test genes have been produced in order to study the species specificity of the rearrangement process and the requirements for accessibility of the targets of the V(D)J recombinase.

One of the first experiments was aimed at determining if the rearrangement signal sequences (RSSs) and associated regions of chicken λ genes can be recognized by the recombination machinery of mouse. The transgenes contained an unrearranged 11.5 kb segment of chicken $V_\lambda 1$, $J_\lambda 1$, $C_\lambda 1$ and flanking sequences (Bucchini et al., 1987). Four transgenic lines expressed and rearranged the transgene in the spleen but not in the thymus, suggesting that transcriptional and rearrangement cues of chicken are recognized by mouse proteins. In one mouse line, rearrangement also occurred in the thymus; this was apparently not due to B cell contamination. It was interpreted as the result of a high copy number (20) of the transgenes.

In another experiment, an unrearranged rabbit κ gene was introduced into the mouse germline (Goodhardt et al., 1987). Eleven transgenic lines were obtained and in all of them the rabbit gene was rearranged in the spleen and in thymus (probably in both B and T cells). Transcription was low. Higher levels of transcripts found with a hybrid gene [V_κ–J_κ1–4 (mouse)–C_κ (rabbit)] suggested that the rabbit κ promoter/enhancer may not be very efficient in mouse cells (M. Goodhardt, personal communication). Transgenic mice with the rabbit gene had mixed antibody molecules, containing both rabbit and mouse κ chains. The results suggest that rearrangement of the rabbit κ gene was not correctly controlled with respect to tissue and stage specificity, and perhaps allelic exclusion.

In a further experiment with rabbit κ transgenes it was found that deletion of the κ intron enhancer eliminated rearrangement in transgenic mice (Kallenbach et al., 1993). It was concluded that while the rabbit κ intron enhancer is defective in transcription in mouse cells, it allows high levels of rearrangement thus showing that transcription and rearrangement are not coupled.

The role of transcription in V(D)J recombination was further addressed in transgenic mice carrying a TCR–Ig hybrid transgene (Ferrier et al., 1990). The V(D)J region was from TCRβ, while the intron and C_μ region were from the Ig locus.

Rearrangement of D to J occurred equally well in T and B cells, whereas VD rearrangement was only seen in T cells. It was suggested that this pattern of rearrangement correlated with the transcription of the unrearranged V_β gene only in T cells.

Transgenic mice were also made with an artificial substrate for the V(D)J recombinase, pHRD (Engler et al., 1991b). The pHRD test gene contains the RSSs of a V_κ and J_κ gene and is rearranged efficiently when carried in stably transfected pre-B cell lines. Contrary to expectations, the test gene was not rearranged in the lymphoid organs of several different transgenic lines. In an attempt to determine the reason for the presumed inaccessibility of the transgene to the V(D)J recombinase, the methylation status of the transgene DNA was analyzed in these mice. It was found that all transgene copies were completely methylated in these mice. Since some other transgenes had shown differential expression depending on the mouse strain (Allen et al., 1990), it was postulated that the methylation of pHRD may be controlled in a mouse strain-specific way. The transgenic mice were crossed with a number of different mouse strains. Crosses to SJL or DBA/2 showed that the C57BL/6 strain was responsible for the methylation of the pHRD transgene. Undermethylation was achieved by crossing the transgene for at least two generations into SJL or DBA/2. Crosses with C57BL/6 × DBA/2 (BXD) recombinant inbred mice mapped the modifier gene, now named Ssm-1, to the distal end of chromosome 4, near the Friend virus susceptibility gene Fv-1 (Engler et al., 1991b). Further crosses showed that the following strains are Ssm-1^+: B6, A, Balb/c, C57L, LP. The strains DBA/2, SJL, CBA, C3H and SM are Ssm-1^-. Presence of the test gene in an Ssm-1^+ strain completely prevents its rearrangement. Undermethylation and lymphoid tissue-specific rearrangement is achieved only when both copies of Ssm-1^+ are lost after crossing for at least two generations into an Ssm-1^- strain, i.e. Ssm-1^+ is dominant. Presumably, this novel modifier may be involved in the control of chromatin organization during development and gene expression. Other mouse strains may have different modifier genes that interact with different target genes.

In mice carrying seven unmethylated copies of the artificial pHRD transgene recombination occurs randomly between any two RSSs within the transgene array (Engler et al., 1993). This finding suggests that the V(D)J recombinase may be limiting in pre-B cells and rules out a simple mechanism of one-dimensional tracking of a recombinase complex with binding sites for both the 12-mer and the 23-mer spacer RSSs.

Since the pHRD artificial substrate does not encode Igs, its rearrangement products cannot be selected for. It was therefore interesting to determine the composition of the joints formed in fetal liver with those present in adult spleen, lymph nodes and bone marrow, which are presumably all formed in the bone marrow (Engler et al., 1992). It was found that the fetal joints lacked untemplated nucleotides (N-regions), but that the adult joints had a high proportion of N-regions. This suggested that the rearrangement mechanism is different at the two developmental stages, most likely due in part to the greater amount of terminal transferase in the bone marrow pro-B/pre-B cells (Gregoire et al., 1979).

SOMATIC HYPERMUTATION OF IG TRANSGENES

Somatic hypermutation is a major mechanism to increase the repertoire of the variable regions of Ig genes. There are no known diseases or mouse models of defective somatic hypermutation and the molecular mechanism of somatic hypermutation is unknown. Since no cell line is as yet available that reliably reproduces the somatic mutations seen *in vivo*, the use of Ig transgenic mice was introduced some years ago as a model to study the molecular mechanism (O'Brien *et al.*, 1987). The experimental model now in use to select cells with mutations is to hyperimmunize light chain transgenic mice with an antigen resulting in mutations in known heavy chain genes, and then to sequence the transgenic light chains in hybridomas that show mutations in the endogenous heavy chain genes (O'Brien *et al.*, 1987). In this way it was clearly shown that κ transgenes are subject to somatic hypermutation in several different chromosomal integration sites and that *cis*-acting sequences present within about 15 kb of κ transgenes (including about 4 kb upstream of the V_κ promoter 5' and the 3'κ enhancer 3') were sufficient to target the mutator. It was further shown that the extent of the mutations in terms of frequency of point mutations per mutated sequence and in the location of the point mutations within the gene is the same as in endogenous κ genes (O'Brien *et al.*, 1987). Thus, the peak of the mutation frequency is over the VJ region, whereas the C region and 3' untranslated region are unmutated (Hackett *et al.*, 1990). Furthermore, a 5' boundary appears to exist within the leader intron (Rogerson, 1994).

Evidence for somatic hypermutation has now also been found in heavy chain transgenes (Giusti and Manser, 1993; Sohn *et al.*, 1993).

Recently, in addition to splenic hybridomas, the Peyer's patches of Ig transgenic mice have been used as a source of B cells that have undergone somatic mutation (Gonzales-Fernandes and Milstein, 1993). B cells that bind high levels of peanut agglutinin (PNA^{hi}) were found to be higher in mutation frequency than PNA^{lo} cells. This procedure can apparently dispense with the need for immunization of the mice. It suffers, however, from the problem that the polymerase chain reaction is required to amplify the single DNA sequence representing each transgene copy and that, unlike with hybridomas, additional samplings cannot be taken from the same cell whose genes were originally cloned.

The relationship between somatic mutation and transcription of the targeted transgenes is obscure. Kappa transgenes lacking the 3' enhancer have been found to be unmutated or mutated only at a very low rate (Sharpe *et al.*, 1990, 1991; Carmack *et al.*, 1991; Betz *et al.*, 1994). However, the control of expression of Ig transgenes in the centroblasts of the germinal center, i.e. the B cells in which the mutator process is presumed to be active, has not been established. The transgene described by Carmack *et al.*, is expressed at a high rate in plasma cells, and so is another unmutated gene that lacks the κ-intron enhancer (Betz *et al.*, 1994). Are these genes perhaps not transcribed in the centroblasts or do these enhancers have additional roles in the mutation process? Clearly, the 3' enhancer is not essential for the mutation process itself; seemingly specific mutations do occur, however, at a very low rate in κ transgenes lacking the 3' enhancer (Betz et al., 1994).

One of the major questions concerning somatic mutation is how the exquisite specificity of targeting to the V region is achieved. The target cannot be simply the V region, because sequences flanking the V region, such as J and the leader intron as well as the JC intron, are also subject to somatic mutation. Replacing the V promoter by a β-globin promoter was reported to permit mutations of the VJ region at almost the same frequency as with transgenes containing the V promoter (Betz et al., 1994). This would suggest that the V region promoter is not a specific targeting sequence, although a comparison of the two promoters may show sequence similarities. A transgenic construct containing only the V promoter and heavy chain intron enhancer was found to permit specific somatic mutation of an associated CAT (chloramphenicol acetyl transferase) gene, albeit at a low frequency (Azuma et al., 1993). Finally, a TCRβ gene expressed as a transgene in B cells under the control of the heavy chain intron enhancer was not mutated to any appreciable extent (Hackett et al., 1992). Thus, the intron enhancer by itself is not sufficient to confer mutability. However, the intron enhancer used in the TCR construct contained only about 700 bp of the 997-bp enhancer fragment used in the CAT construct; thus it was lacking some of the matrix attachment sequences present in the full sequence (Cockerill et al., 1987). Also, while expression of the TCR transgene was high in spleen and thymus (Hackett et al., 1992), the TCRβ promoter may not cooperate with the heavy chain enhancer in transcriptional activation of the transgene in germinal centers.

Thus, the question of the targeting sequences, as well as of the requirement for transcriptional competency, remains open. An unexpected finding was the uneven distribution of mutated copies in arrays of multiple transgenes in two different transgenic lines (O'Brien et al., 1987). A careful analysis with one particular transgenic line showed that in a three-copy array one copy was preferentially targeted, in some cases the middle copy, in some one of the flanking copies (Rogerson et al., 1991). The targeted copy showed multiple point mutations, suggesting that the same copy had been bombarded multiple times by the mutator process. Since there was no evidence for differential expression of the three copies, it appeared unlikely that only one, the mutated copy, was expressed. These findings led us to propose a model linking somatic hypermutation with DNA replication (Rogerson et al., 1991). The model postulates a mutation initiation region (MIR) located upstream of the V-region that is a binding site for a mutator factor. The factor becomes associated with the leading strand of DNA replication, when replication happens to originate upstream (5') of the V region, and is carried along with the replication complex. Its association reduces the fidelity of DNA copying, thus introducing point mutations. The factor has a limited processivity and with increasing distance from the V region increasingly dissociates from the replication complex. In this way, the C region of κ genes is generally not mutated, since the factor has disappeared from the replication complex by the time it reaches $C_κ$. However, in λ genes, the C regions are mutated at a low frequency (Motoyama et al., 1991), perhaps because the distance between VJ and C is considerably shorter.

This model is based on the known scarcity and irregularity of mammalian origins of replication, which would explain why different transgenes are targeted in different cells. The precursors of mutating B cells are presumably resting in G_0 and enter the cell cycle after antigenic stimulation, different cells using different origins of replication.

It is hoped that further experiments with transgenic mice carrying modified Ig genes will lead to a clear understanding of the targeting sequence(s) and other requirements for somatic hypermutation.

CONCLUSIONS

The transgenic mouse model provides a unique opportunity for the study of the control of expression of Ig genes. While in the normal mouse, B lymphocytes that produce one particular identical Ig are rare, one among thousands of different B cells, in transgenic mice almost all of the B cells express the transgene. As described in this chapter, this situation has been very useful in the analysis of the molecular and cellular mechanisms underlying Ig gene rearrangement, B cell development, feedback inhibition of heavy chain gene rearrangement and V(D)J recombinase production, and somatic hypermutation.

Ig transgenes are in general expressed as expected from the transcriptional promoters and enhancers included in the transgene. However, in the whole animal, the very much increased levels of B cells with a single specificity and of serum antibodies of a particular idiotype may lead to unexpected responses, perhaps under the influence of T cells. The apparent preponderance of Ly-1 B cells in anti-NP transgenic mice, for example, is at present unexplained. Furthermore, due to the strong selective pressure by antigen and other unknown cues, the mature repertoire of B cells may not reflect the situation early in B cell development; mature B cells are probably selected to be functional. It will be extremely helpful to exploit Ig transgenic mice in the investigation of the developmental signals and the molecular mechanisms leading to the establishment of a B cell repertoire.

Acknowledgements

I am grateful for reprints of their work to R. Grosschedl, F. Melchers, C. Milstein, D. Lo, N. Lonberg, M. Neuberger, E. Selsing and A. Strasser. Studies from my laboratory were supported by NIH grants HD23089, AI24780 and GM38649.

REFERENCES

Allen, N., Norris, M. and Surani, M. (1990). *Cell* **61**, 853–861.
Alt, F., Rosenberg, N., Lewis, S., Thomas, E. and Baltimore, D. (1981). *Cell* **27**, 381–390.
Azuma, T., Motoyama, N., Fields, L. and Loh, D. (1993). *Int. Immunol.* **5**, 121–130.

Bernier, G. and Cebra, J. (1964). *Science* **144**, 1590–1591.
Betz, A., Milstein, C., Gonzalez-Fernandes, R., Pannell, R., Larson, T. and Neuberger, M. (1994). *Cell* **77**, 239–248.
Bogen, B. and Weiss, S. (1991). *Eur. J. Immunol.* **21**, 2391–2395.
Brinster, R., Ritchie, K., Hammer, R., O'Brien, R., Arp, B. and Storb, U. (1983). *Nature* **306**, 332–336.
Brinster, R., Chen, H., Trumbauer, M., Yagle, M. and Palmiter, R. (1985). *Proc. Natl Acad. Sci. USA* **82**, 4438–4442.
Brombacher, F., Kohler, G. and Eibel, H. (1991). *J. Exp. Med.* **174**, 1335–1346.
Brueggeman, M., Caskey, H., Teale, C., Waldmann, H., Williams, G., Surani, A. and Neuberger, M. (1989). *Proc. Natl Acad. Sci. USA* **86**, 6709–6713.
Bucchini, D., Reynaud, C., Ripoche, M., Grimal, H., Jami, J. and Weill, J. (1987). *Nature* **326**, 409–411.
Carmack, C., Camper, S., Mackle, J., Gerhard, W. and Weigert, M. (1991). *J. Immunol.* **147**, 2024–2033.
Cockerill, P., Yuen, M. and Garrard, W. (1987). *J. Biol. Chem.* **262**, 5394–5397.
Coleclough, C., Perry, R.P., Karjalainen, K. and Weigert, M. (1981). *Nature* **290**, 372–378.
Doglio, L., Kim, J.Y., Bozek, G. and Storb, U. (1994). *Dev. Immunol.* **4**, 1–14.
Durdik, J., Gerstein, R., Rath, S., Robbins, P., Nisonoff, A. and Selsing, E. (1989). *Proc. Natl Acad. Sci. USA* **86**, 2346–2350.
Eccles, S., Sarner, N., Vidal, M., Cox, A. and Grosveld, F. (1990). *New Biologist* **2**, 801–811.
Ehlich, A., Schaal, S., Gu, H., Kitamura, D., Muller, W. and Rajewsky, K. (1993). *Cell* **72**, 695–704.
Einat, P., Bergman, Y., Yaffe, D. and Shani, M. (1987). *Genes Dev.* **1**, 1075–1084.
Engler, P., Haasch, D., Pinkert, C., Doglio, L., Glymour, M., Brinster, R. and Storb, U. (1991b). *Cell* **65**, 939–947.
Engler, P., Roth, P., Kim, J.Y. and Storb, U. (1991a). *J. Immunol.* **146**, 2826–2835.
Engler, P., Klotz, E. and Storb, U. (1992). *J. Exp. Med.* **176**, 1399–1404.
Engler, P., Weng, A. and Storb, U. (1993). *Mol. Cell. Biol.* **13**, 571–577.
Ferrier, P., Krippl, B., Blackwell, T.K., Furley, A.J., Suh, H., Winoto, A., Cook, W.D., Hood, L., Costantini, F. and Alt, F.W. (1990). *EMBO J.* **9**, 117–125.
Giusti, A. and Manser, T. (1993). *J. Exp. Med.* **177**, 797–809.
Gollahon, K.A., Hagman, J., Brinster, R.L. and Storb, U. (1988). *J. Immunol.* **141**, 2771–2780.
Gonzales-Fernandes, A. and Milstein, C. (1993). *Proc. Natl Acad. Sci. USA* **90**, 9862–9866.
Goodhardt, M., Cavelier, P., Akimenko, M.A., Lutfalla, G., Babinet, C. and Rougeon, F. (1987). *Proc. Natl Acad. Sci. USA* **84**, 4229–4233.
Goodnow, C., Crosbie, J., Jorgensen, H., Brink, R. and Basten, A. (1989). *Nature* **342**, 385–391.
Gram, H., Zenke, G., Geisse, S., Kleuser, B. and Buerki, K. (1992). *Eur. J. Immunol.* **22**, 1185–1191.
Green, L., Hardy, M., Maynard-Currie, C., Tsuda, H., Louie, D., Mendez, M., Abderrahim, H., Noguchi, M., Smith, D., Zeng, Y., David, N., Sasai, H., Garza, D., Brenner, D., Hales, J., McGuinness, R., Capon, D., Klapholz, S. and Jacobovits, A. (1994). *Nature Genet.* **7**, 13–21.
Gregoire, K., Goldschneider, I., Barton, R. and Bollum, F. (1979). *J. Immunol.* **123**, 1347–1352.
Grosschedl, R., Weaver, D., Baltimore, D. and Costantini, F. (1984). *Cell* **38**, 647–658.
Grosveld, F., van Assendelft, G., Greaves, D. and Kollias, G. (1987). *Cell* **51**, 975–985.
Hackett, J., Rogerson, B., O'Brien, R. and Storb, U. (1990). *J. Exp. Med.* **172**, 131–137.
Hackett, J., Stebbins, C., Rogerson, B., Davis, M. and Storb, U. (1992). *J. Exp. Med.* **176**, 225–231.
Hagman, J., Lo, D., Doglio, L.T., Hackett, J., Jr, Rudin, C.M., Haasch, D., Brinster, R. and Storb, U. (1989). *J. Exp. Med.* **169**, 1911–1929.
Hagman, J., Rudin, C., Haasch, D., Chaplin, D. and Storb, U. (1990). *Genes Dev.* **4**, 978–992.
Hardy, R.R., Carmack, C.E., Shinton, S.A., Kemp, J.D. and Hayakawa, K. (1991). *J. Exp. Med.* **173**, 1213–1225.
Herzenberg, L. and Stall, A. (1989). *Cold Spring Harb. Symp. Quant. Biol.* **54**, 219.
Iglesias, A., Lamers, M. and Kohler, G. (1987). *Nature* **330**, 482–484.
Jenuwein, T. and Grosschedl, R. (1991). *Genes Dev.* **5**, 932–943.
Kabat, E., Wu, T., Perry, H., Gottesman, K. and Foeller, C. (1991). 'Sequences of Proteins of Immunological Interest'. US Dept. of Health and Human Services, Bethesda, MD.

Kallenbach, S., Babinet, C., Pournin, S., Cavelier, P., Goodhardt, M. and Rougeon, F. (1993). *Eur. J. Immunol.* **23**, 1917–1921.

Kitamura, D., Roes, J., Kuhn, R. and Rajewski, K. (1991). *Nature* **350**, 423–426.

Koehler, G. and Milstein, C. (1976). *Eur. J. Immunol.* **6**, 511–519.

Lo, D., Pursel, V., Linton, P.J., Sandgren, E., Behringer, R., Rexroad, C., Palmiter, R.D. and Brinster, R.L. (1991). *Eur. J. Immunol.* **21**, 1001–1006.

Lonberg, N. (1994). 'Handbook of Experimental Pharmacology, vol. 113, (M. Rosenberg and J.P. Moore, eds), The Pharmacology of Monoclonal Antibodies' Springer-Verlag, Berlin. pp. 49–101.

Lonberg, N., Taylor, L., Harding, F., Trounstine, M., Higgins, K., Schramm, S., Kuo, C., Mashayekh, R., Wymore, K., McCabe, J., Munoz-O'Regan, D., O'Donnell, S., Lapachet, E., Bengoechea, T., Fishwild, D., Carmack, C., Kay, R. and Huszar, D. (1994). *Nature* **368**, 856–859.

Ma, A., Fischer, P., Dildrop, R., Oltz, E., Rathburn, G., Achacoso, P., Stall, A. and Alt, F.W. (1992). *EMBO. J.* **11**, 2727–2734.

Manz, J., Denis, K., Witte, O., Brinster, R. and Storb, U. (1988). *J. Exp. Med.* **168**, 1363–1381.

Motoyama, N., Okada, H. and Azuma, T. (1991). *Proc. Natl Acad. Sci. USA* **88**, 7933–7937.

Muller, W., Ruther, U., Vieira, P., Hombach, J., Reth, M. and Rajewsky, K. (1989). *Eur. J. Immunol.* **19**, 923–928.

Nemazee, D. and Buerki, K. (1989). *Nature* **337**, 562–566.

Neuberger, M.S., Caskey, H.M., Petterson, S., Williams, G.T. and Surani, M.A. (1989). *Nature* **338**, 350–352.

Nussenzweig, M.C., Shaw, A.C., Sinn, E., Danner, D.B., Holmes, K.L., Morse, H.C., III and Leder, P. (1987). *Science* **236**, 816–819.

Nussenzweig, M., Shaw, A., Sinn, E., Campos-Torres, J. and Leder, P. (1988). *J. Exp. Med.* **167**, 1969–1974.

O'Brien, R., Brinster, R. and Storb, U. (1987). *Nature* **326**, 405–409.

Offen, D., Spatz, L., Escowitz, H., Factor, S. and Diamond, B. (1992). *Proc. Natl Acad. Sci. USA* **89**, 8332–8336.

Palmiter, R. and Brinster, R. (1986). *Annu. Rev. Genet.* **20**, 465–499.

Pernis, B., Chiappino, G., Kelus, A. and Gell, P. (1965). *J. Exp. Med.* **122**, 853–876.

Perry, R.P., Kelley, D.E., Coleclough, C., Seidman, J.G., Leder, P., Tonegawa, S., Matthyssens, G. and Weigert, M. (1980). *Proc. Natl Acad. Sci. USA* **77**, 1937–1941.

Perry, R.P., Coleclough, C. and Weigert, M. (1981). *Cold Spring Harb. Symp. Quant. Biol.* **XLV**, 925–933.

Petterson, S., Sharpe, M., Gilmore, D., Surani, A. and Neuberger, M. (1989). *Int. Immunol.* **1**, 509–516.

Ritchie, K.A., Brinster, R.L. and Storb, U. (1984). *Nature* **312**, 517–520.

Roehm, N., Herron, I., Cambier, C., DiGiusto, D., Haskins, K., Kappler, J. and Marrack, P. (1984). *Cell* **38**, 577–589.

Rogerson, B. (1994). *Mol. Immunol.* **31**, 83–98.

Rogerson, B., Hackett, J., Peters, A., Haasch, D. and Storb, U. (1991). *EMBO J.* **10**, 4331–4341.

Roth, P., Doglio, L., Manz, J., Kim, J.Y., Lo, D. and Storb, U. (1993). *J. Exp. Med.* **178**, 2007–2021.

Roth, P., Kurtz, B., Lo, D. and Storb, U. (1995). *J. Exp. Med.* **181**, 1059–1070.

Rudin, C., Hackett, J. and Storb, U. (1991). *J. Immunol.* **146**, 3205–3210.

Rusconi, S. and Kohler, G. (1985). *Nature* **314**, 330–334.

Rusconi, S. (1994) In 'The Impact of Gene Transfer Technique in Eukaryotic Cell Biology' (Shell P.and Starlinger, P. eds), pp. 134–152. Springer Verlag, Heidelberg.

Sharpe, M., Neuberger, M., Pannell, R., Surami, A. and Milstein, C. (1990). *Eur. J. Immunol.* **20**, 1379–1385.

Sharpe, M., Milstein, C., Jarvis, J. and Neuberger, M. (1991). *EMBO J.* **10**, 2139–2145.

Sohn, J., Gerstein, R., Hsieh, C., Lemer, M. and Selsing, E. (1993). *J. Exp. Med.* **177**, 493–504.

Storb, U., Wilson, R., Selsing, E. and Walfield, A. (1981). *Biochemistry* **20**, 990–996.

Storb, U., Ritchie, K., Hammer, R., O'Brien, R., Manz, J., Arp, B. and Brinster, R. (1985). In 'Genetic Manipulation of the Early Mammalian Embryo' (Constantini, F. and Jaenish, R. eds), pp. 197–209. Cold Spring Harbor Laboratory, Cold Spring Harbor.

Storb, U., Pinkert, C., Arp, P., Engler, P., Gollahon, K., Manz, J., Brady, W. and Brinster, R.L. (1986). *J. Exp. Med.* **164**, 627–641.
Strasser, A., Harris, A., Corcoran, L. and Cory, S. (1994). *Nature* **368**, 457–460.
Tsang, H., Pinkert, C., Hagman, J., Lostrum, M., Brinster, R.L. and Storb, U. (1988). *J. Immunol.* **141**, 308–314.
Tsao, B.P., Ohnishi, K., Cheroutre, H., Mitchell, B., Teitell, M., Mixter, P., Kronenberg, M. and Hahn, B.H. (1992). *J. Immunol.* **149**, 350–358.
Vasicek, T.J., Levinson, D.A., Schmidt, E.V., Campostorres, J. and Leder, P. (1992). *J. Exp. Med.* **175**, 1169–1180.
Walfield, A., Selsing, E., Arp, B. and Storb, U. (1981). *Nucleic Acids Res.* **9**, 1101–1109.
Weaver, D., Costantini, F., Imanishi-Kari, T. and Baltimore, D. (1985). *Cell* **42**, 117–127.
Weiler, E. (1965). *Proc. Natl Acad. Sci. USA* **54**, 1765–1772.
Yamamura, K.-I., Kudo, A., Ebihara, T., Kamino, K., Araki, K., Kumahara, Y. and Watanabe, T. (1986). *Proc. Natl Acad. Sci. USA* **83**, 2152–2156.
Yancopoulos, G.D. and Alt, F.W. (1985). *Cell* **40**, 271–281.
Zou, Y., Gu, H. and Rajewsky, K. (1993). *Science* **262**, 1271–1274.

16

B lymphocyte tolerance in the mouse

David Nemazee

Division of Basic Sciences, Department of Pediatrics, National Jewish Center for Immunology and Respiratory Medicine, Denver, CO, USA

Introduction 365
Clonal deletion as a mechanism for peripheral tolerance 367
Receptor editing as a mechanism for central tolerance 371
Implications for autoimmunity 373
Conclusion 375

INTRODUCTION

B lymphocytes with specificity for self-antigens pose a threat to the organism because of their remarkable potential to rapidly proliferate and to differentiate to high-rate antibody secretion. This potential is perhaps best illustrated by model immune responses to foreign antigens in mice, in which it has been shown that very small numbers of antigen-reactive B-cells, on the order of 20 individual cells, can eventually give rise to tens of milligrams of antibody. It also appears that autoreactive B lymphocytes may potentially play a key role in antigen presentation to class II-reactive T lymphocytes in autoimmunity and therefore autoreactive B-cells may have the potential to stimulate and expand quiescent autospecific T helper cells. For these reasons, there has been considerable interest in understanding how potentially harmful B-cell responses are normally avoided and in determining whether or not defects in B-cell tolerance play a primary role in the development of autoimmunity. In this chapter, I review some of the key findings and outstanding questions about B-cell tolerance.

A multiplicity of B lymphocytes

As discussed elsewhere in this book, B lymphocytes go through a number of developmental stages after acquiring their surface antigen receptors. B-cells are probably screened for tolerance at all of these stages. In addition, there are probably developmentally distinct lineages of B-cells, such as B-1 cells (see Hardy and Hayakawa, 1994) that may have unique responses to self and foreign antigens. Distinct lineage cells may in addition differ in the details of their development and in their susceptibility to tolerance. For example, the apparently self-renewing sIgM$^+$ B-1 cell population may need only be screened for tolerance early in development, whereas the so-called conventional B lymphocyte lineage is continually renewed from sIg$^-$ bone marrow precursors, and tolerance induction must be ongoing throughout the life of the individual to be effective.

One can further define at least three distinct stages of development during which B-cells are susceptible to self-tolerance (1) immature, sIgM$^+$ cells in the primary lymphoid organs (e.g. adult bone marrow, fetal liver); (2) small, recirculating sIgM$^+$/sIgD$^+$ cells, which normally make up the majority of B-cells in the peripheral lymphoid organs; and (3) germinal center B-cells that have recently been activated and have altered their antigen specificity through somatic V-region hypermutation. These three types of B-cell are phenotypically, functionally and anatomically distinct, but have in common the expression of sIg, which can interact with the extracellular environment. There are now data suggesting that self-tolerance can be induced during all of these developmental stages, although much work remains to be done to ascertain precise mechanisms and to establish differences in tolerance susceptibility. It is likely that different self-antigens induce tolerance primarily at one or another of these developmental stages, depending on the pattern of tissue-specific expression, the developmental timing of expression, or the accessibility of the antigen in question. Finally, it has long been known that B-cells from neonatal and adult mice can differ greatly in tolerance susceptibility (reviewed in Nossal, 1983), and a number of the key parameters that play a role in this difference are being actively investigated (e.g. Brines and Klaus, 1992; Yellen-Shaw and Monroe, 1992).

The specificity of B-cell tolerance

Before discussing mechanisms, it is of interest to recall what is known about the specificity of self-tolerance in B-cells. There is a very large body of literature in which the specificities of individual spleen B-cells or their clonal progeny have been analyzed by scoring antibody secretion. Using these methods, very high frequencies of autoreactive B-cells have been detected (reviewed by Nossal, 1989). These autoreactive B-cells have antibody reactivity to fixed cells, serum proteins, denatured DNA and cytoskeletal proteins. These results have often been interpreted as disproof of B-cell tolerance. On the other hand, detailed studies examining tolerance to experimental tolerogens or cell surface proteins revealed a rather different picture, indicating that tolerance could greatly diminish the frequency and affinity of reactive B-cells that could be induced to secrete antibody. These studies were done by assessing the frequency of cells or clones reactive to foreign antigen detected with and without pretreatments that induce tolerance (Chiller *et al.*, 1971; Nossal and Pike, 1975; Metcalf and Klinman, 1977; Etlinger and Chiller, 1979). This experimentally induced tolerance is highly specific: for example, tolerance induced to the hapten dinitrophenol (provided in tissue culture as a multimeric conjugate with carrier protein) is profound and fails to induce tolerance in clones reactive to the structurally similar hapten trinitrophenol (TNP) (Teale and Klinman, 1980). Similar results have been observed in comparisons of the antibody repertoires of humans of different ABO blood groups (Galili *et al.*, 1987; Rieben *et al.*, 1992). An additional argument in favor of the idea of B-cell self-tolerance is that, in the absence of B-cell tolerance to self, it is indeed hard to imagine how the science of serology could have emerged, with its routine demonstration of non-self specificity.

How are we then to resolve the apparent contradictions between the results with experimental tolerogens or immunogens and the studies indicating high frequencies of self-reactive B-cells? Perhaps the best attempt to address this question was that of McHeyzer-Williams and Nossal (1988), who devised a culture system capable of non-specifically activating and inducing high-rate antibody production in a high proportion of B-cells. These cultures produced IgM in the absence, and IgG_1 in the presence, of added *interleukin* (II)-4-containing supernatant. Culture supernatants were screened for antibody binding to methanol fixed or unfixed thymoma cells. The striking finding was that B-cells producing IgM antibodies to intracellular antigens were extremely frequent (1 in 37 B-cells), whereas the IgG_1-form of the same antibodies were rare (one B-cell in 3×10^6). Both IgM and IgG_1 antibodies that bound *unfixed* cells, and were therefore specific for the cell surface, were undetectable (less than one B-cell in 3×10^6). The straightforward interpretation of these results is that there exists a very stringent tolerance to cell surface antigens whereas tolerance to intracellular antigens may spare only low-affinity self-reactive B-cells, whose binding requires highly multivalent contacts to be detectable. These differences might reflect the relative inaccessibility of intracellular compared to extracellular antigens.

CLONAL DELETION AS A MECHANISM FOR PERIPHERAL TOLERANCE

A basic mechanistic distinction emerging from recent studies is that B-cell tolerance in the primary lymphoid organ ('central' tolerance) does not necessarily alter cell lifespan, whereas peripheral tolerance, either by clonal elimination or functional inactivation ('anergy'), involves a more or less rapid cell death. I will first discuss peripheral tolerance, and then central tolerance, the mechanism of which is more intimately related to the subject of this book, Ig genes.

It has long been known from anti-IgM suppression experiments *in vivo* that mimics of antigen can lead to the loss of peripheral B-cells, although a direct antibody-mediated cytotoxic effect was not ruled out (Lawton and Cooper, 1974). Ig transgenic mice have allowed more detailed analyses of peripheral tolerance mechanisms *in vivo*. In a model system for tissue-specific B-cell tolerance, Russell *et al.* (1991) showed that a hepatocyte-expressed major histocompatibility complex (MHC) class I K^b molecule could efficiently and profoundly reduce anti-K^k/K^b specific transgenic B-cells in the peripheral lymphoid organs in genetic crosses, rendering the mice virtually B-cell-less in the lymph nodes and eliminating serum anti-K^b titers. These results are most easily explained by tolerance-induced cell death. A series of experiments from the group of T. Honjo have shown that intra-peritoneal injection of antigen, or forms of anti-Ig treatments that cause extensive crosslinking of sIg receptors, can induce rapid apoptosis of B-1 (Murakami *et al.*, 1992; Nisitani *et al.*, 1993) and conventional B-cells in the peritoneum (Tsubata *et al.*, 1994). These investigators have further shown that enforced expression of aberrantly high levels of the *bcl*-2 proto-oncogene protein in B-cells can substantially block this deletion *in vivo* (Nisitani *et al.*, 1993). Moreover, peritoneal B-cells from autoimmune-prone strains of mice are resistant to deletion by such treatments (Tsubata *et al.*, 1994). The experiments of Carsetti *et al.* (1993), in which anti-TNP transgenic mice were challenged with TNP–carrier conjugates, similarly indicated that rapid B-cell depletion could be induced by acute antigen challenge. Using a clonal, splenic focus assay, Teale and Klinman (1980, 1984) showed that B-cell tolerance was an active process requiring new RNA and protein synthesis, and which was reversible in the presence of T-cell help. Recently, Parry *et al.* (1994) found that plastic-bound, but not soluble, anti-Ig reagents induced apoptosis in purified, resting B-cells. Furthermore, anti-CD40 and IL-4 acted synergystically to prevent this death. These results are consistent with the interpretation that *in vivo* peripheral tolerance proceeds by the mechanism of apoptosis.

It is likely that T-cell help, generated through the specific recognition of class II-restricted antigen and apparently delivered through multiple interacting pairs of molecules on the surfaces of B- and T-cells (Clarke and Ledbetter, 1994), can reverse such toxic effects of antigen, at least in a proportion of the responding B-cells. What is less clear at this point is to what extent other physiological stimuli acting through B-cell membrane proteins can provide such survival 'help' *in vivo*. It is presumed that T-independent type II antigens, which can elicit antigen-specific antibody responses in the apparent absence of T-cells, must somehow provide such secondary signals.

Tolerance studies in B-cell lymphomas

Certain B-cell tumors, such as the intensively studied mouse lymphoma WEHI-231, may be reasonably good models for the analysis of antigen-induced clonal elimination in peripheral B-cells and the rescue of B-cell viability and function through secondary signals. There is an extensive literature indicating that apoptosis can be induced by sIg crosslinking in WEHI-231 and a small subset of other sIg^+M tumors (reviewed by Scott, 1993). Many studies have analyzed the nature of the proximal sIg-mediated signals needed and the nature of secondary signals that can protect from anti-Ig-mediated suicide. The details of the signaling pathways involved in normal peripheral tolerance will likely emerge from the study of such tumor models and their mutants. An important recent paper has shown that the sIg-mediated death signal in WEHI-231 cells can be reversed by simultaneous stimulation through CD40 (Tsubata *et al.*, 1993). This experimental model system is a promising avenue of future research into the biochemical nature of the two-signal model for B-cell activation and tolerance (Bretscher and Cohn, 1970).

Tolerance in memory B-cells

B-cells can change their antigen specificity upon T-dependent antigen stimulation through Ig V-region somatic hypermutation in germinal centers. This process has the potential to generate autoreactive B-cells whose reactivity to self must be controlled. Evidence for the existence of a second tolerance window in memory B-cells during or after somatic V-gene hypermutation has been presented (Walker and Weigle, 1985; Linton *et al.*, 1991). These experiments were done by attempting to induce tolerance in hapten–carrier immunized cells by treatment with the initial hapten on a different carrier, thereby providing sIg crosslinking signals, but no T-cell help. Tolerance was assessed by antibody secretion either in whole animals or in splenic fragment culture. Recent studies by Nossal's group (Nossal *et al.*, 1993; Pulendran *et al.*, 1994) have shown that the disruption of ongoing T-cell help by high dose, deaggregated carrier protein is by far a more important means of memory B-cell tolerance induction than direct sIg crosslinking in these B-cells. These results may indicate that antigen-activated B-cells that lose affinity or specificity for the immunizing antigen are at a competitive disadvantage compared to B-cells that maintain specificity. One might extend this idea to suggest that hypermutating germinal center B-cells that acquire self-specificity might be eliminated because this change in specificity would indirectly restrict T-cell help. In any case, the strong evidence that the autoantibody-producing B-cell clones in naturally occurring autoimmune disease are highly selected and the product of repeated somatic V-gene hypermutation (Radic and Weigert, 1994) emphasizes the possibility that defects in important aspects of memory or germinal center B-cell biology may play a role in autoimmunity. In addition, there is evidence obtained from the study of human germinal center cells that sIg crosslinking is required for B-cell survival and *bcl*-2 upregulation (Liu *et al.*, 1991). While these studies fail to strongly support a role for sIg-mediated tolerance in memory B-cells, they indicate a requirement for sustained T-cell help for germinal center B-cell survival.

The relevance of anergy in B-cells

The earliest experiment using Ig transgenic mice to study B-cell tolerance was devised by Goodnow, Basten and colleagues, who bred anti-hen egg lysozyme transgenic mice with mice expressing a transgene-encoded, soluble form of lysozyme. These mice generated substantial numbers of splenic B-cells that, when analyzed in a number of functional assays, were 'anergic' (i.e. hyporesponsive); these animals also manifested an unusual depletion of B-cells from the marginal zone of the splenic lymphoid follicle (Goodnow et al., 1988; Adams et al., 1990; Mason et al., 1992). In addition, the anergic phenotype was characterized by an inability to signal through sIg, and functional hyporesponsiveness that could be reversed by stimulation with lipopolysaccharide but not by T-cell help (Cooke et al., 1994). These results have until recently presented an enigma, because it was assumed that these anergic B-cells, being present in the spleen in numbers comparable to those of the antigen-free control mice, were similarly long-lived, implying that they might either have an affirmative function in the immune system or pose a threat of autoimmunity should their functional inactivation be reversible. It was later shown that the anergic B-cells have an extrordinarily short lifespan *in vivo*, thus greatly diminishing the effective distinction between anergy and deletion (Fulcher and Basten, 1994). This information also suggests that one need not invoke a function for the anergic B-cells, which, had they been longer-lived, might have played a suppressive role as tolerogenic, self-antigen-specific presenting cells (discussed in Nemazee, 1992). Other groups have provided evidence for anergy *in vivo* in normal (Nossal and Pike, 1980; Pike et al., 1982; Gause et al., 1987) and in Ig transgenic mice (Erikson et al., 1991). It remains to be seen whether or not anergy in these systems is also associated with short cell lifespan *in vivo*, but it is striking that in many of these experiments virtually the entire cohort of B-cells exhibited tolerance, and the lifespans of the cells in the face of interclonal competition was not tested. A B-cell–B-cell competition dependence of anergic B-cell lifespan implies that a lymphopenic environment could support greater numbers of autoreactive anergic cells, thus increasing the potential risk of the reversal of tolerance.

The implications for signaling

Despite the possible transient nature of B-cell anergy, which may be followed by cell death, it is interesting to consider the possibility that prior to interaction with self-antigens B-cells have multiple potential fates, depending on the nature of the antigenic signal eventually encountered. A similar idea has been proposed to explain antigen-induced activation of B-cells with conjugates of differing molecular size and structure (Dintzis et al., 1976). In the absence of secondary signals, such as T-cell help, strongly crosslinking signals induce rapid cell death, whereas weaker signals may induce anergy that allows a somewhat longer window of opportunity for interactions with other cell types or factors. Regardless of the functional significance of anergy compared to deletion, it is interesting and perhaps somewhat surprising that the same sIg receptor can lead to qualitatively different responses, depending on the form of antigen interacting with the receptor. This implies that the B-cell is capable of interpreting structural information about antigen that may be useful for determining the class of the response, be it positive or negative. How this could occur at the level of intracellular signal transduction through a single cell surface receptor is, at present, an enigma, but may be explained by various quantitative parameters such as the amplitude or duration of signaling (Brunswick *et al.*, 1989). By extension from the emerging data on the functional importance of receptor affinity for T-cell responses (De Magistris *et al.*, 1992; Evavold *et al.*, 1993), it would appear possible that differences in the affinity of the self-antigen–B-cell receptor interaction could result in qualitatively distinct signals and cell responses.

RECEPTOR EDITING AS A MECHANISM FOR CENTRAL TOLERANCE

Recent Ig transgenic mouse experiments indicate that, in the bone marrow, antigens that mediate clonal deletion probably do not induce rapid programmed cell death of sIgM$^+$ B-cells, but instead induce a reversible 'developmental' block (Nemazee and Bürki, 1989a, 1989b; Hartley *et al.*, 1993). Interestingly and unexpectedly, these immature sIg$^+$ bone marrow B-cells encountering deleting antigen express V(D)J recombinase activator gene (*RAG*) mRNA (Tiegs *et al.*, 1993) and undergo secondary Ig light chain gene rearrangement, allowing a modification in receptor specificity that often renders the cells no longer autoreactive and spares them from death (Gay *et al.*, 1993; Radic *et al.*, 1993; Tiegs *et al.*, 1993; Chen *et al.*, 1994). This mechanism of tolerance has been termed 'receptor editing'.

An intriguing element of the receptor editing hypothesis is how it alters our picture of the forces driving the evolution of the murine Ig genes (Fig. 1). The absence of D regions in light chain genes, combined with their presence in heavy chain genes, should facilitate efficient nested secondary V rearrangements in light chain genes and tend to suppress them in heavy chain genes, because of the constraints of the 12/23-bp spacer rule in recombination signal sequences. The retention through evolution of both κ and λ loci should increase the chances of finding allowable light chains. Furthermore, the unique features of the κ locus become more understandable in the context of editing. Most primary V_κ–J_κ rearrangements are inversional (Harada and Yamagishi, 1991, and references therein) although only ~40% of V_κ genes are thought to be in reading frames of opposite orientation to the J_κ genes (Shapiro and Weigert, 1987). This indicates that the developing B-cell takes some pains to retain the entire remaining V_κ genes, probably in order to maximize the repertoire of V_κ genes that can be used in secondary rearrangements. Finally, the existence of RS (the so-called κ deleting element downstream of C_κ), which mediates the V(D)J recombinase-dependent deletion of Cκ in a large fraction of λ light chain-expressing B-cells, has no obvious function other than to kill active, or potentially active, κ light chain loci. (The old notion that prior RS rearrangements might be necessary for λ rearrangements has been ruled out by C_κ intron enhancer knockout experiments; Chen *et al.*, 1993; Zou *et al.*, 1993.) Thus, the Ig genes demonstrate features that seem more appropriate for the mechanisms of tolerance and receptor selection than for the simple maximization of diversity.

In the Ig heavy chain transgenic mice developed by Weigert's group, in which the transgenic heavy chain, when paired with a wide variety of endogenous light chains, results in an autoreactive (anti-double stranded DNA) receptor, a highly restricted set of light chains are used among splenic B-cells. These active light chains often bear the signs of having resulted from secondary rearrangements; for example, downstream J_κ

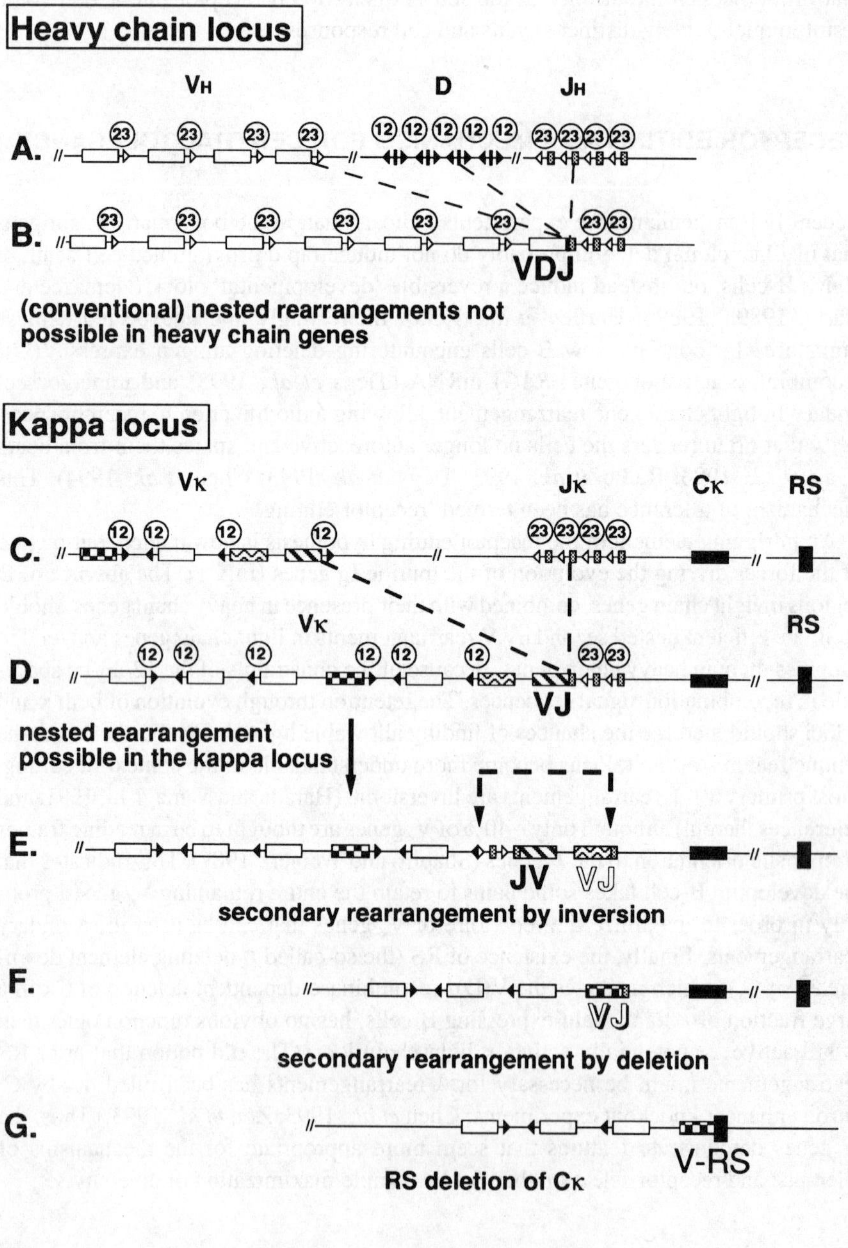

Heavy chain locus

A.

B.
VDJ

(conventional) nested rearrangements not possible in heavy chain genes

Kappa locus

C.

D.
VJ

nested rearrangement possible in the kappa locus

E.
JV VJ

secondary rearrangement by inversion

F.
VJ

secondary rearrangement by deletion

G.
V-RS

RS deletion of $C\kappa$

genes are used at an abnormally high frequency and recurrent, rare V_κ–J_κ combinations arise independently at high frequency (Radic et al., 1993; Chen et al., 1994). These results argue strongly that in mice in which light chain rearrangement and repertoire development are allowed to proceed normally, the production of autoreactive receptors results in nested rearrangements in the κ locus.

The extent to which a receptor editing tolerance mechanism may obtain in normal, non-transgenic B-cells remains to be explored, but is consistent with the high degree of secondary κ rearrangements in normal B-cells, which frequently excise in-frame, potentially functional primary rearrangements (Harada and Yamagishi, 1991). In addition, there is evidence that a significant proportion of immature sIg⁺ B-cells in normal mouse bone marrows express *RAG* mRNA (Ma et al., 1992).

IMPLICATIONS FOR AUTOIMMUNITY

A key question that has not yet been satisfactorily answered is whether or not a defect in B-cell tolerance plays a role in autoimmunity. Theoretically, such a defect could alone cause systemic autoimmune disease, as we have argued elsewhere (Nemazee et al., 1991a). The increasing evidence for an intrinsic B-cell defect in murine autoimmunity is reviewed here.

Fig. 1. The implications of the receptor editing hypothesis for mouse κ locus organization. Unrearranged heavy (A) and κ (C) loci undergo primary functional rearrangements (B and D, respectively). Cells at the pre-B to B-cell transitional stage of development in the bone marrow are undergoing κ rearrangement. The appearance of IgM on the cell surface normally turns off further recombination but, if the newly emerging receptors are autoreactive and their surface IgM is crosslinked by membrane-bound autoantigen, recombinase levels remain high. The structure of the Ig genes favors secondary, nested rearrangements in the κ locus and suppresses them in the H locus because the recombination signal sequences of V and J are compatible in the κ locus and incompatible in the H locus. (Elements with 23-bp spacers join to elements with 12-bp spacers as indicated.) If secondary κ chain rearrangement is successful, the previously active V_κ–J_κ is rendered inactive by nested rearrangement and deletion (F), or by inversion (E), which places the V_κ–J_κ far away from C_κ. Rearrangement may continue until a new, functional VJ is generated, or until the C_κ loci are inactivated by using up the available J_κ or by RS recombination (G). The functional light chain must fail to result in an autoreactive receptor or the editing process may be resumed. It is also possible at this stage that a new light chain that pairs preferentially with the cell's μ heavy chain can in some cases extinguish the B-cell's anti-self specificity without eliminating the active, autospecific VJ on the other chromosome. In addition, some cells undergo RS recombination, which deletes the C_κ locus. Such cells appear to frequently rearrange and express λ loci. In the mouse, further editing of active λ light chain loci that encode autoantibodies is not generally possible, given the organization of the λ genes; however in other species, such as the human, nested secondary $V_\lambda J_\lambda$ rearrangements that can eliminate active $V_\lambda J_\lambda$ may be possible.

Tetraparental chimera studies

There are a number of well-known mouse strains that spontaneously develop lupus-like autoimmunity: MRL/lpr, (NZB × NZW)F1 and BXSB male mice, among others. With increasing age these mice produce large amounts of many kinds of autoantibodies, such as anti-DNA, and suffer from a variety of manifestations of immune complex formation, including kidney failure. Various studies over the years have shown that the autoimmune phenotype can be transferred by bone marrow and is dependent on T-cells (reviewed in Theofilopoulos and Dixon, 1985; Theofilopoulos et al., 1989). However, there is a considerable body of evidence that these mice have intrinsic B-cell defects as well (Gershwin et al., 1979; Jyonouchi et al., 1983, 1985; Mihara et al., 1988). These mice have been analyzed to determine if their intrinsic defect is in self-tolerance in the T helper cells alone, the assumption being that an autoreactive T-cell can recruit normal B-cells into the response against self tissue. These experiments were done by producing mixed bone marrow chimeras or mixed embryo aggregation chimeras, in which T-cells and allotype-marked B-cells of the autoimmune-prone and normal genotypes developed together. In this situation, if normal B-cells could be recruited into the autoimmune response, then the fraction of autoantibody produced from B-cells of the autoimmune-prone and normal genotypes should be the same, or at least proportional to the relative amounts of total serum Ig of the parental allotypes. The striking and consistent finding in all of these studies has been that B-cells of the autoimmune-prone genotype secrete virtually *all* of the autoantibody in these mixed chimeras (Merino et al., 1991; Nemazee et al., 1991b; Sobel et al., 1991). This would indicate that an intrinsic B-cell defect exists in the autoimmune-prone mouse strains. It should be emphasized that there is little evidence for a defect in T-cell thymic deletion by self-superantigen in autoimmune-prone mice (Kotzin et al., 1989; Theofilopoulos et al., 1989).

Autoimmunity in *bcl*-2 mice

The *bcl*-2 oncogene is frequently activated in human B-cell follicular lymphomas as a result of chromosomal translocations that bring it into proximity with the Ig heavy chain enhancer. Overexpression of *bcl*-2 has been shown to inhibit programmed cell death in IL-3-dependent B-cell lines deprived of IL-3 (Vaux et al., 1988). Two groups have described transgenic mice expressing *bcl*-2 under the control of the Ig heavy chain enhancer (McDonnell et al., 1989; Strasser et al., 1990). These mice manifest B-cell hyperplasia and prolonged B-cell survival *in vivo* and *in vitro*. Interestingly, one of these transgenic mouse lines also manifests early mortality as a result, not of B-cell tumors, but of lupus-like glomerulonephritis (Strasser et al., 1991). Since this transgenic mouse line expresses *bcl*-2 in the B-cells, but not detectably in T-cells, these results suggest that a defect in B-cells alone might be responsible for the autoimmunity.

Autoimmunity in anti-erythrocyte transgenic mice

Honjo's group has extensively investigated a transgenic mouse line expressing heavy and light chain genes encoding an erythrocyte autoantibody. Surprisingly, even on a non-autoimmune background (C57BL6) a proportion of these transgenic mice develop hemolytic anemia. This group has established that B-cells of these mice become tolerant to self red blood cells (RBCs) by clonal deletion, but that some B-cells manage to make their way to the peritoneum, where apparently low levels of antigen allow their persistence, expansion and progression to antibody secretion (Okamoto et al., 1992). Intraperitoneal injection of RBCs leads to rapid apoptosis of these B-cells and regression of clinical signs of hemolytic anemia (Murakami et al., 1992). Interestingly, this inducible peripheral deletion could be blocked by bcl-2 transgene overexpression. These results show that in this case induction of antigen-mediated B-cell tolerance is sufficient to prevent autoimmune disease. This in turn implies that defects in B-cell tolerance alone may in some cases induce autoimmune disease. Recent experiments measuring the *in vivo* tolerance susceptibility of B-cells in autoimmune-prone mice have provided additional support for this view (Tsubata et al., 1994).

Development of lupus in mice with severe combined immune deficiency (SCID) reconstituted with (NZB/NZW)F1 mouse pre-B cell clones

Reininger et al. (1992) showed that pre-B cell clones derived from lupus-prone NZB/NZW F1(B/W) mice (but not those derived from normal mice) can differentiate in SCID mice and secrete high titer anti-nuclear antibodies, including anti-DNA. Most important, in a subset of recipients these cells gave rise to lupus-like autoimmune disease in the apparent absence of T-cells, suggesting again that intrinsic B cell defects can be responsible for autoimmune disease.

Collectively, these studies provide strong evidence that intrinsic B cell defects, possibly including self-tolerance defects, may be at the root of some autoimmune diseases.

CONCLUSION

B lymphocyte self-tolerance is mediated by multiple mechanisms and probably plays a role in the prevention of at least some aspects of autoimmunity or hyperreactivity. The processes of B-cell tolerance must in addition play an enormously important role in the development of the B-cell repertoire, which could directly affect both B- and T-cell responses to foreign antigens. A better understanding of B cell tolerance mechanisms will focus our attention on important gene products and processes in B-cells, which, in disease, may be mutated or subverted by parasites.

Acknowledgements

This work was supported by a grant from the Arthritis Foundation and NIH grants RO1 GM 44809, RO1 AI 33608, KO4 AI 01161.

REFERENCES

Adams, E., Basten A. and Goodnow C.C. (1990). *Proc. Natl Acad. Sci. USA* **87**, 5687–5691.
Bretscher, P.A. and Cohn, M. (1970). *Science* **169**, 1042–1049.
Brines, R.D. and Klaus, G.G.B. (1992). *Int. Immunol.* **4**, 765–771.
Brunswick, M., June, C.H., Finkelman, F.D., Dintzis, H.M., Inman, J.K. and Mond, J.J. (1989). *Proc. Natl Acad. Sci. USA* **86**, 6724–6728.
Carsetti, R., Köhler, G. and Lamers, M.C. (1993). *Eur. J. Immunol.* **23**, 168–178.
Chen, C., Radic, M.Z., Erikson, J., Camper, S.A., Litwin, S., Hardy, R.R. and Weigert, M. (1994). *J. Immunol.* **152**, 1970–1982.
Chen, J., Trounstine, M., Kurahara, C., Young, F., Kuo, C.-C., Xu, Y., Loring, J.F., Alt, F.W. and Huszar, D. (1993). *EMBO J.* **12**, 821–830.
Chiller, J.M., Habitch, G.S. and Weigel, W.O. (1971). *Science* **171**, 813–815.
Clarke, E.A. and Ledbetter, J.A. (1994). *Nature* **367**, 425–428.
Cooke, M.P., Heath, A.W., Shokat, K.M., Zeng, Y., Finkelman, F.D., Linsley, P.S., Howard, M. and Gordnow, C.C. (1994). *J. Exp. Med.* **179**, 425–438.
De Magistris, M.T., Alexander, J., Coggeshall, M., Altman, A., Gaeta, F.C.A., Grey, H.M. and Sette, A. (1992). *Cell* **68**, 625–634.
Dintzis, H.M., Dintzis, R.Z. and Vogelstein, B. (1976). *Proc. Natl Acad. Sci. USA* **73**, 3671–3675.
Erikson, J., Radic, M.Z., Camper, S.A., Hardy, R.R., Carmack, C. and Weigert, M. (1991). *Nature* **349**, 331–334.
Etlinger, H.M. and Chiller, J.M. (1979). *J. Immunol.* **122**, 2558–2563.
Evavold, B.D., Sloan-Lancaster, J. and Allen, P.M. (1993). *Immunol. Today* **14**, 602–608.
Fulcher, D.A. and Basten, A. (1994). *J. Exp. Med.* **179**, 125–134.
Galili, U., Buehler, J., Shohet, S.B. and Macher, B.A. (1987). *J. Exp. Med.* **165**, 693–704.
Gause, A., Yoshida, N., Kappen, C. and Rajewsky, K. (1987). *Eur. J. Immunol.* **17**, 981–990.
Gay, D., Saunders, T., Camper, S. and Weigert, M. (1993). *J. Exp. Med.* **177**, 999–1008.
Gershwin, M.E., Castles, J.J., Erickson, K. and Ahmed, A. (1979). *J. Immunol.* **122**, 2020–2025.
Goodnow, C.C., Crosbie, J., Adelstein, S., Lavoie, T.B., Smith-Gill, S.J., Brink, R.A., Pritchard-Briscoe, H., Wotherspoon, J.S., Loblay, R.H., Raphael, K., Trent, R.T. and Basten, A. (1988). *Nature* **334**, 676–682.
Harada, K. and Yamagishi, H. (1991). *J. Exp. Med.* **173**, 409–415.
Hardy, R.R. and Hyakawa, K. (1994). *Adv. Immunol.* **55**, 297–339.
Hartley, S.B., Cooke, M.P., Fulcher, D.A., Harris, A.W., Cory, S., Basten, A. and Goodnow, C.C. (1993). *Cell* **72**, 325–335.
Jyonouchi, H., Kincade, P.W., Good, R.A. and Gershwin, M.E. (1983). *J. Immunol.* **131**, 2219–2225.
Jyonouchi, H., Kincade, P.W. and Good, R.A. (1985). *J. Immunol.* **134**, 858–864.
Kotzin, B.L., Babcock, S.K. and Herron, L.R. (1988). *J. Exp. Med.* **168**, 2221–2229.
Lawton, A.R. and Cooper, M.D. (1974). *Contemp. Top. Immunobiol.* **3**, 193–225.
Linton, P.-J., Rudie, A. and Klinman, N.R. (1991). *J. Immunol.* **146**, 4099–4104.

Liu, Y., Mason, D.Y., Johnson, G.D., Abbot, S., Gregory, C.D., Hardie, D.L., Gordon, J. and Maclennan, I.C.M. (1991). *Eur. J. Immunol.* **21**, 1905–1910.
Ma, A., Fisher, P., Dildrop, R., Oltz, E., Rathbun, G., Achacoso, P., Stall, A. and Alt, F,W. (1992). *EMBO J.* **11**, 2727–2734.
McDonnell, T.J., Deane, N., Platt, F.M., Nunez, G., Jaeger, U., McKearn, J.P. and Korsmeyer, S.J. (1989). *Cell* **57**, 79–88.
McHeyzer-Williams, M.G. and Nossal, G.J.V. (1988). *J. Immunol.* **141**, 4118–4123.
Mason, D.Y., Jones, M. and Goodnow, C.C. (1992). *Int. Immunol.* **4**, 163–175.
Merino, R., Fossati, L., Lacour, M. and Izui, S. (1991). *J. Exp. Med.* **174**, 1023–1029.
Metcalf, E.S. and Klinman, N.R. (1977). *J. Immunol.* **118**, 2111–2116.
Mihara, M., Ohsugi, Y., Saito, K., Miyai, T., Togashi, M., Ono, S., Murakami, S., Dobashi, K., Hirayama, F. and Hamaoka, T. (1988). *J. Immunol.* **141**, 85–90.
Murakami, M., Tsubata, T., Okamoto, M., Shimizu, A., Kumagai, S., Imura, H. and Honjo, T. (1992). *Nature* **357**, 77–80.
Nemazee, D. (1992). *Res. Immunol.* **143**, 272–275.
Nemazee, D. and Bürki, K. (1989). *Proc. Natl Acad. Sci. USA*, **86**, 8039–8043.
Nemazee, D. and Bürki, K. (1989). *Nature* **337**, 562–566.
Nemazee, D., Russell, D., Arnold, B., Haemmerling, G., Allison, J., Miller, J.F.A.P., Morahan, G. and Buerki, K. (1991a). *Immunol. Rev.* **122**, 117–132.
Nemazee, D., Guiet, C., Buerki, K. and Marshak-Rothstein, A. (1991b). *J. Immunol.* **147**, 2536–2539.
Nisitani, S., Tsubata, T., Murakami, M., Okamoto, M. and Honjo, T. (1993). *J. Exp. Med.* **178**, 1247–1254.
Nossal, G.J.V. (1983). *Annu. Rev. Immunol.* **1**, 33–62.
Nossal, G.V.J. (1989). In 'Fundamental Immunology', 2nd edn (W.E. Paul, ed.), pp. 571–586. Raven Press, New York.
Nossal, G.J.V. and Pike, B.L. (1975). *J. Exp. Med.* **141**, 904–915.
Nossal, G.J.V. and Pike, B.L. (1980). *Proc. Natl Acad. Sci. USA* **77**, 1602–1606.
Nossal, G.J.V., Karvelas, M. and Pulendran, B. (1993). *Proc. Natl Acad. Sci. USA* **90**, 3088–3092.
Okamoto, M., Murakami, M., Shimizu, A., Ozaki, S., Tsubata, T., Kumagai, S-i. and Honjo, T. (1992). *J. Exp. Med.* **175**, 71–79.
Parry, S.L., Holman, M.J., Hasbold, J. and Klaus, G.G.B. (1994). *Eur. J. Immunol.* **24**, 974–979.
Pike, B.L., Boyd, A.W. and Nossal, G.J.V. (1982). *Proc. Natl Acad. Sci. USA* **79**, 2013–2017.
Pulendran, B., Karvelas, M. and Nossal, G.J.V. (1994). *Proc. Natl Acad. Sci. USA* **91**, 2639–2643.
Radic, M.Z. and Weigert, M. (1994). *Annu. Rev. Immunol.* **12**, 487–520.
Radic, M.Z., Erikson, J., Litwin, S. and Weigert, M. (1993). *J. Exp. Med.* **177**, 1165–1173.
Reininger, L., Radaszkiewicz, T., Kosco, M., Melchers, F. and Rolink A.G. (1992). *J. Exp. Med.* **176**, 1343–1353.
Rieben, R., Tucci, A., Nydegger, U.E. and Zubler, R.H. (1992). *Eur. J. Immunol.* **22**, 2713–2717.
Russell, D.M., Dembic, Z., Morahan, G., Miller, J.F.A.P., Bürki, K. and Nemazee, D. (1991). *Nature* **354**, 308–311.
Scott, D.W. (1993). *Adv. Immunol.* **54**, 393–425.
Shapiro, M.A. and Weigert, M. (1987). *J. Immunol.* **139**, 3834–3839.
Sobel, E.S., Katagiri, T., Katagiri, K., Morris, S.C., Cohen, P.L. and Eisenberg, R.A. (1991). *J. Exp. Med.* **173**, 1441–1449.
Strasser, A., Harris, A.W., Vaux, D.L., Webb, E., Bath, M.L., Adams, J. and Cory, S. (1990). *Curr. Top. Microbiol. Immunol.* **166**, 175–181.
Strasser, A., Whittingham, S., Vaux, D.L., Bath, M.L., Adams, J., Cory, S. and Harris, A.W. (1991). *Proc. Natl Acad. Sci. USA* **88**, 8661–8665.
Teale, J.M. and Klinman, N.R. (1980). *Nature* **288**, 385–387.
Teale, J.M. and Klinman, N.R. (1984). *J. Immunol.* **133**, 1811–1817.
Theofilopoulos, A.N. and Dixon, F.J. (1985). *Adv. Immunol.* **37**, 269–390.
Theofilopoulos, A.N., Kofler, R., Singer, P. and Dixon, F.J. (1989). *Adv. Immunol.* **46**, 61–109.
Tiegs, S.L., Russell, D.M. and Nemazee, D. (1993). *J. Exp. Med.* **177**, 1009–1020.

Tsubata, T., Wu, J. and Honjo, T. (1993). *Nature* **364**, 645–648.
Tsubata, T., Murakami, M. and Honjo, T. (1994). *Curr. Biol.* **4**, 8–17.
Vaux, D.L., Cory, S. and Adams, J.M. (1988). *Nature* **335**, 440–442.
Walker, S.M. and Weigle, W.O. (1985). *Cell. Immunol.* **90**, 331–338.
Yellen-Shaw, A.J. and Monroe, J.G. (1992). *J. Exp. Med.* **176**, 129–137.
Zou, Y.-R., Takeda, S. and Rajewsky, K. (1993). *EMBO J.* **12**, 811–820.

17

Immunoglobulin variable region gene segments in human autoantibodies

Kathleen N. Potter and J. Donald Capra

Department of Microbiology, University of Texas Southwestern Medical Center, Dallas, Texas, USA

Introduction 379
V(D)J rearrangement of immunoglobulin genes......................380
Generation of antibody diversity..381
V(D)J gene segment utilization by autoantibodies 382
Autoantibody-associated idiotypes 384
Auto-anti-idiotypic antibodies...385
Cold agglutinins..386
Rheumatoid factors 389
Anti-DNA antibodies..391
Conclusions 392

INTRODUCTION

In order to carry out its function of protecting the host against invasion by pathogenic microbial agents, the immune system must distinguish between foreign and self-antigens in order to avoid self-destruction. This state of balance between the immune system and the host is referred to as self-tolerance. There are many unanswered questions, however, regarding the mechanisms involved in establishing and maintaining tolerance and how tolerance is broken. The concept of self-tolerance predicts that autoreactive B cell clones will either be inactivated or removed from the immunological repertoire, a process called clonal deletion (Nossal, 1992). It has become apparent that the immune system frequently produces autoantibodies directed against self-antigens, and there has been considerable speculation that these autoantibodies are required for maintaining the healthy state of the individual (Avrameas, 1991). The data also demonstrate that autoantibodies are derived from gene segments representing every V_H family and many are germline encoded not requiring somatic mutation for binding activity (reviewed in Pascual and Capra, 1991a). The finding that pathogenic autoantibodies can be germline encoded, and that most of us contain the same or very similar gene segments (see below), suggests that the immunoglobulin repertoire alone is not a risk factor in autoimmune disease. Indeed most of the evidence would suggest that individuals with autoimmune disease do not have or express a unique set of immunoglobulin genes.

In spite of the ubiquitous presence of autoreactive antibodies, autoimmune disease is not common. The question of whether there are any characteristic differences between the affinity or specificity and molecular composition of natural autoantibodies found in healthy individuals and pathogenic antibodies in patients with autoimmmune diseases is still an intriguing one. Structural analysis using X-ray crystallography (Bentley *et al.*, 1990; Ban *et al.*, 1994), electron microscopy (Roux *et al.*, 1987) and anti-idiotypic antibodies, which are discussed below, are attempting to correlate structural features of antibodies with their biological function. In this chapter we discuss gene segment usage by autoantibodies as compared with the normal repertoire, and the functional significance of the idiotypes formed in the expressed immunoglobulins.

V(D)J REARRANGEMENT OF IMMUNOGLOBULIN GENES

The variable regions of immunoglobulin light (L) and heavy (H) chains are assembled in an ordered fashion from three multigene families of DNA segments referred to as the variable (V), diversity (D; in H chains only) and joining (J) gene segments (reviewed in Tonegawa, 1983). The majority of the human immunoglobulin H chain gene segments are on the long arm of chromosome 14q in band 32.3, which consists of approximately 1.1 megabases (Mb) of DNA. A complete map of the human immunoglobulin V_H locus on the telomeric region of chromosome 14q has been determined (Matsuda *et al.*, 1993; Cook *et al.*, 1994). The locus contains 87 V_H gene segments, of which only 50 are functional, approximately 30 D segments, nine J_H segments consisting of six functional and three pseudogenes, and a constant (C) region consisting of nine functional and two pseudogenes. The V_H genes are grouped into seven families based on greater than 80% homology at the DNA level and 75% at the amino acid level. The V_H7 family is closely related to the V_H1 family, and together they contain approximately 25 members (van Dijk *et al.*, 1993). The V_H2 family has approximately four members, V_H3 is the largest family with approximately 55 members, although only 50% are functional, V_H4 has approximately 12 members, V_H5 has three members (two are functional), and V_H6 has one member and it is functional.

There are two isotypes of light chains, κ (Cox *et al.*, 1994) and λ (Vasicek and Leder, 1990.) In humans, approximately 60% are κ and 40% are λ. The κ and λ loci are located on chromosomes 2 and 22 respectively. The κ locus spans over 2 Mb of

DNA and consists of two large duplicated blocks that are inversely oriented. The locus contains approximately 76 κ gene segments, of which 32 are potentially functional, five tandemly arrranged J_κ gene segments and one C_κ gene. There are no D segments in the L chain complex. Approximately 25 V_κ orphon gene segments, most of which are pseudogenes, have been mapped to other chromosomes (reviewed in Zachav, 1993). The λ locus contains seven unique C_λ gene segments each preceded by a single J_λ gene segment. It is estimated that there are approximately 70 V_λ gene segments. The V_λ gene segments have been classified into nine families based on amino acid and nucleotide sequences (Williams and Winter, 1993).

Variable regions of heavy chains are formed from the union of a single D segment with a single J_H segment to form DJ_H followed by the joining of V_H to DJ_H, both by site-specific recombination events (Yancopoulos and Alt, 1986; Alt et al., 1992). The union of the rearranged variable region with the C region is accomplished through splicing at the RNA level (Early et al., 1980). $V_L J_L$ recombination occurs by the same mechanisms as for the H chain locus.

B cells in the preimmune repertoire express germline-encoded antibodies of the IgM isotype. Subsequent to stimulation by antigen, IgM-bearing B cells may differentiate into plasma cells, which secrete antibody of either the IgG, IgA or IgE class. Class-switch recombination is a regulated intrachromosomal event involving deletion of intervening DNA segments (reviewed in Coffman et al., 1993). Antibodies with the original specificity are produced but the effector functions are now characterized by the newly acquired C region. The isotype of antibodies may be significant in the generation of pathogenic antibodies as IgG antibodies are usually somatically mutated in the variable region and have higher affinities for antigen than IgM antibodies.

GENERATION OF ANTIBODY DIVERSITY

The diversity of antibody specificities is regulated by several different mechanisms.

1. The multiplicity of germline genes for V, D and J gene segments provides the raw material from which the immune system draws.
2. The various combinations of different $V_H DJ_H$ and $V_L J_L$ gene segments are the initial substrates for generating a multitude of structures that is further expanded by D–D fusions and D inversions (Meek et al., 1989; Gu et al., 1991).
3. N-segment additions (template-independent insertions by the action of terminal deoxynucleotidyl transferase; TdT) (Alt and Baltimore, 1982) and template-encoded P-nucleotide additions (Lafaille et al., 1989) generate diversity at the junction of the V, D and J gene segments. N-segments are rarely detected in prenatal and neonatal repertoires but commonly found in $V_H DJ_H$ junctions in adult lymphocyte repertoires (Meek, 1990). Processes that either add or delete nucleotides during the joining of D and J_H coding regions result in the generation

of new amino acid residues in the V_H complementarity determining region 3 (CDR3) and significantly affect antibody specificity.
4. Somatic mutation occurs in antigen-activated B-cell populations resulting in the production of surface receptors having increasingly high affinity for the activating antigen (Gearhart and Bogenhagen, 1983; Tonegawa, 1983).
5. Particular combinations of H and L chains increase the antibody repertoire since, in theory, any H chain can pair with any L chain. Analyzing a large database of amino acid sequences of antibodies with various specificities, Kabat and Wu (1991) determined that the $CDR3_H$ has the major influence on antibody specificity and diversity. L chains have been shown to modulate antibody binding even when the H chain dominates the binding interaction (Radic et al., 1991).

Using the polymerase chain reaction, Tomlinson et al. (1992) sequenced 74 human germline V_H segments from a single individual and built a directory of 122 V_H segments with different nucleotide sequences. Their analysis indicates the structural diversity of the germline repertoire for antigen binding is fixed at about 50 groups of V_H segments. Sequence comparisons also indicate that there is little polymorphism in the smaller human V_H gene families (Williams et al., 1991; Tomlinson et al., 1992), although Snyder et al. (1992) concluded that the V_H4 germline genes exhibit considerable polymorphism. The diversity of the variable regions of H and L chains is to generate an expansive number of shapes required to interact with the equally vast number found in invading microorganisms. This large diversity of reactivities, however, puts the host's tissues in jeopardy as similar structures are used by both the host and the microbial invaders. Although expression of autoantibodies is inherent in the human immune response, there are several mechanisms that limit the production of autoreactive antibodies. Using mice transgenic for genes encoding an anti-DNA antibody, Chen et al. (1994) found a significant reduction of peripheral B cells and anti-DNA antibody producing B cells could not be recovered from the spleens. The remaining B cells escaped deletion by revising their antigen receptors in several ways, including elimination of the transgenic H chain gene via intrachromosomal recombination followed by rearrangement and expression of endogenous V_H genes, ongoing rearrangement of endogeous κ chain genes to generate a non-DNA-binding antibody, a process known as 'receptor editing' (Gay et al., 1993; Radic et al., 1993a; Tiegs et al., 1993) and expression of rare endogenous L chains to generate a non-DNA-binding antibody.

V(D)J GENE SEGMENT UTILIZATION BY AUTOANTIBODIES

Studies on V_H gene segment utilization in humans comes from analysis of the H chain repertoire expressed by normal B cells, fetal B lymphocytes, B cell malignancies and autoantibodies. The results indicate that there is preferential expression of particular

V_H gene segments during ontogeny. The reasons for this bias, however, are unknown. It would appear that the human expressed repertoire draws predominantly from fewer than a dozen germline genes (reviewed in Pascual et al., 1992a). This notion is substantiated by Cox et al. (1994) who found that of 50 V_κ germline gene segments no more than 24 were found in rearranged sequences, and only about 11 sequences were frequently used. This indicates that any analysis of biased usage will have to take into account the limited expressed repertoire. Nonetheless, the V_H3, D_HQ52, DXP1, J_H3 and J_H4 gene segments are recurrently used in human fetal liver (Schroeder and Wang, 1990; Pascual et al., 1993), while in human adult peripheral B cells the V_H3 family is overrepresented and the V_H1 family is underrepresented and J_H4, J_H5, J_H6 and DXP are overutilized while D_HQ52 is underutilized (Wasserman et al., 1992).

It is postulated that the development of certain autoimmune diseases is related to infectious microorganisms. Based on the limited information available, the antibody response to foreign proteins does not appear to be restricted to specific V_H or V_L gene segments. Antibodies to the capsular polysaccharide of *Haemophilus influenzae* type b predominantly are derived from the V_H3 family, with some V_H1 usage (Silverman and Lucas, 1991). The V_H3 encoded H chains are structurally related to the 20P1 and 30P1 V_H genes characteristic of the early human repertoire. An analysis of the human anti-rabies virus response indicates that the majority (seven out of nine) of the antibodies are V_H3 encoded, with the remaining two being derived from the V_H1 and V_H4 families (Ikematsu et al., 1993). Human anti-insulin antibodies preferentially use V_H3 family gene segments not frequently used in fetal repertoires (Thomas, 1993). A review of anti-human immunodeficiency virus antibodies indicates that these antibodies are derived from a number of different gene segments from different gene families (Andris and Capra, 1995). The same gene usage has been observed between a human rheumatoid factor and anti-herpes viruses (Rioux et al., 1994). There appears to be an overlap in gene segment usage between anti-pathogen and autoantibody responses. Most of the V regions studied have undergone somatic mutation typical of an antigen-driven response.

Natural autoantibodies are self-reactive antibodies isolated from healthy donors in low concentration (Guilbert et al., 1982). They are not necessarily the result of antigen induction as they are produced by B cells isolated from human embryonic tissue and cord blood (Levinson et al., 1987). Natural antibodies are encoded by unmutated or minimally mutated germline gene segments with all V_H families being represented (Sanz et al., 1989). They tend to be low-avidity IgM antibodies, although other isotypes are found, and exhibit polyspecificity. The polyspecificity of antibodies could either reflect structures common to several seemingly unrelated antigens or be a property of their antibody-combining sites which can interact with unrelated molecules. Studying human anti-Rh monoclonal IgM antibodies, Blancher et al. (1991) found that the majority of the induced antibodies which cross-reacted with intracellular self-antigens were of the IgM isotype, whereas the IgG antibodies rarely exhibited cross-reactivity. This suggests that polyreactive antibodies are part of the immune response against foreign antigens. This result also indicates that the cross-reactivity of IgM antibodies is related to the properties of the antibody-combining site, which permits interaction with unrelated molecules. One should exercise care when referring to an antibody as monospecific as this may merely indicate that a large enough panel of antigens has not been screened to detect additional reactive antigens.

Natural antibodies do not appear to be pathogenic, although it has not been excluded that they could play a role in autoimmune disease, as some of the antigens recognized by natural autoantibodies are also targets for disease-associated antibodies in autoimmune diseases. Autoantibodies to intracellular antigens often recognize highly conserved determinants that correspond to sites critical to the function of the antigenic molecule (Boitard, 1992). It is postulated that natural autoantibodies function in the first line of defense against pathogens, are involved in the clearance of senescent cells and play a role in repertoire selection through V region-dependent interactions with immunoglobulin receptors on lymphocytes (Varela and Coutinho, 1991). There is evidence for natural IgM autoantibodies regulating the production of natural IgG autoantibodies through V region-dependent interactions resulting in low serum levels of autoreactive IgG antibodies (Hurez *et al.*, 1993).

Like the natural autoantibodies, many disease-associated autoantibodies are also polyreactive. Over 100 antibodies from a wide variety of autoimmune disorders have been sequenced and analysed (reviewed by Capra and Natvig, 1993). In general, the population of V_H genes used to encode autoantibodies reflects the normal expressed repertoire. Over 75% of V_H gene segments utilized are derived from approximately a dozen germline genes with essentially no restriction. There is an exception with respect to the almost exclusive use of the $V_H 4$–21 gene segment in cold agglutinins (CAs). However, even in CAs, there appears to be an apparent random utilization of J_H, V_κ, J_κ, V_λ and J_λ gene segments (reviewed in Bona *et al.*, 1993; Victor and Capra, 1994). These results suggest that typical antigen-driven processes most likely occur in most autoimmune dieseases. Autoantibodies associated with autoimmune disease processes generally exhibit polyclonal responses. This suggests that there is probably no involvement of superantigens in most autoimmune diseases, as this concept implies induction of antibodies composed of only one or very few V_H regions (Pascual and Capra, 1991b; Goodglick and Braun, 1994; Silverman, 1994).

AUTOANTIBODY-ASSOCIATED IDIOTYPES

An idiotope (Id) refers to a single V region determinant unique to an antibody or group of antibodies. The term idiotype refers to the sum of idiotopes expressed on the V region of an antibody and represents a class of serologically defined determinants. Idiotopic determinants are associated with any portion of the V_H and V_L regions. Private idiotopes are associated predominantly with antibody CDRs. Cross-reactive idiotopes (CRIs) are public idiotopes shared by different antibody molecules from unrelated individuals and, in some cases, represent markers of V region-encoded germline genes. CRIs were first described by Williams et al. (1968) who demonstrated that many human CAs from unrelated individuals reacted with absorbed polyclonal rabbit antisera. As with the CAs, anti-DNA antibodies from unrelated patients with systemic lupus erythematosus (SLE) (Shoenfeld et al., 1983), anti-acetylcholine receptor antibodies in myasthenia gravis (Dwyer et al., 1983) and anti-thyroglobulin antibodies in thyroiditis (Zanetti et al., 1984) all exhibit CRI.

Anti-idiotypic (anti-Id) antibodies serve as probes of antibody V region determinants and are used to examine questions of clonal origin and relatedness, trace functional B cell populations and study repertoire development in humans. Some anti-Ids recognize conformational determinants dependent on combined V_H and V_L interactions, while others are V_H and V_L specific Ids. The characterization of the structural correlates of idiotopes is, therefore, of fundamental value in understanding the structure–function relationships of Id–anti-Id interactions and of antigen–antibody interactions in general.

Understanding the structures of the antibodies involved in anti-self reactions will presumably provide vital information for the control of some of the disease processes (Schwartz, 1993). There are a number of Ids that are associated with autoantibodies, although their presence does not strictly imply autospecificity. This finding correlates with the lack of restriction in the usage of V_H gene segments in autoantibodies. Approaches to the study of idiotypy include serological analysis and site-directed mutagenesis. Serological approaches involve generating both anti-peptide and anti-whole antibody antisera. Mutagenesis can be used to determine single amino acids involved in antigen binding and has been applied to determining amino acids involved in anti-Id binding. Mutagenesis that results in the loss of anti-Id binding can either mean that the mutagenized area is itself the anti-Id binding site, or that this area is responsible for the display of the actual determinant at a region distal to the mutagenized site.

AUTO-ANTI-IDIOTYPIC ANTIBODIES

Many investigators have concluded that idiotypes may play a role in the self-regulation of the immune system. This is based on the demonstration that an individual can spontaneously produce antibodies which react with its own antibodies and arise during a variety of immune responses. Auto-anti-idiotypic antibodies (AAIAs) are normally present in the sera of healthy individuals. AAIAs of the IgG class have been detected that react with natural autoantibodies (Rossi et al., 1989). Anti-Ids have been shown to bind directly to Id-bearing B cells and inhibit the induction of Id-positive antibodies by specific antigens. T cells with Id specificity have also been detected (King et al., 1993). The spontaneous appearance of AAIAs to anti-DNA antibodies associated with SLE has been demonstrated with the inverse correlation between serum levels of anti-DNA antibodies and their anti-Id in the same lupus patient (Abdou et al., 1981; Zoluai and Eyquem, 1983). This inverse correlation is thought to be indicative of the biological significance of anti-idiotypes. The idiotypic network theory postulated by Jerne (1974) views the immune system within an individual as an organized structure of antibody molecules and B and T lymphocytes interconnected by complementary Id–anti-Id structures leading to a suppressed immune state. The introduction of antigen potentially alters this balance leading to a newly established equilibrium. The network could potentially be disrupted through mimicry of self-antigen by AAIA resulting in a regulatory breakdown, the outcome of which could be autoimmune disease. There is, however, little direct proof that auto-anti-idiotypes regulate the production of autoantibodies. It is interesting to note that the anti-anti-Id response can generate a subset of antibodies specific not only for the Id but also bind with low affinity to the original antigen that elicited the Id-bearing antibody (Tincani et al., 1993). These dual recognition antibodies are called 'epibodies' (Fischel and Eilat, 1992). Mendlovic et al. (1988) induced a lupus-like disease in normal mice by immunization with a human monoclonal antibody bearing the public Id 16/6. High levels of murine anti-16/6 and anti-anti-16/6 antibodies associated with anti-DNA activity were detected in the sera of the immunized mice. Bona et al. (1986) suggest that epibodies may have the physiological function of contributing to the stability of immune complexes thereby allowing them to be more efficiently taken up by cells of the reticuloendothelial system.

In the following sections, autoantibodies from specific autoimmune diseases will be discussed in more detail.

COLD AGGLUTININS

CAs are autoantibodies that agglutinate red blood cells in the cold. Some CAs are considered pathogenic, playing a direct role in the development of the clinical symptoms of cold agglutinin disease (CAD), while some are non-pathogenic and found in the sera of healthy people. CAs are also found in patients with B cell tumors (Stevenson et al., 1993) and patients with infectious mononucleosis (Chapman et al., 1993). The CAs were the first autoantibodies for which CRIs were described (Williams et al., 1968). Virtually all CAs from patients with CAD show reactivity with the rat monoclonal antibody 9G4 (Stevenson et al., 1986). In addition, it has been shown that CAs which are I/i specific and produced under chronic lymphoproliferative conditions almost uniformly are encoded by the V_H4-21 gene segment (Leoni et al., 1991; Pascual et al., 1991a, 1992b; Silberstein et al., 1991; Grillot-Courvalin et al., 1992). Essentially all V_H4-21 derived antibodies tested are 9G4 reactive, whereas antibodies encoded by gene segments from other V_H families are 9G4 unreactive (Pascual et al., 1991a; Silberstein et al., 1991; Thompson et al., 1991; Logtenberg et al., 1992). The 9G4 Id, therefore, is a marker for V_H4-21 gene segment-encoded antibodies. 9G4 reacts with 10% of all B cells in the normal adult human population and the fetal spleen at 14 weeks' gestation (Stevenson et al., 1989). The V_H4-21 gene is overrepresented in the autoimmune repertoire (reviewed in Pascual and Capra, 1992). In addition to antibodies with I/i specificity, this gene segment is also utilized by antibodies having reactivity against structurally distinct blood group antigens (Thompson *et al.*, 1991; Bye *et al.*, 1992), as well as antibodies with anti-DNA (van Es *et al.*, 1991) and rheumatoid factor (RF) activities (Silberstein *et al.*, 1991; Pascual *et al.*, 1992b) and exhibiting polyreactivity (Thompson *et al.*, 1992). If the frequent detection is due to preferential utilization of the V_H4–21 gene segment, this implies a greater frequency of recombination between V_H4–21 and the DJ_H than other V_H gene segments. Alternatively, the apparent overutilization of the V_H4 21 gene segment may be the result of polyclonal or oligoclonal expansion. This question can best be addressed by examining the non-productive gene segment rearrangement in B cells, as this recombination event is not antigen selected.

Using mutational analysis and expression of recombinant antibodies in the baculovirus system, we localized the 9G4 Id to the H chain FR1 (1993). We have also shown by capture ELISA that transfer of the V_H4–21 FR1 to a 9G4-unreactive antibody converts it to 9G4 reactive (Potter, K.N., Li, Y.-C. and Capra, J.D., unpublished data). This indicates that the 9G4 Id is definitely FR1 associated and is not located at a distal site that is structurally influenced by FR1. Combinatorial antibodies composed of a V_H4–21-encoded H chain paired with a variety of L chains were all 9G4 reactive. The H chain alone did not react with 9G4 indicating that the determinant requires a specific H chain conformation that is stabilized by the L chain. The structural relationship between the 9G4-reactive site and the antigen-binding site is under investigation. It is known that interaction between CAs and I/i antigen partially blocks the Id (Williams *et al.*, 1968) and that the interaction between CAs and 9G4 blocks hemagglutination (Stevenson *et al.*, 1989; Thompson *et al.*, 1991). The localization of the 9G4 Id to FR1 suggests that this inhibition of I/i binding is possibly due to steric hindrance or an altered structure resulting from the 9G4 binding. We have preliminary evidence indicating that I binding activity is lost when the FR1 of a V_H4–21-encoded CA is replaced by the FR1 encoded by the closely related V71–2 gene segment, which is not usually associated with CA activity (Li, Y.-C., Spellerberg, M.B., Stevenson, F.K., Capra, J.D. and Potter, K.N., unpublished data).

The V_H4–21 gene segment is found both in germline configuration and with mutations when expressed as the IgM isotype (Pascual *et al.*, 1991, 1992b; Silberstein *et al.*, 1991). Germline V_H4–21-encoded antibodies are found in both anti-I and anti-i antibodies (Pascual *et al.*, 1992b) indicating that in addition to requiring the V_H4–21 FR1 to FR3-encoded portion of the variable region for antigen binding, the D and J gene segments, which contribute to the CDR3 amino acids, are involved in determining the fine specificity of a CA. Amino acid sequence comparisons indicate that there is no similarity in the CDR3 sequences between CAs with respect to length and amino acid composition. Studies using monoclonal CAs against synthetic carbohydrates reveals that each antibody is binding to a discrete epitope on the I/i antigen (Feizi *et al.*, 1979). The idiotypic and structural heterogeneity of CA light chains was studied by Jefferies *et al.* (1992) using nine monoclonal anti-idiotypic antibodies. They found that, in contrast with the conserved Id associated with the CA H chain, there is heterogeneity of L chain idiotype expression. This could be due either to utilization of different gene segments or somatic mutation. These authors suggest that the L chain idiotypes may relate to fine binding specificity. It is postulated, therefore, that the V_H4–21-encoded portion of the variable region of the H chain is required for the global interaction or 'docking' of the antibody with the l/i antigen, while the H chain CDR3 and the L chain modulate the fine specificity of the antibody (Potter et al., 1993). Experiments are ongoing to test this hypothesis.

As mentioned earlier, CAs can either be pathogenic or benign. CAs isolated from patients with CAD are almost always V_H4–21 encoded antibodies. Pascual et al. (1992b) found that the best antibody inducing hemagglutination in a series of eight CAs was a V_H4–21 germline-encoded antibody. The other CAs were encoded by mutated V_H4–21 gene segments with 1–11 amino acid differences from the germline sequence. Replacement to substitution analysis indicates that the mutations are not indicative of an antigen-driven response. Serum levels of CAs increase following some infections, including Epstein–Barr virus (EBV)-induced infectious mononucleosis. Chapman et al. (1993) examined V_H4–21-encoded antibodies following a polyclonal response after EBV infection. Five antibodies were germline in sequence, with the sixth having a single amino acid difference in FR4. There is no evidence, therefore, for antigen-driven affinity maturation of these autoantibodies. The use of the V_H4–21 gene segment in response to a viral infection demonstrates that, at least in some cases, the autoantibody gene repertoire is the same as that used in the immune response to external antigens. Naturally occurring anti-l antibodies were found to be encoded by V71–2 and V2–1 gene segments, in addition to V_H4–21 (Schutte et al., 1993) and Jefferies et al. (1993) report that naturally occurring anti-I/i CAs can be encoded by different V_H3 genes as well as the V_H4–21 gene segment. These results may suggest a distinction between pathogenic and benign anti-l/i autoantibodies.

As a number of 9G4-reactive antibodies are autoantibodies, mechanisms that control their overproduction must be in effect. This control could involve an idiotypic network involving the 9G4 determinant located in FR1. There are many cells displaying the 9G4-reactive FR1 at any one time. This suggests that a potential superantigen, of either bacterial or viral origin, could simultaneously interact with all of the V_H4–21-encoded antibodies regardless of their specificities, resulting in polyclonal activation. This proliferation increases the possibility that B cells will be activated that are able to produce autoantibodies against a self-structure. CAD is an example of a monoclonal gammopathy, in which the CA produced is derived from a single clone that has been over-stimulated. Polyclonal stimulation followed by antigen selection could explain this phenomenon.

It remains unexplained as to why antibodies to the l/i antigen are virtually always encoded by the single V_H4–21 gene segment. Every other autoantigen analyzed has antibodies encoded by several gene segments and usually gene segments from several different V_H families (Pascual et al., 1992b). This exclusivity perhaps signifies that the l/i antigen represents a unique structural category, the dimensions of which are most compatable with V_H4-21-encoded antibody structures. It will be interesting to know whether the I/i antigen will remain the exception, or with continued research becomes the prototype of other antigens that are bound by antibodies derived predominantly from a single V_H gene segment.

RHEUMATOID FACTORS

RFs are antibodies that bind the Fc region of IgG immunoglobulins (reviewed in Randen et al., 1992). RFs are found in patients with B cell dyscrasias, rheumatoid arthritis (RA) and patients with inflammatory and infectious diseases. RFs are also produced in healthy people following immunization. This suggests that RF may play a physiological role in normal immune function such as in the clearance of circulating immune complexes.

RA is a non-organ specific disease involving the destruction of joints resulting from a chronic inflammatory process involving the synovium. RFs can form immune complexes that bind complement and perpetuate inflammation (Molines et al., 1986). While higher titers of RFs corrrelate with a more aggressive form of the disease (Panush et al., 1971; Withrington et al., 1984), the event initiating RA is unknown. Some theories involve altered glycosylation of IgG molecules or an infectious agent. RFs in patients with RA show broad cross-reactivity with low affinity for individual antigens.

The initial classification of human monoclonal IgM paraproteins with RF activity was into three serologically determined CRI groups named Wa, Po (Kunkel et al., 1973, 1974) and Bla (Agnello et al., 1980), an RF that also exhibits anti-histone activity. These RFs are found in the sera of patients with B cell dyscrasias such as mixed cryoglobulinemia and Waldenström's macroglobulinemia (Kunkel et al., 1973), diseases that are unrelated to RA. Approximately 60% of mixed cryoglobulin RF of the IgM κ isotype express the Wa CRI, approximately 15% express the Po CRI and approximately 15% express the Bla CRI. Amino acid and nucleotide sequence analysis have shown that the Wa group is composed of antibodies with $V_κ3b$-derived L chains and V_H1-encoded H chains, Po antibodies are predominantly composed of $V_κ3a$ L chains paired with V_H3 H chains and Bla CRI are composed of $V_κ3a$ L chains associated with V_H4 H chains.

The mouse monoclonal antibody 17.109 reacts with $V_κ3b$-encoded L chains that are derived from the germline gene segments *Humkv325* and *Humkv305* (Radoux et al., 1986; Liu et al., 1989), while the mouse monoclonal antibody 6B6.6 recognizes $V_κ3a$-encoded L chains that are *Humkv328* derived (Liu et al., 1989). Mouse monoclonal antibody LC1 was raised against a human monoclonal antibody with RF activity. LC1 detects a CRI on the H chain of many non-pathogenic IgM antibodies and monoclonal IgM autoantibodies, many of which exhibit RF activity. Silverman et al. (1990) demonstrated that LC1-reactive V_H4 H chains preferentially pair with κ3a (6B6.6 Id positive) L chains in antibodies with RF activity, and represents approximately 13–22% of RFs. The LC1 CRI has been reported in subpopulations of cells in the germinal centers as well as the mantle zones of secondary human B cell follicles (Pratt et al., 1991). LC1 reacts with malignant B cells from over 10% of patients with chronic lymphocytic leukemia (CLL) or small lymphocytic lymphoma (SLL) (Kipps et al., 1990). The LC1 Id is expressed predominantly on $CD5^-$ B lymphocytes, as well as $CD5^+$ cells isolated from cord blood (Dean et al, 1993). We determined that this Id is located in FR1 of all $V_H 4$ family gene segments, excluding $V_H 4$–21 (Potter et al., 1994). These LC1-reactive gene segments are remarkably similar in their FR1 sequences and different from that of the $V_H 4$–21 sequence. This finding explains the mutually exclusive pattern of binding observed between 9G4 and LC1 (Mageed et al., 1991). Using combinatorial antibodies produced in the baculovirus system, we demonstrated that some L chains paired with LC1-reactive H chains result in the loss of LC1 reactivity. This indicates that the L chain influences the display of the LC1 determinant and provides another mechanism, other than mutation, to explain the loss of an H chain idiotope. The monoclonal antibodies G6 and G8 are H chain-specific anti-Id that recognize $V_H 1$-encoded antibodies (Mageed et al., 1986), and the monoclonal antibodies B6 and D12 detect $V_H 3$-associated CRIs. The structural correlates of these Ids have not been determined.

The anti-Ids 17.109, 6B6.6 and G6, which recognize the majority of monoclonal RFs from patients with B cell dyscrasias, react with only 1–2% of polyclonal RFs from patients with RA. The anti-Id 4C9 (Davidson et al., 1992), however, reacts predominantly with an L chain determinant on approximately 23% of IgM RFs from patients with RA and not with RFs from patients with B cell malignancies. IgG RF from RA synovial fluids did not react with 4C9, indicating that either different genes are used in the two isotypes or that somatic mutation has removed the Id.

In order to further characterize what are believed to be pathogenic RFs present in patients with RA, RFs have been derived from the synovial membrane of patients with classical RA and the juvenile polyarticular form of the disease, as well as from the peripheral blood lymphocytes of patients with classical RA and SLE. A comprehensive analysis of monoreactive and polyreactive RFs from patients with RA and RFs from peripheral blood of patients with SLE and Sjögren's syndrome indicates that RFs and polyreactive antibodies derive from a diverse array of V_H and V_L gene segments. Many RFs and polyreactive antibodies are direct or nearly direct copies of germline genes, which indicates that RFs are part of the normal B cell repertoire. Some exhibit somatic mutation and the CDR3 show extensive variability in length and composition (Pascual et al., 1992c). The polyreactive antibodies were exclusively associated with λ L chains and there was an overutilization of $V_H 3$ genes. IgG RFs show somatic mutations typical of an antigen-driven response (Randen et al., 1993). Olee et al. (1992) suggest that potentially pathogenic IgG RFs derive from the natural antibody repertoire in a normal immune system through somatic mutation and antigen selection.

Hay et al. (1991) used IgG binding by octapeptides to determine that the FR2 and FR3 of $V_\kappa 3b$ light chains have the ability to bind IgG. It is possible that RF activity is demonstrated only when an L chain is combined with a particular H chain. The particular structural correlates responsible for RF activity have not been determined and are a topic of current research.

ANTI-DNA ANTIBODIES

Anti-dsDNA antibodies are considered to be markers for the autoimmune disease SLE and may be involved in the pathogenesis of the disease, most notably with renal involvement (reviewed by Isenberg *et al.*, 1994). The biological processes responsible for initiating anti-dsDNA reactivity *in vivo* are not understood, especially as approximately 30% of patients diagnosed as having SLE do not produce detectable levels of anti-DNA antibodies. These patients produce other antinuclear antibodies, however, having anti-Sm, anti-Ro and anti-La specificities.

Anti-DNA antibodies of the IgM isotype have been sequenced and are found to be derived from the V_H1, V_H2, V_H3, V_H4 and V_H6 families of gene segments (reviewed in Pascual and Capra, 1992; Chastagner *et al.*, 1994), while the few anti-DNA antibodies of the IgG isotype studied to date are derived from V_H3 and V_H4 family gene segments (Winkler *et al.*, 1992). Both κ and λ light chains are used in anti-DNA antibodies. Analysis of the V_H gene segments indicate that they are somatically mutated, characteristic of an antigen-driven response (Davidson *et al.*, 1990; van Es *et al.*, 1991; Chastagner *et al.*, 1994).

Ids commonly associated with anti-DNA antibodies are referred to as 9G4, 3I, 0-81 and 16/6. The V_H4-21-derived anti-DNA antibodies express the FR1-positioned 9G4 Id, which is discussed in the section on CAs. Some anti-DNA antibodies are encoded by other members of the V_H4 family, such as V71-2 (Davidson *et al.*, 1990) and presumably express the murine monoclonal LC1 Id that has also been localized to FR1 of V_H4-encoded antibodies, excluding V_H4-21. The 16/6 idiotope is elevated in the sera of patients with active lupus (Isenberg *et al.*, 1984). Antibodies displaying the 16/6 Id are encoded by the V_H3 family member V_H26 (Dersimonian *et al.*, 1987; Spatz *et al.*, 1990).

Sequence analysis of anti-DNA antibodies indicates that arginine residues are important in the CDR3 of the H chain for DNA binding (Radic *et al.*, 1993b). Using site-directed mutagenesis, the substitution of an arginine with a glycine in the CDR3 of the H chain abrogated DNA binding. This indicates that arginine influences the electrostatic potential of the contact residues in the antibody. Winkler *et al.* (1992) examined six IgG monoclonal anti-DNA antibodies that were somatically mutated. Five were V_H3 encoded and one was V_H4 encoded. The mutation pattern indicated that basic amino acids were being selected in all three CDRs of both the H and L chains. However, analysis of antibodies with DNA specificity indicates that antibodies without enrichment of basic residues in the CDRs can still bind DNA. This would indicate that the different epitopes on the DNA surface are not homogeneous in their electrostatic properties, and would not all induce antibodies having the same amino acid requirements. Nonetheless, comparisons of sequences of anti-DNA antibodies indicate a conserved core in the CDR3 of the H chain strongly suggesting that the CDR3 residues determine the fine DNA-binding specificity (Isenberg et al., 1994).

Individual patients appear to express a restricted range of antibody specificities and some of the antibodies can be polyreactive. Isenberg et al. (1994) suggest that anti-dsDNA antibodies are the result of polyclonal B cell activation, which includes an anti-dsDNA response followed by antigen-specific stimulation. This hypothesis is very similar to that proposed to explain the data derived from analysis of CAs.

CONCLUSIONS

It is apparent that the normal immune system frequently produces autoantibodies; however, autoimmunity is a rare event. This implies that autoimmunity results from a disturbance of the control mechanisms. Genetic analysis of autoantibodies indicates that the gene segments used by autoantibodies derive from the same repertoire as those used by the normal immune system to exogenous antigens. Most non-pathogenic, natural IgM autoantibodies are encoded by V_H gene segments expressed in germline configuration. They are present in the serum in low concentration, are polyreactive and bind antigen with low affinity. The V_H gene segments used by autoantibodies from patients with autoimmune diseases reflect the normal expressed repertoire. Natural IgM autoantibodies, therefore, may become pathological when B cells are stimulated to produce excessive quantities of these otherwise innocuous antibodies. The production of somatically mutated, high-affinity IgG autoantibodies appears to be selected by an antigen-driven response. Autoimmunity, therefore, does not seem to be the sole result of heredity, but probably depends on the interaction of multiple gene products and environmental factors. This analysis of autoantibodies, their gene segment usage and idiotypes has potential clinical significance as the Ids on neoplastic B cells provide tumor antigens for vaccination of patients with B cell tumors.

Acknowledgements

We acknowledge the National Institutes of Health (AI-12127 and CA-44016) for financial support. J.D. Capra holds the Edwin L. Cox Distinguished Chair in Immunology and Genetics at the University of Texas Southwestern Medical Center at Dallas.

REFERENCES

Abdou, N.I., Wall, H., Lindsley, H.B., Halsey, J.F. and Suzuki, T. (1981). *J. Clin. Invest.* **67**, 1297–1304.
Agnello, V., Arbetter, A., de Kasep, G.I., Powell, R., Tan, E.M. and Joslin, F. (1980). *J. Exp. Med.* **151**, 1514–1527.

Alt, F.W. and Baltimore, D. (1982). *Proc. Natl Acad. Sci. USA* **79**, 4118–4122.
Alt, F.W., Oltz, E.M., Young, F., Gorman, J., Taccioli, G. and Chen, J. (1992). *Immunol. Today* **13**, 306–314.
Andris, J.S. and Capra, J.D. (1995). *J. Clin. Immunol.* **15**, 17–26.
Avrameas, S. (1991). *Immunol. Today* **12**, 154–158.
Ban, N., Escobar, C., Garcia, R., Hasel, K., Day, J., Greenwood, A. and McPherson, A. (1994). *Proc. Natl Acad. Sci. USA* **91**, 1604–1608.
Bentley, G.A., Boulot, G., Riottot, M.M. and Poljak, R.J. (1990). *Nature* **348**, 254–257.
Blancher, A., Roubinet, F., Oksman, F., Terynch, T., Broly, H., Chevaleyre, J., Vezon, G. and Ducos, J. (1991). *Vox Sang.* **61**, 196–204.
Boitard, C. (1992). *Curr. Opin. Immunol.* **4**, 741–747.
Bona, C.A., Yang, C.-Y., Kohler, H. and Monestier, M. (1986). *Immunol. Rev.* **90**, 115–128.
Bona, C.A., Kelso, G., Pascual, V. and Capra, J.D. (1993). In 'The molecular pathology of autoimmune diseases' (Bona, C.A., Siminovich, K.A., Zanetti, M. and Theophillopoulos, A.N. eds). Harwood Academic Press, New York.
Bye, J.M., Carter, C., Cui, Y., Gorick, B.D., Songsivilai, S., Winter, G., Hughes-Jones, N.C. and Marks, J.D. (1992). *J. Clin. Invest.* **90**, 2481–2490.
Capra, J.D. and Natvig, J.B. (1993). *Immunologist* **1**, 16–19.
Chapman, C.J., Spellerberg, M.B., Smith, G.A., Carter, S.J., Hamblin, T.J. and Stevenson, F.K. (1993). *J. Immunol.* **151**, 1051–1061.
Chastagner, P., Demaison, C., Theze, J. and Zouali, M. (1994). *Scand. J. Immunol.* **39**, 165–178.
Chen, C., Radic, M.Z., Erikson, J., Camper, S.A., Litwin, S., Hardy, R.R. and Weigert, M. (1994). *J. Immunol.* **176**, 1970–1982.
Coffman, R.L., Lebman, D.A. and Rothman, P. (1993). *Adv. Immunol.* **54**, 229–270.
Cook, G.P., Tomlinson, I.M., Walter, G., Riethman, H., Carter, N.P., Buluwela, L., Winter, G. and Rabbitts, T.H. (1994). *Nature Genet.* in press.
Cox, J.P.L., Tomlinson, I.M. and Winter, G. (1994). *Eur. J. Immunol.* **24**, 827–836.
Davidson, A., Manheimer-Lory, A., Aranow, C., Peterson, R., Hannigan, N. and Diamond, B. (1990). *J. Clin. Invest.* **85**, 1401–1409.
Davidson, A., Lopez, J., Sun, D. and Prus, D. (1992). *J. Immunol.* **148**, 3873–3878.
Deane, M., Mackenzie, L.E., Stevenson, F.K., Youinou, P.Y., Lydyard, P.M. and Mageed, R.A. (1993). *Scand J. Immunol.* **38**, 348–358.
Dersimonian, H., Schwartz, R.S., Barrett, K.J. and Stollar, B.D. (1987). *J. Immunol.* **139**, 2496–2501.
Dwyer, D.S., Bradley, R.J., Urquhart, C.K. and Kearney, J.F. (1983). *Nature* **301**, 611–614.
Early, P., Rogers, J., Davis, M., Calame, K., Bond, M., Wall, R. and Hood, L. (1980). *Cell* **20**, 313–319.
Feizi, T., Childs, R.A., Watanabe, K. and Hakomori, S.I. (1979). *J. Exp. Med.* **149**, 975–980.
Fischel, R. and Eilat, D. (1992). *J. Immunol.* **149**, 3089–3096.
Gay, D., Saunders, T., Camper, S. and Weigert, M. (1993). *J. Exp. Med.* **177**, 999–1008.
Gearhart, J.J. and Bogenhagen, D.F. (1983). *Proc. Natl Acad. Sci. USA* **80**, 3439–3443.
Goodglick, L. and Braun. J. (1994). *Am. J. Pathol.* **144**, 623–636.
Grillot–Courvalin, C., Brouet, J.C., Piller, R.F., Rassenti, L.Z., Lahaume, S., Silverman, G.J., Silberstein, L. and Kipps, T.J. (1992). *Eur. J. Immunol.* **22**, 1781–1788.
Gu, H., Kitamura, D. and Rajewsky, K. (1991). *Cell* **65**, 47–54.
Guilbert, B., Dighiero, G. and Avameas, S. (1982). *J. Immunol.* **128**, 2779–2787.
Hay, F.C., Soltys, A.J., Tribbick, G. and Geysen, H.M. (1991). *Eur. J. Immunol.* **21**, 1837–1841.
Hurez, V., Kaveri, S-V. and Kazatchkine, M.D. (1993). *Eur. J. Immunol.* **23**, 783–789.
Ikematsu, H., Harindranath, N., Ueki, Y., Notkins, A.L. and Casali, P. (1993). *J. Immunol.* **150**, 1325–1337.
Isenberg, D.A., Shoefeld, Y., Nadaio, M.P., Rauch, J., Reichin, M., Stollar, B.D. and Schwartz, R.S. (1984). *Lancet* **ii**, 418–422.
Isenberg, D.A., Ehrenstein, M.R., Longhurst, C. and Kalsi, J.K. (1994). *Arthritis Rheum.* **37**, 169–180.
Jefferies, L.C., Silverman, G.J., Carchidi, C.M. and Silberstein, L.E. (1992). *Clin. Immunol. Immunopathol.* **65**, 119–128.

Jefferies, L.C., Carchidi, C.M. and Silberstein, L.E. (1993). *J. Clin. Invest.* **92**, 2821–2833.
Jerne, N.K. (1974). *Ann. Immunol.* **125C**, 373–389.
Kabat, E.A. and Wu, T.T. (1991). *J. Immunol.* **147**, 1709–1719.
King, C.A., Wills, M.R., Hamblin, T.J. and Stevenson, F.K. (1993). *Cell. Immunol.* **147**, 411–424.
Kipps, T.J., Robbins, B.A., Tefferi, A., Meisenholder, G., Banks, P.M. and Carson, D.A. (1990). *Am. J. Pathol.* **136**, 809–816.
Kunkel, H.G., Agnello, V., Joslin, F.G., Winchester, R.J. and Capra, J.D. (1973). *J. Exp. Med.* **137**, 331–342.
Kunkel, H.G., Winchester, R.J., Joslin, F.G. and Capra, J.D. (1974). *J. Exp. Med.* **139**, 128–136.
Lafaille, J.J., DeCloux, A., Bonneville, M., Takagaki, Y. and Tonegawa, S. (1989). *Cell* **59**, 859–870.
Leoni, L., Ghiso, J., Goni, F. and Frangione (1991). *J. Biol. Chem.* **266**, 2836–2842.
Levinson, A.I., Dalal, N.F., Haidar, M. and Tar., L. (1987). *J. Immunol.* **139**, 2237–2241.
Liu, M.-F., Robbins, D.L., Crowley, J.J., Sinha, S., Kozin, F., Kipps, T.J., Carson, D.A. and Chen, P.P. (1989). *J. Immunol.* **142**, 688–694.
Logtenberg, T., Schutte, M.E.M., Elbeling, S.B., Gmelig-Meyling, F.H.J. and van Es, J.H. (1992). *Immunol. Rev.* **128**, 23–47.
Mageed, R.A., Dearlove, M., Goodall, D.M. and Jefferis, R. (1986). *Rheumatol. Int.* **6**, 179–183.
Mageed, R.A., MacKenzie, L.E., Stevenson, F.K., Yuksel, B., Shokri, F., Maziak, B.R., Jefferis, R. and Lydyard. P.M. (1991). *J. Exp. Med.* **174**, 109–113.
Matsuda, F., Shi, E.Y., Nagaoka, H., Matsumura, R., Haino, M., Fukita, Y., Takaishi, S., Imai, T., Riley, J.H., Anand, R., Soeda, E. and Honjo, T. (1993). *Nature Genet.* **3**, 88–94.
Meek, K. (1990). *Science* **250**, 820–823.
Meek, K.D., Hasemann, C.A. and Capra, J.D. (1989). *J. Exp. Med.* **170**, 39–57.
Mendlovic, S., Brocke, S., Shoenfeld, Y., Ben-Bassat, M., Meshorer, A., Bakimer, R. and Mozes, E. (1988). *Proc. Natl Acad. Sci. USA* **85**, 2260–2264.
Molines, T.E., Lea, T., Mellbye, O.J., Pahle, J., Grand, O. and Harboe, M. (1986). *Arthritis Rheum.* **29**, 715–721.
Nossal, G.V.J. (1992). *Adv. Immunol* **52**, 283–331.
Olee, T., Lu, E.W., Huang, D.-F., Soto-Gil, R.W., Deftos, M., Kozin, F., Carson, D.A. and Chen., P.P. (1992). *J. Exp. Med.* **175**, 831–842.
Panush, R.S., Bianco, N.E. and Schur, P.H. (1971). *Arthritis Rheum.* **14**, 737–747.
Pascual, V. and Capra, J.D. (1991a). *Adv. Immunol.* **49**, 1–74.
Pascual, V. and Capra, J.D. (1991b). *Current Biol.* **1**, 315–317.
Pascual, V. and Capra, J.D. (1992). *Arthritis Rheum.* **35**, 11–18.
Pascual, V., Victor, K.J., Lelsz, D., Spellerberg, M.B., Hamblin, T.J., Thompson, K.M., Randen, I., Natvig, J., Capra, J.D. and Stevenson, F.K. (1991). *J. Immunol.* **146**, 4385–4391.
Pascual, V., Widhopf, G. and Capra, J.D. (1992a). *Int. Rev. Immunol.* **8**, 147–157.
Pascual, V., Victor, K., Lelsz, D., Spellerberg, M.B., Hamblin, T.J., Stevenson, F.K. and Capra, J.D. (1992b). *J. Immunol.* **149**, 2337–2344.
Pascual, V., Victor, K., Randen, I., Thompson, K., Steinitz, M., Forre, O., Fu, S.-M., Natvig, J.B. and Capra, J.D. (1992c). **36**, 349–362.
Pascual, V., Verkruyse, L., Casey, M.L. and Capra, J.D. (1993). *J. Immunol.* **151**, 4164–4172.
Potter, K.N., Li, Y.-C., Pascual, V., Williams, R.C., Byres, L.C., Spellerberg, M., Stevenson, F.K. and Capra, J.D. (1993). *J. Exp. Med.* **178**, 1419–1428.
Potter, K.N., Li, Y.-C. and Capra, J.D. (1994). *Scand. J. Immunol.* **40**, 43–49.
Pratt, L.F., Szubin, R., Carson, D.A. and Kipps, T.J. (1991). *J. Immunol.* **147**, 2041–2046.
Radic, M.Z., Mascelli, M.A., Erikson, J., Shan, H. and Weigert, M. (1991). *J. Immunol.* **146**, 176–182.
Radic, M.Z., Erikson, J., Litwin, S. and Weigert, M. (1993a). *J. Exp. Med.* **177**, 1165–1173.
Radic, M.Z., Mackle, J., Erickson, J., Mol, C., Anderson, W.F. and Weigert, M. (1993b). *J. Immunol.* **150**, 4966–4977.
Radoux, V., Chen, P.P., Sorge, J.A. and Carson, D.A. (1986). *J. Exp. Med.* **164**, 2119–2124.
Randen, I., Thompson, K.M., Pascual, V., Victor, K., Beale, D., Coadwell, J., Forre, O., Capra, J.D. and Natvig, J.B. (1992). *Immunol. Rev.* **128**, 49–71.

Randen, I., Pascual, V., Victor, K., Thompson, K.M., Forre, O., Capra, J.D. and Natvig, J.B. (1993). *Eur. J. Immunol.* **23**, 1220–1225.
Rioux, J.D., Rauch, J., Silvestri, L. and Newkirk, M.M. (1994). *Scand. J. Immunol.* **40**, 350–354.
Rossi, F., Dietrich, G. and Kazatchkine, M.D. (1989). *Immunol. Rev.* **110**, 135–149.
Roux, K.H., Monafo, W.J., Davie, J.M. and Greenspan, N.S. (1987). *Proc. Natl Acad. Sci. USA* **84**, 4984–4988.
Sanz, I., Casali, P., Thomas, J.W., Notkins, A.L. and Capra, J.D. (1989). *J. Immunol.* **142**, 4054–4061.
Schroeder, H.W. Jr and Wang, J.Y. (1990). *Proc. Natl Acad. Sci. USA* **87**, 6146–6150.
Schutte, M.E.M., van Es, J.H., Silberstein, L.E. and Logtenberg, T. (1993). *J. Immunol.* **151**, 6569–6576.
Schwartz, R.S. (1993). In 'Fundamental Immunology', 3rd edn, pp. 1033–1097 (W.E. Paul, ed.). Raven Press, New York.
Shoenfeld, Y., Isenberg, D.A., Rauch, J., Madaio, M.P., Stollar, B.D. and Schwartz, R.S. (1983). *J. Exp. Med.* **158**, 718–730.
Silberstein, L.E., Jefferies, L.C., Goldman, J., Friedman, D., Moore, J.S., Nowell, P.C., Roelcke, D., Pruzanski, W., Roudier, J. and Silverman, G.J. (1991). *Blood* **78**, 2372–2386.
Silverman, G.J. (1994). *Immunologist* **2**, 51–57.
Silverman, G.J. and Lucas, A.H. (1991). *J. Clin. Invest.* **88**, 911–920.
Silverman, G.J., Schrohenloher, R.E., Accavitti, M.A., Koopman, W.J. and Carson, D.A. (1990). *Arthritis Rheum.* **33**, 1347–1360.
Snyder, J.G., Yu-Lee, L.-Y. and Marcus, D.M. (1992). *Eur. J. Immunol.* **22**, 1075–1082.
Spatz, L.A., Wong, K.K., Williams, M., Desai, R., Goher, J., Berman, J.E., Alt, F.W. and Latov, N. (1990). *J. Immunol.* **144**, 2821–2828.
Stevenson, F.K., Wrightam, N., Glennie, M.J., Jones, D.B., Cattan, A.R., Feizi, T., Hamblin, T.J. and Stevenson, G.T. (1986). *Blood* **68**, 430–436.
Stevenson, F.K., Smith, G.J., North, J., Hamblin, T.J. and Glennie, M.J. (1989). *Br. J. Haematol.* **72**, 9–15.
Stevenson, F.K., Spellerberg, M.B., Treasure, J., Chapman, C.J., Silberstein, L.E., Hamblin, T.J. and Jones, D.B. (1993). *Blood* **82**, 224–230.
Thomas, J.W. (1993). *J. Immunol.* **150**, 1375–1382.
Thompson, K.M., Sutherland, J., Barden, G., Melamed, M.D., Randen, I., Natvig, J.B., Pascual, V., Capra, J.D. and Stevenson, F.K. (1991). *Scand. J. Immunol.* **34**, 509–518.
Thompson, K.M., Sutherland, J., Barden, G., Melamed, M.D., Wright, M.G. and Bailey, S. (1992). *Immunology* **76**, 146–157.
Tiegs, S.L., Russell, D.M. and Nemazee, D. (1993). *J. Exp. Med.* **177**, 1009–1020.
Tincani, A., Balestrieri, G., Allergri, F., Cattaneo, R., Fornasieri, A., Li, M., Sinico, A. and Amico, G.D. (1993). *Clin. Exp. Rheumatol.* **12**, 129–134.
Tomlinson, I.M., Walter, G., Marks, J.D., Llewelyn, M.B. and Winter, G. (1992). *J. Mol. Biol.* **227**, 776–798.
Tonegawa, S. (1983). *Nature* **302**, 575–581.
van Dijk, K.W., Mortari, F., Kirkham, P.M., Schroeder, H.W., Jr and Milner, E.C. (1993). *Eur. J. Immunol.* **23**, 832–839.
van Es, J.H., Gmelig-Meyling, F.H.J., van de Akker, W.R.M., Aanstoot, H., Derksen, R.H.W.M. and Logtenberg, T. (1991). *J. Exp. Med.* **173**, 461–470.
Varela, F. and Coutinho, A. (1991). *Immunol. Today* **12**, 159–166.
Vasicek, T.J. and Leder, P. (1990). *J. Exp. Med.* **172**, 609–620.
Victor, K. and Capra, J.D. (1994). In 'Autoimmunity: Physiology and Disease', pp. 19–34. (A. Coutinho and M.D. Kazatchkine, eds.) Wiley-Liss, Inc, New York.
Wasserman, R., Ito, Y., Galili, N., Yamada, M., Reichard, B.A., Shane, S., Lange, B. and Rovera, G. (1992). *J. Immunol.* **149**, 511–516.
Williams, C., Weigel, L., Sanz, I. and Capra, J.D. (1991). In 'Anti-idiotypic Vaccines. Progress in Vaccinology' Vol. 3, pp. 22–30. (P.-A. Cazenave, ed.). Springer-Verlag, New York.
Williams, R.C., Kunkel, H.G. and Capra, J.D. (1968). *Science* **161**, 379–381.
Williams, S.C. and Winter, G. (1993). *Eur. J. Immunol.* **23**, 1456–1461.
Winkler, T.H., Fehr, H. and Kalden, J.R. (1992). *Eur. J. Immunol.* **22**, 1719–1728.

Withrington, R.H., Teitsson, I., Valdimarssor, H. and Seifert, M.H. (1984). *Ann. Rheum. Dis.* **43**, 679–685.
Yancopoulos, G.D. and Alt, F.W. (1986). *Annu. Rev. Immunol.* **4**, 339–368.
Zachau, H.G. (1993). *Gene* **135**, 167–173.
Zanetti, M., Rogers, J. and Katz, D.H. (1984). *J. Immunol.* **133**, 240–243.
Zouali, M. and Eyquem, A. (1983). *Ann. Immunol. Inst. Pasteur* **134C**, 377–391.

18

Regulation of immunoglobulin gene transcription

Kathryn Calame* and Sankar Ghosh†

* Department of Microbiology, Columbia University College of Physicians
and Surgeons, New York, NY, USA
† Howard Hughes Medical Institute, Yale University School of Medicine,
New Haven, CT, USA

Introduction 397
Functional elements of Ig promoters and enhancers 398
Description of transcriptional regulators affecting Ig gene transcription 400
Roles of individual regulatory elements in heavy chain locus 410
B-cell and developmental-stage specificity of IgH transcription 412
Regulating elements in light chain transcription...413
B-cell and developmental-stage specificity of Ig κ and λ light chain transcription ..415
Perspectives 417

INTRODUCTION

The transcriptional regulation of immunoglobulin (Ig) genes has been actively investigated since Ig gene clones first became available some 15 years ago. Since Ig genes are expressed only in B cells and are differentially expressed at different stages of B-cell development, their regulation allows mechanisms of both tissue and stage specificity to be explored. Ig gene transcription has provided an extremely useful paradigm for understanding general mechanisms of gene regulation in eukaryotes. For example, the first cellular enhancer was discovered in the IgH locus. In addition, several key transcription factor families, including NF-κB/rel proteins, Oct/POU proteins, E2A proteins and TFE3 proteins, were discovered as a result of studying Ig gene transcription. Ig gene transcription is also interesting because of its apparent involvement in other mechanisms that are critically important for humoral immunity: V(D)J recombination, isotype class switching and somatic hypermutation of V genes. In this chapter, we discuss recent advances in understanding the regulatory elements that control heavy and light chain Ig gene transcription. We consider the specific proteins that bind to the regulatory elements and how these elements function within the context of B-cell development. For other recent reviews on this topic see Kadesch (1992), Li *et al.* (1991), Liou and Baltimore (1993), Nelsen and Sen (1992) and Staudt and Lenardo (1991).

FUNCTIONAL ELEMENTS OF IG PROMOTERS AND ENHANCERS

Each functional Ig V gene segment has a transcriptional promoter. In addition, transcriptional enhancers are located within the J–C intron and 3' of C genes in both the heavy and κ gene loci. Although no enhancers have been detected in the J–C introns of the λ locus (Picard and Schaffner, 1984), there are two enhancers that are located at the 3' end of each constant region cluster (Hagman *et al.*, 1990). There have been reports of additional elements that may also regulate Ig transcription (Jenuwein and Grosschedl, 1991; Matthias and Baltimore 1993) but these will not be considered further here because less is known about them. Representative V gene promoters and Ig enhancers for heavy and light chain genes are illustrated in Figs 1

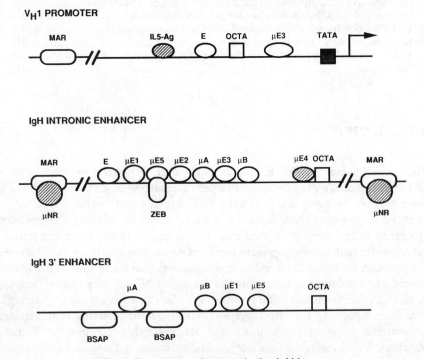

Fig. 1. Regulatory elements in the IgH locus.

and 2, showing the binding sites for transcriptional regulators that have been identified biochemically by *in vitro* binding studies and/or functionally by mutation and transfection. Since activator binding sites within the IgH intronic enhancer are particularly closely packed, these sites are also illustrated at the sequence level in Fig. 3.

Ig LIGHT CHAIN PROMOTERS

Ig κ INTRONIC ENHANCER

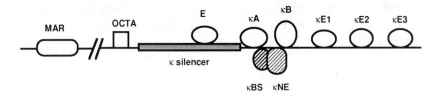

Ig κ 3' ENHANCER

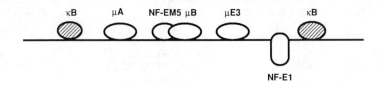

Ig λ 2-4 ENHANCER

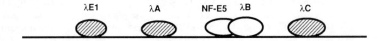

Fig. 2. Regulatory elements in the light chain loci. Activator binding sites are shown above the lines; negative regulator sites are shown below the lines. Matrix attachment regions (MARs) are centered on the lines. Slashed symbols represent sites recognized by proteins that have not been cloned. The positions of the elements are approximate and not to scale.

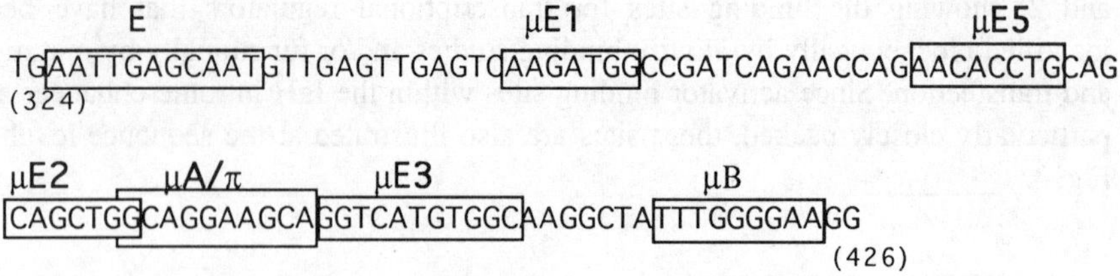

```
        E                      μE1                            μE5
TG AATTGAGCAAT GTTGAGTTGAGTC AAGATGG CCGATCAGAACCAG AACACCTG CAG
(324)

   μE2      μA/π      μE3                 μB
CAGCTGG CAGGAAGCA GGTCATGTGGC AAGGCTA TTTGGGGAA GG
                                                   (426)
```

Fig. 3. Activator sites in the IgH intronic enhancer.

DESCRIPTION OF TRANSCRIPTIONAL REGULATORS AFFECTING IG GENE TRANSCRIPTION

A summary of consensus sequences of established protein binding sites in Ig light and heavy chain genes is presented in Table I. In the discussion below, protein(s) that bind these sites, presented in alphabetical order, are considered in more detail.

Table 1. Ig enhancer and promoter protein-binding sites. The consensus sequences for each site are shown, followed by the protein(s) that are likely to bind the site in B cells and a description of the DNA-binding motif of the protein(s).

Site	Sequence	Proteins	Motif
μE1, NF-E	AAAATGG G T	YY1	C_2H_2 zinc fingers
μE2, μE5 κE2	GCAGNTGT A G	E2A proteins, E12 E47, E2–5, E2–2, HEB	BHLH
μE3, κE3	CATGTG C	TFE3, TFEB, USF I/II Myc proteins	BHLHZP
μA	AGGAAGCA	Ets proteins, ets-1 ERP, erg-3, fli-1	Ets domain
μB	TTTGGGGAA	Ets proteins, PU.1	Ets domain
E, κE1	ATTNTGCAAT A T	C/EBP proteins, NF-IL6, Ig/EP	BZIP
Oct	ATGNATATT AA	Oct-1, Oct-2	Helix turn helix Homeodomain

κB	GGGRATYYC
	    C
	    T
		NF-κB/Rel proteins, p50, p65,
		p52, c-*Rel*
			Rel domain

BSAP	GNNCANTG (3′) A GCGTGACCA (5′) GA GG	Pax proteins, BSAP	Paired domain

C/EBP family proteins

Protein-binding sites, called 'E sites', are functionally important in the intronic IgH enhancer, several V_H promoters, the intronic κ enhancer and the γ1 germline promoter (Peterson et al., 1988; Tsao et al., 1988; Cooper et al., 1992; Xu and Stavnezer, 1992). E sites are recognized by the CCAAT/enhancer (C/EBP) family of proteins (Landschulz et al., 1988; Roman et al., 1990). In addition to Ig genes, binding sites for C/EBP proteins occur in regulatory elements of many other genes including cytokine genes, liver-specific genes, acute response genes, the c-*fos* gene and several viral genes (reviewed in Akira and Kishimoto, 1992).

C/EBP proteins comprise a subfamily of the 'basic-zipper' (B-ZIP) family of proteins. BZIP proteins dimerize via α helices that contain heptad repeats of leucine resides (leucine zippers); dimerization juxtaposes the basic regions so that they directly contact DNA (Vinson et al., 1989; O'Shea et al., 1991). C/EBP proteins are a specific family by virtue of their ability to heterodimerize with each other and to recognize the same DNA consensus sequence, which is distinct from sequences recognized by other BZIP proteins (Akira and Kishimoto, 1992). The family includes C/EBP (Landschulz et al., 1988), NF-IL6 (Akira et al., 1990), LIP, a smaller protein resulting from translation initiation at an internal AUG codon of NF-IL6 mRNA (Descombes and Schibler, 1991), Ig/EBP (Roman et al., 1990), CRP3 (Williams et al., 1991) and CHOP 10 (Ron and Habener, 1992).

C/EBP, NF-IL6 and CRP3 are transcriptional activators that share homology in their N-terminal regions. C/EBP has three activator regions; at least two regions in combination are necessary for transcriptional activation (Nerlov and Ziff, 1994). Based on homology, NF-IL6 and CRP3 are likely to have similar activation domains. Only part of one domain is retained in the smaller proteins LIP (Descombes and Schibler, 1991) and Ig/EBP (Roman et al., 1990). Ig/EBP contains no functional activator region and inhibits C/EBP family activators (Roman et al., 1990), presumably by binding site competition and possibly also by forming inactive heterodimers. LIP, which has a structure similar to Ig/EBP, also inhibits C/EBP activators in non-B cells (Descombes and Schibler, 1991) but is a weak transcriptional activator in B cells (Cooper et al., 1994).

The different C/EBP family proteins are differentially expressed and their expression patterns vary in different tissue lineages. In hepatocytes and adipocytes, alternate expression of C/EBP and NF-IL6 is important for normal differentiation and expression of development-specific genes (Cao et al., 1991; Isshiki et al., 1991; Samuelsson et al., 1991; Alain et al., 1992). However, a unique pattern of C/EBP family expression is seen in B cells where only NF-IL6 and Ig/EBP are expressed at reasonable levels. Although Ig/EBP is ubiquitously expressed, highest levels are found in early B cells and levels decrease 5–20 fold during B-cell development (Roman et al., 1990; Cooper et al., 1994). By contrast, NF-IL6 levels are very low in early B cells but are significantly induced when B-cell lines or normal splenocytes are stimulated by lipopolysaccharide (LPS). Thus NF-IL6 binding to E sites activates IgH and Ig κ transcription in mature B cells and plasma cells but not in earlier B cells

(Cooper et al., 1994). E sites may also be important for germline γ1 gene transcription, which precedes isotype switching to γ1. Mice deficient for NF-IL6 develop normally but are more susceptible to infection, suggesting they may have B cell defects (L. Poli and F. Costantini, personal communication). The role, if any, of high Ig/EBP levels in early B cells is not clear since Ig/EBP alone neither activates nor represses transcription. It may be that by binding E sites Ig/EBP helps modify the chromatin conformation of the IgH and κ loci so that other activators can bind.

In other cell lineages, NF-IL6 is induced in response to IL-1 and IL-6 (Akira and Kishimoto, 1992; Hsu and Chen, 1993) and this is likely to be true in B cells as well since IgH gene transcription is increased upon IL-6 treatment of responsive B cell lines (Raynal et al., 1989). In other lineages the binding and/or activating ability of NF-IL6 is also modulated by phosphorylation in response to different signal transduction paths (Wegner et al., 1992; Nakajima et al., 1993; Trautwein et al., 1993). It is not known if NF-IL6 is phosphorylated during B-cell development.

E2A/E2-2 proteins

The μE5 and μE2 sites in the IgH intronic enhancer, μE5 site in the 3′ IgH enhancer and κE2 site in the intronic κ enhancer and μE2/5 sites in the λ enhancers are recognized by members of the E2A family. (These sites are sometimes called 'E boxes'; however, to distinguish them from E sites recognized by C/EBP proteins they will be called μE2/5 sites here.) E2A family proteins bind the μE5 and κE2 sites with higher affinity than the μE2 site. Three genes encode E2A family activators that are present in B cells: (1) *E12*, *E47* and, *E2–5* are differential splice products of the *E2A* gene (Kadesch, 1992); (2) *E2–2* (Henthorn et al., 1990) and *HEB* (Hu et al., 1992) genes also encode related proteins. There are other family members such as Myo D and myogenin (Tapscott et al., 1988) that are not expressed in B cells but which are important for muscle development and other functions. Family member Id is a transdominant negative regulator of E2A proteins (Benezra et al., 1990; Kreider et al., 1992). For recent reviews on this family see Kadesch (1992) and Olson and Klein (1994).

E2A family proteins contain basic helix loop helix (BHLH) domains, which are responsible for protein dimerization and DNA binding. X-ray crystallographic analysis shows that the BHLH structure is similar to the BHLH zipper (BHLHZIP) structure of TFE3 family proteins (Ellenberger et al., 1994). Family members bind DNA as dimers and heterodimerize with each other.

In vivo footprinting showed that the μE2 site was occupied in a B-cell specific manner (Ephrussi et al., 1985). In contrast to the μE1 and μE3 sites where only ubiquitously expressed proteins bind, there is evidence for B-cell specific complexes binding the μE5 and μE2 sites even though mRNA for E2A family proteins is ubiquitously expressed (Murre et al., 1991). In addition, mutational analyses suggest that proteins which negatively regulate the IgH intronic enhancer in non-B cells bind to the μE5 site (Weinberger et al., 1988). Further studies have confirmed the importance of the μE5 site for conferring B-cell specificity. Activation by adjacently bound TFE3 is repressed in non-B cells but can be relieved by ectopic expression of E2A (Ruezinsky *et al.*, 1991), suggesting that a repressor in non-B cells is replaced by an activator in B cells. Binding and immunoblots show E2A products predominantly present in mature B cells (Bain *et al.*, 1993), a non-DNA binding form of E2A protein in pro-B cells and a unique E2A complex in plasma cells (Jacobs *et al.*, 1993). It is not known if the B-cell specific protein(s) is the product of an unidentified B-cell specific gene or a modified form of a ubiquitously expressed E2A protein. Id, when ectopically expressed, inhibits E2A proteins but it is not expressed in any B cells except possibly in a few very early B cells (Wilson *et al.*, 1991). Embryonic stem cells that are deficient in E2A have no detectable abnormalities (Zhuang *et al.*, 1993) but the effect of E2A gene deficiency on Ig expression has not yet been reported. The E2A gene is translocation to homeobox gene PBS in acute lymphoblastic leukemias (Korsmeyer, 1992).

An interesting zinc finger protein, ZEB, has been identified that binds to μE5 and μE4 sites (T. Kadesch, personal communication); it does not bind μE2. ZEB is ubiquitously expressed but indirect evidence suggests it is a repressor of IgH transcription that may be displaced by a B-cell specific activator.

Ets family proteins including PU.1

The IgH intronic enhancer has two funtionally important ets sites: μA (Nelsen *et al.*, 1993) or π (Libermann and Baltimore, 1993; Rivera *et al.*, 1993) and μB. The 3' IgH enhancer also has two ets sites called μA and μB; there is an ets site in the 3' κ enhancer and in the λ enhancers. The ets family of transcription regulators is large and different family members recognize different DNA sequences (see Wasylyk *et al.*, 1993 for a review). The family includes PU.1, ets-1, ets-2, elf-1, fli-1, erg-1, erg-3, GABPα, ERP (Nye *et al.*, 1992; Leiden, 1993; Wasylyk *et al.*, 1993; Lopez *et al.*, 1994). Since the family is large, it is difficult to know with certainty which proteins bind particular ets sites *in vivo*. *In vitro* PU.1 binds μB sites approximately 10 times better than μA sites; ets-1 binds μA sites better than μB sites and there may be binding cooperativity when two ets sites are nearby (R. Sen, personal communication).

Ets family proteins contain a conserved 84 amino acid 'ets domain' comprised of basic and leucine residues, which mediates DNA binding (Nye *et al.*, 1992). The proteins bind as monomers in the major groove with minor groove contacts. There is an inhibitory domain for binding in ets-1 and ets-2 but not PU.1 (Hagman and Grosschedl, 1992).

One unique aspect of ets family proteins is that they appear to require association with other proteins for function. Ets proteins usually do not transactivate promoters containing only ets-binding sites but do activate, often synergistically, in conjunction with various other proteins. For example, in the IgH intronic enhancer, PU.1 and ets-1 (Nelsen *et al.*, 1993) and erg-3, fli-1 and E12/E47 (Rivera *et al.*, 1993) interactions have been shown to be functionally important. In the 3' κ enhancer, NF-EM5 associates with PU.1, depending upon the phosphorylation state of PU.1 (Pongubala *et al.*, 1992, 1993); a similar complex appears to form on λ enhancer (Eisenbeis *et al.*, 1993). In T-cell receptor (TCR) enhancers there is evidence for interaction between ets proteins and CBF (Leiden, 1993; Wotton *et al.*, 1994) and similar interactions are possible in the IgH intronic enhancer but have not yet been investigated. The mechanism by which ets proteins activate in association with other proteins is not understood.

Some ets family members are expressed in a restricted manner. For example, ets-1 is preferentially expressed in lymphocytes (Bhat *et al.*, 1987), PU.1 is expressed in macrophages and B cells (Klemsz *et al.*, 1990). Others such as ets-2 are more universally expressed. ERP is expressed widely but in B cells is only expressed at high levels in early B cells (Lopez *et al.*, 1994).

The *PU.1* gene has recently been ablated in mice and causes lethality late in embryonic development. Analysis of fetal livers shows that four hematopoietic lineages fail to develop: B cells, T cells, granulocytes and macrophages (H. Singh, personal communication). Thus these mice show that PU.1 plays a critical role in early B-cell development but do not allow the role of PU.1 in Ig transcription to be assessed.

Octamer proteins

Ig promoters are only active in B-lineage cells. All heavy and light chain promoters share a highly conserved octamer motif, ATTTGCAT, that is positioned approximately 70 bp upstream of the transcriptional start site (Parslow *et al.*, 1984); in addition, there is an octamer site in the IgH intronic enhancer. Synthetic promoters containing just an octamer site and a TATA box can direct accurate and B-lineage-restricted transcription, thus highlighting the importance of the octamer element in the tissue-specific expression of Ig promoters (Dreyfus *et al.*, 1987, Wirth *et al.*, 1987). Quite surprisingly however, the same octamer element drives the expression of a number of ubiquitously expressed genes, including the histone *H2B* gene (LaBella *et al.*, 1988). Therefore the specificity of the octamer element in B cells is either due to its promoter context or due to the presence of a specific accessory factor in B cells.

Two transcription factors, Oct-1 and Oct-2, can bind to the octamer element in B cells (Singh *et al.*, 1986; Staudt *et al.*, 1986; Muller *et al.*, 1988; Scheiderheit *et al.*, 1988; Strum *et al.*, 1988). These factors belong to a family of homeodomain proteins known as the POU family (Herr *et al.*, 1988) whose characteristic feature is the presence of a ~160 amino acid bipartite DNA-binding domain containing an N-terminal ~75 amino acid POU region and a C-terminal ~60 amino acid homeodomain. While Oct-1 is ubiquitously expressed, Oct-2 has a more restricted pattern of expression, primarily in lymphoid cells and the central nervous system (Staudt *et al.*, 1986; Strum *et al.*, 1988). Recent studies have suggested that both Oct-1 and Oct-2 function by interacting with a common B-cell factor, OCA-B, which stimulates B-cell-specific transcription (Pierani et al., 1990; Luo et al., 1992). However the presence of both a ubiquitously expressed member and a tissue-specific member in the same cell does not provide an obvious answer to the question of how B-lymphocyte-specific expression is achieved through specific octamer elements. Although it is possible that Oct-2 is more effective than Oct-1 on Ig promoters, *in vitro* transcription studies suggest that both Oct-1 and Oct-2 are equally effective in stimulating transcription in B-cell extracts that have been depleted of octamer proteins. Also, the level of Oct-2 in different cells does not correlate well with the level of heavy chain enhancer activity (LeBowitz et al., 1988; Johnson et al., 1990).

Two recent studies have attempted to determine the potential role of Oct-2 in Ig transcription, by creating mutants of Oct-2 both in a mature B-cell line and in mice (Corcoran et al., 1993; Feldhaus et al., 1993). In the former, the analysis was complicated by the fact that the knock-out protocol did not result in a complete loss of Oct-2 protein from the cell, but in a significant reduction (Feldhaus et al., 1993). However, even with the drastically reduced amounts of Oct-2 protein, there was no diminution in the production of the heavy chain mRNA in these cells. In the other more definitive study, homozygous mice containing disrupted Oct-2 alleles were generated (Corcoran et al., 1993). These mice die one day following birth, indicating an as yet unidentified role for Oct-2 in the viability of these mice. More interestingly, examination of the B lymphocytes from the fetal liver of the homozygous knock-out mice indicates no apparent effect on the expression of the Ig heavy chain or on the normal differentiation of cells in the B lineage. Instead, the only defect in B cells from these mice is their inability to differentiate into Ig-secreting cells when cultured *in vitro* with LPS. Therefore Oct-2 plays a unique role only in the process of B-cell activation or late stages of differentiation, but does not appear to have a specific role in Ig transcription. Thus Oct-2 is probably redundant for Ig heavy chain gene expression.

Pax proteins, BSAP

Examination of different protein-binding sites on the IgH 3' enhancer that are in common with the intronic enhancer failed to identify any site that was specifically occupied in the plasma cell stage (Grant et al., 1992). Instead it appears that late B-cell activity of the 3' enhancer may be due to the binding of a factor, BSAP (B-cell specific activator protein), to specific sites in the 3' enhancer at earlier developmental stages (Singh and Birshtein, 1993). This factor is present in both fetal liver and B lymphocytes, but is not detected in plasma cells, T cells or other cell types, indicating that it plays an important role in B cells (Berberis et al., 1990).

BSAP belongs to a family of homeobox proteins that share a region of homology known as the paired domain (Adams et al., 1992) The paired domain consists of 128 amino acids that have been conserved through evolution. The protein contains a bipartite DNA-binding domain, and when it binds to DNA as a monomer the two domains interact with adjacent major grooves on the same side of the DNA helix

(Czerny et al., 1993). Since both half sites contribute to the binding affinity, a significant amount of degeneracy in the overall sequence can be tolerated. In mammals there are at least nine paired box or Pax genes (Pax 1–9). BSAP, which was cloned as a mammalian homolog of a transcription factor from sea urchin called TSAP (tissue-specific activator protein), was later determined to be the product of the mammalian Pax-5 gene (Adams et al., 1992).

A number of B-cell genes have BSAP sites that function in stimulating transcription. These genes include CD19, V_{pre-B}, 15 and the blk tyrosine kinase gene (Kozmik et al., 1992; Okabe et al., 1992; Singh and Birshtein, 1993). In addition BSAP has been implicated in the regulation of the expression of sterile transcripts of the germline ε locus (Liao et al., 1992). BSAP might also influence heavy chain class switching as there are multiple binding sites in different switch regions (Waters et al., 1989; Rothman et al., 1991; Liao et al., 1992; Xu et al. 1993). Therefore it appears that BSAP is among a small subclass of transcription factors that can serve either as an activator or as a repressor (another example is the YY1 protein), depending on the promoter that is regulated.

NF-κB/Rel proteins

Deletion and mutational analysis carried out on the intronic enhancer in the Ig κ locus indicated that among the various protein-binding sites only the κ B site was regulated in a developmental stage-specific manner (Sen and Baltimore, 1986a,b; Atchison and Perry, 1987; Lenardo et al., 1987). This site was active in transcription only in mature B cells and plasma cells that were expressing the κ gene, but inactive in pre-B cells and other non-lymphoid cells (Sen and Baltimore, 1989b; Lernbecher et al., 1993). Therefore it appeared that the transcription factor binding to this site, NF-κB, was responsible for the regulated expression of the κ intronic enhancer. Although active nuclear NF-κB binding activity can only be detected in mature B cells and macrophages, it is widely expressed in most cell types analyzed where it is present as an inactive cytosolic protein by being bound to an inhibitory molecule known as IκB (Baeuerle and Baltimore, 1988; Baeuerle et al., 1988). Treatment of cells with various agents such as LPS, tumor necrosis factor α (TNF-α), IL-1 or double-stranded RNA cause the dissociation of this cytoplasmic complex and the released NF-κB translocates to the nucleus (Baeuerle and Baltimore, 1988a; Grilli et al., 1993). In non-B cells, NF-κB is responsible for the inducible expression of a wide variety of genes including cytokines, lymphokines, adhesion molecules and acute phase proteins (Grilli et al., 1993).

NF-κB was first characterized as a heterodimer of 50 kDa, and 65 kDa subunits (p50 and p65); however cloning of the genes encoding these subunits revealed that they were part of a larger family of proteins that include the oncogene v-Rel and the proto-oncogene c-Rel (Ghosh et al., 1990; Kieran et al., 1990; Nolan et al., 1991; Ruben et al., 1991; Bours et al., 1992). All of these proteins share a 300 amino acid region of homology at their N terminus that has been termed the Rel-homology domain (RHD). Additional members of the family that have been identified include Rel B and p52 (Schmid *et al.*. 1991; Ryseck *et al.*, 1992). All of these proteins can homo- and hetero-dimerize with one another and bind to DNA as dimers. While homodimeric p50 or p52 are transcriptionally inactive and if overexpressed can function in a dominant negative fashion, other combinations such as p50/p52–p65, p50–c-*Rel*, p65–p65 and p65–c-*Rel* have been demonstrated to activate transcription in various promoters. Of the different family members, p50 and p52 are synthesized as longer precursors of 105 and 100 kDa respectively (p105 and p100) that are processed through an ATP-dependent mechanism to the smaller forms (Fan and Maniatis, 1991).

The longer precursor forms have C-terminal regions that contain repeats of a sequence motif, known as ankyrin repeats. The C-terminal region can fold back and cover the nuclear localizing signal present in the N-terminal RHD, thus ensuring that the precursor forms are exclusively cytosolic in localization (Henkel *et al.*, 1992). In the case of p105, the C-terminal region can also be encoded by a distinct mRNA that is expressed in B cells and the protein product, named IκB-γ, can interact with p50 protein and help to exclude it from the nucleus (Inoue *et al.*, 1992). The importance of ankyrin repeats in specifying interaction with Rel domains became clear when it was found that all the major isoforms of IκB i.e. IκB-α (the homolog in chicken is known as pp40) (Davis *et al.*, 1991; Haskill *et al.*, 1991), IκB-β (S. Ghosh, unpublished data), IκB-γ (Inoue *et al.*, 1992) and Bcl-3 (Franzoso *et al.*, 1992) contain multiple closely spaced ankyrin repeats. It appears that the number of repeats determines which Rel family member a particular IκB isoform can bind (Beg *et al.*, 1992). Thus, IκB-γ and Bcl-3 with seven repeats interact with p50 and p52 whereas IκB-α/pp40 and IκB-β with five or six repeats binds to p65 and c-*Rel* but not p50 or p52. The presence of two isoforms of IκB with similar specificity is probably to increase the potential for regulation, as each different IκB may respond to distinct signaling pathways.

The mechanism by which NF-κB is activated from its cytoplasmic form upon treatment with various inducers is still an active area of research. Studies carried out *in vitro* with purified NF-κB, IκB-α and different protein kinases indicated that while protein kinase C (PKC), cAMP-dependent protein kinase (PKA) and heme regulated kinase (HRI) can cause the dissociation of NF-κB–IκB complex, only PKC and PKA are able to directly phosphorylate IκB-α and prevent it from associating with NF-κB (Ghosh and Baltimore, 1990). Similar studies carried out subsequently on the IκB-β protein revealed an additional regulatory step in addition to phosphorylation. Treatment of the purified protein with alkaline phosphatase prevented it from interacting with NF-κB indicating that basal phosphorylation was necessary for the activity of IκB-β (Link *et al.*, 1992). However, this simple and appealing model of direct phosphorylation of IκB by protein kinases remains to be demonstrated conclusively *in vivo*. An unexpected complication has been the finding that upon stimulation of cells, the IκB-α protein (IκB-β has yet to be tested) is immediately degraded, probably after first being phosphorylated (Beg *et al.*, 1992; Scott *et al.*, 1993; Sun *et al.*, 1993). This degradation probably leads to a more complete release of NF-κB, which upon translocation to the nucleus activates transcription of various genes, including that of IκB-α. The newly synthesized IκB-α then accumulates and shuts off the NF-κB response, thus ensuring a transient response. Such an autoregulatory feedback loop appears to be ideal for regulating a transcription factor whose purpose is to carry out rapid but transient expression of target genes. The degradation of IκB-α seen upon stimulation has however prevented the isolation of sufficient amounts of the phosphorylated IκB-α form for analysis. A recent report demonstrated that TPCK, a chymotrypsin inhibitor, can block the activation of NF-κB, probably by inhibiting the protease that degrades IκB-α. However treatment with TPCK does not lead to an accumulation of phosphorylated IκB-α, suggesting that it may actually be inhibiting an earlier signaling step (H. Yang-Su and S. Ghosh, unpublished data). Further work is clearly necessary to resolve this issue.

TFE3/USF proteins

μE3 protein-binding sites were first identified in the IgH intronic enhancer by *in vivo* footprinting studies that showed B-cell-specific occupation of the site (Ephrussi *et al.*, 1985). There are also μE3 sites in V_H promoters (Peterson and Calame, 1989), the IgH 3' enhancer (Pettersson *et al.*, 1990; Lieberson *et al.*, 1991) and the κ intronic enhancer (κE3). *In vitro* binding studies reveal several ubiquitously expressed proteins that bind μE3 sites but no B-cell-specific μE3-binding proteins have been found. The B-cell-specific occupation of the IgH intronic enhancer site *in vivo* may result from cooperative association between ubiquitously expressed μE3-binding proteins and B-cell-specific proteins bound at nearby sites.

The proteins that bind μE3 sites are members of the TFE3/USF family. This family of proteins contains BHLHZIP domains that mediate DNA binding via the basic regions and protein dimerization via the HLHZIP domains. The crystal structure of the BHLHZIP domains reveals coiled α helices similar to that of the BZIP proteins (Ferre *et al.*, 1993). The zipper regions appear to determine the dimerization specificity of the proteins (Beckmann and Kadesch, 1991) and the family can be subdivided by their ability to heterodimerize. TFE3 (Beckmann *et al.*, 1990; Roman *et al.*, 1992), TFEB (Carr and Sharp, 1990; Fisher *et al.*, 1991), TFEC (Zhao *et al.*, 1993), MI (Hodgkinson *et al.*, 1993) comprise one subfamily that heterodimerize with one another; USF I (Gregor *et al.*, 1990) and II (Sirito *et al.*, 1994), FIP (Blanar and Rutter, 1992) comprise a second subfamily and c-, N- and L-Myc along with Max (Blackwell *et al.*, 1990; Blackwood and Eisenman, 1991) comprise a third subfamily. All of these proteins bind μE3 and related sites (Blackwell *et al.*, 1993). A single base change in the μE3 site (from CATGTG to CACGTG) makes a higher affinity binding site and is the sequence first identified as a c-Myc/Max binding site (Blackwell *et al.*, 1990).

TFE3 and TFEB are strong transactivators in transfection assays. TFE3 has two transcriptional activation domains: an amphipathic α helix in the N terminus (Beckmann *et al.*, 1990; Roman *et al.*, 1991) and a proline-rich region C-terminal to the BHLHZIP region (S. Artandi and K. Calame, unpublished data). Since TFE3 and TFEB are homologous it is likely that these motifs are also activation regions in TFEB. TFE3 and TFEB can form tetramers and the HLHZIP domains appear to mediate tetramer formation, possibly by stacking of α helices. Tetramerization may be the mechanism by which TFE3 can activated from distant, enhancer sites (S. Artandi et al., 1994). USF is a very weak transactivator in transfection assays but the recombinant protein activates transcription *in vitro*. This assay has been used to define a bipartite activation domain in the USF N terminus that has no distinguishing chemical characteristics (Kirschbaum et al., 1992). In Ga14 fusion assays, the N-terminal 143 amino acids of c-Myc activate transcription; this contains glutamine, proline and acidic residues (Kato, 1990). TFEC lacks activation domains and inhibits transactivation by TFE3 (Zhao et al., 1993).

TFE3, TFEB, USF I and II, and c-Myc are widely expressed in many cells including B cells at all developmental stages (Beckmann et al., 1990; Marcu, 1992; Roman et al., 1992; Sirito et al., 1994). TFE3 is LPS inducible in B-cell lines and splenic B cells (C. Roman, K. Merrell and K. Calame, unpublished data); it is not known if other family memers are also LPS inducible. TFE3, like many other transcription factors, is differentially spliced. The short form lacks the N-terminal activation domain and has a transdominant negative effect on activation by the long form (Roman et al., 1991). MI is expressed in developing eye, ear and skin, but not in the B-cell lineage (Hodgkinson et al., 1993); similarly TFEC is apparently not expressed in B cells (Zhao et al., 1993).

The role of individual family members for activating Ig transcription remains unclear. *In vitro* binding studies suggest that USF is the predominant protein binding to µE3 sites in crude nuclear extracts from B cells. However, in cotransfections TFE3 and TFEB activate µE3-dependent reporters much better than USF. B cells that are deficient in TFE3 have defects in cell surface markers such as CD23 and HSA and do not secrete normal levels of Ig (K. Merrell and K. Calame, unpublished data).

YY-1

The µE1 site, like the µE3 site, was shown by *in vivo* footprinting to be occupied in a B-cell-specific manner in the IgH intronic enhancer (Ephrussi et al., 1985) but *in vitro* studies showed a ubiquitously expressed protein binds the site (Peterson and Calame, 1987; Riggs et al., 1991). There is also a µE1 site in the 3' IgH enhancer and in the 3' κ enhancer (NF-E1). A zinc finger protein, YY1, recognizes the µE1 site (Hariharan et al., 1991; Park and Atchison, 1991; Shi et al., 1991; Flanagan et al., 1992). Antisera against YY1 supershifts all µE1-binding activity in B-cell extracts, suggesting there are no other proteins that bind the site (S. Saleque and K. Calame, unpublished data).

YY1 is a particularly intriguing transcriptional regulator because it has different activities in different gene contexts; it can be an activator, a repressor or an initiator of transcription (Seto *et al.*, 1991; Shi *et al.*, 1991). In the IgH intronic enhancer mutational analyses show YY1 is an activator (Lenardo *et al.*, 1987) but in the 3' κ enhancer it appears to be a repressor (Park and Atchison, 1991). YY1 has four C_2H_2 zinc fingers in the C terminus that mediate DNA binding. The N terminus contains an acidic region and an unusual run of histidine residues. The mechanism(s) of YY1 action are not well understood although it has recently been shown to associate with TFIIB to initiate transcription on some promoters in a mechanism independent of TBP (Usheva and Shenk, 1994).

YY1 is widely expressed in many lineages including B cells (Riggs *et al.*, 1991; Safranyl and Perry, 1993). YY1 activity is inhibited by high levels of c-Myc, which associates directly with YY1 and may interfere with the ability of YY1 to contact components of the basal transcription machinery such as TFIIB or other transcription proteins (Shrivastava *et al.*, 1993).

ROLES OF INDIVIDUAL REGULATORY ELEMENTS IN THE HEAVY CHAIN LOCUS

The IgH locus contains two well-defined transcriptional enhancers and a third weaker one that has not been studied in detail (Matthias and Baltimore, 1993). All three enhancers are lymphoid specific. The intronic enhancer is the strongest (Lieberson *et al.*, 1991) and best studied (for recent reviews see Calame, 1989; Staudt and Lenardo, 1991; Nelsen and Sen, 1992). It has been used to target expression of many transgenes to lymphoid cells, for example the c-*myc* gene (Adams *et al.*, 1985). The intronic enhancer is active throughout B-cell development including early stage pro-B and pre-B cells. Activation of rearranged V_H promoters is clearly a primary role of the intronic enhancer. In addition, it contains a promoter for sterile μ transcripts (Su and Kadesch, 1990) and appears to be important for activating sterile μ transcripts, initiated at multiple sites, which precede and may be required for VDJ joining (Ferrier *et al.*, 1990; Capone *et al.*, 1993; Chen *et al.*, 1993). It may also be important for somatic mutation (Umar *et al.*, 1991; Rothenfluh *et al.*, 1993).

The chromatin structure of the IgH locus is in an accessible or open configuration in B cells (Freeman and Garrard, 1992). Using an unusual transgenic model to study accessibility of the IgH locus to T7 polymerase, a 95-bp region of the IgH intronic enhancer (from μE5 to μB) was shown to be sufficient to act as a locus control region (Grosveld *et al.*, 1987), based on its independence of copy number and insertion site (Jenuwein *et al.*, 1993). However, more sequence, including the matrix attachment regions (MARs) (Cockerill *et al.*, 1987; Freeman and Garrard, 1992), was necessary to activate the sterile μ Pol II promoter and, presumably, V_H promoters. MARs provide nucleation sites for base unpairing and affinity for nuclear scaffold. They may define chromatin boundaries and may be important for transcriptional activation by the IgH intronic enhancer (Bode *et al.*, 1992); however, there are no direct data to establish the role of MARs in the IgH enhancer or promoters. Recently MAR-binding proteins have been identified, which should aid in dissecting the function of these elements in transcriptional regulation (Dickinson *et al.*, 1992). In addition, the IgH intronic enhancer is located at or near an origin of chromosomal DNA replication, which is more active in B lymphocytes than fibroblasts and which may also contribute to chromatin accessibility in B cells (Ariizumi *et al.*, 1993).

The continuing expression of the Ig heavy chain in cell lines that lack the intronic enhancer highlights the importance of the 3′ IgH enhancer for heavy chain expression (Klein *et al.*, 1984; Wabl and Borrows, 1984; Auguilera *et al.*, 1985). The activation of the 3′ enhancer in late stages of B-cell differentiation suggests that 3′ enhancer function is probably necessary for the transition from the mature B to plasma cell stage and for the expression of heavy chain in plasma cells (Dariavach *et al.*, 1991). By contrast, the intronic enhancer remains active throughout different stages of development (Gerster *et al.*, 1986). The IgH 3′ enhancer was first discovered in rat (Pettersson *et al.*, 1990) and later described in mouse where it is located 12.5 kb 3′ of the C_α secreted exon (Dariavach *et al.*, 1991; Lieberson *et al.*, 1991). In addition to providing further activation of V_H promoters in late B cells, this enhancer probably plays a role in activating I-region transcripts that precede, and appear to be required for, class switching (Pettersson *et al.*, 1990; Dariavach *et al.*, 1991; Lieberson *et al.*, 1991). Indeed I transcripts and class switching to γ and ε in mice that lack the IgH 3′ enhancer is severely compromised (Cogne *et al.*, 1994). There is evidence that 3′ enhancer activity is position dependent (Mocikat *et al.*, 1993) and that it may be responsible for activation of genes such as c-*myc* when they are translocated to the IgH locus (Marcu, 1992).

Neither the transcriptional activity nor the 5′ sequence of all V_H promoters (Meek *et al.*, 1990) has been systematically compared. Thus, it is not known if differential promoter activity may be related to V_H gene rearrangement or usage (Sheehan *et al.*, 1993). The octamer sites appear to play a dominant role in V_H promoter activity (Jenuwein and Grosschedl, 1991). There is different spacing in different promoters between octamer and heptamer sites and the relative location of other sites in different V_H promoters (Eaton and Calame, 1987; Cooper *et al.*, 1992). A few V_H promoters initiate transcription bidirectionally but the functional significance of this is not known (Nguyen, 1991). A MAR is located in $V_H 1$ (Webb *et al.*, 1991a) and topoII is associated with a protein complex that binds $V_H 1$ after antigen activation (Webb *et al.*, 1991b, 1993); these proteins may help determine chromatin structure of the V_H region. Germline V_H transcripts have been identified (reviewed in Alt *et al.*, 1986; Yancopoulos and Alt, 1986; Ramakrishnan and Rosenberg, 1988) that probably play a role in VD joining. They appear to initiate at the same sites as rearranged transcripts but their regulation is not understood (Berman *et al.*, 1991). V_H promoters appear to be active early in B-cell development but probably not as early as the intronic enhancer since there are Abelson murine leukemia virus (A-MuLV) lines that lack germline V transcripts (Ramakrishnan and Rosenberg, 1988).

B-CELL AND DEVELOPMENTAL-STAGE SPECIFICITY OF IGH TRANSCRIPTION

B-cell specific activators

Several activators appear to play a role in the B-cell specificity of heavy chain promoters and enhancers. These include octamer proteins (for promoters and enhancers), ets family proteins (enhancers) and E2A family proteins (enhancers). The importance of Oct and μB sites for B-cell expression has been confirmed by transgenic mouse studies (Jenuwein and Grosschedl, 1991; Annweiler et al. 1992). The μB site, which binds PU.1, is required in early B cells (Nelsen et al. 1990); later, μB and Oct are redundant in the intronic enhancer. μB cooperates with the μA (π) site, which binds other ets family proteins (Nelsen et al., 1993). This is consistent with the finding that when Oct sites are present, the μA/π site is required only in early B cells (Libermann and Baltimore, 1993).

The precise Oct-binding proteins that confer B-cell specificity are not completely understood. The idea that Oct sites are important primarily in later B cells is consistent with the observation that Oct-2 is expressed at higher levels in later B cells and is induced in response to cytokines (Miller et al. 1991). However, as discussed above, levels of Oct-2 do not correlate with IgH expression and IgH genes are expressed in mice lacking Oct-2 (Corcoran et al., 1993). Oct-1 and Oct-2 both appear to be capable of activating transcription in B cells by virtue of their association with another factor, OCA-B (Luo et al., 1992). OCA-B appears to be B cell specific but its expression pattern during B-cell development has not been reported. It is interesting to note the predominance of Oct motifs in Ig promoters and enhancers in primitive vertebrates such as *Xenopus* (Haire et al. 1991) and catfish (B. Magor and G. Warr, personal communication), suggesting its importance in evolutionarily simpler regulatory elements.

Ubiquitously expressed activators binding to the E and μE3 sites, NF-IL6 and TFE3 respectively, are induced when B cells are activated with LPS (Cooper et al., 1994; C. Roman and K. Merrell, unpublished data). In fact, E sites perform as activator sites only in mature B cells and plasmacytomas. Since NF-IL6 is induced in response to IL-6 in many cells (Akira and Kishimoto, 1992), it is likely that E sites are important for the ability of IL-6 to induce IgH transcription (Raynal et al. 1989).

Negative regulators

There is good evidence for negative regulators that inhibit the activity of both IgH enhancers in non-B cells and in B cells at particular stages of development. In the intronic enhancer, four binding sites for a protein called NF-μNR, are located in AT-rich regions near the MARs (Scheuermann and Chen, 1989; Scheuermann, 1992). Deletions suggest that NF-μNR is a negative regulator in non-B cells. cDNA encoding this protein has not been cloned and its structure is not known. There is a binding site for NF-μNR in the V_H1 promoter but it is not known if NF-μNR functions as a negative regulator in this location (N. Avitahl and K. Calame, unpublished data). Mutational studies also showed that the μE5 site binds a negative regulator in non-B cells (Weinberger et al., 1988) and cotransfections have shown that the negative activity can be overcome by E2A family activators (Ruezinsky et al., 1991). As discussed above, ZEB binds this site and is ubiquitously expressed. ZEB may function as a ubiquitous negative regulator that can be displaced by B-cell-specific E2A activators. Treatment with anti-μ causes B cells to shut off Ig expression. The mechanism by which this occurs is not fully understood but anti-μ has been shown to affect the intronic enhancer by increasing binding to μE5 and decreasing binding at Oct (Chen et al., 1991; Moore et al., 1993).

Somatic cell hybrids between B cells and T cells or fibroblasts have shown that non-B cells contain transdominant regulators that extinguish IgH transcription; both the intronic enhancer and V_H promoters are inhibited (Zaller et al., 1988). The octamer motif in the intronic enhancer appears to be necessary for this repression, which appears to proceed by a mechanism distinct from other negative regulatory mechanisms (Yu et al., 1989; Shen et al., 1993).

In the 3' enhancer, negative regulation in early B cells appears to be mediated by the paired domain protein BSAP, which is expressed in early B cells (Barberis et al., 1990; Adams et al., 1992; Singh and Birshtein, 1993). It is not clear why this enhancer is not active in early B cells, but this could be related to its apparent role in activating I region transcription, which precedes isotype switching later in B-cell development.

REGULATORY ELEMENTS IN LIGHT CHAIN TRANSCRIPTION

Regulatory elements in the κ light chain locus

Expression of the κ light chain gene is due to the activity of the promoter and two enhancer elements, one located in the intron between the joining and constant region segments, the other located approximately 9 kb downstream of the constant region exon (Emorine et al., 1983; Queen and Baltimore, 1983; Bergman et al., 1984; Falkner and Zachau, 1984; Picard and Schaffner, 1984; Gopal et al., 1985; Queen et al., 1986; Meyer and Neuberger, 1989). All three elements are B-cell specific but the developmental stage-specific expression of the κ gene is due to the regulated activity of the two enhancers. The intronic enhancer was discovered first and has been studied extensively. This enhancer is inactive in early stages of B-cell development, but becomes active during the transition from a pre-B cell to a mature B cell (Sen and Baltimore, 1986a,b; Atchison and Perry, 1987; Lenardo et al., 1987). The enhancer is highly conserved between human, mouse and rabbit genes and also shares some regulatory sites with the heavy chain intronic enhancer. The stage restriction was primarily believed to be due to the binding of NF-κB to a 10-bp site, termed the κB site (Sen and Baltimore, 1986a; Atchinson and Perry, 1987). Mutation of the κB site abolishes the activity of the intronic enhancer and also its ability to be induced in pre-B cells by inducers such as LPS or phorbol esters (Lenardo et al., 1987). Therefore the constitutive activation of NF-κB in mature B cells and plasma cells appeared to be a critical regulatory event in the differentiation from a pre-B to a mature B cell. However, as indicated above, it is unclear if NF-κB itself is responsible for activating this site or if there are additional binding activities in these cells. Besides the κB site, there are three µE2/5 site sequences in this enhancer that bind to different members of the HLH-transcription factor family (Gimble and Max, 1987; Hromas et al., 1988). Mutational studies indicate that the κE1, 2 and 3 sites are crucial for the activity of this enhancer (Lenardo et al., 1987).

The search for the 3' enhancer in the κ locus was driven by the observation that high levels of κ light chain protein continue to be expressed in the plasmacytoma S107, which lacks the transcription factor NF-κB (Atchison and Perry, 1987). This additional enhancer is located 8.5 kb downstream of the constant region exon and is as active as the intronic enhancer (Meyer and Neuberger, 1989; Meyer et al., 1990; Fulton and Van Ness, 1993). It is also active in mature B cells and plasma cells but not in pre-B cells, T cells or other cell types (Pongubala and Atchison, 1991). The majority of the activity of this enhancer can be localized to a 132-bp core that does not contain κB sites, thereby providing an explanation for the expression of κ gene in S107 cells. This 132-bp core contains an µE2 sequence that can bind HLH proteins. The involvement of HLH proteins in the regulation of this enhancer was further substantiated when the activity of the enhancer could be repressed by expressing the HLH-inhibitory protein Id (Benezra et al., 1990; Pongubala and Atchison, 1991). Further characterization of this enhancer revealed that it binds Pu.1, a B-cell restricted ets family member (Pongubala et al., 1992). Surprisingly, activation through Pu.1 requires the binding of another B-cell-restricted factor, NF-EM5 to an adjacent site. The association between these proteins also requires phosphorylation of the Pu.1 protein (Pongubala and Atchison, 1991). A similar situation is also thought to occur in the λ enhancers, but it is not clear if the activity of Pu.1 in all instances requires an obligatory interaction with another transcription factor (Eisenbeis et al., 1993).

Regulatory elements in the λ light chain locus

The λ locus does not appear to contain enhancer elements in the intron between the joining and constant regions (Picard and Schaffner, 1984). Instead two enhancers have been identified at the 3' end of each constant region cluster. These enhancers, named $E_\lambda 2$-4 (located 15.5 kb downstream of C12C14) and $E_\lambda 3$-1 (located 35 kb downstream of Cl3Cl1) are separated by 100 kb of the mouse genome (Hagman et al., 1990). The two enhancers are specific for B cells and appear to be homologous in their functional elements. Two homologous domains, λA and λB, appear to be responsible for imparting the majority of the activity to these enhancers (Rudin and Storb, 1992). These domains bind to both B-cell-specific and non-specific nuclear factors and complex protein–protein interaction appears to be responsible for imparting full activity. The proteins binding to the λA region have not yet been defined, but the λB region contains a binding site for Pu.1 that is regulated by the binding of NF-EM5 to an adjacent site, reminiscent of the κ 3' enhancer (Eisenbeis et al., 1993). Since both the κ 3' enhancer and the λ enhancers are active at similar stages of development, it is reasonable that they are regulated by common elements.

B-CELL AND DEVELOPMENTAL-STAGE SPECIFICITY OF IG κ AND λ LIGHT CHAIN TRANSCRIPTION

The B-cell and developmental stage-specific expression of the κ light chain gene is considered to be primarily due to the transcription factor NF-κB binding to the Ig κ intronic enhancer. Mutational studies indicated that the κB site was crucial for inducibility of the intronic enhancer in pre-B cells (Lenardo et al., 1987). The other characterized binding sites in the intronic enhancer, κE1, κE2 and κE3, bind factors that are either expressed ubiquitously or expressed at earlier stages of B-cell development and therefore cannot explain the developmental-stage specificity of this enhancer (Lenardo et al., 1987). The strong correlation in cell lines between the activity of NF-κB and the expression of the κ gene led to the suggestion that NF-κB acted like a developmental switch in the transition from a pre-B cell to a mature B cell. Examination of the composition of the constitutively active NF-κB complex in mature B cells indicated that it was composed of p50–c-*Rel* heterodimers and not the proto typical p50–p65 heterodimers (Miyamoto et al., 1994; R.J. Phillips and S. Ghosh, unpublished data). A separate study has found that mature B cells contain a constitutive complex of p50–RelB heterodimers (Lernbecher et al., 1993). It is unclear how a portion of the Rel complexes can become constitutively activated in B cells (and macrophages) but not in any other cell type. It has been demonstrated recently that the rate of degradation of IκB-α protein is greater in mature B cells than in pre-B cells suggesting that the pathway leading to the modification and degradation of IκB-α is stimulated in these cells (Miyamoto et al., 1994). Although the importance of the κB site in the κ intronic enhancer is well established and the question of how NF-κB becomes constitutively active only in mature B cells remains to be answered, two separate studies have provoked a rethinking about the role of NF-κB as a regulator of the developmental-stage-specific Ig κ light chain gene.

1. Mutant mice lacking the NF-κB p50 gene have been generated by gene targeting (W. Shaw and D. Baltimore, personal communication). These mice are born and develop normally and contain normal numbers of B cells. The number of B cells expressing the κ gene are similar to wild-type mice. The only defects that have been noticed stem from the role of NF-κB in inducing expression of genes such as cytokines and lymphokines. Although it is possible that some other member of

the Rel-family, e.g. p52, substitutes for p50, the results are still striking since all the κB-binding complexes detected in B cells appear to contain p50. Additional knock-outs of other members of this family will be necessary to reach an unequivocal conclusion.
2. The majority of studies that defined a developmental-stage-specific activation of NF-κB in the transistion from a pre-B to a mature B cell were carried out with A-MuLV-transformed murine pre-B cell lines. However it appears that the *abl* oncogene itself can inhibit the activation of NF-κB and therefore the absence of nuclear NF-κB in transformed pre-B cells is likely to be an artifact of the transforming event (Chen *et al.*, 1994; Klug *et al.*, 1994). In addition, pre-B cells isolated from the bone marrow and amplified by *in vitro* culture with IL-7 contain active NF-κB but do not express the κ light chain gene. Therefore the correlation between activity of NF-κB and expression of the κ light chain gene does not appear to hold and at present it is unclear if NF-κB plays as critical a role in κ gene expression as originally envisioned.

Instead it has become evident that negative regulatory elements play an important role in determining the stage specificity of the intronic enhancer. In the Ig κ intronic enhancer, a 200-bp fragment that lies upstream of the NF-κB binding site was found to contain negative regulatory activity (Pierce *et al.* 1991). This silencer element contains multiple negative elements that prevent the enhancer from functioning in non-B cells, e.g. T cells that have nuclear NF-κB. Further analysis has shown that there is an added negative regulatory element of 27 bp, termed κNE, that lies just upstream of the κB site, but is not part of the previously characterized κ silencer. This sequence is homologous to mouse B1 repetitive elements and is conserved between mouse and human κ intronic enhancers. The κNE itself is not B-cell specific, but acquires specificity when paired to a contiguous sequence termed κBS (Saksela and Baltimore, 1993). A specific nuclear factor complex was identified with the κNE sequence and termed B1-NF. Mutational studies suggested that this complex has a functional role, although further characterization of this factor, including cloning, will be necessary before more definitive conclusions can be drawn about its function.

Although it remains unclear which, if any, factor specifies the developmental-stage specificity of the κ intronic enhancer, it is likely that the octamer sites in the light chain promoters and the E sites in the intronic enhancer are also critical for providing B-cell specificity. As discussed earlier, which octamer-binding protein imparts B-cell specificity is unknown, but the associating protein OCA-B probably plays an important role (Luo *et al.*, 1992). Also, even though the μE2/5 sites bind to ubiquitously expressed regulators they function as positive regulators only in B cells. An additional level of control is also imparted by the κ 3' enhancer. Both Pu.1 and NF-EM5 are B-cell restricted in their expression and the enhancer is only active in B cells (Pongubala *et al.*, 1992). Although the enhancer is not active in pre-B cells both Pu.1 and NF-EM5 are present and therefore an additional level of regulation must exist to impart the developmental-stage specificity to this enhancer. Such a regulation is probably provided by negative elements that flank the 132-bp core activating region in this enhancer (Park and Atchison, 1991). A specific protein was found to bind to one of the regulatory sequences and was termed NF-E1. Cloning of the corresponding cDNA revealed that it was identical to the previously characterized transcription factor, YY1. Therefore, similar to BSAP in the IgH 3' enhancer (Singh and Birshtein, 1993), both positive and negative regulation can be provided by the same protein.

Both the enhancers in the λ locus are B-cell specific, but their patterns of activity in different developmental stages have not been reported. Since these enhancers also bind to Pu.1 and NF-EM5, it is likely that they are regulated similarly to the κ 3' enhancer (Eisenbeis et al., 1993). In addition these enhancers contain μE2/5 sites, which may also contribute to B-cell specificity.

PERSPECTIVES

Although recent progress in understanding proteins that regulate Ig gene transcription has been impressive, as the length of this chapter indicates, key questions remain. Which factors regulate B-cell lineage and developmental-stage specificity? What role do negative regulators play in these events? How are appropriate levels of key regulators achieved? How are the regulators affected by B-cell growth or differentiation signals? The importance of these questions is underscored by recent demonstrations that expression of heavy and light chain Ig genes is necessary and sufficient, in *RAG*-1 or *RAG*-2 deficient mice, to drive ordered B-cell differentiation (Spanopoulou et al., 1994; Young et al., 1994).

This chapter was completed in June 1994.

REFERENCES

Adams, B., Dorfler, P., Aguzzi, A., Kozmik, Z., Urbanek, P., Maurer, F.I. and Busslinger, M. (1992). *Genes Dev.* **6**, 1589–1607.
Adams, J.M., Harris, A.W., Pinkert, C.A., Corcoran, L.M., Alexander, W.S., Cory, S., Palmiter, R.D. and Brinster, R.L. (1985). *Nature* **318**, 533–538.
Akira, S. and Kishimoto, T. (1992). *Immunol. Rev.* **127**, 25–50.
Akira, S., Isshiki, H., Sugita, T., Osamu, T., Kinoshita, S., Nishio, Y., Nakajima, T., Hirano, T. and Kishimoto, T. (1990). *EMBO J.* **9**, 1897–1906.
Alam, T., An, M.R. and Papaconstantinou, J. (1992). *J. Biol. Chem.* **267**, 5021–5024.
Alt, F., Blackwell, K., DePinho, R., Reth, M. and Yancopoulos, G. (1986). *Immunol. Rev.* **89**, 5–30.
Annweiler, A., Muller, U. and Wirth, T. (1992). *Nucleic Acids Res.* **20**, 1503–1509.
Artandi, S., Cooper, C., Shrivastava, A. and Calame, K. (1994). *Mol. Cell. Biol.* **14**, 7704–7716.
Ariizumi, K., Wang, Z. and Tucker, P.W. (1993). *Proc. Natl Acad. Sci. USA* **90**, 3695–3699.
Atchison, M.L. and Perry, R.P. (1987). *Cell* **48**, 121–128.
Auguilera, R., Hope, T. and Sakano, H. (1985). *EMBO J.* **4**, 3689–3693.
Baeuerle, P.A. and Baltimore, D. (1988). *Cell* **53**, 211–217.
Baeuerle, P.A., Lenardo, M., Pierce, J.W. and Baltimore, D. (1988). *Cold Spring Harb. Symp. Quant. Biol.* **53**, 789–798.

Bain, G., Gruenwald, S. and Murre, C. (1993). *Mol. Cell. Biol.* **13**, 3522–3529.
Barberis, A., Widenhorn, K., Vitelli, L. and Busslinger, M. (1990). *Genes Dev.* **4**, 849–859.
Beckmann, H. and Kadesch, T. (1991). *Genes Dev.* **5**, 1057–1066.
Beckmann, H., Su, L.-K. and Kadesch, T. (1990). *Genes Dev.* **4**, 167–179.
Beg, A.A., Ruben, S.M., Scheinman, R.I., Haskill, S., Rosen, C.A. and Baldwin, A.S. (1992). *Genes Dev.* **6**, 1899–1913.
Benezra, R., Davis, R., Locksohn, D., Turner, D. and Weintraub, H. (1990). *Cell* **61**, 49–59.
Berberis, A., Widenhorn, K., Vitelli, L. and Busslinger, M. (1990). *Genes Dev.* **4**, 849–859.
Bergman, Y., Rice, D., Grosschedl, R. and Baltimore, D. (1984). *Proc. Natl Acad. Sci. USA* **81**, 7041–7045.
Berman, J.E., Humphries, C.G., Barth, J., Alt, F.W. and Tucker, P.W. (1991). *J. Exp. Med.* **173**, 1529–1535.
Bhat, N., Fisher, R., Fujiwara, S., Ascione, R. and Papas, T. (1987). *Proc. Natl Acad. Sci. USA* **84**, 3161–3165.
Blackwell, T.K., Kretzner, L., Blackwood, E.M., Eisenman, R.N. and Weintraub, H. (1990). *Science* **250**, 1149–1151.
Blackwell, T.K., Huang, J., Ma, A., Kretzner, L., Alt, F., Eisenman, R. and Weintraub, H. (1993). *Mol. Cell. Biol.* **13**, 5216–5224.
Blackwood, E.M. and Eisenman, R.N. (1991). *Science* **251**, 1211–1217.
Blanar, M. and Rutter, W. (1992). *Science* **256**, 1014–1018.
Bode, J., Kohwi, Y., Dickinson, L., Joh, T., Klehr, D., Mielke, C. and Kohwi, S.T. (1992). *Science* **255**, 195–197.
Bours, V., Burd, P.R., Brown, K., Villalobos, J., Park, S., Ryseck, R.P., Bravo, R., Kelly, K. and Siebenlist, U. (1992). *Mol. Cell. Biol.* **12**, 685–695.
Calame, K. (1989). *Trends Genet.* **5**, 395–399.
Cao, Z., Umek, R. and McKnight, S. (1991). *Genes Dev.* **5**, 1538–1552.
Capone, M., Watrin, F., Fernex, C., Horvat, B., Krippl, B., Scollay, R. and Ferrier, P. (1993). *EMBO J.* **12**, 4335–4346.
Carr, C.S. and Sharp, P.A. (1990). *Mol. Cell. Biol.* **10**, 4384–4388.
Chen, J., Young, F., Bottaro, A., Stewart, V., Smith, R.K. and Alt, F.W. (1993). *EMBO J.* **12**, 4635–4645.
Chen, U., Scheuermann, R.H., Wirth, T., Gerster, T., Roeder, R.G., Harshman, K. and Berger, C. (1991). *Nucleic Acids Res.* **19**, 5981–5989.
Chen, Y.-Y., Wang, L.C., Huang, M.S. and Rosenberg, N. (1994). *Genes Dev.* **8**, 688–697.
Cockerill, P., Yuen, M.-H. and Garrard, W. (1987). *J. Biol. Chem.* **262**, 5394–5397.
Cogne, M., Lansford, R., Bottaro, A., Zhang, J., Gorman, J., Young, F., Cheng, H.-L. and Alt, F. (1994). *Cell* **77**, 737–747.
Cooper, C., Johnson, D., Roman, C., Avitahl, N., Tucker, P. and Calame, K. (1992). *J. Immunol.* **149**, 3225–3231.
Cooper, C., Berrier, A., Roman, C. and Calame, K. (1994). *J. Immunol.* **153**, 5049–5058.
Corcoran, L.M., Karvelas, M., Nossal, G.J., Ye, Z.S., Jacks, T. and Baltimore, D. (1993). *Genes Dev.* **7**, 570–582.
Czerny, T., Schaffner, G. and Busslinger, M. (1993). *Genes Dev.* **7**, 2048–2061.
Dariavach, P., Williams, G.T., Campbell, K., Pettersson, S. and Neuberger, M.S. (1991). *Eur. J. Immunol.* **21**, 1499–1504.
Davis, N., Ghosh, S., Simmons, D.L., Tempst, P., Liou, H.-C., Baltimore, D. and Bose, H.R. (1991). *Science*, **253**, 1268–1271.
Descombes, P. and Schibler, U. (1991). *Cell* **67**, 569–579.
Dickinson, L.A., Joh, T., Kohwi, Y. and Kohwi, S.T. (1992). *Cell* **70**, 631–645.
Dreyfus, M., Doyen, N. and Rougeon, F. (1987). *EMBO J.* **6**, 1685–1690.
Eaton, S. and Calame, K. (1987). *Proc. Natl Acad. Sci. USA* **84**, 7634–7638.
Eisenbeis, C.F., Singh, H. and Storb, U. (1993). *Mol. Cell. Biol.* **13**, 6452–6461.
Ellenberger, T., Fass, D., Arnaud, M. and Harrison, S. (1994). *Genes Dev.* **8**, 970–980.
Emorine, L., Kuehl, M., Weir, L., Leder, P. and Max, E. (1983). *Nature* **304**, 447–449.

Ephrussi, A., Church, G., Tonegawa, S. and Gilbert, W. (1985). *Science* **227**, 134–140.
Falkner, F.G. and Zachau, H.G. (1984). *Nature* **310**, 71–74.
Fan, C. and Maniatis, T. (1991). *Nature* **354**, 395–398.
Feldhaus, A.L., Klug, C.A., Arvin, K.L. and Singh, H. (1993). *EMBO J.* **12**, 2763–2772.
Ferre, D.A., Prendergast, G.C., Ziff, E.B. and Burley, S.K. (1993). *Nature* **363**, 38–45.
Ferrier, P., Krippl, B., Blackwell, K., Furley, A.J., Suh, H., Winoto, A., Cook, W., Hood, L., Costantini, F. and Alt, F. (1990). *EMBO J.* **9**, 117–125.
Fisher, D., Carr, C., Parent, L. and Sharp, P. (1991). *Genes Dev.* **5**, 2342–2352.
Flanagan, J.R., Becker, K.G., Ennist, D.L., Gleason, S.L., Driggers, P.H., Levi, B.Z., Appella, E. and Ozato, K. (1992). *Mol. Cell. Biol.* **12**, 38–44.
Franzoso, G., Bours, V., Park, S., Tomita-Yamaguchi, M., Kelly, K. and Siebenlist, U. (1992). *Nature* **359**, 339–342.
Freeman, L.A. and Garrard, W.T. (1992). *Crit. Rev. Eukaryotic Gene Expr.* **2**, 165–209.
Fulton, R. and Van Ness, B. (1993). *Nucleic Acids Res.* **21**, 4941–4947.
Gerster, T., Picard, D. and Schaffner, W. (1986). *Cell* **45**, 45–52.
Ghosh, S. and Baltimore, D. (1990). *Nature* **344**, 678–682.
Ghosh, S., Gifford, A.M., Rivere, L.R., Tempst, P., Nolan, G.P. and Baltimore, D. (1990). *Cell* **62**, 1019–1029.
Gimble, J.M. and Max, E.E. (1987). *Mol. Cell. Biol.* **7**, 15–25.
Gopal, T.V., Shimada, T., Baur, A.W. and Nienhuis, A. (1985). *Science* **229**, 1102–1104.
Grant, P.A., Arulampalam, V., Ahrlund, R.L. and Pettersson, S. (1992). *Nucleic Acids Res.* **20**, 4401–4408.
Gregor, P., Sawadogo, M. and Roeder, R. (1990). *Genes Dev.* **4**, 1730–1740.
Grilli, M., Chiu, J.J. and Lenardo, M.J. (1993). *Int. Rev. Cytol.* **143**, 1–62.
Grosveld, F., Blom van Assendelft, G., Greaves, D. and Kollias, G. (1987). *Cell* **51**, 975–985.
Hagman, J. and Grosschedl, R. (1992). *Proc. Natl Acad. Sci. USA* **89**, 8889–8893.
Hagman, J., Rudin, C.M., Haasch, D., Chaplin, D. and Storb, U. (1990). *Genes Dev.* **4**, 978–992.
Haire, R.N., Ohta, Y., Litman, R.T., Amemiya, C.T. and Litman, G.W. (1991). *Nucleic Acids Res.* **19**, 3061–3066.
Hariharan, N., Kelley, D.E. and Perry, R.P. (1991). *Proc. Natl Acad. Sci. USA* **88**, 9799–9803.
Haskill, S., Beg, A., Tompkins, S.M., Morris, J.S., Yurochko, A., Sampson-Johannes, A., Mondal, K., Ralph, P. and Baldwin, A.S. (1991). *Cell* **65**, 1281–1289.
Henkel, T., Zabel, U., van, Z., Muller, J., Fanning, E. and Baeuerle, P. (1992). *Cell* **68**, 1121–1133.
Henthorn, P., Kiledjian, M. and Kadesch, T. (1990). *Science* **247**, 467–470.
Herr, W., Strum, R., Clerc, R., Corcoran, L., Baltimore, D., Sharp, P., Ingraham, H., Rosenfeld, G., Finney, G., Revkun, G. and Horvitz, H. (1988). *Genes Dev.* **2**, 1513–1516.
Hodgkinson, C.A., Moore, K.J., Nakayama, A., Steingrimsson, E., Copeland, N.G., Jenkins, N.A. and Arnheiter, H. (1993). *Cell* **74**, 395–404.
Hromas, R., Pauli, U., Marcuzzi, A., Lafrenz, D., Nick, H., Stein, J., Stein, G. and Van Ness, B. (1988). *Nucleic Acids Res.* **16**, 953–967.
Hsu, W. and Chen, K.S. (1993). *Mol. Cell. Biol.* **13**, 2515–2523.
Hu, J.S., Olson, E.N. and Kingston, R.E. (1992). *Mol. Cell. Biol.* **12**, 1031–1042.
Inoue, J., Kerr, L.D., Kakizuka, A. and Verma, I.M. (1992). *Cell* **68**, 1109–1120.
Isshiki, H., Akira, S., Sugita, T., Nishio, Y., Hashimoto, S., Pawlowski, T., Suematsu, S. and Kishimoto, T. (1991). *New Biol.* **3**, 63–70.
Jacobs, Y., Vierra, C. and Nelson, C. (1993). *Mol. Cell. Biol.* **13**, 7321–7333.
Jenuwein, T. and Grosschedl, R. (1991). *Genes Dev.* **5**, 932–943.
Jenuwein, T., Forrester, W.C., Qiu, R.G. and Grosschedl, R. (1993). *Genes Dev.* **7**, 2016–2032.
Johnson, D.G., Carayannopoulos, L., Capra, J.D., Tucker, P.W. and Hanke, J.H. (1990). *Mol. Cell. Biol.* **10**, 982–990.
Kadesch, T. (1992). *Immunol. Today* **13**, 31–36.
Kato, G., Barrett, J., Villa-Garcia, M. and Dang, C. (1990). *Mol. Cell. Biol.* **10**, 5914–5920.
Kieran, M., Blank, V., Logeat, F., Vandekerckhove, J., Lottspeich, F., LeBail, O., Urban, M.B., Kourilsky, P., Baeuerle, P.A. and Israel, A. (1990). *Cell* **62**, 1007–1018.

Kirschbaum, B.J., Pognonec, P. and Roeder, R.G. (1992). *Mol. Cell. Biol.* **12**, 5094–5101.
Klein, S., Sablitsky, F. and Radbruch, A. (1984). *EMBO J.* **3**, 2473–2476.
Klemsz, M., McKercher, S.R. and Maki, R. (1990). *Cell* **61**, 113–124.
Klug, C.A., Gerety, S.J., Shah, P.C., Chen, Y.-Y., Rice, N.R., Rosenberg, N. and Singh, H. (1994). *Genes Dev.* **8**, 678–687.
Korsmeyer, S.J. (1992). *Annu. Rev. Immunol.* **10**, 785–807.
Kozmik, Z., Wang, S., Dorfler, P., Adams, B. and Busslinger, M. (1992). *Mol. Cell. Biol.* **12**, 2662–2672.
Kreider, B.L., Benezra, R., Rovera, G. and Kadesch, T. (1992). *Science* **255**, 1700–1702.
LaBella, F., Sive, H.L., Roeder, R.G. and Heintz, N. (1988). *Genes Dev.* **2**, 32–39.
Landschulz, W., Johnson, P., Adashi, E., Graves, B. and McKnight, S. (1988). *Genes Dev.* **2**, 786–800.
LeBowitz, J.H., Kobayash, T., Staudt, L., Baltimore, D. and Sharp, R.A. (1988). *Genes Dev.* **2**, 1227–1237.
Leiden, J.M. (1993). *Annu. Rev. Immunol.* **11**, 539–570.
Lenardo, M., Pierce, J. and Baltimore, D. (1987). *Science* **236**, 1573–1577.
Lernbecher, T. and Muller, U., Wirth, T. (1993). *Nature* **365**, 767–770.
Li, S.C., Rothman, P., Boothby, M., Ferrier, P., Glimcher, L. and Alt, F.W. (1991). *Adv. Exp. Med. Biol.* **292**, 245–251.
Liao, F., Giannini, S.L. and Birshtein, B.K. (1992). *J. Immunol.* **148**, 2909–2917.
Libermann, T.A. and Baltimore, D. (1993). *Mol. Cell. Biol.* **13**, 5957–5969.
Lieberson, R., Giannini, S., Birshtein, B. and Eckhardt, L. (1991). *Nucleic Acids Res.* **19**, 933–937.
Link, E., Kerr, L.D., Schreck, R., Zabel, U., Verma, I.M. and Baeuerle, P.A. (1992). *J. Biol. Chem.* **267**, 239–246.
Liou, H.C. and Baltimore, D. (1993). *Curr. Opin. Cell Biol.* **5**, 477–487.
Lopez, M., Oettgen, P., Akbarali, Y., Dendorfer, U. and Libermann, T. (1994). *Mol. Cell. Biol.* **14**, 3292–3309.
Luo, Y., Fujii, H., Gerster, T. and Roeder, R.G. (1992). *Cell* **71**, 231–241.
Marcu, K. (1992). *Annu. Rev. Biochem.* **61**, 809–860.
Matthias, P. and Baltimore, D. (1993). *Mol. Cell. Biol.* **13**, 1547–1553.
Meek, K., Rathbun, G., Reininger, L., Jaton, J.C., Kofler, R., Tucker, P.W. and Capra, J.D. (1990). *Mol. Immunol.* **27**, 1073–1081.
Meyer, K.B. and Neuberger, M.S. (1989). *EMBO J.* **8**, 1959–1964.
Meyer, K.B., Sharpe, M.J., Surani, M.A. and Neuberger, M.S. (1990). *Nucleic Acids Res.* **18**, 5609–5615.
Miller, C.L., Feldhaus, A.L., Rooney, J.W., Rhodes, L.D., Sibley, C.H. and Singh, H. (1991). *Mol. Cell. Biol.* **11**, 4885–4894.
Miyamoto, S., Chiao, P.J. and Verma, I.M. (1994). *Mol. Cell. Biol.* **14**, 3276–3282.
Mocikat, R., Harloff, C. and Kutemeier, G. (1993). *Gene* **136**, 349–353.
Moore, B.B., Ariizumi, K., Tucker, P.W. and Yuan, D. (1993). *J. Immunol.* **150**, 3366–3374.
Muller, M.M., Ruppert, S., Schaffner, W., Matthias, P. (1988). *Nature* **336**, 544–551.
Murre, C., Voronova, A. and Baltimore, D. (1991). *Mol. Cell. Biol.* **11**, 1156–1160.
Nakajima, T., Kinoshita, S., Sasagawa, T., Sasaki, K., Naruto, M., Kishimoto, T. and Akira, S. (1993). *Proc. Natl Acad. Sci. USA* **90**, 2207–2211.
Nelsen, B. and Sen, R. (1992). *Int. Rev. Cytol.* **133**, 121–149.
Nelsen, B., Kadesch, T. and Sen, R. (1990). *Mol. Cell. Biol.* **10**, 3145–3154.
Nelsen, B., Tian, G., Erman, B., Gregoire, J., Maki, R., Graves, B. and Sen, R. (1993). *Science* **261**, 82–86.
Nerlov, C. and Ziff, E. (1994). *Genes Dev.* **8**, 350–362.
Nguyen, Q.T., Doyen, N., d'Andon, M.F. and Rougeon, F. (1991). *Nucleic Acids Res.* **19**, 5339–5344.
Nolan, G.P., Ghosh, S., Liou, H.-C., Tempst, P. and Baltimore, D. (1991). *Cell* **64**, 961–969.
Nye, J., Petersen, J., Gunther, C., Jonsen, M. and Graves, B. (1992). *Genes Dev.* **6**, 975–990.
Okabe, T., Watanabe, T. and Kudo, A. (1992). *Eur. J. Immunol.* **22**, 37–43.
Olson, E. and Klein, W. (1994). *Genes Dev.* **8**, 1–8.
O'Shea, E.K., Klemm, J.D., Kim, P.S. and Alber, T. (1991). *Science* **254**, 539–544.
Park, K. and Atchison, M.L. (1991). *Proc. Natl Acad. Sci. USA* **88**, 9804–9808.
Parslow, T.G., Blair, D.L., Murphy, W.J., Granner, D.K. (1984). *Proc. Natl Acad. Sci. USA* **81**, 2650–2654.

Peterson, C. and Calame, K. (1987). *Mol. Cell. Biol.* **12**, 4194–4203.
Peterson, C. and Calame, K. (1989). *Mol. Cell. Biol.* **9**, 776–786.
Peterson, C., Eaton, S. and Calame, K. (1988). *Mol. Cell. Biol.* **8**, 4972–4980.
Pettersson, S., Cook, G.P., Bruggemann, M., Williams, G.T. and Neuberger, M.S. (1990). *Nature* **344**, 165–168.
Picard, D. and Schaffner, W. (1984). *Nature* **307**, 80–82.
Pierani, A., Heguy, A., Fujii, H. and Roeder, R. (1990). *Mol. Cell. Biol.* **10**, 6204–6215.
Pierce, J.W., Gifford, A.M. and Baltimore, D. (1991). *Mol. Cell. Biol.* **11**, 1431–1437.
Pongubala, J.M. and Atchison, M.L. (1991). *Mol. Cell. Biol.* **11**, 1040–1047.
Pongubala, J.M., Nagulapalli, S., Klemsz, M.J., McKercher, S.R., Maki, R.A. and Atchison, M.L. (1992). *Mol. Cell. Biol.* **12**, 368–378.
Pongubala, J.M., Van, B.C., Nagulapalli, S., Klemsz, M.J., McKercher, S.R., Maki, R.A. and Atchison, M.L. (1993). *Science* **259**, 1622–1625.
Queen, C. and Baltimore, D. (1983). *Cell* **33**, 741–748.
Queen, C., Foster, J., Stauber, C. and Stafford, J. (1986). *Immunol. Rev.* **89**, 49–68.
Ramakrishnan, L. and Rosenberg, N. (1988). *Mol. Cell. Biol.* **8**, 5216–5223.
Raynal, M.-C., Liu, Z., Hirano, T., Mayer, L. and Chen-Kiang, S. (1989). *Proc. Natl Acad. Sci. USA* **86**, 8024–8028.
Riggs, K., Merrell, K., Wilson, G. and Calame, K. (1991). *Mol. Cell. Biol.* **11**, 1765–1769.
Rivera, R.R., Stuiver, M.H., Steenbergen, R. and Murre, C. (1993). *Mol. Cell. Biol.* **13**, 7163–7169.
Roman, C., Platero, J.S., Shuman, J. and Calame, K. (1990). *Genes Dev.* **4**, 1404–1415.
Roman, C., Cohn, L. and Calame, K. (1991). *Science* **254**, 94–97.
Roman, C., Matera, A.G., Cooper, C., Artandi, S., Blain, S., Ward, D.C. and Calame, K. (1992). *Mol. Cell. Biol.* **12**, 817–827.
Ron, D. and Habener, J. (1992). *Genes Dev.* **6**, 439–453.
Rothenfluh, H.S., Taylor, L., Bothwell, A.L., Both, G.W. and Steele, E.J. (1993). *Eur. J. Immunol.* **23**, 2152–2159.
Rothman, P., Li, S., Gorham, B., Glimcher, L., Alt, F. and Boothby, M. (1991). *Mol. Cell. Biol.* **11**, 5551–5561.
Ruben, S.M., Dillon, P.J., Schreck, R., Henkel, T., Chen, C., Maher, M., Baeuerly, P.A. and Rosen, C.A. (1991). *Science* **251**, 1490–1493.
Rudin, C., M. and Storb, U. (1992). *Mol. Cell. Biol.* **12**, 309–320.
Ruezinsky, D., Beckmann, H. and Kadesch, T. (1991). *Genes Dev.* **5**, 29–37.
Ryseck, R.P., Bull, P., Takamiya, M., Bours, V., Siebenlist, U., Dobrzanski, P. and Bravo, R. (1992). *Mol. Cell. Biol.* **12**, 674–684.
Safrany, G. and Perry, R.P. (1993). *Proc. Natl Acad. Sci. USA* **90**, 5559–5563.
Saksela, K. and Baltimore, D. (1993). *Mol. Cell. Biol.* **13**, 3698–3705.
Samuelsson, L., Stromberg, K., Vikman, K., Bjursell, G. and Enerback, S. (1991). *EMBO J.* **10**, 3787–3793.
Scheiderheit, C., Cromlish, J.A., Gerster, T., Kawakami, K., Balmaceda, C.-G., Currie, R.A. and Roeder, R.G. (1988). *Nature* **336**, 551–557.
Scheuermann, R.H. (1992). *J. Biol. Chem.* **267**, 624–634.
Scheuermann, R. and Chen, U. (1989). *Genes Dev.* **3**, 1255–1266.
Schmid, R.M., Perkins, N.D., Duckett, C.S., Andrews, P.C. and Nabel, G.J. (1991). *Nature* **352**, 734–736.
Scott, M.L., Fujita, T., Liou, H.-C., Nolan, G.P. and Baltimore, D. (1993). *Genes Dev.* **7**, 1266–1276.
Sen, R. and Baltimore, D. (1986a). *Cell* **47**, 921–928.
Sen, R. and Baltimore, D. (1986b). *Cell* **46**, 705–716.
Seto, E., Shi, Y. and Shenk, T. (1991). *Nature* **354**, 241–245.
Sheehan, K.M., Mainville, C.A., Willert, S. and Brodeur, P.H. (1993). *J. Immunol.* **151**, 5364–5375.
Shen, L., Lieberman, S. and Eckhardt, L.A. (1993). *Mol. Cell. Biol.* **13**, 3530–3540.
Shi, Y., Seto, E., Chang, L.S. and Shenk, T. (1991). *Cell* **67**, 377–388.
Shrivastava, A., Saleque, S., Kalpana, G., Artandi, S., Goff, S. and Calame, K. (1993). *Science* **262**, 1889–1892.

Singh, H., Sen, R., Baltimore, D. and Sharp, P.A. (1986). *Nature* **319**, 154–158.
Singh, M. and Birshtein, B.K. (1993). *Mol. Cell. Biol.* **13**, 3611–3622.
Sirito, M., Lin, Q., Maity, T. and Sawadogo, M. (1994). *Nucleic Acids Res.* **22**, 427–433.
Spanopoulou, E., Roman, C., Corcoran, L., Schlissel, M., Silver, D., Nemazee, D., Nussenzweig, M., Shinton, S., Hardy, R. and Baltimore, D. (1994). *Genes Dev.* **8**, 1030–1042.
Staudt, L. and Lenardo, M. (1991). *Annu. Rev. Immunol.* **9**, 373–398.
Staudt, L.M., Singh, H., Sen, R., Wirth, T., Sharp, P.A. and Baltimore, D. (1986). *Nature* **323**, 646–648.
Strum, R.A., Das, G. and Herr, W. (1988). *Genes Dev.* **2**, 1582–1599.
Su, L.K. and Kadesch, T. (1990). *Mol. Cell. Biol.* **10**, 2619–2624.
Sun, S.C., Ganchi, P.A., Ballard, D.W. and Greene, W.C. (1993). *Science* **259**, 1912–1915.
Tapscott, S., Davis, R., Thayer, M., Cheng, P., Weintraub, H. and Lassar, A. (1988). *Science* **242**, 405–411.
Trautwein, C., Caelles, C., van der Geer, P., Hunter, T., Karin, M. and Chojkier, M. (1993). *Nature* **364**, 544–547.
Tsao, B., Wang, X.-f., Peterson, C. and Calame, K. (1988). *Nucleic Acids Res.* **16**, 3239–3253.
Umar, A., Schweitzer, P.A., Levy, N.S., Gearhart, J.D. and Gearhart, P.J. (1991). *Proc. Natl Acad. Sci. USA*, **88**, 4902–4906.
Usheva, A. and Shenk, T. (1994). *Cell*, **76**, 1115–1121.
Vinson, C., Sigler, P. and McKnight, S. (1989). *Science* **246**, 911–916.
Wabl, M. and Borrows P., B. (1984). *Proc. Natl Acad. Sci. USA* **81**, 2452–2455.
Wasylyk, B., Hahn, S.L. and Giovane, A. (1993). *Eur. J. Biochem.* **211**, 7–18.
Waters, S., Saikh, K. and Stavnezer, J. (1989). *Mol. Cell. Biol.* **9**, 5594–5601.
Webb, C.F., Das, C., Eneff, K.L. and Tucker, P.W. (1991a). *Mol. Cell. Biol.* **11**, 5206–5211.
Webb, C.F., Das, C., Eaton, S., Calame, K. and Tucker, P.W. (1991b). *Mol. Cell. Biol.* **11**, 5197–5205.
Webb, C.F., Eneff, K.L. and Drake, F.H. (1993). *Nucleic Acids Res.* **21**, 4363–4368.
Wegner, M., Cao, Z. and Rosenfeld, M. (1992). *Science* **256**, 370–373.
Weinberger, J., Jat, P. and Sharp, P. (1988). *Mol. Cell. Biol.* **8**, 988–992.
Williams, S., Cantwell, C. *et al.*, (1991). *Genes Dev.* **5**, 1553–1596.
Wilson, R.B., Kiledjian, M., Shen, C.P., Benezra, R., Zwollo, P., Dymecki, S.M., Desiderio, S.V. and Kadesch, T. (1991). *Mol. Cell. Biol.* **11**, 6185–6191.
Wirth, T., Staudt, L. and Baltimore, D. (1987). *Nature* **329**, 174–178.
Wotton, D., Ghysdael, J., Wang, S., Speck, N.A. and Owen, M.J. (1994). *Mol. Cell. Biol.* **14**, 840–850.
Xu, L., Gorham, B., Li, S.C., Bottaro, A., Alt, F.W. and Rothman, P. (1993). *Proc. Natl Acad. Sci. USA* **90**, 3705–3790.
Xu, M.Z. and Stavnezer, J. (1992). *EMBO J.* **11**, 145–155.
Yancopoulos, G.D. and Alt, F.W. (1986). *Annu. Rev. Immunol.* **4**, 339–368.
Young, F., Ardman, B., Shinkai, Y., Lansford, R., Blackwell, K., Mendelsohn, M., Rolink, A., Melchers, F. and Alt, F. (1994). *Genes Dev.* **8**, 1043–1057.
Yu, H., Porton, B., Shen, L. and Eckhardt, L. (1989). *Cell* **58**, 441–448.
Zaller, D., Yu, H. and Eckhardt, L. (1988). *Mol. Cell. Biol.* **8**, 1932–1939.
Zhao, G.Q., Zhao, Q., Zhou, X., Mattei, M.G. and de Crombrugghe, (1993). *Mol. Cell. Biol.* **13**, 4505–4512.
Zhuang, Y., Kim, C., Bartelmez, S., Cheng, P., Groudine, M. and Weintraub, H. (1993). *Proc. Natl Acad. Sci. USA* **89**, 12132–12136.

Index

Figures are indicated by *italic* type, major page references in **bold**.

Abelson lines (A-MulV-transformed cells), 207
    class switching, 236, 245–246, 248, 252
    light chain rearrangements, 197, 198, 199
    μ heavy chain transgenic mice, 351
    RS (recombining sequence) DNA, 200–201
Accessory (A) cells, 34
    α factors, 39
    MHC Class II molecule expression, 35
Adhesion molecules, 107
Affinity maturation, 8, 9, 57, 84
    amphibians, 326
    analytic methodology, 58–62, *61*
    B-cell subsets, 93
    chondrichthyans, 317
    germinal centre reaction 43, 64–67
    primary response, 62–64
    selection, 74–77
    somatic hypermutation, 58
Agnathans
    Ig genes, 333–334
    immune system tissues, 335
ali, 295
Alicia rabbits, 295, 297
Allelic exclusion, 229, 350
    chicken B-cells, 272–275
    rabbit, 289
    Raja, 321
    regulated model, 350
    stochastic model, 350
Allotypes 4
α5β1 integrin, 107
α factors, 42
Alternative splicing
    heavy chain secreted/membrane-bound form, 130, 131, 164–165, 237–238
    Raja, 322
    IgD isotype switching, 240
    teleost IgM, 325
Alu sequences, kappa locus, 185, 186
Amblystoma mexicanum, 326
Aminopeptidase N, 12

Amphibians
    heavy chain genes, 326
    hematopoetic bone marrow, 336
    light chain genes, 331
Anergy
    B-cell development, 52–53
    B-cell tolerance, 369–370
Anti-DNA antibodies, **391–392**
    V_H4-21 usage, 387
    V_H gene usage, 391
Anti-erythrocyte transgenic mice, autoimmunity, 375
Anti-idiotype antibodies, 385
Antibody-forming cells (AFC) *see* Plasma cells
Antigen internalization, **139–140**
Antigen receptor activation motif (ARAM) *see* Tyrosine activation motif (TAM)
Antigen receptor homology 1 (ARH1) *see* Tyrosine activation motif (TAM)
Antigens, T-dependent/T-independent B cell responses, 36
Apoptosis
    B-cell development, 9, 51, 53
    B-cell lymphomas, 368
    B-cell tolerance induction, 368
    bcl-2 rescue, 66, 353, 368
    CD40, 66, 251
    germinal centre processes, 43, 66, 75
    helper T cell rescue, 368
    SIg cross-linkage, 368
Auto-anti-idiotype antibodies, 385–386
Autoantibodies, **379–392**
    anti-DNA antibodies, **391–392**
    cold agglutinins, **386–389**
    disease-associated, 384
    germline origins 155, 380
    idiotypes, 385
    infectious microorganism gene usage responses, 383
    natural, 380, 383–384
    rheumatoid factors, **389–391**
    V(D)J gene segment utilization, 155, 382–384

424 Index

Autoimmune disease
  cross-reactive idiotype, 385
  kappa chain repertoire, 187
  V_H segment polymorphisms associations, 152
Autoimmunity
  B-cell anergy, 370
  B-cell tolerance, 365, **373–375**
    anti-erythrocyte transgenic mice, 375
    autoimmune-prone chimeric mice, 374
    bcl-2 mice, 374
    scid mice with lupus-prone pre-B-cell clones, 375
  germinal centre memory cells, 369
  V-gene hypermutation, 369
Avian Ig genes
  heavy chain genes, 327
  light chain genes, 331–332
  see also Chicken

B220/6B2
  B-cell progenitors, 86
  B-cell subsets, 91
  follicular B-cells, 91
  pre-B cells, 104
  pro-B cells, 103
B29, 130, 136
  promoters, 136
B7-2, 35
  activated B-cells, 38
B-1 cells, **90–91**, 366
  affinity maturation, 93
  CD5 expression, 91
  distribution, 90
  isotype switching, 92
  λ transgenic mice, 355
  origins, 95–96, 97
  phenotype, 90, 91
  repertoire, 94–95
  self-replenishment, 91–92, 95
  SIg cross-linkage-induced apoptosis, 368
  without CD5 expression, 91
B-1a cells, 38, 64, 84, 87, 91
  origins, 95–96, 97
B-1b cells, 91, 98
B-2 cells (conventional B-cells), 84, 87
  distribution, 87
  λ transgenic mice, 355
  marginal zone (MZ) subset, 90
  phenotypic characteristics, 87, 88, 89, 91
B-cell activation/development, 9, 34, 45–48, *46*, 83–84, **84–86**, *85*
  affinity maturation, 93

anergy, 51–53
antibody response maturation, **57–77**
antigen-independent early phases, 45, *46*
antigenic stimulation response, **92–93**
CD antigens, **10–16**, 10–11, 16–17
cell cycle continuation/completion, **39–41**, 54
  G_2 phase, 40
  S phase, 39–40
clonal selection, 8, 42, 51–53
cytokines/regulatory factors, 112–113, **114–121**
  negative regulators, 121
developmental stages, **6–10**
germinal centre reaction, 9, 65, 66
helper T-cell dependent responses, 8–9, 34, 35, 36–38, *37*, *40*, 64, 92–93
  CD40, 251
helper T-cell independent responses, 8, 34, 35–36, *36*, *40*, 64, **92–93**, 92
hypermutation in sheep, **279–286**
Ig molecule suppressive mechanisms, 44–45
Ig transgenic mice, 350–353
immunodeficiency diseases, **22–25**
immunoglobulin class switching *see* Class switch recombination (CSR)
immunoglobulin gene rearrangements, 6, 48, 62, 84, 86, 351
  regulation, 228–229
immunoglobulins expression, 7–8
mature B-cell stage, 8, *46*, 47
  tolerance, 366
memory cells, 9–10, 92
MHC restriction, 34, 35
phenotypic changes, *85*
plasma cell stage, 9, 10
primary immune response, 92, 93
repertoire expression, 62–63
screening for tolerance, 43–44, 366
stromal cell interactions, **110–114**
surface immunoglobulin (SIg) expression, 50–51
surface-expressed molecules, 39
B-cell lineages, **87**, 98–99
  conventional/follicular B-cells (B-2), 87–90
  origins, **95–98**
  tolerance, 366
  V(D)J recombination control, 228
B-cell malignancies
  rabbit V_H gene usage, 294
  rheumatoid factors, 389
  see also Lymphoma
B-cell receptor (BCR), 5, 8, 129, 130
  antigen internalization, **139–140**
  antigen specificity, 5

Index   425

CD19 cross linkage, 13
Igα/Igβ heterodimer
   C_H complex, *136*, 136, 238
   signalling function, 130
   membrane-bound heavy chain, 130, 238
   signal transduction, 5–6, 137
      protein tyrosine kinase (PTK) activation, 137–138
B-cell subsets, **86–87**
   B-1 cells *see* B-1 cells
   development, **83–99**
   isotype switching, 92–93
   marginal zone (MZ) B-cells, 90
   repertoire, 94–95
   self-replenishment, 91–92
   T-dependent/independent responses, 94
B-cell tolerance, **365–376**, 382
   anergy, 369–370
   autoimmunity, **373–375**
      anti-erythrocyte transgenic mice, 375
      autoimmune-prone chimeric mice, 374
      bcl-2 mice, 374
      scid mice with lupus-prone pre-B-cell clones, 375
   B-cell developmental stages, 366
   B-cell lineages, 366
   central, **371–373**
   clonal deletion, **367–371**, 379
      B-cell lymphomas, 368-369
      memory B-cells, 369
      signalling, 370-371
   Ig transgenic mice, 368
   receptor editing, **371–373**
   specificity, 366–367
B cells, **3-26**
   antigen receptors, 8, **129–140**
   chicken precursor generation, **270–272**, 272
   differentiation antigens, **3–17**
      overview, 16–17
   immature *see* Immature B-cells
   ontogeny, 17
B-ZIP proteins, 401
Basic fibroblast growth factor (bFGF), 107
Basic helix loop helix (BHLH) domains, 402
Bathyraja aleutica, 321
*bcl-*2, 43
   apoptosis rescue, 66, 368
   autoimmunity, 374
   B-cell tolerance, 374
   transgenic mice, 353
BHLHZIP domains, 408, 409
Biased usage
   DXP1, 383

J_H3–6, 383
Q52 V_H family, 163, 383
rabbit D gene segments, 299–300
V_H1, 298–299, 383
V_H3, 383
7183 V_H family, 163
$V_H$segment, 154–155, 383
   autoantibodies, 155
   fetal V_H segment, 155
   human, 154–155
   mouse, **163–164**
   ontogeny, 154, 155, 163
   pre-B cell, 163, 164
$V_κ$ gene segment, 383
BiP/GRP78, 6, 7
   IgM binding, 131
blk, 137
   BSAP binding site, 406
BlyF, 136
Bone marrow stromal antigen-1 (BST-1), 117
   pre-B cell development, 115, 116
Bone marrow stromal cells *see* Stromal cells
BP-1 aminopeptidase A, 12
Bruton's tyrosine kinase (BTK), 22, 23
BSAP (B-cell specific activator protein), 242, 247
   Ig gene transcription regulation, 405–406, 413
Burkitt's lymphoma, 14

C3 gene, 39
6C3/BP-1, 106
C3b receptor *see* CD35
C4-binding protein, 14
c-fos mutants, 120–121
c-kit, 118–119
c-Myc, 408, 409
C/EBP family proteins, 401–402
Cα genes, 164
   rabbit, 290, 304, 306–309, 311
      evolution, 307–309
      IgA *in vitro* expression, 306–309
      IgA *in vivo* differential expression, 307
      Leporidae, 308–309
      Ochotonidae, 309
      organization, 306
   structure, 237
$C_δ$ genes, 164
   deletion in class switch recombination, 240
   rabbit, 309
   structure, 237
$C_ε$ genes
   rabbit, 305–306
   structure, 237

C_γ gene
  pseudogene, 166
  rabbit, 290, 304, 305–306, 311
    recombinant haplotypes, 310
  structure, 237
C_H genes, 145, 164
  alternative splicing processes, 130, 131, 237–238
    membrane exons expression, 164–165
    class switch recombination (CSR), 235–236
    intrachromosomal deletion, 240–241
  deletion mutations, 166
  germline repertoire, 146
  germline transcription, 245–246
    I_α promoter, 248
    I_ϵ promoter, 247
    I_γ promoters, 247, 248
    LPS inducible promoters, 248
    transcripts, 246
  hinge region, 237
  human, 166, 237
  intracytoplasmic segments, 164, 165
  lower vertebrates
    amphibians, 326
    dipnoi, 325
    Heterodontus, 319
    Hydrolagus, 323
    Raja, 320
    teleosts, 324
  membrane-bound Ig expression, 131, 164–165, 237–238
  mouse, 166, 237
  nucleotide sequences, 165
  polymorphisms, 146, 165
  rabbit, 290, **304–311**
    allogroups/haplotypes, 296
    allotypic specificities, 289, 296, 304–305
    d allotype, 304, 305
    e allotype, 304, 305
    f allotype, 304, 306, 307
    g allotype, 304, 306, 307
    Ms allotype, 304
    n allotype, 304
    organization/expression, 305–311
    recombinant haplotypes, 310–311
    V_H *trans* association, 310
  rat, 166
  secretory Ig expression, 131, 164, 237, 238
  structure, 164–165, **237–238**
  Tm exons, 130, *131*, 131–132
C_H locus, 145, 146, **164–167**, 237
  pseudogenes, 237
Caiman crocodylus, 326

C_κ deleting element (kappade), 179
C_κ deletion, **200–202**, 211
  in human, 201
C_κ gene, 173, 381
  proximal (p) copy, 179
C_L genes
  chicken, 332
  chondrichthyans, 328
  teleosts, 330
C_λ genes, 194, 381
  human, 195
C_λ-like genes, 195
CALLA *see* CD10
C_μ genes
  chicken heavy chain, 269
  deletion in class switch recombination, 240
  rabbit, 304, 305–306
  recombinant haplotypes, 310
  structure/organization, 145, 237
  teleosts, 324
Carcharhinus plumbeus, 329
CD4, 38
CD5
  B-cell subsets, 86
  T cell subsets, 86
CD5 B-cells *see* B-1a cells
CD10, B-cell differentiation, **12–13**
CD11b *see* MAC-1
CD13, 1, 12
CD19, 24
  B-cell differentiation, 11, **13–14**, 39
  BSAP binding site, 406
  signal transduction, 13, 22
    X-linked agammaglobulinemia (XLA), 22–23
CD20, B-cell differentiation, **14**
CD21 (CR2), 13
  B-cell cell cycle continuation, 39, 40
  B-cell differentiation, **14–15**
CD22
  B-cell differentiation, **15**
  B-cell subsets, 91
  spleen immature B-cells, 90
CD23
  B-cell subsets, 91
  maturing B-cell expression, 9, 86
CD26, 12
CD28, 35
  helper T-cell expression, 38
CD34, B-cell differentiation, 11
CD35 (CR1), 14
  B-cell cell cycle continuation, 40
CD38, 64

B-cell differentiation, **15–16**
B-cell maturation, 9
CD40, 35
  apoptosis rescue signalling, 66, 369
  B-cell cell cycle continuation, 41
  B-cell differentiation, **11**, 13, 38
  CD40L interaction, 251
  class switching regulation, 251
  helper T-cell expression, 38
  hyper-IgM immunodeficiency (HIM), 23
  mutation in X-linked hyper-IgM syndrome, 42
  T-dependent B-cell responses, 42, 251
CD40L gene defect, hyper-IgM immunodeficiency (HIM), 23
CD43 (leukosialin), 24
  B-1 cell expression, 90
  B-cell progenitor expression, 86
  B-cell subsets, 91
CD44, 107
CD45
  B-cell cell cycle continuation, 39
  B-cell differentiation, **11–12**
  isoforms, 12
CD45 knock-out mice, 12
CD45R, 12
CD72, 13
CD77, 9
CD79a *see* Igα
CD79b *see* Igβ
CD81, 13
CDR1, 18
  diversity, 69–70
  somatic hypermutation, 69–70, *70*
CDR2, 18
  diversity, 69, 70
CDR3, 18, 20, 207, 382
  anti-DNA antibodies, 391, 392
  diversity generation, 69, 70
  fetal repertoire, 20, 21, 22
  somatic hypermutation, 69, *70*
  teleosts, 324
  see also heavy chain CDR3
CDRs (complementarity determining regions)
  fetal antibodies, 18
  hypermutation in sheep ileal Peyer's patches (IPP), *283*, 282–284
  private idiotypes, 384
Cell cycle
  B cell activation, **39–41**, 54
    G_2 phase, 40
    S phase, 39–40
  lymphokines activity, 39, 40
Central B-cell tolerance, **371–373**

Chelydra serpentina, 327
Chicken
  allelic exclusion, 272–275
    silencer elements, 278
  antibody repertoire, 267
    pre-immune, 303
  B-cell precursor generation, **270–272**, *272*
    DJ_H rearrangement, 271
  bursal Ig sequence selection, 275–276
  constant region isotypes, 327
  D elements, *268*, 269–270, *270*
  gene conversion for Ig diversification, 186, **269–279**, *277*, 303, 327
  heavy chain locus, *268*, 269, 327, 338
  heavy chain rearrangements, 270, 271, *272*, 278
    selection in bursal/non-bursal sites, 275
  light chain locus, *268*, 269, 331–332
  light chain rearrangements, 270, 271, 278
  silencer/anti-silencer regulation, 273, *274*, 275
Chlamydoselachus anguineus, 321
Chondrichthyans, 317
  H chain isotypes, 321
  Ig genes, 316
    germline-joined, *318*, 319–320
    heavy chain, 323, 338
    light chain, 328–330
  IgX, 321
CHOP 10, 401
Class switch recombination (CSR), 10, 67, 84, **235–259**, 236, 381
  amphibians, 326
  B-1 cells, 92
  B-cell activation/differentiation, 8, 9, 10
  B-cell ontogeny, 17
  B-cell subsets, 92–93
  common recombinase, 243
  competition model, 256
  control models, **251–256**
  cytokines/growth factors, 42, 243, 247, **248–250**, 248
    IL-4, **249**
    IL-5, **249–250**
    interferon-γ, **250**
    transforming growth factor β (TGF-B), **250**
  directed, 242–243, 247, 248
  DNA replication, 241
  experimental study systems, 257–259
    gene targeting in mice, 257–258
    in vitro switch substrate constructs, 257
  germline transcription, *245*, 245–246, 247, 248
    accessibility model, **251–254**
    I exon protein factor binding, 254

Class switch recombination (CSR) (*continued*)
  germline transcription (*continued*)
    Iα promoter, 248
    Iε promoter, 247
    Iγ promoters, 247, 248
    LPS inducible promoters, 248
    regulatory role, **242–248**
    transcription process as regulator, 252–253
    transcripts as regulators, 246, 253–254
  IL-4, 243, *244*, 246, 247
  intrachromosomal deletion pathway, 240–241
  isotype-specific recombinases, 243
  looping-out model 240
  LPS-responsive, 243, *244*, 246
    IgH 3' enhancer, 255–256
    local promoter elements, 255
    locus control mechanism, 254–256
  memory B-cells, 67
  molecular mechanisms, **239–242**
  protein factors, 242
  RAG-2-deficient blastocyst complementation system, 252, *258*, 259
  RNA splicing events, 241
  sequential switching, 241
  switch regions, 239–240
  T cells in regulation, 92–93, **248–251**
    T cell/B-cell interactions, 250–251
Clonal deletion
  B-cell development, 8, 42, 51–53
  B-cell tolerance, **367–371**, 379
Cluster of differentiation antigens, 10–11
  B-cell development, **10–16**, *16*, 16–17
Coding end hairpins, 220
Cold agglutinin disease, 386, 388
Cold agglutinins, 384, **386–389**
  V_H4-21, 386–388
Collagen, stromal cell extracellular matrix (ECM), 107
Common variable immunodeficiency (CVID), 25
Constant region genes, **237–239**
CR1 (complement receptor 1) *see* CD35
CR2 (complement receptor 2) *see* CD21
Cross-reactive idiotype, 384–385
  rheumatoid factors, 389
CRP3, 401
Cytokines
  B-cell development, 112–113, **114–121**, 118, 119
    negative regulators, 121
  class switch regulation, 243, **248–250**
  receptor subunit sharing, 120

D (diversity) gene segments, 205, *206*, 380
  chicken, *268*, 269–270, *270*
  rabbit, 290, *299*, 299
    somatic hypermutation, 303
    usage, 299–300, 311
D_H, 145, 380
  chicken, 327
  fetal repertoire, 20, 21
  gene rearrangements/combinatorial joining, 4
    chicken B-cell precursor cells, 271
    deletional joining, 218
    J_H segment recombination, 218, 381
  human chromosomes 15 and 16, 154
  lower vertebrates
    amphibians, 326
    Heterodontus, 317, 320
    Latimeria chalumnae, 325
    Raja, 320
  mouse, **162–163**
  physical mapping, **152–153**, *152*
D_HQ52, 383
Decay accelerating factor (DAF), 14–15
Deletion, class switch recombination (CSR), 240–241
Deletional joining, **216–218**, *217*
Dextrans-activated B-cells, 96
  B-1 cells, 94, 96
Dinitrophenol tolerance, 367
Dipeptidyl peptidase IV, 12
Dipnoi heavy chain genes, 325–326
DNA replication
  activated B-cell S phase, 39–40
  class switch recombination (CSR), 241
Double-strand break repair (DSBR)
  Ku autoantigen, 227–228
  V(D)J recombination, 225, 227
DXP1 biased usage, 383

E2-2 gene, 402
E2A gene, 402
E2A/E2-2 proteins, 397, 402–403
  B-cell specific activation, 412
  basic helix loop helix (BHLH) domains, 402
  ZEB binding, 403, 413
E sites, 401–402
Early B-cell factor (EBF), 136–137
elf-1, 403
Elops saurys, 324
E_μ, 305–306
E_μ-*myc* transgenic rabbits, 305
E_μ/E_κ-*myc* transgenic rabbits, 294
Enhancers

Index 429

B-cell specific activation, 412
    functional elements, 398–400
    heavy chain locus *see* Heavy chain locus enhancer
    Ig transgene expression, 349
    light chain locus, *399*
    protein-binding sites, 400
Epigonal organs, 336
Epstein–Barr virus (EBV) receptor, 14, 15
Eptatretus stoutii, 335
erg-1, 403
erg-3, 403
ERP, 403
Error-prone DNA replication/polymerase, 73
    class switch recombination (CSR), 241
    ileal Peyer's patches (IPP) V genes, 284–285
ets proteins, 403–404
    B-cell specific activation, 412
ets-1, 403
ets-2, 403
Evolutionary aspects, *338*
    heavy chain genes, 337, 338
    immunoglobulin, 337
    immunoglobulin gene superfamily, 337
    κ locus, 371
    λ locus, 194, 371
    light chain genes, 316, 337, 338
    rabbit Cα genes, 307–309
    TCR genes, 334–335, 337
    V_H genes, 332–333
    V_H locus, 157–158, 332–333
    Vκ genes, 186–187
        orphons, 186
Extracellular matrix (ECM), stromal cells, 107, 108–109

Fc receptor-mediated B-cell suppression, 45
FcεRIII, 45
Fcγ$RIIb_1$, 45
Fcγ$RIIb_2$, 45
FcγRIII, 45
Fetal liver kinase-2 (FLK-2)/STK-1, 117
Fetus
    B-1a cell progenitors, 95, 96, 97
    CDR3 repertoire, 20–21, 22
    immunoglobulin repertoire, 22
    N nucleotide addition, 20–21
    V_H gene repertoire, 19
    V_H segment biased usage, 155
Fibronectin, extracellular matrix (ECM), 107
FIP, 408
FL (fetal liver kinase ligand), 116, 117

Flat antibody-binding site hypothesis, 22
fli-1, 403
FLP recombination system, 213
Follicular B-cells, 87, 98–99
Follicular dendritic cells (FDC), 43, 65
    germinal centre reaction, 64
FR1, 18
FR2, 18
FR3, 18
FR4, 18
FRs, fetal antibodies, 18
fyn, 137, 138

G-CSF, 121
GABPα, 403
Gadus morhua, 330
γ1 germline promoter, 401, 402
γ2b transgenic mice, 351–352
    c line, 353
γc chain gene, mutations in XSCID, 23–24
Gene conversion
    chicken Ig diversification, 186, 267, **269–279**, *277*, 303, 327
    rabbit Ig diversification, 300, *302*, 302, 311
    silencer elements, 278–279
    V_H segment, 156–157
    $V_\lambda 1$, 276
Germinal centre reaction, 64–67
Germinal centres, 9, 43, *65*, 336
    affinity maturation, 9, 64, 93
        selection process, 75
    apoptosis, 75
    B-cell tolerance, 366, 369
    CD38 marker, 64
    dark zone, 9, 43, 64, 66
    follicular mantle zone, 9, 43, 64
    light zone, 9, 43, 64, 66
    memory cells, 369
    somatic hypermutation, 9, 64, 65, 66
    T-dependent B-cell responses, 9, 43
Ginglymostoma cirratum, 329
GM-CSF
    B-cell regulation, 121
    stromal cell extracellular matrix (ECM) binding, 107
GM-CSF receptor, 120
gp130, 120
gp130/gp35–65 surrogate L chain complex, 48
Growth factors
    B-cell development, 112–113, **114–121**, 118, 119
    class switch regulation, 243

Gut-associated lymphoid tissue (GALT), 337
  agnathans, 335
  elasmobranchs, 336
  teleosts, 336
  see also Ileal Peyer's patches

Haemophilus influenzae response, 20, 187
  V_H gene usage, 383
Hagfish, 333, 335
Haptens, T-cell dependent B-cell activation, 36, 37, 38
Heavy chain CDR3 (HCDR3), 20
  controlled diversification, 20–21
Heavy chain gene rearrangements, 4, 371, *372*, 381
  µ chain inhibition in transgenic mice, 351–353
  pre-B-cells, 104
  pro-B-cell, 84
Heavy chain genes, 380
  assembly, *206*
  avians, 327
    chicken, 338
  class switching *see* Class switch recombination (CSR)
  constant region, 5
  evolutionary aspects, 337, 338
  lower vertebrates, 316
    amphibians, 326
    chondrichthes, 338
    dipnoi, 325–326
    germline-joined genes, *318*, 319–320
    Heterodontus, 317–320, *318*
    Hydrolagus, 323–324
    Latimeria, 338
    osteichthes, 338
    Raja, 320–323
    reptilians, 326–327
    teleosts, 324–325
  pre-B-cells expression, 104
  pro-B-cell transcription, 6
  pseudo-constant region, 5
  rabbit, **289–312**
  repertoire diversification, 18–22
  Tm (transmembrane) exons, 130, *131*, 131–132
  transcriptional regulation
    B-cell specific activators, 412
    negative regulators, 412–413
    see also Heavy chain locus enhancer
  transgenic mice, 348
    µ transgenes, 347–348
Heavy chain locus, **145–167**, 145
  chicken, *268*, 269, 327
  protein-binding sites, 400

Heavy chain locus enhancer, 397, *398*, 398–399
  activator sites, *400*
  BSAP binding, 405
  class switch recombination (CSR), 255–256
  E sites, 401
  ets sites, 403
  matrix attachment regions (MARs), 410, 411
  µE2/5 sites, 402
  octamer site, 404
  regulatory element function, 410–411
  TFE3/USF protein binding sites, 408
  YY-1 binding, 409
Heavy chain locus promoter, *398*
Heavy chain protein
  constant region, 4–5
  membrane-bound form, 5, 237
  pre-B-cell synthesis, 199
  secreted form, 5, 237, 238
HEB gene, 402
Herpes virus, gene usage response, 383
Heterodontus, 339
  heavy chain genes, 317–320
    constant region, 319, 321
    diversity generation, 318–319
    germline-joined genes, *318*, 319–320
    µ-type gene clusters, 317, 320–321
    N-additions, 319
    recombination signal sequences (RSSs), 318
    TCR-like characteristics, 320
    V_H gene clusters, 317–318
  light chain genes, 328, 329
Heterodontus francisci, 317
Heterodontus japonicus, 336
Hinge region, C_H genes, 164
HIV virus, gene usage response, 383
HLA-disease associations, 25
HLH proteins, 414
Holocephalans
  heavy chain genes, 323
  immune system tissue, 336
Homology-mediated joining, 220–221
HSA
  B-cell progenitor expression, 86
  B-cell subsets
    B-1 cells, 90, 91
    marginal zone (MZ) B-cells, 90, 91
    splenic immature B-cells, 90, 91
Hybrid joins, 215–215, *216*
Hybridomas
  affinity maturation analysis, 58–59
  immunoglobulin class switching, 236
  µ heavy chains transgenic mice, 351
Hydrolagus

heavy chain genes, 323–324
  C_H genes, 323
  VDJ-joined gene, 323
  V_H gene clusters, 323, 324
  light chain genes, 328, 330
Hydrolagus colliei, 323, 329
  germline-joined heavy chain genes, 320
Hyper-IgM immunodeficiency (HIM), 23

I regions, 253
  protein factor binding, 254
  transcripts as regulators, 253–254
Iα promoter, 248
Iϵ promoter, 247, 251, 255
Iγ promoter, 247
Iγ1 promoter, 252
Iγ2b promoter, 248
  replacement experiment, 252, 255
Iγ3 promoter, 248
I/i antigen, V_H4–21 interaction, 388–389
Ictalurus punctatus, 324, 330
Idiotype, 384
Ig transgenic mice, **346–350**
  B-cell development, 51–53, 350–353
  B-cell tolerance, 51–53, 368, 369–370
  endogenous Ig gene effects, 350–356
  Ig gene rearrangements, 353–356
  heavy chain transgenes, light chain rearrangements, 372–373
  human antibody production, 349
  Ig gene expression, 349–350
  κ transgenes, 346–347
  λ transgenes, 346–347
  μ transgenes, 347–348
  somatic hypermutation, 358–360
Ig/EBP, 401
IgA, 235
  C_H genes, 237
  chicken, 327
  class switch regulation, 243
    IL-5, 249
    transforming growth factor β (TGF-B), 250
  effector functions, 239
  heavy chain constant region, 4
  infant levels, 17
  rabbit, 304
    Cα gene *in vitro* expression, 306–307
    IgA–f isotype *in vitro* expression, 306
    IgA–g isotype *in vitro* expression, 306
    isotype function, 306–307
    isotype *in vivo* differential expression, 307
    isotypes, 304

  secreted form, 238
  surface expressed form, 132
IgA deficiency, 25
Igα, 5, 7, 8, 45, 165, 238
  plasma cell expression, 10
  pre-B-cells, 48
Igα/Igβ heterodimer, 5, 130, **133–137**
  antigen receptor homology motif, 5–6
  B-cell receptor (BCR), 8, 130
  C_H transmembrane part (Tm) interaction, 132, 238
  characteristics, 133
  genes, 136–137
  membrane Ig association, 130, 131, 132–133, 136
  protein sequence, 133–135, *134*
  protein tyrosine kinase (PTK) interaction, 130, **137–139**
  SH2 domains, 139
  signal transduction, 5–6, 8, 137, 238
  tyrosine activation motif (TAM), *135*, 135–136, 138–139
Igβ, 5, 7, 8, 45, 165, 238
  pre-B-cells, 48
IgD, 235
  alternative splicing, 240
  B lineage cell expression, 8
  B-1 cell expression surface, 90, 91
  B-2 cells (conventional B-cells) surface expression, 87, 91
  B-cell development, 9
  C_H genes, 237
  effector functions, 238
  heavy chain constant region, 4
  marginal zone (MZ) B-cells surface expression, 90, 91
  maturing B-cell surface expression, 8, 86
  membrane-expressed form, 8, 132, 133
  rabbit, 304–305, 309
  secreted form, 238
IgD knock-out mice, 8
IgE, 235
  C_H genes, 237
  class switching
    IL-4, 249
    IL-13, 249
    regulation, 243, 249
    sequential switching, 241
    T cell/B-cell interaction, 251
  effector functions, 239, 242
  heavy chain constant region, 4
  membrane-expressed form, 132
  rabbit, 304

IgG, 235
  autoantibodies
    anti-DNA antibodies, 391
    natural, 383, 384
  chicken, 327
  effector functions, 238–239
  fetus/infant, 17
  heavy chain constant region, 4
  isotypes, 235
  membrane-expressed form, 132, 133
  primary response, 236
  rabbit, 304
    allotypic specificities, 304
    trans IgG heavy chains, 310
  secondary response, 236
  transplacental passage, 17
IgG$_1$, 235
  B-cell tolerance, 267
  C$_H$ genes, 237
  class switch regulation, 243
    IL-4, 249
    IL-5, 249–250
    T cell/B-cell interaction, 251
  effector functions, 239
IgG$_{2a}$, 235
  C$_H$ genes, 237
  class switch regulation, 243
    interferon-γ, 250
  effector functions, 239, 242
IgG$_{2b}$, 235
  C$_H$ genes, 237
  class switch regulation, 243
    IL-4, 249
  effector functions, 238
IgG$_3$, 235
  C$_H$ genes, 237
  class switch regulation, 243
    IL-4, 249
  effector functions, 238
IgG$_4$, class switch regulation, 249
IgM, 235
  activated B-cell secretion, 42
  alternative splicing processes, 130, 131
  autoantibodies
    anti-DNA antibodies, 391
    natural, 383, 384
  B lineage cell expression, 8
  B-1 cell surface expression, 90, 91
  B-2 cells (conventional B-cells) surface expression, 87, 91
  B-cell receptor (BCR), 8
  B-cell tolerance, 367
  BIP binding, 131

C$_H$ genes, 237
chicken, 327
effector functions, 238
heavy chain constant region, 4
immature B-cell surface expression, 8, 86
infant levels, 17
lower vertebrates
  amphibians, 326
  Heterodontus, 317
  teleosts, 324, 325
marginal zone (MZ) B-cells surface expression, 90, 91
mature B-cell surface expression, 8
membrane-bound form, 129–130, 132, 133
  Igα/Igβ heterodimer association, 130, 132
pentameric form, 131
plasma cell expression, 10
primary immune response, 92, 93, 236
rabbit, 304
  allotypic specificities, 304
rheumatoid factors, 389
secondary response, 236
secretory form, 129–130, 238
sheep ileal Peyer's patches (IPP) surface expression, 279
IgX
  amphibians, 326
  Raja, 321
  transcripts, 322–323
  V$_H$ region elements, 321
IgY
  amphibians, 326
  chicken, 327
IL-1, B-cell regulation, 114–115, 121
IL-2, 249
  B-cell cell cycle continuation, 40
  B-cell maturation, 35
  class switching regulation, 250
  helper T-cells secretion, 38, 42
IL-2R, γc chain mutations in XSCID, 23, 24, 120
IL-3
  B-cell regulation, 117–118, 121
  stromal cell extracellular matrix (ECM) binding, 107
IL-3R, 120
IL-4, 249
  B-cell regulation, 114–115, 121
  class switch regulation, 42, 243, *244*, 246, 247, **249**
    local promoter elements, 255
    T cell/B-cell interaction, 251
  helper T-cells secretion, 38
IL-4R, 24, 120

IL-5, 249
    B-cell cell cycle continuation, 40
    B-cell maturation, 35
    class switch regulation, 42, **249–250**, 250
    helper T-cells secretion, 38
IL-5R, 120
IL-6, 412
    B-cell development regulation, 115
    T-dependent B-cell responses, 42–43
IL-7, 14
    pre-B cell development, 104, 115
        cytokine co-factors, 115, 116
    pro-B cell development, 117
    X-linked agammaglobulinemia (XLA), 22–23
IL-7R, 24, 120
    B-cell development, 119–120
IL-9R, 24
IL-10, B-cell development regulation, 115
IL-11, B-cell development regulation, 115
IL-13, class switch regulation, 249
IL-15 receptor, 24
Ileal Peyer's patches (IPP), sheep, 279
    as bursal equivalent, 279
    hypermutation, 281, 282–284
        CDRs, *283*, 283–284, 284
        molecular mechanism, 284–285
    Ig diversification, 280, **281–282**, 286
    Ig gene rearrangements in B-cell progenitors, 279
    selection of B-cells, 281, 285–286
    V_λ rearrangements, *280*, 281
    V_λ selection processes, 281
Immature B-cells, 8, 86, 104
    bone marrow lineages, 110
    in vitro culture, 111
    splenic B-cell phenotype, 91
    surface Ig expression, 8, **50–51**, 53, 54, 235
    tolerance, 366
Immune system tissues, lower vertebrates, 335–337
Immunodeficiency diseases, **22–25**
Immunoglobulin
    B lineage cell expression, 7–8
    evolutionary aspects, 337
    fetus, 17, 22
    lower vertebrates, 334–335
    neonate, 17–18
    pre-B cell production, 6–7
    repertoire development, **17–22**
        V_H antibodies, 18–20
    soluble form, 5
    structure, 4
    transmembrane form *see* **Membrane-bound immunoglobulin**
    transplacental passage, 17
Immunoglobulin class switching *see* Class switch recombination (CSR)
Immunoglobulin fold, 334, 337
Immunoglobulin gene rearrangements, 4, 34, 48
    B-cell development, 84, 86
    B-cell repertoire, 62–63, *63*
    CD19 regulation, 14
    chicken B-cell precursors, 270
    IL-7 regulation, 14
    light chain transgene effects, 353–356
    non-random nature, 62
    regulation in pre-B cells, 50
    see also Class switch recombination (CSR); V(D)J recombination
Immunoglobulin gene superfamily, 334
    C domains, 334, 337
    evolutionary aspects, 337
    invertebrates, 334
    lower vertebrates, 334–335
    V domains, 334, 337
Immunoglobulin genes, 4–5
    germline repertoire, 4
    transcription, **397–417**
Immunoglobulin isotypes
    effector functions, 238–239
    rabbit, 304–305
Immunoglobulin therapy, 22
Immunologic memory, 8–9
Insects, immunoglobulin gene superfamily, 334
Insulin-like growth factor-1 (IGF-1), pre-B-cell development, 115, 116
Interferon-γ, 249
    class switch regulation, 42, 243, 246, **250**
    stromal cell extracellular matrix (ECM) binding, 107
Inversional joining, **216–218**, *217*
Invertebrates, immunoglobulin gene superfamily, 334
Isotype class switching *see* Class switch recombination (CSR)

J chain, 238
J (joining) gene segments, 205, *206*, 380
    chicken, *268*, 269
    FR4, 18
    hypermutation target area, 68
J11d^{high}, 93
J11d^{low}, 93
J558 V_H family, 158, 160
    organization, 160, 162
J_H gene segments, 145, 380
    chicken, *268*, 269, 327

J$_H$ gene segments (*continued*)
  chromosomal location, 145
  D^H recombination, 381
  fetal repertoire, 20, 21
  lower vertebrates
    amphibians, 326
    Heterodontus, 317
    Raja, 320
    teleosts, 324
  mouse, **162–163**
  physical mapping, **152–153**, *152*
  pseudogenes, 152
  rabbit, 290, *301*
    usage, 300, 311
  rearrangements/combinatorial joining, 4, 381
J$_H$3, 383
J$_H$4, 383
J$_H$5, 383
J$_H$6, 383
J$_κ$ gene segments, 173, 381
  V$_κ$–J$_κ$ junctions, 184
  V$_κ$–J$_κ$ rearrangements, 181–182
J$_L$ gene segments
  chicken, *268*, 269, 332
  chondrichthyans, 328
  rearrangements/combinatorial joining, 4
  teleosts, 331
J$_λ$ gene segments, 194, 381
  human, 195
  V$_λ$ recombinations, 196

κ knockout mice, L-chain expression, 198–199, 201
κ light chain genes, 173–188
  rearrangements
    in Abelson lines, 198, 199
    H-chain production signalling, 199
    order of recombination, 197–200
    in transgenic mouse, 356
  somatic hypermutation, 74
κ light chains, 5
  species differences, 5
κ locus, 173–175, 380–381
  B3–J$_κ$–C$_κ$–κ de region, 174, **178–179**
    allelic polymorphism, 180
    biomedical implications, 187
    chromosome 2 pericentric inversion, 187
    distal (d) copy, 174, 179, 182, 186
      homozygous absence, 187
    duplication, 174
    evolutionary aspects, 371
    haplotypes, 180
    organization, *174*, 174–175, *176*
    polymorphism, **180–181**
      duplication differentiating (DDP), 180
      haplotype 11, 181
    proximal (p) copy, 174, 182, 186
    pseudogenes, 186
    rearrangements in tolerance induction, 372–373
    regulatory elements, 413–414
    repetitive sequences, 185
    signal joints, 181–182
    structural organization in germline DNA 179
    transcriptional regulation, 415–417
    unique sequences/sequence-tagged sites (STS), 186
  V$_κ$ genes, **175–178**
κ locus enhancer, 398, *399*
  E sites, 401
  ets site, 403
  μE2/5 sites, 402
  NF-κB binding, 415
  NK-κB binding, 406, 413–414
  PU.1, 404, 414
  YY-1 repression, 410
κ transgenic mice, 346–347
  endogenous κ gene rearrangement inhibition, 354–355
  light chain rearrangements, 356
  somatic hypermutation, 358
  kde (κ-deleting element), 201
  RS (recombining sequence) DNA homology, 201
Ku autoantigen, 227–228

Lagomorph C$_κ$ gene evolution, 307–309
λ5, 7, 48, 50, 86, 104
  C$_λ$-like gene human equivalent, 195
  pre-B-cell differentiation, **194–195**
λ light chain genes, **193–202**
  evolution, 194
  human, 195
  mouse, 193–195
    low expression, 195–196
    lowλ1 phenotype, 195–196
  organization, **193–195**, *194*
  pseudogenes, 194, 195
  rearrangements, **196**
    in Abelson lines, 197, 198, 199
    H-chain production signalling, 199
    in κ knockout mice, 198–199
    non-random, 196
    order of recombination, 197–200
    ordered model, 197, 198

stochastic model, 197
λ light chain locus, 380, 381
  evolutionary aspects, 371
  regulatory elements, 414–415
  sheep, 281
λ light chains, 5
λ locus enhancer, 417
  λ transgene expression, 347
  NF-EM5 binding, 415, 417
  PU.1 binding, 404, 414, 415, 417
λ transgenic mice, 346–347
  endogenous κ gene rearrangement inhibition, 355
Lampreys, 333, 335
Latimeria chalumnae, 325
Latimeria heavy chain genes, 338
lck, 137
5′-Leader–VDJ–C$_\delta$–C$_\delta$–3′, 8
Leporidae, C$_a$ genes, 308
Leu, 13 13
Leukemia inhibitory factor (LIF) knockout mice, 119
Leukocyte common antigen *see* CD45
Leukosialin *see* CD43
Leydig's organ, 336
Light chain genes, 5, 380–381
  avians, *268*, 269, 331–332
  evolutionary aspects, 316, 337, 338
  lower vertebrates, 316, 327–332, *328*
    amphibians, 331
    chondrichthyans, 328–330
    isotype classification, 332
    teleosts, 330–331
  rearrangements 4, 371, *372*, 381
  transgenic mice
    κ transgenes, 346–347
    λ transgenes, 346–347
Light chain locus
  chicken, *268*, 269
  enhancer, *399*
  promoter, *399*
  protein-binding sites, 400
  transcriptional regulation, 415–417
Light chain transcriptional regulation, 413–415
  κ light chain locus, 413–414
  λ light chain locus, 414–415
Light chain transgenes, endogenous Ig gene rearrangement effect, 353–356
Light chains
  allotype/isotype exclusion, 197
    κ knockout mice, 198–199, 201
    ordered model, 197, 198, 199
    stochastic model, 197, 199
  constant region, 4

immature B-cell expression, 104
low λ chain expression in mice, 195–196
LINE1 sequences, 185
LIP, 401
Locus control region (LCR), 256
Lower vertebrates, **315–339**
  agnathans (jawless vertebrates), 333–334
  allelic exclusion, 321
  heavy chain genes
    amphibians, 326
    dipnoi, 325–326
    Heterodontus, 317–320
    Hydrolagus, 323–324
    Raja, 320–323
    reptilians, 326–327
    sarcopterygians, 325
    teleosts, 324–325
  immune system organs/tissues, 335–337
  immunoglobulin, 334–335
  immunoglobulin gene superfamily, 334–335
  light chain genes, 327–332, *328*
    amphibians, 331
    chondrichthyans, 328–330
    isotype classification, 332
    teleosts, 330–331
LPS inducible promoters (I$_r$2b and I$_r$3), 248
LPS-activated B-cells, 34, *36*, 41
  B-1 cells, 94
  class switching regulation, 242–243, *244*, 246
    IgH 3′ enhancer, 255–256
    locus control mechanism, 254–256
  Ig molecule suppression, 45
  IgG isotypes, 238
LPS-responsive factor, 242
Ly-1 B-cells *see* B-1 cells
Lymph node B-cells, 87, 91
Lymph nodes, 336
Lymphoid follicles, 87
Lymphokines
  B-cell cell cycle progression, 39, 40
  B-cell proliferation, 35, 54
  helper T-cells secretion, 38
  Ig class switching, 42
Lymphoma
  B-cell tolerance, 368–369
  immunoglobulin class switching, 236, 246
  κ chain repertoire, 187
  recombination signal sequences (RSSs), 213
lyn, 137, 138
Lyt-1 *see* CD5

M-CSF, osteopetrotic (*op/op*) mice defect, 119

MAC-1 (CD11b), 90, 91
Marginal zone (MZ) B-cells, 90, 91, 98–99
MAT switch, 276–278
Matrix attachment regions (MARs), 410, 411
Max, 408
mb-1, 130, 131, 136
   pre-B-cell expression, 104
   promoters, 136
      BlyF, 136
Membrane cofactor protein (MCP), 15
Membrane-bound immunoglobulin, 5, 129, **130–133**
   C_H genes, 129, 164
   heavy chain protein, 237, 238
   Igα/Igβ heterodimer association, 136, 238
   intracytoplasmic part, 131–132, 237, 238
   mature B-cells, 132
   memory B-cells, 132
   Raja, 322
   Tm (transmembrane) exons, 130, *131*, 131–132
   Tm (transmembrane) part, 132, 237, 238
   transport to cell surface, 132–133
Memory B-cells, 9–10, 38, **42–43**, 86
   B-cell development, 8, 9
   B-cell tolerance, 369
   germinal centres, 64, 66
      hypermutation/selection process, *76*, 76–77
   high affinity response, 66–67, 93
   life-span, 43
   membrane-bound immunoglobulin, 132
   phenotypes, 10
   T cell-dependent development, 92
MHC Class II molecules
   accessory (A) cells, 35
   MHC restriction of B-cell responses, 34, 35
   T-cell dependent B-cell responses, 37, 38
   TCR binding, 38
MHC restriction 34, 35
MI, 408, 409
Mixed cryoglobulinaemia, 389
Molluscs, immunoglobulin gene superfamily, 334
μ heavy chains
   B lineage cell expression, 8
   B-cell development, 350
   BiP interaction, 6, 7
   heavy chain rearrangement inhibition in transgenic mice, 351–353
   pre-B-cell, 48, 86
   ψ light chain complex, 7–8
μ transgenic mice, 347–348
μ-ψLC receptor, 8
Myelomas
   c-*myc* translocation, 243
   immunoglobulin class switching, 236

N-nucleotide addition, 4, **218–219**, 381
   addition mechanism, 215, 216
   CDR3 generation, 20
   fetal antibodies, 18, 20–21
   Heterodontus, 319
   rabbit, 300, 303
   terminal deoxynucleotidyl transferase (TdT), 219, 224
   V(D)J recombination, 221
   V_κ gene rearrangements, 184
N-syndecan, 107
Nasopharyngeal carcinoma, 14
Neonatal immunoglobulins, 17
NF-EM5, 416, 417
   λ enhancer binding, 415
NF-IL6, 401, 402, 412
NF-κB, 397, 414
   κ enhancer binding, 415
NF-mNR, 412, 413
NK-κB, 406–408
NSF-B_1, 242
Nucleotide substitution bias, **71**, 71–72

OCA-B, 404, 412, 416
Ochotonidae, C_κ genes, 308, 309
Oct-1, 404, 405, 412
Oct-2, 404, 405, 412
   mouse mutants, 405
Octamer proteins, 397, 404–405
   B-cell specific activation, 412
Octamer-binding proteins, 242, 416
Oncorhynchus mykiss, 324, 330
Open-and-shut joins, 215–216, *216*
Orphons
   V_H segments, 153, 154, 157
   expression, 156
   V_κ genes, 175, 179, 184, 381
   evolutionary aspects, 186
   pseudogenes, 185
   $V_\kappa I$, 184
   W groups, 185, 187
   Z family, 184–185
Osteichthyes, Ig genes, 316, 338
Osteopetrotic (*op/op*) mice, 119

P-nucleotide addition, **219–220**, 381
   addition mechanism, 215
   scid mutation, 226
   V(D)J recombination, 221
Pax genes, 406
Pax proteins, 405–406

Pax-5, 242
Periarteriolar lymphoid sheaths (PALS), 64, 93
Peritoneal B-cells, 90
Petromyzon marinus, 335
Phosphorylcholine (PC) B-1 cell activation, 94
pHRD transgenic mice, 357
Plasma cells, 9, 10, 17, 64, 86
   germinal centre reaction, 66
   primary immune response, 93
Pleiotrophin, 107
POU proteins, 397, 404, 412
Pre-B cell colony enhancing factor (PBEF), 115, 116
Pre-B cell growth stimulating factor (PBSF), 115, 116
Pre-B cells, 6–7, 17, 34, *46*, 47, 104
   bone marrow lineages, 110
   CD antigens, **16**
   cytokines/regulatory factors in development, 115
   fetal lineages, 98
   fibronectin receptor ($\alpha 5\beta 1$ integrin), 107
   Ig gene rearrangements, 351
      heavy chain, 84, 86, 104
      light-chain, 197–198, 199
   IL-7 effects, 115
   in vitro culture, 111, 114
   $\lambda 5$ expression, 194–195
   light chain allotype/isotype exclusion, 197–198, 199
   μH chain expression, 48
   μH chain-surrogate L chain receptor, 48–50
   phenotype, *85*, 86
   receptor function in growth control, **33–54**
   stromal cell extracellular matrix (ECM) interaction, 107
   stromal hyaluronidate binding, 107
   V_H segment biased usage, 163, 164
Pre-B-I cells, 48
Pre-B-II cells, 48, *49*
Primary response, 92, 93, 236
   immunoglobulin class switching, 236
Pro-B cells, 6, *46*, 47, 103
   CD antigens, *16*
   cytokines regulating development, 117
   developmental lineages, 84, 97–98
   Ig gene rearrangements, 84, 351
   in vitro culture, 111, 114
   phenotype, *85*, 86
Pro-B-I cells, 48
Promoters, 411
   B cell specific activation, 412
   functional elements, 398–400

   heavy chain locus, *398*
   light chain locus, *399*
   protein-binding sites, 400
Protein tyrosine kinase (PTK)
   Igα/Igβ heterodimer interaction, 130, **137–139**
   SH2 domains, 137, 139
Protein-binding sites, transcriptional regulation, **400**
Protopterus aethiopicus, 325
Pseudo-light chain proteins *see* Surrogate L chain
Pseudogenes
   C_ρ, 166
   C_H locus, 237
   chicken V_H genes, 327
      $V_H 1$, 269
   chicken V_λ, 269
   J_H segment, 152
   κ locus, 186
   λ locus, 194, 195
   sheep V_λ, 281
   V_H locus, 148, 149, 151, **156–157**, 380
      human chromosomes 15 and 16 homologies, 154
   V_κ, 174, 175
      orphon segments, 381
ψ light chain *see* Surrogate L chain
PU.1, 403–404, 416, 417
   gene ablation in mice, 404
   κ enhancer binding, 404, 414
   λ enhancer binding, 404, 414, 415, 417

Q52 V_H family, 158, 160, 162
   biased usage, 163

Rabbit
   allelic exclusion, 289
   antibody diversity generation, 300, 302–304
   antibody repertoire, 286
      pre-immune, 303
   C_H genes, **304–311**
   D gene usage, 299–300
   gene conversion, 300, *302*, 302, 311
   heavy chain genes, **289–312**
   heavy chain recombinant haplotypes, 310–311
   Ig allotypes, 289
   J_H gene usage, 300
   N-nucleotide addition, 300, 303
   somatic hypermutation, 300, 303
   trans IgG heavy chains, 310
   V_H genes, 289–290, **290–304**

Rabies virus, V_H gene usage response, 383
RAG-1, 14, 51, 52, 228, 229
  chromosomal organization, 222
  class switch recombination (CSR), 242
  Rch-1 interaction, 222
  V(D)J recombination, 207, **222–223**
RAG-1 mutant mice, 49, 207, 223
RAG-2, 14, 51, 52, 228, 229
  chicken bursa B cells, 223, 278
  chromosomal organization, 222
  class switch recombination (CSR), 242
  V(D)J recombination, 207, **222–223**
RAG-2 mutant mice, 49, 50, 207, 223
RAG-2-deficient blastocyst complementation system, *258*
  class switch recombination (CSR), 252, 259
Raja
  heavy chain genes, 320–323
    allelic exclusion, 321
    germline joining, *318*, 321
    IgX, 321, 322
    isotypes, 321
    μ-type gene clusters, 320–321
    secreted/transmembranous expression, 322
    transcriptional control, 321–322
  light chain genes, 328, 329, 330
Raja eglanteria, 322
Raja erinacea, 320
Rch-1, 222
Rearrangement test transgenes, 356–357
Receptor editing, 382
  B cell tolerance, **371–373**
Recombination signal sequences (RSSs)
  adjacent nucleotide sequence effects, 213
  chicken λ genes, rearrangement in transgenic mouse, 356
  coding sequence joining, *210*, **213–216**, *214*
    hybrid/open-and-shut joins, 215–216, *216*
  heptamer sequences, 211
    isolated, 211, 213
  in inversional/deletional joining, 216–218, *217*
  lower vertebrates
    Heterodontus, 318
    Latimeria chalumnae, 325
    Raja IgX V_H region, 321
    reptilians, 327
    teleosts, 324
  N-region addition, 219
  nonamer sequences, 211
  nucleotide composition, 209, 211
  rabbit V_H gene preferential usage, 298–299
  V(D)J recombination, **209–213**, *212*

Recombination substrates, V(D)J recombination models, *208*, 208–209
Rel proteins, 397, 406–408
  light chain transcriptional regulation, 415, 416
Repetitive sequences
  class switch recombination (CSR) switch regions, 239–240
  κ locus, 185
  V_H segment, 157, 158
Reptilian heavy chain genes, 326–327
Rheumatoid arthritis
  rheumatoid factors, 389, 390
  $V_κ$ gene allelic polymorphism, 180
Rheumatoid factors, **389–391**
  B cell dyscrasias, 389
  B cell malignancies, 389
  CRI groups, 389
  gene usage, 383
    V_H4–21, 387
    V_H segment, 155
  LC1 reactivity, 389–391
  rheumatoid arthritis, 389, 390
RS (recombining sequence) DNA 200–201
  κ-deleting element, 201, 371

Salmo salar, 324
Sarcopterygian heavy chain genes, 325
scid mice *see* Severe combined immunodeficiency (SCID) mouse mutants
Secondary response
  amphibians, 326
  immunoglobulin class switching, 236
Secreted immunoglobulins, 5, 129–130, *130*, 131
  alternative splicing, 131, 164, 237
  Raja, 322
  C_H genes, 129, 164, 237
Selection
  affinity maturation, 74–77
  chicken bursal Ig sequences, 275–276
Sequence-tagged sites (STS), κ locus, 186
Severe combined immunodeficiency (SCID) mouse mutants, 207, 353
  double-strand break repair (DSBR) defect, 225
  with lupus-prone pre-B cell clones, 375
  P-nucleotides, 219–220, 226
  V(D)J recombination defect, **225–226**
SH2 domains in PTKs, 137, 139
Sheep
  antibody repertoire, 267, 286
  B cell hypermutation, **279–286**
  ileal Peyer's patches (IPP) *see* Ileal Peyer's patches (IPP)

pre-immune antibody repertoire, 303
Short consensus repeat (SCR), 14
Sialophorin *see* CD43
Signal joints, V_κ–J_κ rearrangements, 181–182
Silencer/anti-silencer regulation
  chicken gene conversion, 278–279
  chicken light chain rearrangement, 273, *274*, 275
  yeast mating type (MAT) switch, 276–277
SJL λ1 defect, 195–196
Sjögren's syndrome, 389
Sl (steel) mice, stromal cell environment defect, 118
SLF (steel locus factor), 119
  pre-B cell development, 104, 115
  pro-B cell development, 117
Somatic hypermutation, **67–74**, 382
  affinity maturation, 8, 9, 58, 93
  analytic methodology, 58–62, **61**
  B cell activation, 42
  B-1 cells, 93
  chondrichthyans, 317
  error prone repair model, *72*, 72, 73
  germinal centres, 43, 64, 65, 66
  hot spots, 69
  Ig transgenes, 358–360
  mechanisms, 73–74
  memory B-cells, 10, 67
  non-random distribution, **68–71**
  nucleotide substitution bias, 71–72
  passenger transgenes, 68
  rabbit, 300
    V_H genes, 303
  rate, 67
  self-reactive mutants, 44
  sheep antibody repertoire, 267
  sheep ileal Peyer's patches (IPP), 281, 282–284, 303
    CDRs, *283*, 283–284
    molecular mechanism, 284–285
  strand polarity, 72
  target areas, 68
  transitions, 71
  transversions, 71
  V_κ genes, 183
Spheroides nephelus, 324
Spleen, lower vertebrates, 335–336, 337
Splenic B-cells, 87, 90
  B-1 cells, 90, 91
  immature B-cell subset, 90, 91
  marginal zone (MZ) B-cells, 90, 91
  phenotypes, *88*, *89*, 91
src family kinases, 13

B-cell receptor (BCR) activation, 137–138
  SH2 domains, 139
  X-linked agammaglobulinemia (XLA), 22
Steel locus factor *see* SLF
STK-1, 117
Stromal cells, **104–106**
  B cell interactions, **110–114**
    cytokines/growth factors, 118, 119
    in vitro studies, 110–111
    in vivo studies, 110
    LIF knockout mice, 119
    osteopetrotic (*op/op*) mice, 119
    Sl (steel) mice, 118
    W (white spotting) mice, 118–119
  cytokines/regulatory factors, 112–113, **114–121**
  hemopoetic cell relationship, 105–106
  in vitro culture, 111, 114
  origin, 104–105
  phenotypic characteristics, **106–107**, 108–109
    antigen expression, 106, **108–109**
    extracellular matrix (ECM), 107, 108–109
Surface immunoglobulin (SIg)
  B lineage cell expression, 8, 41
  B-cell tolerance, 366, 367
  cross-linkage
    apoptosis, 368
    C19, 13
    T-independent antigens, 36
  IgD, 8
  IgM, 235
  immature B-cell expression, 8, **50–51**, 53, 54, 86, 235
  mature B-cell expression, 8, 33, 34
  pre-B cells, 7
  rabbit, 304–305
  T-cell dependent haptenic group binding, 36–37, 38
  T-cell independent B-cell responses, 36, 38
Surface immunoglobulin (SIg)-associated proteins, 5–6
Surrogate L chain, 48
  cell surface expression, 7
  μ chain associations, 7
  μH chain pre-B receptor, 48–50, 53
  pre-B cells, 7, 86, 104
  receptors, 7
Switch recombination substrate constructs, 257
syk, 137
  SH2 domains, 139
Systemic lupus erythematosus (SLE)
  anti-DNA antibodies, 391
  rheumatoid factors, 389

T helper-cells
A cell processed antigen recognition, 35
B-cell responses, **35–38**, *37*
activation, 34, 35, *40*, 250–251
apoptosis rescue, 368
CD40, 251
maturation, 8–9
class switch recombination (CSR) regulation, **248–251**
cytokine production, 38, **248–250**
germinal centre reaction, 64
memory B-cell development, 42
subsets, 248, 249
T-dependent antigens, 36
T-cell development, 12
TAPA-1, 39
TCR, 5
evolutionary aspects, 334–335, 337
MHC class II molecule binding, 38
TCR-Ig hybrid transgenic mice, 356–357
Teleosts
heavy chain genes, 324–325
immune system tissue, 336
light chain genes, 330–331
Terminal deoxynucleotidyl transferase (TdT), 4
fetal expression, 18, 21
IL-7 regulation, 14
mouse mutants, 207, 219
N-region addition, 219, 224
pre-B cell regulation, 50
V(D)J recombination, **223–225**
V_κ gene rearrangements, 184
TFE3 proteins, 397, 408–409, 412
TFEB, 408, 409
TFEC, 408, 409
T_H1 cells, 248, 249
T_H2 cells, 248, 249
Thy-1, pro-B cell expression, 103
Thymus, lower vertebrates, 335–336, 337
Tm (transmembrane) exons, 130, *131*, 131–132, 165
effect of mutation, 132
Heterodotus, 319
Igα/Igβ heterodimer interaction, 132
Raja, 322, 323
splicing in teleosts, 324–325
Tolerance *see* B-cell tolerance
TPCK, 408
trans acting factors, somatic hypermutation, 74
Trans IgG heavy chains, 310
Transcriptional regulation, **397–417**
activators, 401
BSAP, 405–406

C/EBP family proteins, 401–402
E2A/E2-2 proteins, 402–403
ets family proteins, 403–404
heavy chain locus
B-cell specific activators, 412
negative regulators, 412–413
see also Heavy chain locus enhancer
light chain locus, 413–415
B-cell specificity, 415–417
developmental stage specificity, 415–417
κ light chain locus, 413–414
NK-κB/Rel proteins, 406–408
octamer proteins, 404–405
Pax proteins, 405–406
promoter functional elements, 398–400
protein-binding sites, 400
PU.1, 403–404
TFE3/USF proteins, 408–409
YY-1, 409–410
Transforming growth factor β (TGF-B)
B-cell regulation, 121
class switch regulation, 42, 243, 246, 247, 248, **250**
local promoter elements, 255
Transgenic mice, **345–360**
affinity maturation analysis, 60, 62
Ig transgenes *see* Ig transgenic mice
rearrangement test transgenes, 356–357
Transgenic rabbits, 294
Trinitrophenol tolerance, 367
Tyrosine activation motif (TAM)
BCR/TCR internalization, 140
Igα/Igβ heterodimer, *135*, 135–136
protein tyrosine kinase (PTK) activation, 138–139
Tyrosine kinase
B-cell receptor (BCR) signal transduction, 5,6
see also Protein tyrosine kinase (PTK)

USF proteins, 408–409
USFI, 408, 409
USFII, 408, 409

Vaccine response, neonate/infant, 17–18
VCAM-1, stromal cell expression, 107
V(D)J recombination, 4, 145, 181, **205–230**, *206*, **380–381**, *381*
B-cell development, 6, 351
CD10 expression, 13
CDR3 generation, 18, 20
fetus, 20–21

chicken, 278
coding end modification, **218–221**
  coding end hairpins, 220
  N-region addition, 218–219
  P-nucleotide addition, 219–220
coding sequence joining, *210*, **213–216**, *214*, 226
control, 228–230
diversity generation, 381–382
double-strand break repair (DSBR) mutation effects, 225, 227
double-strand cleavage, 214–215
fetal antibodies, 18
Heterodontus, 318
homology-mediated joining, 220–221
Ku autoantigen, 227–228
λ genes, **196**
light-chain genes, 196
  in Abelson lines, 197, 198, 199
  allotype/isotype exclusion, 197
  C_κ deletion, **200–202**
  order of recombination, 197–200
  RS (recombining sequence) DNA, 200–201
mechanism overview, *210*
model systems, 207–209
  cell lines, 207–208
  mouse mutants, 207–208
  recombination substrates, *208*, 208–209
N-region addition, 221
non-random nature, 182–183, 196, 298, 299–300
P-nucleotide addition, 221, 226
rabbit, 298, 299–300
  D gene segments usage, 299–300
RAG 1, 207, **222–223**
RAG 2, 207, **222–223**
rearrangement test transgenes, 356–357
receptor editing in tolerance induction, 371, *372*
recombination signal sequences (RSSs), **209–213**, *212*
  adjacent nucleotide sequence effects, 213
  coding sequence joining, *210*, **213–216**, *214*
  hybrid/open-and-shut joins, 215–216, *216*
  inversional/deletional joining, **216–218**, *217*
  isolated heptamer sequences, 211, 213
  severe combined immunodeficiency (*scid*) mutation, 207, 219–220, **225–226**
terminal deoxynucleotidyl transferase (TdT), **223–225**
transcriptional orientation, 149
V_H gene assembly, 48, 145
V_κ genes, **181–184**
  V_κ–J_κ, 181–182

VGAM 3–8, 158, 162
V_H promoters, 411
  E sites, 401
V_H segments, 18, 145, 146, 380
  Alicia rabbits, 295, 297
  autoantibodies utilization, 19, 155, 380, 382–383, 385
    disease-associated autoantibodies, 384
    natural autoantibodies, 383
  biased usage, 19, 383
    autoantibodies, 155
    human, **154–155**
    mouse, **163–164**
    ontogeny, 154–155, 163
  chicken, 327
    pseudogenes, 327
    $V_H 1$, 327
  evolution, 157–158, 332–333
  fetal repertoire, 19–20
    $V_H 3$, enrichment, 19–20
  gene conversion, 156–157
  gene families, 18, 19, 146–148
    conserved regions, 148
    human, 146–148
    mouse, 158–160
    phylogenetic tree, 146–147, *147*
  gene rearrangements/combinatorial joining *see* V(D)J recombination
  germline repertoire, 146, 382
  human chromosomes 15 and 16, 153–154, *154*, 157
  human locus, **146–158**
  hypermutation boundary, 68
  lower vertebrates
    amphibians, 326
    dipnoi, 325
    Heterodontus, 317, *318*, 318–319, 320
    Hydrolagus, 323, 324
    Latimeria chalumnae, 325
    Raja, 320–321
    teleosts, 324
  mapping, 146
  memory B-cell somatic mutation, 10
  mouse locus, **158–164**
  number, 151
  orphon segments, 153, 154
    expression, 156
  physical mapping, 148–151, *150*
    human, *151*
    mouse, 160–162, *161*
  polymorphisms, 146, 151–152
    disease associations, 152
  pseudogenes, 148, 149, 151, **156–157**, 380

V_H segments (continued)
  rabbit, 289–290, **290–304**, 311
    allogroups/haplotypes, 289, 296
    C_H trans association, 310
    organization, 290, 290–292, 291
    recombinant haplotypes, 310–311
    somatic hypermutation, 303
    usage, **292–299**, 294
    w group, 296
  repertoire formation, 18–20
    human, **154–155**
    mouse, **163–164**
  repetitive sequences, 157, 158
  V_Ha^- allotype, **295–298**
    x locus, 295, 296
    y locus, 295, 296
V_H1, 19, 147, 269, 311, 327, 380
  ali mutant/V_H1–a2 deletion, 295
  anti-DNA antibodies, 391
  mouse homologue, 158
  NF-mNR binding, 413
  preferential rearrangement, 298–299, 383
  pseudogenes, 269
  rabbit, 291, 294
    usage in B-cell tumours, 294
    usage in normal B-cells, 294–295
    usage in total VDJ gene repertoire, 298
  rearrangements in bursal precursor cells, 270, 271
  recombinant haplotypes, 310
V_H2, 147, 380
  anti-DNA antibodies, 391
  mouse homologue, 158
V_H3, 19, 148, 380
  anti-DNA antibodies, 391
  biased usage, 383
  mouse homologues, 158, 159
V_H3609N, 162
V_H4, 147, 148, 380
  anti-DNA antibodies, 391
  mouse homologue, 158
V_H4–21
  anti-DNA antibodies, 391
  cold agglutinins, 386–388
  I/i antigen interaction, 388–389
V_H5, 147, 148, 380
V_H6, 19, 147, 148, 380
  anti-DNA antibodies, 391
  mouse homologues, 158
  physical mapping, 149
V_H7, 147–148
V_H10, 162
V_H11, 162
V_H12, 162

V_H21, cold agglutinins, 384
3660 V_H family, 158, 162
3609 V_H family, 158, 160, 162
7183 V_H family, 158, 160
  biased usage, 163
V_Ha allotypes, 289, 291
  V_H1–a2 deletion effect, 295
  V_H gene usage, 292–293
V_Ha1, 289, 292, 293
V_Ha2, 289, 292, 293
V_Ha3, 289, 292, 293
V_HI, *Heterodontus*, 317, 318
V_HII, *Heterodontus*, 317, 318
V_HSM7, 162
V_Hx32, 295, 296, 297, 298
  usage in total VDJ gene repertoire, 298
V_Hy33
  usage in total VDJ gene repertoire, 298
V_Hz, 297, 298
  usage in total VDJ gene repertoire, 298
V_κ genes, 173–174, **175–178**, 176, 381
  allelic polymorphisms, 180
  biased usage, 383
  conserved sequence elements, 177–178
  evolution, 186–187
  mouse, 163–164
  orphons, 175, 179, 184–185, 186, 381
  pseudogenes, 174, 175, 185, 381
  rearrangement, **181–184**
    non-random nature, 182–183
    V_κ–J_κ, 181–182
    V_κ–J_κ junctions, 184
  somatic mutation, 183
  subgroups, 176–177
V_κ–J_κ junctions, 184
V_κ–J_κ rearrangements, 181–182
  signal joints, 181–182
$V_\kappa3b$, rheumatoid factors, 389, 390
V_L genes
  chicken, 268, 269, 332
  chondrichthyans, 328
  memory B-cell somatic mutation, 10
  rearrangements see V(D)J recombination
  teleosts, 330, 331
VLA-4, stromal cell VCAM-1 binding, 107
V_λ genes, 194, 381
  sheep, 281
    hypermutation, 281, 282–284
    pseudogenes, 281
    rearrangements, 280, 281
    selection processes, 281
$V_\lambda1$, 269
  gene conversion, 276

hypermutation, 68, 71
rearrangements in bursal precursor cells, 270, 271
V_{pre-B}, 7, 48, 50, 86, 104
BSAP binding site, 406

W κ locus orphons, 185, 187
W (white spotting) mice, 118, 119
hematopoietic stem cell defect, 118
Waldenström's macroglobulinaemia, 389
WASP positional cloning, 24
Wiskott–Aldrich syndrome, 24

X-linked agammaglobulinemia (XLA), 22–23
tyrosine kinase gene mutation, 42
X-linked hyper-IgM syndrome, CD40 mutation, 42

X-linked severe combined immunodeficiency (XSCID), 23–24
B-cell defect, 120
IL-2R γ chain defect, 120
Xenopus, 331
Xenopus laevis, 326
XR-1 CHO line, 227
xrs-6 CHO line, 227

y33-encoding gene, 294, 295, 296, 298
Yeast mating type (MAT) switch, 276–278
YxxL motif *see* Tyrosine activation motif (TAM)
YY-1, 409–410, 417

Z $V_κ I$ orphons, 184–185
ZEB, 403, 413
Zinc finger proteins, 403, 409